A DICTIONARY OF USEFUL AND EVERYDAY PLANTS
AND THEIR COMMON NAMES

A DICTIONARY OF
USEFUL AND EVERYDAY PLANTS
AND THEIR
COMMON NAMES

BASED ON MATERIAL CONTAINED IN J. C. WILLIS:
A DICTIONARY OF THE FLOWERING PLANTS AND FERNS
(6TH EDITION, 1931)

F. N. HOWES
Formerly Keeper of the Museum
Royal Botanic Gardens, Kew

CAMBRIDGE UNIVERSITY PRESS

CAMBRIDGE

LONDON · NEW YORK · MELBOURNE

Published by the Syndics of the Cambridge University Press
The Pitt Building, Trumpington Street, Cambridge CB2 1RP
Bentley House, 200 Euston Road, London NW1 2DB
32 East 57th Street, New York, NY 10022, USA
296 Beaconsfield Parade, Middle Park, Melbourne 3206, Australia

© Cambridge University Press 1974

Library of Congress catalogue card number: 73–91701

ISBN 0 521 08520 9

First published 1974
Reprinted with corrections 1975

Printed in Great Britain
at the
University Printing House, Cambridge
(Euan Phillips, University Printer)

CONTENTS

PREFACE

When the revised and enlarged edition of Willis' *Dictionary of the Flowering Plants and Ferns* (7th edition, 1966) appeared, disappointment was expressed in some quarters that it had been found necessary to leave out much of the more general information contained in the earlier editions. This had to be done by the reviser for reasons of space as there was so much new material, new genera etc., to be incorporated. The information that was left out was concerned with economic uses, ornamental plants, and common names.

After due consideration it was decided that this kind of information might well be supplied, for those requiring it, in the form of a separate book and this volume was therefore prepared. The information of this sort given in the older editions of the *Dictionary* has been greatly extended and brought up to date. The total number of entries has been more than doubled and many new features have been added. With regard to common names, the emphasis has been mainly on 'everyday' plants in countries where English is spoken, or used to some extent. Trade names and the names of economic or commercial plant products, including timbers, have been included. For reasons of space and in accordance with a modern trend for works of this kind authorities for specific names have not been given.

I wish to express my appreciation to the Director of the Royal Botanic Gardens, Kew, and the Bentham–Moxon Trustees for facilities kindly provided, also to former colleagues for assistance or advice freely given.

F. N. HOWES

PUBLISHER'S NOTE

Dr Howes died in February 1973 shortly after completing his work on the typescript. Many botanists have helped with the additional work necessary in preparing the text for the publisher. In particular, David Mabberley, currently of the Botany School, University of Oxford, has been of great assistance in reading the proofs against Dr Howes's typescript.

The publisher will be grateful to receive corrections of errors that could be included in a reprint. Please send to: Cambridge University Press, The Pitt Building, Trumpington Street, Cambridge CB2 1RP, England.

September 1975

ABBREVIATIONS

Af.	Africa *or* African	Hemisph.	Hemisphere
agric.	agricultural	herbac.	herbaceous
Am.	America *or* American	Himal.	Himalayas
Amaz.	Amazon region	hort.	horticultural(ly)
ann.	annual	imp.	important
Argen.	Argentina	inc.	including
As.	Asia *or* Asian	Ind.	India *or* Indian
Aus.	Australia *or*	Indon.	Indonesia
	Australian	inflor.	inflorescence
b.	bark	Jam.	Jamaica
Br.	Britain, British *or*	Jap.	Japan *or* Japanese
	British Isles	l.	leaf
Braz.	Brazil *or* Brazilian	Madag.	Madagascar *or*
C.	Central		Malagasy
Calif.	California *or*	med.	medicinal(ly)
	Californian	Medit.	Mediterranean
Can.	Canada *or* Canadian		region
cli.	climate *or* climates	Mex.	Mexico *or* Mexican
cosmop.	cosmopolitan	mt.	mountain
cult.	cultivated *or* cultiva-	N.	North
	tion	N.Z.	New Zealand
cv.	cultivar	orn.	ornament(ally)
decor.	decorative	pantrop.	pantropical
diam.	diameter	peren.	perennial
E.	East	pl.	plant
E.I.	East Indies	r.	root
ed.	edible	rhi.	rhizome
esp.	especially	S.	South
esst.	essential	sp.	species (*plural* spp.)
Eur.	Europe *or* European	subsp.	subspecies (*plural*
Euras.	Eurasia		subspp.)
f.	form	subtemp.	subtemperate
fl.	flower *or* flowering	subtrop.	subtropics *or* sub-
fol.	foliage		tropical
fr.	fruit	Tasm.	Tasmania
gdn.	garden	temp.	temperate
h.	hour	trop.	tropics *or* tropical

ABBREVIATIONS

U.S.	United States of America	W.	West
usu.	usually	w.	with
v.	very	W.I.	West Indies *or* West Indian
var.	variety	×	indicates a plant of hybrid origin
veg.	vegetable (esculent)	+	indicates a plant of graft-hybrid origin
vol.	volume		

Note. The many mentions made in this dictionary to 'Willis' refer to Willis, J. C. *Dictionary of the Flowering Plants and Ferns,* seventh (1966) or eighth (1973) edition.

A

Aandblom (S.Af.) Evening fl., name used for various night-scented spp. of *Hesperantha*, *Freesia* and *Gladiolus*.

Abaca *See* Manila hemp.

Abele *Populus alba*, white poplar.

Abelia Orn. fl. shrubs, widely cult.

Abies *See* fir.

Abir A scented powder used in Hindu worship, the principal ingredient being the rhizome of *Hedychium spicatum*.

Abiu (Braz.) *Pouteria cainito*, trop. Am. tree w. ed. fr., wild and cult.

Abobra *A. tenuifolia*, S.Am., cult., quick-growing climber, fls. green, fragrant.

Abortifacients A large number of pls. are known to have been used in different parts of the world as abortifacients, inc. spp. of *Arum*, *Aristolochia*, *Calotropis*, *Celastrus*, *Colubrina*, *Dalbergia*, *Daphne*, *Euphorbia*, *Genista*, *Kibatalia*, *Rubia*, *Urginea* and *Veratrum*.

Abricock Old English name for apricot.

Abroma *A. augusta*, trop. As., b. fibre resembles hemp.

Absinthe *Artemisia absinthium*, cult. herb, used for flavouring the liqueur absinthe and vermouth.

Abura *Mitragyna ciliata*, W.Afr., a commercial timber, used for furniture.

Abutilon Many spp. cult. for orn., esp. *A. megapotamicum*, Braz.

Acacia This large genus, w. some 800 spp., is well represented in Aus. and Af. and has numerous economic uses. Many spp. cult. for orn. (*see Acacia* in Willis). **Apple ring** —, *A. albida*, widespread in Af.: **Benin rope** —, *A. ataxacantha*: **seyal** —, *A. seyal*: **Congo** —, *A. macrophylla*: **false** —, *Robinia pseudoacacia*, cult.: **rose** —, *R. hispida*: **sweet** —, *Acacia farnesiana*. *See* wattle.

Acaena Cult. orn. pls., many creeping; stand shade well.

Acajou (W.I.) *Guarea trichilioides*, forest tree w. purplish-red wood; name also used for cashew, also mahogany (in French).

Acalypha Cult. orn. shrubs in warm cli., esp. *A. wilkesiana* and *A. hispida:* some are weeds, e.g. *A. ciliata*, trop.

Acanthopanax Orn. trees and shrubs, mainly E.As., cult., orn. fol., usually armed.

Acanthospermum *A. hispidum*, a trop. weed w. spiny frs.

Acanthus Cult. orn. perens., ls. or fol. of *A. mollis*, Medit. considered to have been copied by ancient sculptors in their designs.

Acaroid resin Or gum accroides: the resin obtained from Australian grass trees (5 spp. of *Xanthorrhoea*): originally used by aborigines for fixing spearheads to shafts: the resin is present on the trunk or between the persistent old l. bases. The main use of the resin has been as a varnish or lacquer for metals, also for leather varnish.

Aceituna (C.Am.) *Quassia* (= *Simarouba*) *glauca*, seeds yield oil.

Acer *See* maple.

Acerola Or Barbados cherry, *Malpighia glabra*, trop. Am., small tree, ed. fr. has high vitamin content, used for syrups.

1

ACHILLEA

Achillea Cult. orn. fls., many cultivars.

Achimenes Cult. orn. pls., warm cli., *A. grandiflora* a popular house pl.

Achira (Peru) *Canna edulis*, tubers ed.

Acidanthera Orn. (= *Gladiolus*) spp., mainly S.Af., fls. scented: some good pot pls.

Acokanthera Shrubs, mainly Af., poisonous to stock (arrow poisons), orn., cult.

Acom (W.I.) *Dioscorea bulbifera*, a yam.

Aconite *Aconitum* spp., N. temp., poisonous, some med., esp. *A. napellus*, dried rs. collected in C.Eur., formerly used in liniments in the treatment of neuralgia, sciatica and rheumatism. **Nepal** —, *A. spicatum:* **winter** —, *Eranthis hyemalis* and other spp., cult.

Acorn The seed of oaks (*Quercus*): *Dioscorea bulbifera*, W.I., a yam.

Actinidia Climbing shrubs, E.As., cult. for orn., some w. ed. frs., esp. *A. chinensis*, the Chinese gooseberry or 'Kiwi fruit'.

Ada *A. aurantiaca*, Columbia, cult. orchid of easy cult., orange fls.

Adam and Eve *Aplectrum hyemale*, N.Am., terrestrial orchid, greenish-brown fls. —'s **hood,** *Ficus* spp., W.Af.: —'s **needle,** *Yucca* spp. cult.: —'s **flannel,** *Verbascum thapsus.*

Adder's meat *Stellaria holostea:* — **mouth,** *Malaxis*, N.Am.: — **tongue fern,** *Ophioglossum:* **dwarf adder's tongue,** *Ophioglossum lusitanicum:* **yellow** — —, *Erythronium americanum*, N.Am., yellow flowers, cult.

Adenophora *A. lilifolia*, cult. orn. blue fls., called Campanula in E.Af.

Adhatoda *Justicia adhatoda* (= *Adhatoda vasica*), Ind., ls. med., wood used for beads.

Adiantum *Adiantum* spp., maidenhair fern, widely cult.

Adlay Ed. forms of Job's tears, *Coix lacryma-jobi.*

Adonis Cult. orn. pls., *A. aestivalis* (pheasant's eye) one of the best known.

Adrue (W.I.) *Cyperus articulatus*, r. med.

Aechmea Orn. bromeliads, trop. Am., cult. *A. fasciata*, showy fls. and ls.

Aerial roots Adventitious rs. arising above ground, often forming buttresses (*Palmae, Pandanaceae*), pillars (*Araceae, Ficus*), clasping and climbing organs (*Araceae, Hedera, Orchidaceae, Tecoma*), water absorbing organs (*Orchidaceae, Velloziaceae*), assimilating organs (*Orchidaceae, Podostemaceae*), thorns (*Acanthorrhiza*), or parasitic suckers (*Cuscuta, Viscum*).

Aerides Cult. orn. orchids, As., fls. often in long arching racemes.

Aeschynomene Trop. shrubs, ls. of some spp. sensitive to touch, like the sensitive pl. *A. aspera* yields shola pith, used for sun helmets.

Aesculus See horse chestnut, buckeye.

Afara *Terminalia superba*, W.Af., a commercial timber, used for furniture in Br.

African blackwood *Dalbergia melanoxylon*, a trade timber: — **bladder nut,** *Diospyros lucida:* — **bowstring hemp,** *Sansevieria:* — **boxthorn,** *Lycium horridum:* — **breadfruit,** *Treculia africana:* — **canarium,** *Canarium schweinfurthii*, a trade timber: — **celtis,** *Celtis adolfi-fridericii, C. soyauxii, C. zenkeri*, all trade timbers: — **corn lily,** *Ixia:* — **cypress,** *Callitris:* — **ebony,** *Diospyros* spp., trade timber: — **elemi** *Boswellia freriana:* — **fleabane,** *Tarchonanthus camphoratus*, S.Af.: — **harebell,** *Wahlenbergia, Roella ciliata:* — **hemp,** *Sparmannia africana:* — **kino,** *Pterocarpus erinaceus:* — **lily,** *Agapanthus umbellatus:* — **lotus,** *Zizyphus lotus:* — **mahogany,** *Khaya* spp., esp. *K.*

2

anthotheca, K. grandifoliola, K. ivorensis, K. nyasica, K. senegalensis, all trade timbers: — **marigold,** *Tagetes erecta:* — **oak,** *Lophira, Oldfieldia:* — **padouk,** *Pterocarpus soyauxii,* W.Af., a trade timber: — **peach,** *Nauclea latifolia* (= *Sarcocephalus esculentus*): — **pencil cedar,** *Juniperus procera:* — **pepper,** *Xylopia aethiopica:* — **rosewood,** *Pterocarpus erinaceus:* — **rubber,** *Landolphia:* — **satin-bush,** *Podalyria sericea:* — **satin-wood,** *Fagara macrophylla,* a trade timber: — **teak,** *Oldfieldia africana:* — **toad flower,** *Stapelia:* — **tulip tree,** *Spathodea campanulata:* — **violet,** *Saintpaulia ionantha:* — **walnut,** *Lovoa klaineana,* a trade timber: — **whitewood,** *Enantia chlorantha:* — **yellow-wood,** *Podocarpus* spp.

Afrikander (S.Af.) *Gladiolus* spp.

Afrormosia *A. elata,* W.Af., a commercial timber, resembles teak.

Afzelia *A. africana, A. bipindensis. A. pachyloba* from W.Afr. and *A. quanzensis* from E.Af. all yield commercial timbers.

Aganosma Shrubs or climbers, As., cult., large showy fls.

Agapanthus Cult. orn. fls., S.Af., many hybrids and cultivars.

Agapetes Cult. orn. evergreen shrubs, As.

Agastache *Agastache* spp., N.Am., ls. used for flavouring.

Agathis *See* kauri pine.

Agathosma Aromatic heath-like shrubs, S.Af., cult.

Agave *See* century plant, *Agave americana.* Several spp. yield commercial fibres, e.g. sisal, cantala, henequin, lecheguilla etc. *See* separate headings and *Agave* in Willis.

Agba *Gossweilerodendron balsamiferum,* W.Af., a trade timber used for furniture.

Ageratum Several spp. cult. as orn. fls. esp. for edging and bedding, e.g. *A. houstonianum: A. conyzoides* is a common weed in warm cii.

Aglaia *A. odorata,* fragrant fls. used by Chinese for scenting tea.

Agrimony *Agrimonia* spp., N.Hemisph. **Common** —, *A. eupatoria* a popular home remedy from early times, esp. for liver complaints: **hemp** —, *Eupatorium cannabinum:* **scented** —, *Agrimonia odorata.*

Agrostis Genus which includes many imp. agric. and lawn grasses.

Aguacate Lat. Am. name for avocado pear.

Ague weed *Gentiana quinquefolia,* E.N. Am.

Air potato *Dioscorea bulbifera.*

Ajowan *Trachyspermum ammi* (= *Carum copticum*), As., As. Minor, aromatic seeds, med. and a spice, a source of thymol, cult., ann.

Ajuga Several spp. cult. gdn. pls., often procumbent.

Akeake (N.Z.) *Dodonaea viscosa; Olearia avicenniifolia.*

Akebia Orn. climbing shrubs, E.As., cult. *A. lobata* and *A. quinata* with ed. frs.

Akee or **ackee apple** *Blighia sapida,* W.Af., long cult. in W.I.: the creamy white aril, eaten cooked w. fish, is a common Jam. breakfast dish: now canned for export. The fresh aril can cause 'vomiting sickness' due to the presence of hypoglycine, which is soluble in water and is leached out when the akees are boiled or parboiled, as is usual.

Akund fibre (Ind.) *Calotropis procera,* Ind., trop.Af., shrub or small tree yielding a b. fibre and seed floss similar to kapok.

3

Al (Ind.) *Morinda citrifolia*, a red dye is obtained from the r., other spp. yield yellow dyes used locally, esp. *M. tinctoria*.

Alan *Shorea albida*, Sarawak, Brunei, a commercial timber.

Albardine *Lygeum spartum*, Medit., esparto-like grass.

Alberta *Alberta magna*, S.Af. handsome shrub, reddish fls., cult.

Albizia Several spp. grown as shade for trop. crops: many spp. exploited for timber, e.g. *A. procera*, Malaysia, and *A. odoratissima*, Ind., Sri Lanka. In W.Af. the following yield commercial timbers – *A. adiantifolia*, *A. ferruginea* and *A. zygia*.

Alchemilla Several spp. cult. as gdn. pls. (lady's mantle).

Alcohol Ordinary or potable alcohol (ethyl alcohol) is obtained by the fermentation of saccharine liquids, w. or without subsequent distillation, depending upon the product required. A wide range of veg. products provide the raw material for the commercial production of alcohol, notable among them being potatoes, sugar beet, sugar cane, the grape, various cereals, esp. maize, rice, barley and rye. In warm countries, the juice of several palms, obtained by tapping, is used as a local source of alcohol such as spp. of *Arenga*, *Borassus*, *Cocos*, *Caryota*, *Corypha*, *Elaeis*, *Nypa* and *Phoenix*. In C.Am. *Agave* is much used, as are some trop. r. crops in certain other countries.

Alder *Alnus* spp. The **common** or **Eur.** —, *A. glutinosa*, favours damp situations and the wood withstands damp conditions: used for piles, soles of clogs, artificial limbs etc., many cult. vars. **Am.** —, *A. serrulata:* **Am. green** —, *A. crispa:* **berry bearing** —, *Frangula alnus* (= *Rhamnus frangula*): — **buck-thorn**, *Frangula alnus:* **Caucasian** —, *Alnus subcordata:* **green** —, *A. viridis:* **grey** —, *A. incana:* **hazel** —, *A. rugosa:* **Himalayan** —, *A. nitida:* **Italian** —, *A. cordata:* **Jap.** —, *A. japonica:* **mountain** —, *A. viridis:* **Oregon** —, *A. oregona:* **red** —, *A. rubra:* **seaside** —, *A. maritima:* **smooth** —, *A. serrulata:* **W.I.** —, *Conocarpus erectus:* **white** —, *Alnus incana*.

Alder bark B. of the common Eur. alder, *Alnus glutinosa*, has been used for tanning leather in some countries, esp. the Balkans and Turkey, mainly in small-scale or home tanneries. Used alone it imparts a dark or reddish colour and has a tendency to make the leather brittle.

Alecost *See* costmary.

Alehoof *Glechoma hederacea*, derivation of name uncertain.

Aleppo galls *See* oak galls.

Alerce *Fitzroya cupressoides* (= *F. patagonica*), Chile, a commercial timber; *Libocedrus tetragona*, Chile: *Tetraclinis articulata*, N.Af.

Alexanders *Smyrnium olusatrum*, once a salad plant and pot-herb, later superseded by celery.

Alfa Or esparto grass, *Stipa tenacissima*.

Alfalfa *See* lucerne, *Medicago sativa*.

Algaroba *Ceratonia siliqua*, Eur., carob: *Prosopis* spp., trop. Am.

Algarobilla *Caesalpinia brevifolia*, S.Am., the dry pods of this shrub are used in tanning in Chile and may be exported. They resemble divi-divi in tanning properties and have been much used for furs and fine skins: tannin content 45–50 %.

Algerian fibre *Chamaerops humilis*, *see* vegetable hair: — **grass**, esparto.

Algodoncillo *Abutilon integerrimum*, S.Am., cult. orn. tree, large yellow fls.

Alizarin The dye-stuff of madder, *Rubia tinctorum.*

Alkali grass (Am.) *Distichlis, Puccinellia, Zigadenus.*

Alkaloids Compounds of great complexity chemically and widely distributed among pls. Some are poisonous, often virulently so, others are imp. as drugs. The name often indicates the pl. or genus from which the alkaloid is derived, e.g. aconitin (*Aconitum*), cinchonin (*Cinchona*), digitalin (*Digitalis*), hyoscyamin (*Hyoscyamus*), solanin (*Solanum*), strophanthin (*Strophanthus*), strychnine (*Strychnos*) etc.

Alkanet or **alkannin** *Alkanna tinctoria*, Medit., red dye from rs.: *Pentaglottis* (=*Anchusa*) *sempervirens*, SW.Eur. **Bastard** —, *Lithospermum arvense.*

Allamanda Mainly trop. evergreen climbers w. large handsome tubular fls. *A. cathartica*, yellow fls. and *A. violacea*, widely cult.

Allanblackia Seeds of *A. oleifera*, W.Af. and *A. stuhlmannii*, E.Af. yield ed. fats.

Alleghany blackberry *Rubus allegheniensis.*

Allgood *Chenopodium bonus-henricus*, a pot-herb (spinach).

Allheal *Valeriana officinalis, Stachys palustris:* **W.I.** —, *Micromeria obovata.*

Alligator apple *Annona* spp.: — **bonnet**, *Nymphaea* spp., N.Am.: — **juniper**, *Juniperus pachyphlaea:* — **pear**, *Persea americana* (avocado pear): — **wood**, *Guarea trichilioides*, W.I.: — **pepper**, *Aframomum melagueta*, W.Af., a spice.

Allium Very many spp. w. showy fls., cult.: many are vegs.

Allseed *Radiola linoides* (=*Millegrana radiola*), seed freely produced: **four leaved** —, *Polycarpon tetraphyllum.*

Allspice Word in common use for the spice pimento (*Pimenta dioica*) because the odour is considered to be suggestive of a combination of different spices; cinnamon, nutmeg, cloves, etc. Name also used for *Chimonanthus praecox* (=*C. fragrans*), wintersweet, early fl. shrub w. fragrant fls. **Californian** —, *Calycanthus occidentalis:* **Carolina** —, *Calycanthus floridus*, aromatic b., med.: — **jasmine**, *Gelsemium:* **wild** —, *Lindera* spp., aromatic b.

Allthorn *Koeberlinia spinosa*, S. U.S., Mex., a leafless parasite w. thorny twigs.

Almeidina *Euphorbia tirucalli*, Angola, name used for the rubbery extract, also for rubber waste or last tappings of plantation rubber. *See* Tirucalli.

Almond *Prunus amygdalus* (=*Amygdalus communis*). This dessert nut is produced in greater quantity than any other nut, S.Eur., Calif., Aus. etc.: its popularity never wanes: much used in confectionery, cakes, macaroons etc., oil freely used in cosmetics. **Barbados** —, *Terminalia catappa:* **bitter** —, *Prunus amygdalus* var. *amara:* **country** — (Ind.), *Terminalia catappa:* **Cuddapah** —, *Buchanania lanzan* (=*B. latifolia*), Ind.: **dog** —, *Andira inermis*, W.Af.: **earth** —, *Cyperus esculentus*, tubers ed.: **Hottentots'** —, *Brabeium stellatifolium:* **Indian** —, *Terminalia catappa:* **Java** —, *Canarium commune*, cult. trop.: **Russian dwarf** —, *Prunus tenella*, cult.: — **tree** (W.I.), *Terminalia catappa*, cult. trop.: **wild** —, *Brabeium stellatifolium*, S.Af.

Almondette *Buchanania lanzan* (=*B. latifolia*), Ind. tree, kernels used like almonds.

Alnus *See* alder.

Aloe Cult. orn. pls., esp. in dry subtrop., mainly Af. and esp. S.Afr., *A. variegata* (partridge-breasted aloe) is an accommodating room or house pl., grown since Victorian times, in company with *Aspidistra*. **Am.** —, *Agave americana*, century

ALOES

pl.: **Bombay** —, *Agave vivipara*, yields fibre: **candelabra** —, *Aloe candelabrum*, S.Af.: **Cape** —, *A. ferox:* **Curaçao** —, *A. barbadensis.*

Aloes The drug aloes or bitter aloes is the dried and powdered l. juice of several spp. of *Aloe*, the juice having been drained from the cut ls. The more imp. kinds are Cape aloes, *Aloe ferox*, and Curaçao or Barbados aloes, *A. barbadensis* (= *A. vera*). The main use is as a purgative. — **wood**, *Aquilaria agallocha*, *see* eagle wood; *Cordia sebestina*, trop. orn. tree.

Alonsoa *A. warscewiczii*, Peru, and other spp., cult. orn. pls. (pot pls.).

Alpam root *Apama wallichii*, Ind., used for snake bite.

Alpinia Cult. orn. fls., often v. showy, rs. may have smell and taste of ginger, used med. locally.

Alsike *Trifolium hybridum*, a cult. clover.

Alsomitra *A.* (= *Macrozanonia*) *macrocarpa*, Polynesia. A cucurbit with a large gourd-like fr. which, on ripening, releases seeds w. large wings. These are carried long distances by wind. Their structure interests the aeronautically minded.

Alsophila *See Cyathea.*

Alstonia *A. congensis*, *A. boonei*, trop. Af., yield trade timber, soft and light.

Alternanthera Cult. orn. pls. favoured as a bedding pl.

Alum root *Heuchera* spp., N.Am. cult. gdn. pls., esp. *H. sanguinea.*

Aluminium plant *Pilea cadierei*, aluminium-like markings on ls., a house pl.

Alva marina Sea wrack or eel-grass, *Zostera marina.*

Alyce clover *Alysicarpus vaginalis*, cult. warm cli., fodder or hay.

Alyssum *A. saxatile* (= *Aurinia saxatilis*) and other spp., cult. orn. fls. edging and rock gdn. pls. **False** —, *Berteroa incana*, a weed: **hoary** —, *Alyssum alyssoides:* **sea** or **sweet** —, *Lobularia maritima* (= *Alyssum maritimum*).

Amaraboya *A. princeps*, Colombia, shrub, showy carmine fls., cult.

Amaranth, Globe *Gomphrena globosa*, cult. orn. fl.

Amaranthus Some are cult. orn. pls., some are weeds or spinach pls.

Amaryllis *See under* lily.

Amatungulu (S.Af.) *See* Natal plum.

Amazon lily *Eucharis grandiflora*, cult., fls. 10–12 cm across.

Ambal (Ind.) *Phyllanthus emblica*, b., ls. and frs. used in local tanning.

Ambarella (Sri Lanka) *Spondias dulcis*, cult. trop., subacid frs. for preserves.

Ambatch *Aeschynomene* (*Herminiera*) *elaphroxylon*, trop. Af., soft, light wood for rafts.

Amber Fossil resin from the Baltic region believed to be from various conifers now extinct. It is mainly recovered from extensive beds of blue earth near the eastern shores of the Baltic sea by mining and is best known for its ornamental value.

Amboyna wood *Pterocarpus indicus* and allied spp., E.I., a trade timber, burrs used for panelling, veneers and inlay work.

Ambrette seed *See* musk-mallow.

Ambrevade *Cajanus cajan*, pigeon pea, name used in Madag. and Mauritius.

Amburana *A. cearensis*, Braz., good timber, seeds aromatic, contain coumarin.

Amelanchier Small trees or shrubs, mainly N.Am., cult. for fls. and autumn fol., esp. *A. canadensis* (shad bush) and *A. ovalis* (snowy mespilus).

6

American barberry *Berberis canadensis:* — **basswood,** *Tilia americana:* — **beech,** *Fagus ferruginea:* — **bluebell,** *Campanula americana:* — **century plant,** *Agave americana:* — **chestnut,** *Castanea dentata:* — **columbine,** *Aquilegia canadensis:* — **cowslip,** *Dodecatheon:* — **cress,** *Barbarea praecox:* — **currant,** *Ribes sanguineum:* — **dewberry,** *Rubus canadensis:* — **ebony,** *Brya ebenus:* — **elder,** *Sambucus canadensis:* — **elemi,** *Bursera:* — **ephedra,** *Ephedra americana:* — **flytrap,** *Apocynum androsaemifolium:* — **grass** (W.I.), *Thysanolaena agrostis:* — **holly,** *Ilex opaca:* — **hornbeam,** *Carpinus caroliniana:* — **laurel,** *Kalmia:* — **laurel, great,** *Rhododendron maximum;* — **lotus,** *Nelumbo lutea:* — **mandrake,** *Podophyllum peltatum:* — **mastic,** *Schinus molle:* — **mistletoe,** *Phorodendron:* — **pennyroyal,** *Hedeoma pulegioides:* — **raspberry,** *Rubus strigosus:* — **senna,** *Cassia marylandica:* — **spikenard,** *Aralia racemosa:* — **swamp lily,** *Saururus cernuus:* — **vervain,** *Verbena hastata:* — **waterweed,** *Elodea canadensis:* — **wild gooseberry,** *Ribes cynosbati:* — **wild olive,** *Osmanthus americanus:* — **wisteria,** *Wisteria frutescens:* — **witch elder,** *Fothergilla:* — **wormseed,** *Chenopodium ambrosioides.*

Amherstia *A. nobilis,* Burma, cult. trop., remarkably attractive pink fls. and inflor.: has been described as the most beautiful of all fl. trees.

Ammi *A. visnaga,* Medit., As. Minor, fr. long used in folk medicine.

Ammobium *A. alatum,* Aus., an 'everlasting', silvery fl. heads 3–4 cm across.

Ammoniacum Or 'gum ammoniacum', a med. gum-resin obtained from *Dorema ammoniacum* and other spp. in As. Minor.

Amorphophallus Fleshy-rooted pls., Old World trop., some v. large, e.g. *A. titanum* Sumatra (this and *Rafflesia,* also of Sumatra, may be the world's largest 'flowers'), l. stalk up to 4 m long, l. 4 m across, spathe 8 m long and spadix-over 1 m long. *A. campanulatus* is a r. crop in S.Ind. and elsewhere. *A. rivieri,* China, cult. as a room pl. (The *Amorphophallus* 'flower' is, of course an inflor.).

Ampelopsis Name used hort. for three different climbing pls., viz. *Parthenocissus quinquefolia, P. thomsonii* and *P. tricuspidata* (*Ampelopsis veitchii* or 'Virginia creeper').

Amphicome *A. arguta* and *A. emodi,* Himal., cult. perens. w. showy fls.

Anabasis *A. aphylla,* As. Minor, shrub, has been used as insecticide.

Anacharis Submerged aquatic pls., N.Am., used in aquaria.

Anchovy pear *Grias cauliflora,* cult. trop., ed. fr., large orn. ls. 1 m × 23 cm wide.

Anchusa Cult. orn. fls., esp. *A. italica,* many vars., fls. attract hive bees.

Andra-assy oil *Joannesia princeps,* S.Am., seed oil a drastic purgative.

Andes rose *Bejaria* (=*Befaria*) *racemosa* and other spp.

Andiroba (Braz.) *Carapa guianensis,* an imp. S.Am. timber, seeds yield oil.

Andromeda, Marsh *A. polifolia,* N.Hemisp., cult. orn. shrub, many vars.

Andropogon Tall trop. and subtrop. grasses, some used for thatching or sand fixation.

Androsace Primula-like pls., cult.

Anemone Genus includes cult. orn. fls., e.g. *A. hupehensis* (Japanese anemone), and some med. pls.: **wood —,** *A. nemorosa.* See *Anemone* in Willis.

Anemopaegma Mainly climbing fl. shrubs, S.Am., cult. warm cli.

Angelica *Angelica archangelica* (=*Archangelica officinalis*), Eur., naturalized

7

in Br., culinary herb: green stems or petioles preserved in sugar, like ginger, and much used in confectionery and cooking: ls. used for flavouring sauces and wines. — **tree,** *Aralia,* China, cult.: **wild** —, *Angelica sylvestris,* common in Br.

Angelin *Andira inermis,* trop. Am., W.I., a trade timber (partridge wood).

Angelonia Orn. perens. or shrubs, C. & S.Am., related to *Alonsoa,* cult.

Angel's eyes *Veronica chamaedrys:* — **tears,** *Narcissus triandrus:* — **trumpet,** *Datura suaveolens,* widely cult. shrub, warm cli., large scented white fls.: — **wand,** *Chamaelirium luteum,* N.Am.

Angico *Piptadenia rigida,* S.Am., a timber tree, also yields gum.

Angostura *Galipea officinalis, Cusparia febrifuga,* S.Am., b. has aromatic and stimulant properties, used for angostura bitters.

Anguloa Large S.Am. orchids, some spp. cult.

Anigozanthos W.Aus. pls. w. handsome, if strange, woolly fls., cult. *See* kangaroo paw.

Anise *Pimpinella anisum,* dried frs. a much-used spice and flavouring material, distilled for oil which is used med. and for liqueurs (anisette). **Chinese** —, *Illicium verum:* **Jap. star** —, *I. anisatum:* **purple** —, *I. floridianum:* **star** —, *I. verum,* tree of SW. China, frs. distilled for oil, used like anise: — **tree,** *I floridianum.*

Anjan *Hardwickia binata,* Ind., a durable wood for constructional work.

Ankalaki butter *Polygala butyracea,* W.Af. ed. fat from seeds.

Annatto *Bixa orellana,* trop. Am. shrub or small tree often cult. for orn. in warm cli.: seed coat is the source of a reddish yellow dye used mainly in colouring foodstuffs, esp. cheese and butter, as it is quite harmless.

Annual Completing the life cycle within a year: — **rings,** the growth rings seen in sectioning or cutting across the stems of woody pls., esp. the trunks of deciduous trees. Interpreted w. caution they may give an indication of age. Some trees from the tropics, where growth may be continuous, do not show clearly defined growth or annual rings.

Anodendron *A. paniculatum,* Sri Lanka, climber, strong bast fibre, used for fishing nets.

Anoectochilus Some spp. cult., orn. foliage, mainly trop. As.

Ant-orchid (Aus.) *Chiloglottis formicifera:* — **plants,** *see* myrmecophilous pls.

Antelope grass (W.Af.) *Echinochloa pyramidalis,* seeds made into meal: — **horn,** *Asclepias* (= *Asclepiodora*) *viridis,* N.Am.: — **orchid,** *Dendrobium canaliculatum,* Aus.

Anthericum Cult. orn. pls.; some are good border pls.

Anthurium Many spp. cult. for orn. fol. or spathe.

Anthyllis Sun-loving orn. pls., cult.

Antiaris *Antiaris africana, A. welwitchii,* trop. Af., commercial timber.

Antigonon *See* coral vine.

Antirrhinum *See* snapdragon.

Antrophyum Cult. orn. ferns w. unusual simple fleshy fronds: sunken sori.

Anu *Tropaeolum tuberosum,* a starchy r. crop of the Andes.

Apache beads *Anemopsis californica,* aquatic, rs. made into cylindrical beads.

Aphelandra Trop. and subtrop. orn. pls., Am., *A. squarrosa* var. *louisae*, a popular house pl., yellow inflor., orn. ls.

Apio *Arracacia xanthorrhiza*, a S.Am. r. crop.

Apitong *Dipterocarpus* spp. commercial timber, also called 'bagac', Philippines: *D. grandiflorus, D. lasiopodus, D. vernicifluus.*

Apocynum Some spp. cult. for orn., e.g. *A. androsaemifolium*, E. N.Am., some are poisonous, e.g. *A. venetum*, S.Eur., dogbane.

Aponogeton Genus which includes many cult. orn. aquatics.

Apple *Malus pumila* (=*Pyrus malus*). The apple has been cult. from ancient times and the numerous cultivars are considered to be derived from the wild sp., *Malus pumila*, found wild over much of Eur. and As. It is considered that two other spp., *Malus sylvestris* and *M. prunifolia*, may have been ancestors of some of the cultivated apples. Most of the vars. now cult. in Br. have originated or been bred in Br., many of them during the last two centuries. In Br. there is a marked distinction between 'eating' and 'cooking' apples but this does not apply in many other countries. Special vars. are cult. for cider manufacture. The wood of the apple is hard and heavy but not extensively used, with supplies often more or less restricted to old orchard trees. It is close-grained and a good carving wood, although not as good as pear for this purpose: may be used for gun stocks, tool handles, mallets etc.; once used for long case (grandfather) clocks.

Numerous other frs. or pls. have acquired the name apple for one reason or another, e.g.: akee —, *Blighia sapida:* **alligator** —, *Annona palustris:* **balsam** —, *Momordica balsamina:* **bush** —, *Heinsia:* **Chinese** —, *Malus prunifolia:* **crab** —, *Malus* spp. many cult., orn.: **custard** —, *Annona* spp.: **Dingaan's** —, (S.Af.), *Doryalis caffra:* **elephant** —, *Feronia elephantum:* **emu** —, *Owenia acidula:* **golden** —, *Spondias cytherea, Aegle marmelos:* **kangaroo** —, *Solanum aviculare:* **Kei** — (S.Af.), *Doryalis caffra:* **love** —, tomato: **Malay** —, *Eugenia malaccensis:* **mamee** —, *Mammea americana:* **mammee** — (Af.), *Ochrocarpus africana:* **may** —, *Podophyllum peltatum:* **monkey** —, *Annona, Strychnos, Anisophyllea:* **Otaheite** —, *Spondias cytherea:* **prairie** —, *Psoralea esculenta:* **rose** —, *Eugenia jambos:* **sand** —, *Parinari*, S.Af.: **— of Sodom**, *Calotropis procera, Solanum* spp., E.Af.: **star** —, *Chrysophyllum cainito:* **sugar** —, *Annona:* **thorn** —, *Datura stramonium:* **velvet** —, *Diospyros:* **wood** —, *Feronia elephantum.*

Appleberry (Aus.) *Billardiera scandens*, evergreen climber, cream fls. cult.

Apricot *Prunus armeniaca* (=*Armeniaca vulgaris*), well known fr. originated in China: introduced to Britain about 1500, being called at first abricock or apricock: an imp. commercial fr. in many countries, being eaten raw, dried, candied, canned and as jam: used for cordials, wine, apricot brandy and liqueur: oil from kernels used in cosmetics. **Af.** —, *Ochrocarpus africanus:* **Briancon** —, *Prunus brigantina:* **Jap.** —, *P. mume:* **Santa Domingo** —, *Mammea americana:* **— vine**, *Passiflora incarnata:* **wild** — (S.Af.), *Landolphia capensis.*

Aquarium plants Many water pls. and bog pls., from temperate and warm countries, are regularly grown in aquaria. They may be intended for orn., to aerate the water, or for the benefit of the fish in other ways. The successful choice of suitable aquarium pls. depends upon a number of factors, such as

9

the size of the aquarium, the amount of light (esp. sunlight) the aquarium is likely to be subjected to and the nature of the water, hard (alkaline) or soft. Too much sunlight is likely to result in the profuse growth of algae which causes the death of some aquarium pls. Common trop. aquarium pls. include spp. in the following genera – *Alternanthera*, *Aponogeton*, *Ambulia*, *Anubias*, *Bacopa*, *Blyxa* (esp. *B. echinosperma*), *Cabomba* (esp. *C. aquatica*), *Ceratopteris* (*C. thalictroides*, a fern), *Cryptocoryne*, *Echinodorus* (esp. *E. berteroi*, 'cellophane plant'), *Eichhornia*, *Floscopa*, *Heteracantha* (esp. *H. zosterifolia*), *Hydrilla* (esp. *H. verticillata*), *Hygrophylla* (esp. *H. polysperma*), *Pistia*, *Rotala*, *Salvinia* (water fern), *Synnema* (*S. triflorum*, 'water wisteria'), *Vallisneria*, *Wolffia*.

Pls. used for cold aquaria include many temp. aquatics, some of them common Eur. or Br. and Am. water pls. They include spp. of – *Alisma*, *Azolla* (fern), *Callitriche*, *Cardamine* (esp. *C. lyrata*), *Ceratophyllum*, *Luronium*, *Elodea*, *Lemna*, *Limnobium*, *Ludwigia*, *Myriophyllum*, *Nymphoides* (esp. *N. peltata*), *Sagittaria*, *Vallisneria* (esp. *V. spiralis*). [de Wit, H. C. D. (1964) *Aquarium plants*, 255 pp.]

Aquilaria *A. agallocha*, *see* eagle wood.

Aquilegia *A. vulgaris*, columbine, and other spp. cult. orn. fls., many hybrids.

Arabian coffee *Coffee arabica*, mts. of Ethiopia, taken to Arabia, long cult. *See* coffee.

Arabis *A. albida* (=*A. caucasica*) and other spp., cult. orn. pls., esp. rock gdn. and edging.

Arachis *See* groundnut.

Arachnis As. orchids w. showy fls. and persistent distichous ls., cult.

Aralia Trees, shrubs or herbac. pls., N.Am., As., Aus., several spp. cult.

Aramina fibre *Urena lobata*, a jute-like bast fibre, used for coffee bags, cultivated commercially in Braz. and Congo (Congo jute).

Arar *Tetraclinis articulata*, NW.Af., wood used since ancient times, tree yields sandarac.

Araroba (Braz.) *Andira araroba*, a brown powder form cavities in trunk ('Goa powder') used med.

Araucaria Cult. orn. or timber trees, e.g. bunya-bunya, Chile, Parana and Moreton Bay pines. (*See also Araucaria* in Willis.)

Araujia *A. sericofera*, S.Am., vigorous orn. climber, white fls., cult. warm cli.

Arboloco *Montanoa moritziana*, Columbia, timber much used locally.

Arbor-vitae Name applied to *Thuja* spp. (=*Thuya*), conifers native to E.As. and N.Am.: some cult. for ornament and for hedges, wood light, soft, fragrant w. many uses. **Am.** —, *T. occidentalis*, many vars.: **Chinese** —, *T. orientalis:* **Jap.** —, *T. japonica:* **western** —, *T. plicata* N.Am. also called western Red Cedar, an imp. commercial wood.

Arbour A place to sit, usu. protected by trees or climbers, often part of a pergola. — **vine**, Spanish (W.I.), *Ipomoea tuberosa*.

Arbutus Evergreen trees or shrubs w. orn. bark, cult., e.g. strawberry tree: **trailing** —, *Epigaea repens*, N.Am., cult. (Mayflower).

Archangel, yellow *Lamiastrum* (= *Lamium*) *galeobdolon*, Eur., inc. Br.

Archontophoenix *A. cunninghamiana*, Illawarra palm, Queensland, cult.

Arctostaphylos Small orn. trees or shrubs, mainly Am., cult.

Arctotis Or Af. daisy, orn. fls., cult., many spp. and hybrids.

Ardisia *A. crispa*, E.I., widely cult., scarlet, persistent frs., a pot pl.

Areca nut *See* betel nut.

Arenaria Some spp. cult. mainly in rockeries, others are weeds.

Argan *Argania spinosa* (=*A. sideroxylon*), Morocco, seeds yield ed. oil, like olive oil.

Arghan fibre *Bromelia magdalene*, Colombia, l. fibre once exported.

Argyle apple (Aus.) *Eucalyptus cinerea*; **Duke of** —'s **tea tree**, *Lycium barbarum*.

Argyreia Elegant climbers of Old World trop., showy fls., cult.

Arisaema Cult. orn. tuberous pls. allied to *Arum*, As., Am.

Aristolochia Many cult. orn. climbers, trop. and subtrop., e.g. *A. gigas*, pelican fl.; *A. sipho*, Dutchman's pipe; *A. elegans*, calico fl.; *A. ornithocephala*, bird's head; many spp. have local med. uses: several used for snake bite.

Aristotelia *A. macqui*, Chile, small tree w. ed. frs., esteemed for wine.

Arizona poppy *Kallstroemia grandiflora*, attractive desert ann.

Armeria Cult. orn. tufted perens. many alpine.

Arnatto *See* annatto.

Arnica Or mountain tobacco, *A. montana*, Eur., N.Am., rhis. and fls. med.

Arrabidaia *A. magnifica*, Colombia, climbing shrub w. large showy fls., 9 cm across.

Arracacha *Arracacia xanthorrhiza*, Andean r. crop, resembles parsnip, strong flavour.

Arrack The potable spirit prepared from the juice of various palms in trop. Asia, e.g. *Arenga, Borassus, Cocos, Caryota, Corypha* etc.

Arrowgrass Common —, *Triglochin palustris*: **marsh** —, *T. maritima*, Br.

Arrowhead *Sagittaria sagittifolia* and other spp., cosmop. aquatics, starchy rs. eaten.

Arrow poisons In many parts of the world primitive peoples make use of poisoned arrows or blow-pipe darts for hunting or combating their enemies. The poisons used for tipping such arrows or darts are usually of a mixed nature and derived from more than one sp. of pl. They may also have mixed with them an adhesive such as a veg. gum or may have animal products mixed with them but these may not be particularly toxic or may be of magical significance only.

An important arrow poison in E. & N.Af., used by many tribes, is *Acokanthera schimperi*. Almost any part of the pl. may be used and toxicity may vary with the time of year. The poison, a cardiac glucoside, is extremely powerful and can kill in a few minutes. Another common E.Af. arrow poison is derived from *Adenium obesum*. It contains alkaloids of the digitalis type, acting on the cardiovascular and nervous systems.

The Amerindians of trop. S.Am. commonly use spp. of *Strychnos* for arrow and dart poisons, notably *Strychnos toxifera*, one of the sources of the famous *curare* poison. The poison causes paralysis of the lungs. The Old World spp. of *Strychnos* contain alkaloids of the strychnine class, causing tetanus-like convulsions. In Braz. the long spines of the palm *Jessenia batana* are used for blow-pipe darts, a cotton-like seed floss being wrapped round the end so that it fits correctly into the bore of the blow-pipe. A hollow stem of some sort, about 2 m in length, sometimes of bamboo, is used for the blow-pipe.

Arrowroot True arrowroot, also called **W.I., St Vincent** or **Bermuda arrowroot**, consists of the starch or starch granules extracted from the rhizomes of *Maranta arundinacea*, a native of C.Am. and cult., trop. and subtrop. It has the general properties of starch, is considered to be readily digestible and is commonly used as an invalid or infant food and as gruel in the treatment of diarrhoea. Many starches have been called 'arrowroot' or used as substitutes for true arrowroot, e.g. **Af.** —, *Tacca leontopetaloides* (= *T. pinnatifida): **Bombay** —, *Curcuma angustifolia:* **Braz.** —, *Manihot esculenta* (cassava): **Chinese** —, *Nelumbium:* **E.I.** —, *Curcuma angustifolia:* **English** —, *Solanum tuberosum* (potato!): **Fiji** —, *Tacca leontopetaloides:* **Florida** —, *Zamia:* **Guyana** —, *Dioscorea alata:* **Hawaiian** —, *Tacca leontopetaloides:* **Ind.** —, *Curcuma angustifolia:* **Jap.** —, *Pueraria thunbergii:* **marble** —, *Myrosma cannifolia:* **Para** —, *Manihot esculenta* (cassava): **Portland** —, *Arum maculatum:* **Queensland** —, *Canna edulis:* **Rio** —, *Manihot esculenta:* **Tahiti** —, *Tacca leontopetaloides.*

Arrow-weed *Tessaria*, N.Am.

Arrow-wood *Viburnum dentatum*, *V. acerifolium*, N.Am., young straight shoots used for arrows.

Artabotrys Some spp. cult. for fragrant fls. or ed. fr.

Artemisia Several spp. cult. for orn. fol., others med.

Artichoke Common, globe, or **French** —, *Cynara scolymus*, young fl. heads a much used veg., also blanched shoots, many vars.: **Chinese** or **Jap.** — *Stachys sieboldii*, ed. tubers: **Jerusalem** —, *Helianthus tuberosus*, ed. tubers, also fed to livestock.

Artillery plant *Pilea muscosa* (= *P. microphylla*), trop. Am., cult., also called gunpowder or pistol pl.: puffs of pollen ejected by exploding stamens: may be induced by touching or shaking the pl.

Artocarpus Several trees provide useful timber others ed. frs., e.g. bread-fruit, jak. See *Artocarpus* in Willis.

Arum Several spp. cult., mainly for orn. fol.: **dragon** —, *Dracunculus vulgaris* (= *Arum dracunculus*), Medit., cult.

Arum lily *Zantedeschia aethiopica*, S.Af., the common white sp. widely cult. much used for wreaths. **Black throated** — —, *Z. albomaculata:* **golden** — —, *Z. pentlandii:* **pink** — —, *Z. rehmannii:* **yellow** — —, *Z. elliottiana*, *Z. albomaculata.*

Aruncus *A. sylvester*, N.Hemisph., cult. gdn. pl.

Arundina Terrestrial orchids, trop. As., showy fls. in terminal racemes.

Arundinaria Bamboos widely cult. for orn. poles or stakes.

Asafoetida Oleo-gum-resin w. med. uses obtained from spp. of *Ferula*, notably *F. foetida* and *F. rubricaulis:* may be used in veterinary medicine.

Asarabacca *Asarum europaeum*, Eur. naturalized in Br., med., purgative.

Asarum Several spp. cult., dark resin-scented fls. visited by flies.

Asclepias Cult. orn. pls., e.g. *A. curassavica*, w. red. fls., a trop. weed.

Ascorbic acid Or vitamin C, is widespread in the veg. kingdom, rich sources being citrus frs., guavas, rose-hips, and blackcurrants: may be prepared synthetically: used in treating the undernourished, bottle-fed infants etc.

Ash *Fraxinus excelsior*, **common** or **Eur. ash**, wild and cult., long esteemed for valuable timber, strong, elastic, straight-grained and bends well, favoured for wheels and shafts, motor vehicle bodies, agricultural implements and sports goods. Other commercially imp. kinds of ash are – **Am.**, **Can.**, or **white** —, *F. americana*, *F. pennsylvanica*, *F. nigra:* **Jap.** —, *F. mandshurica*, often highly figured and used for furniture. Other trees or plants that have acquired the name of 'ash' include – **Alpine** —, *Eucalyptus delegatensis*, Aus., *Fraxinus nigra*, N.Am.: **black** or **brown** —, *F. nigra:* **blue** —, *F. quadrangularis:* **blueberry** —, *Elaeocarpus* spp.: **Blue Mt.** —, *Eucalyptus oreades:* **Canary** —, *Beilschmiedia bancroftii:* **Cape** —, *Ekebergia capensis:* **crow's** —, *Flindersia australis:* **flowering** —, *Fraxinus ornus:* **grey satin** —, *Eugenia gustavioides:* **manna** —, *Fraxinus ornus*, cult. in Sicily for the sweet exudate or 'manna': **mountain** —, *Sorbus aucuparia:* **Oregon** —, *Fraxinus oregona:* **prickly** —, *Zanthoxylum* spp.; *Aralia spinosa:* **— pumpkin,** *Benincasia hispida:* **red** —, *Fraxinus pubescens; Alphitonia* spp.: **silver** —, *Flindersia pubescens*, Aus.: **white** —, *Fraxinus americanus:* **yellow** —, *Emmenosperma alphitonioides*, *Eucalyptus luehmanniana.*

Ashanti blood (Ghana) *Mussaenda erythrophylla*, a common W.Af. shrub: one sepal of fl. enlarged and bright red making the fl. v. conspicuous: fls. attractive to humming birds in W.I. **— pepper,** *Piper guineense.*

Ashplant *Leucophyllum frutescens*, *L. texanum*, SW. U.S., used in desert landscaping.

Ashwagandha (Ind.) *Withania somnifera*, an ancient drug used in Ayurvedic med.

Asoka tree (Ind.) *Saraca indica*, cult. orn. tree, clusters of orange-red fls.

Asparagus *Asparagus officinalis*, gdn. asparagus, the tender young shoots a much esteemed veg., must be grown in well-prepared beds of rich soil: extensively canned. **Bath** —, *Ornithogalum pyrenaicum:* **— bean,** *Dolichos unguiculata* (= *D. sesquipedalis*): **— bush,** *Dracaena mannii:* **Cape** —, *Aponogeton distachys*, young shoots ed.: **— fern,** *Asparagus plumosus*, used by florists, many vars.: **— pea,** *Psophocarpus tetragonolobus*, *Tetragonolobus purpureus:* **— tree,** *Dracaena mannii:* **wild** — (S.Af.), *Asparagus africanus*, young shoots eaten: **white** —, *A. alba*, N.Af., shoots ed.

Aspasia Cult. C. & S.Am. orchids.

Aspen *Populus tremula*, Eur., wood much used for matches, chip-baskets for fr. etc. **Can.** —, *P. tremuloides* has similar uses.

Asperula Many cult. orn. pls., esp. for rock gdn. and fl. border.

Asphodel *Asphodelus microcarpus* (= *A. aestivus*), Medit., starchy tubers may be eaten, also used for gum. **Bog** —, *Narthecium ossifragum:* **giant** —, *Eremurus* spp.: **Scottish** —, *Tofieldia pusilla:* **yellow** —, *Asphodeline lutea.*

Aspidistra *A. lurida*, China, long cult. as an indoor fol. pl., esp. in Victorian times, easily grown and tolerant of neglect.

Aspilia Some spp. trop. weeds, e.g. *A. africana*, *A. helianthoides.*

Asplenium Includes many cult. orn. ferns, easy to grow as house pls., e.g. *A. bulbiferum* has young pls. developing on fronds, *A. nidus* (bird's nest fern).

Assai palm *Euterpe edulis*, Guyana, young crown or bud ed., palm cabbage.

Assagai wood *Curtisia dentata* (= *C. faginea*), S.Af., a strong durable wood, used by Zulus for spear or assagai shafts and formerly in wagon-making by early settlers.

Assam indigo *Strobilanthes flaccidifolius*, yields a dye.

Aster *Aster* spp. Many cult. gdn. fls. belong to the genus including the 'Michaelmas daisies', mainly of N.Am. origin. The popular ann. gdn. or **China** —, is *Callistephus chinensis*. **Mex.** —, *Cosmos:* **S.Af.** —, *Felicia* spp.

Astilbe The genus includes many good gdn. pls., wide colour range.

Astrocaryum Several spp. of these trop. Am. palms yield fibre and oil from the frs., used locally, or are cult. for orn.

Asystasia Cult. orn. pls., trop. and subtrop., v. showy fls.

Athyrium Cult. orn. ferns.

Atlantic cedar *Cedrus atlantica*, Atlas Mts., NW.Af., several hort. vars.

Atractylis *A. gummifera*, Medit., yields a 'gum' used to adulterate mastic.

Atraphaxis Cult. orn. shrubs. often spiny, adapted for dry cli.

Atropine An alkaloid obtained from *Duboisia* spp. (Aus.), *Hyoscyamus* or other Solanaceous pls. or by synthesis: used in ophthalmology to dilate the pupil and has many other med. uses: may cause poisoning, e.g. belladonna frs. with children.

Attar of rose *See* otto of rose.

Aubergine Name for brinjal, egg-plant or garden egg, *Solanum melongena*.

Aubrietia *Aubrieta deltoidea*, E.Medit., popular wall and rock gdn. pl., numerous cultivars.

Aucuba *A. japonica*, Jap. laurel, popular evergreen shrub.

Aunt Lucy *Ellisia nyctelia*, N.Am.

Auricula Two main types – (1) **'show'** —, *Primula auricula* and (2) **'alpine'** —, *P. pubescens*. The former have the characteristic mealy fls., usu. cult. under glass for protection.

Australian blackwood *Acacia melanoxylon:* — **bluebell**, *Wahlenbergia gracilis:* — **chestnut**, *Castanospermum australe*, cult., orn. and street tree: — **currant**, *Leucopogon:* — **gum**, *Eucalyptus:* — **daisy**, *Vittadinia:* — **desert kumquat**, *Eremocitrus glauca:* — **feather palm**, *Ptychosperma:* — **fuchsia**, *Correa*, *Epacris:* — **grass tree**, *Xanthorrhoea:* — **heath**, *Epacris:* — **honeysuckle**, *Banksia:* — **kino**, *Eucalyptus rostrata:* — **mint**, *Prostanthera:* — **red berry**, *Rhagodia:* — **red cedar**, *Cedrela toona* var. *australis:* — **sandalwood**, *Fusanus*.

Austrocedrus *A.* (= *Libocedrus*) *chilensis*, Chilean cedar.

Autumn crocus *Colchicum autumnale*, Eur., source of colchicine. — **snow-flake**, *Leucojum autumnale*.

Auxins *See* hormones.

Avaram bark *Cassia auriculata*, Ind., an imp. tanning material in Ind. at one time: basal coppice shoots are cut and the b. peeled off. The thin b. dries in quill form and has a tannin content of about 18 %.

Avens, mountain *Dryas octapetala*, cult., trailing pl., several vars.: **water** —, *Geum rivale:* **white** —, *G. canadense:* **wood** —, *G. urbanum:* **yellow** —, *G. japonicum*.

Avignon berry *Rhamnus infectoria*, SW.Eur., frs. formerly used in dyeing.

Avocado pear *Persea americana* (= *P. gratissima*), C.Am., a well known 'hors d'oeuvres' fr. widely cult. in warm cli.: a notable feature is its richness in protein and fat or oil for a dessert fr., eaten in various ways, often w. other food: oil used in cosmetics. **Coyo** —, *Persea schiedeana:* **Mex.** —, *P. drymifolia.*

Avodiré *Turraeanthus africanus,* W.Af., a commercial timber, used for furniture in Br.

Awarra palm *Astrocaryum tucumoides,* Guyana, ls. used for mats.

Awlwort *Subularia aquatica,* has awl-like ls.

Awusa nut (W.Af.) *See* conophor nut.

Ayan *Distemonanthus benthamianus,* W.Af., a commercial timber.

Ayapana *Eupatorium ayapana,* Braz., ls. used as med. tea, cult.

Azalea A subgenus of *Rhododendron:* cult. orn. fl. shrubs: ls. deciduous: **Alpine** —, *Loiseleuria procumbens:* **flame** —, *Rhododendron calendulaceum,* N.Am.: **Florida flame** —, *R. austrinum,* N.Am.: **Ind.** —, *R. simsii:* **mock** —, *Adenium coetaneum:* **swamp** —, *Rhododendron viscosum* N.Am.: **trailing** —, *Loiseleuria procumbens,* arctic.

Azaleodendron Hybrids between subgenus *Azalea* and other *Rhododendron* spp.

Azara *Azara* spp., Chile, cult. evergreen shrubs w. small fls.

Azarole *Crataegus azarolus,* S.Eur., long cult. for ed. apple-flavoured frs.: also for hedges.

Azolla Floating aquatic ferns, may increase rapidly: *A. caroliniana* is an alien in Br. *See Azollaceae* in Willis.

Aztec clover *Trifolium amabile:* — **pine,** *Pinus teocote:* — **tobacco,** *Nicotiana rustica.*

B

Babassu palm *Orbignya speciosa,* Braz., kernels yield oil.

Babe-in-a-cradle orchid *Epiblema grandiflorum,* SW.Aus.

Babiana *Babiana* spp., S.Af. cult orn., corms eaten by baboons.

Baboon root *Babiana plicata,* S.Af., corms eaten by baboons and early settlers.

Babul bark *Acacia arabica,* Ind., formerly an imp. tanning material in Ind., later supplanted by wattle b. and extract: tannin content about 12%: tree imp. for fuel.

Baby's breath *Gypsophila paniculata,* cult. gdn. pl.; *Galium sylvaticum,* N.Am.

Bacaba palm *Oenocarpus distichus,* S.Am., beverage prepared from frs., also cooking oil.

Baccaurea *B. dulcis* and other spp., Malaya, fr. ed., used for beverages.

Baccharis Several spp. med. in S.Am.

Bachelor's buttons Name used for several pls., e.g. double fl. forms of *Bellis perennis, Centaurea cyanus, Ranunculus acris,* also *Gomphrena globosa, Lychnis* and *Craspedia uniflora,* Aus. **Yellow** — —, *Polygala lutea, P. rugelii,* N.Am.: — **pears,** frs. of *Solanum mammosum.*

Backra yam (Jam.) Good variety of yam (*bakra* may be used for anything good or superior).

Bacon and eggs *Daviesia* spp., *Nemcia capitata,* W.Aus.

BACOPA

Bacopa Aquatic perens., mostly Am., cult.

Bacu or **Baku** *Tieghemella heckelii* (= *Dumoria, Mimusops*), W.Af., ed. fat from kernels; mahogany-like wood.

Bacury *Platonia insignis*, Braz. ed. fr. canned for export.

Badger's bane *Aconitum lycoctonum*, N.As., cult., many gdn. vars.

Badinjan (W.I.) Brinjal, *Solanum melongena*.

Baeckea *B. frutescens*, trop. As., ls. used as tea.

Bael fruit *Aegle marmelos*, trop. As., widely cult., fr. pulp ed., used in conserves: tough durable wood.

Bag flower *Clerodendron thomsoniae*, W.Af., striking red and white fls., cult.

Bagac Or apitong, *Dipterocarpus grandiflorus*, Philippines, imp. commercial timber.

Bagasse Crushed sugar cane after extraction of juice, mainly used as fuel for cane mills, also for fibre board, paper and for mulching.

Bagpod *Sesbania vesicarum* (= *Glottidium*) N.Am., a weed poisonous to livestock.

Bahama grass *Cynodon dactylon:* — **hemp**, *Agave sisalana:* — **whitewood**, *Canella alba*.

Bahia fibre Name for piassaba fibre: — **grass**, *Paspalum notatum:* — **piassaba**, *Attalea funifera:* — **rosewood**, *Dalbergia nigra:* — **wood**, *Caesalpinia brasiliensis*.

Baib grass *Eulaliopsis binata* (= *Ischaemum angustifolius*), Ind., considered for paper.

Bajri (Ind.) *Pennisetum typhoides*, bulrush millet.

Baked apple berry (N.Am.) *Rubus chamaemorus*.

Bakerantha *B.* (= *Bakeria*) *tillansoides*, S.Am., cult. orchid, pink fls. in large inflor.

Bakupari *Rheedia brasiliensis*, S.Am., ed. tree fr., esteemed in Braz.

Bakury *Platonia insignis*, trop. S.Am., ed. tree fr., seeds yield oil.

Balanites *B. aegyptiaca* and other spp., seeds yield oil, fr. ed.

Balata Non-elastic rubber obtained by tapping *Manilkara bidentata* (= *Mimusops balata*), trop. Am., used mainly for machinery belting and substitute for chicle. *Ecclinusa balata*, Braz., yields a substitute or inferior balata.

Balau *Shorea* spp., esp. *S. glauca* and *S. maxwelliana*, Malaya, heavy commercial hardwood. **Red —**, *S. guiso* and *S. kunstleri*.

Bald money *Meum athamanticum*, Eur. inc. Br., aromatic frs.

Balisier (W.I.) *Calathea discolor, Helicornia bihai*, large ls. w. many uses – baskets etc.

Balustine flowers Name used for dried pomegranate fls., used med.

Ball moss Or bunch moss, *Tillandsia recurvata*, N.Am.: — **mustard**, *Neslia:* — **nettle**, *Solanum*, N. Am.

Balloon flower *Platycodon grandiflorus*, China, large blue fls. balloon-like in bud, good border pl.: — **pea**, *Sutherlandia frutescens*, S.Af.: — **vine**, *Cardiospermum*, trop.

Balm *Melissa officinalis*, S.Eur., peren. herb, sweet- or lemon-scented, popular in gdns., used in seasoning to impart lemon taste, also used in scents and liqueurs. In beekeeping it is claimed that rubbing the inside of skeps or boxes with crushed ls. encourages bee swarms to occupy them. **Bastard —**, *Melittis*

16

melissophyllum: **bee** —, *Monarda didyma,* N.Am., cult. gdn. pl., red fls., many vars.: — **of Gilead,** *Abies balsamea, Populus tacamahaca, P. gileadensis, Commiphora opobalsamum* (= *C. gileadensis):* **horse** —, *Collinsonia canadensis:* **mountain** —, *Eriodictyon glutinosum,* N.Am.: **western** —, *Monardella parviflora.*

Balmony *Chelone glabra,* N.Am.

Balsa *Ochroma pyramidale* (= *O. lagopus*), trop. Am., cult., a quick-growing small tree yielding the world's lightest commercial wood, lighter than cork, w. numerous special uses, insulation, model aeroplanes etc. Other lightweight trop. Am. woods may be called balsa.

Balsam apple *Momordica balsamina,* cult. vine, trop., fr. a veg.: — **bog,** *Azorella caespitosa,* grows in tufts constituting the balsam-bog of the Falkland Isles: **broad-leaved** —, *Oreopanax capitatum,* W.I.: **Canada** —, *Abies balsamea,* N.Am., used in optical and microscopical work: **Copaiba** —, *Copaifera langsdorfii* and other spp., S.Am., med.: — **fig,** *Clusia rosea,* W.I.: — **fir,** *Abies balsamea,* N.Am.: **gdn.** —, *Impatiens balsamina,* Ind., numerous vars.: — **of Gilead,** *Commiphora opobalsamum:* **Gurjan** —, *Dipterocarpus:* **Ind.** —, *Impatiens glandulifera,* naturalized in Br.: **Mecca** —, *Commiphora opobalsamum* (= *C. gileadensis):* **orange** —, *Impatiens capensis:* — **pear,** *Momordica charantia,* trop. vine, fr. a veg.: **Peru** —, *Myroxylon balsamum* var. *pereirae,* med.: — **poplar,** *Populus balsamifera, P. tacamahaca,* N.Am.: **rock** —, *Clusia, Peperomia:* — **root,** *Balsamorhiza sagittata,* N.Am.: **seaside** —, *Croton eluteria:* **Tolu** —, *Myroxylon balsamum* var. *balsamum,* med.: — **tree,** *Copaifera mopane,* trop. & S.Af.: **wild** —, *Impatiens noli-tangere,* Br.

Balucanat *Aleurites trisperma,* Philippines, drying oil from seeds.

Bambarra groundnut *Voandzeia subterranea,* imp. Af. food crop, many vars.

Bamboo The bamboos are an imp. group of grasses, with numerous economic uses, esp. in As. countries. They vary in size from dwarf to v. large with culms or stems over 30 m in length and 15–30 cm in diam. There are some 45 genera, mainly trop. and subtrop. and some temp. (*see* Willis). The more imp. economic genera include *Arundinaria, Phyllostachys, Bambusa* and *Dendrocalamus.*

The number of uses to which the bamboos are put is legion, especially in As. The stems are hollow, with cross partitions at the nodes and the wood is elastic and very hard, owing to the deposition of silica in the cell walls. The stems are consequently very light and strong and are also easily split. They are largely used in building, entire as posts, and split as roofing tiles. Split bamboo woven into a kind of matting and attached to the uprights may serve as walls. Bridges are often made of bamboo. It also furnishes water pipes, water vessels, gutters, floats, beehives, walking sticks, pipes, flutes, masts, furniture, household utensils, agric. tools etc. Finely split bamboos are made into mats, blinds, rigging, baskets, fans, hats, coarse clothing, umbrellas, ropes, brushes etc. In Ind. and elsewhere bamboo is used in the manufacture of paper pulp. In the stems of *Bambusa arundinacea* curious concretions of silica are found, known as tabashir or bamboo manna, used in the Orient as a medicine. The so-called 'split cane' fishing rods, manufactured from 'Tongking cane' (*Arundinaria amabilis*) are popular with trout fishermen because of their great strength and lightness. Tongking canes of various dimensions are widely used in other ways, especially for gdn. purposes. Bamboo canes have been much used for walking sticks. A popular umbrella handle for mens' umbrellas is made from the base of the Ind. bamboo *Thyrsostachys siamensis*. The largest of the bamboos is the giant bamboo of Burma (*Dendrocalamus giganteus*), cult. in other parts of the trop. The stems or culms may reach a height of over 30 m with a diam. of 25 cm or more near the base, clumps being 15 m in diam. The culms, sawn across at a joint or node, make useful pl. pots or buckets. The young shoots of a number of different bamboos are eaten in the Orient. They must be used while they are v. young, before they become fibrous. Soil is usu. heaped over them beforehand, otherwise they are prone to be bitter. In China ed. bamboo shoots are canned and freely exported to other countries, *Bambusa beecheyana* being much used. Bamboos are extensively cult. for orn. purposes and for screens, windbreaks and hedges in temp. as well as warm countries. It is only the smaller stemmed bamboos that are suitable for the colder cli. [Lawson, A. H. (1968) *Bamboos: A gardener's guide to their cultivation in temperate climates*, 192 pp.]. **Berry-bearing** —, *Melocanna bambusoides*: **black** —, *Phyllostachys nigra*: **Calcutta** —, *Dendrocalamus strictus*: **common** —, *Bambusa vulgaris*: **dwarf** —, *Arundinaria pygmaea*: **fishpole** —, *Phyllostachys aurea*: **Heavenly** —, *Nandina domestica*: **madake** —, *Phyllostachys bambusoides*: **male** —, *Bambusa arundinacea*: **palm** —, *Raphia* spp., *Chrysalidocarpus lutescens*: **red-berried** — (N.Am.), *Smilax walteri*: **savannah** or **solid** —, *Oxytenanthera abyssinica*: **spiny** —, *Bambusa arundinacea*: **spotted** —, *Ochlandra stridula* var. *maculata*: **Terai** —, *Melocanna bambusoides*: **Tongking** —, *Arundinaria amabilis*: **— vines** (N.Am.), *Smilax* spp.: **wild** —, *Smilax auriculata*. See also Bambuseae in Willis.

Banana *Musa* spp. One of the most popular and widely grown frs. of the trop. and subtrop. produced in enormous quantities for export. There are numerous vars. differing in size, shape, colour, flavour of fr. and suitability for handling and export. In Ind. all bananas are called plantain but the word is usu. used for the large-fruited, starchy or cooking varieties imp. for human food in many countries, esp. trop. Afr. All are derived from hybrids of *Musa acuminata* and *M. balbisiana* and may be called *M.* × *sapientum*. Banana or plantain flour may be made from the former and banana figs (peeled dried fr.) from the latter. The **Cavendish** or **Canary** — ('*Musa nana*', '*M. cavendishii*'), of Chinese origin and of v. fine flavour, is favoured in many countries, esp. in the subtrop. There are many cultivars, mainly dwarf. In the Old World trop. there are many wild bananas. The frs. have seeds. Cult. bananas do not. Wild bananas have been used in breeding work. Some are grown as orn. pls. **Abyssinian** —, *Ensete ventricosum*, see ensete: **flowering** —, *Musa coccinea*: **Jap.** —, *Musa basjoo*: **— orchid** (Aus.), *Philidota pallida*: **wild** — (Malaya), *Musa acuminata*: **wild** — (S.Af.), *Strelitzia nicolai* (= *S. angusta*).

Ban rhea *Villebrunea integrifolia*, Ind., shrub or small tree, yields good fibre.

Banak *Virola koschnyi* (= *V. merendonis*) C.Am., a commercial timber.

Bandakai (Ind.) Okra, *Hibiscus esculentus*.

Band plant *Vinca major*, Medit., naturalized in Br.

Baneberry *Actaea spicata*, N.Hemisph., cult. orn. pl., frs. poisonous.

Banksia Aus. evergreen trees and shrubs often w. striking fol. or frs., cult. **Acorn** —, *B. prionotes:* **heath-leaved** —, *B. ericifolia:* **saw** —, *B. serrata:* **silver** —, *B. marginata:* **swamp** —, *B. robur.*

Banyan tree *Ficus benghalensis*, Ind., aerial rs. form supporting pillars and trees may reach immense size.

Baobab *Adansonia digitata*, trop. Af., tree w. unusually thick trunk and characteristic appearance typical of many savannah areas: frs. w. local med. and other uses. **Aus.** —, *Adansonia gregori.*

Baptisia *B. tinctoria* and other spp. once used as a dye.

Barb grass *Monerma cylindrica*, Medit.: **coast** — —, *Parapholis incurva*, Medit., useful colonizer in salt areas (Aus.): **Barbed wire grass** (Aus.), *Cymbopogon refractus*, inflor. resembles barbed wire.

Barbadine (W.I.) *Passiflora quadrangularis.*

Barbados almond (W.I.) *Terminalia catappa:* — **cedar**, *Juniperus bermudiana:* — **cherry**, *Malpighia punicifolia:* — **gooseberry**, *Pereskia aculeata:* — **lilac**, *Melia azedarach:* — **lily**, *Hippeastrum equestre:* — **mastic tree**, *Mastichodendron sloaneanum:* — **pride**, *Caesalpinia pulcherrima:* — **snowdrop**, *Zephyranthes tubispatha.*

Barbara's buttons *Marshallia*, N.Am.

Barbary fig *Opuntia ficus-indica*, Am., barrier hedge, fr. ed., widely naturalized Medit. and elsewhere.

Barbasco Name used in S.Am. for almost any pl. used for poisoning fish, esp. *Lonchocarpus* spp.

Barbatimao *Styphnodendron barbatimao*, Braz., b. (tannin 20–35 %) used for tanning.

Barberry *Berberis* spp. Cult. orn. fl. shrubs usu. w. spines, mainly As., frs. of many ed. or preserved, wood and b. yield yellow dye. **Alleghany** or **Am.** —, *B. canadensis:* **common Eur.** —, *B. vulgaris:* **Darwin's** —, *B. darwinii*, widely cult.: **Ind.** —, *B. aristata:* **Magellan** —, *B. buxifolia:* **Nepal** —, *B. nepalensis:* **stenophylla** —, *B.* × *stenophylla*, a popular hybrid and hedge plant.

Barberton daisy *Gerbera jamesonii*, S.Af., widely cult., many vars.

Barcelona nut *See* hazel nut.

Bareet grass *Leersia hexandra*, As., trop. fodder grass.

Barilla Name for ash of certain pls., mainly sodium carbonate, once used in soap and glass making. *Halogeton*, *Salsola* and *Suaeda* were largely used.

Bark The b. of trees or other pls. varies enormously in physical and chemical properties and may supply many useful products for mankind. Some trees have a thick, soft or spongy b. such as the giant sequoia (*Sequoiadendron giganteum*) where is may be 20 cm thick or more. It is claimed to afford fire protection to the tree. In contrast to this many trees have a very thin or very hard b., such as the so-called 'ironbarks' of Aus. (*Eucalyptus* spp.). Certain bs. such as birch b. (*Betula* spp.) are quite impervious to water. For this reason they have been used for roofing temporary structures and for making canoes by the Amerindians. The b. of the cork oak or commercial cork is in a somewhat special category and is much used (*see* cork). Some bs. yield imp. drugs such

19

as cinchona (quinine) and cascara; others afford spices such as cinnamon and cassia or cinnamon cassia. Tan bs. have long been imp. in the tanning of leather and for supplying tannin extracts, used also for certain other purposes. One of the oldest and best known is oak b., now little used. Black wattle b. and extract (mimosa extract) is however extensively used. There are many other tan bs. Some tree bs. contain imp. products which are obtained by tapping and not by removing the b. from the tree. Rubber or rubber-like substances from various trees and resins afford good examples.

B. is an imp. source of commercial fibre or bast fibres such as hemp or jute and the many jute substitute fibres. The inner b. or bast of the lime tree (*Tilia*) is fibrous and used for matting in W.Eur. The inner fibrous b. of several other trees (notably *Ficus* spp.) after being beaten and treated is used as b.-cloth for clothing in some parts of the world. The lace b. tree (*Lagetta lintearia*) of the Caribbean reg. provides an orn. inner b. resembling lace and used locally for fancy articles.

Barleria Orn. shrubs. cult. trop. and subtrop.

Barley *Hordeum* spp. **Common —**, *H. vulgare*, is a cereal of age-old cult., having been grown by the Ancient Egyptians and grains have been found in the lake dwellings of Switzerland. As human food in Eur. countries it is used less now. It is much used in malting, for beer and whisky, and for animal feeding. The so-called 'pearl barley', much used for soups and stews, is the barley grain rubbed down to a more or less spherical shape. Barley water is an aqueous infusion w. med. uses. The numerous vars. of cult. barley may be grouped according to the nature of the ear – six-rowed, four-rowed and two-rowed (*see Hordeum* in Willis). **Abyssinian —**, *Hordeum irregulare:* **Egyptian —**, *H. trifurcatum:* — **grass**, *H. leporinum*, Medit.: **meadow —**, *H. secalinum:* **wild —**, *H. murinum:* **wood —**, *Hordelymus europaeus.*

Barna tree, sacred *Crateva nurvala*, Ind. cult. orn. tree, white fls.

Barnyard grass *Echinochloa crus-galli*, cosmop. weed.

Barosma *See* buchu.

Barrel cactus *Ferocactus, Echinocactus*, SW. U.S.

Barren wort *Epimedium alpinum.*

Barrier plants Barrier pls. or defensive hedges are often grown as a cheap form of fencing or to keep out trespassers or wild and domestic animals from crops. Usually such pls. are of a thorny or prickly nature, sometimes heavily armed and it is this that accounts for their effectiveness. Among pls. commonly used in temp. countries are hawthorn, holly, barberry and myrobalan.

In warm cli. many more spp. are in use as defensive hedges, especially in the less developed countries of the trop. and subtrop. The following are some of them – *Acacia*, several spp., e.g. *A. armata*, kangaroo thorn; *A. sphaerocephalata*, bull's horn acacia: *Aegle marmelos*, bael fruit: *Agave americana*, century plant and other spp.: *Atalantia buxifolia: Azima tetracantha: Balanites aegyptiaca: Bambusa spinosa, B. arundinacea*, spiny bamboo: *Barleria prionitis: Bougainvillea glabra* (needs training): *Bromelia pinquin: Caesalpinia coriaria*, divi-divi; *C. (Poinciana) pulcherrima*, Barbados pride (sparingly armed): *C. sepiaria*, Mysore thorn, *C. crista*, bonduc and other spp.: *Capparis sepiaria*, hedge caper; *C. spinosa* caper: *Carissa grandiflora*, Natal plum; *C. carandas*, Ceylon damson: *Dichrostachys cinerea: Doryalis caffra*, Kei apple: *Duranta repens*, pigeon berry: *Erythrina corallodendrum*, coral bean and other spp.: *Euphorbia antiquorum* and other spp.: *Fouquieria splendens*, ocotillo: *Furcraea gigantea*, Mauritius hemp: *Gleditschia triacanthos*, honey locust: *Maytenus* (=*Gymnosporia*) *emarginata: Haematoxylon campechianum*, logwood: *Hakea acicularis; H. gibbosa*, (can become a nuisance): *Lantana camara*, many vars. some dwarf: *Lycium afrum*, African boxthorn: *Malpighia glabra*, Barbados cherry: *Mimosa rubicaulis: Opuntia ficus-indica* and other spp.: *Paliurus spina-christi*, Christ's thorn: *Pandanus tectorius*, and other spp.: *Parkinsonia aculeata*, Jerusalem thorn: *Pereskia aculeata*, Barbados gooseberry or 'Pereskia bleo': *Pisonia aculeata*, cockspur: *Pithecolobium dulce*, Madras thorn (trop. Am.); *P. unguis-cati: Polyscias filicifolia: Poncirus trifoliata*, trifoliate orange: *Prosopis juliflora*, mesquite: *Pyracantha coccinea* var. *lalandii* and other spp.: *Rosa bracteata*, Macartney rose and other spp.: *Scutia myrtina: Triphasia trifolia*, myrtle lime: *Ximenia americana*, wild olive: *Zizyphus jujuba*, jujube; *Z. mucronata*, buffalo thorn. [Howes, F. N. (1946) *Fence and barrier plants in warm climates. Kew Bull.* **1**, 51–87.]

Bartonia *Mentzelia lindleyi* (=*Bartonia aurea*), cult. ann., showy fls.

Bartsia *Bartsia* spp., r. parasites. **Alpine** —, *B. alpina:* **red** —, *B. odontites*, common, may occur in gdns.: **yellow** —, *Parentucellia viscosa*.

Barus camphor *See* borneo camphor, *Dryobalanops aromatica*.

Barwood *Pterocarpus soyauxii*, W.Af., a trade hardwood, once used for dyeing.

Basil *Ocimum basilicum*, a much used flavouring herb, esp. in Medit.: used in casseroles, salads, sauces, pizza and other Italian dishes, also for liqueurs. **Am.** —, *O. basilicum:* **bush or dwarf** —, *O. minimum:* **duppy** —, *O. micranthum*, W.I.: **hoary** —, *O. canum*, a pot-herb: **holy** —, *O. sanctum:* **shrubby** —, *O. gratissimum:* **sweet** —, *O. basilicum:* — **thyme**, *Calamintha acinos*, Eur. & Br.

Basiloxylon *B. excelsum*, S.Am., yields good timber.

Basket flower *Hymenocallis narcissiflora:* — **hoop**, *Croton lucidus*, W.I.

Basketry, plants used for In the making of baskets innumerable pls. are used throughout the world. Bamboos, grasses, sedges and palm ls. are freely used, as are the tough stems of many twining or climbing pls. (*Paullinia, Smilax, Aristolochia, Lonicera, Rubus* etc.). For larger baskets the supple shoots of hazel, willow and other shrubs are employed. Thin strips of light wood, such as poplar and willow are used in making trugs or gardeners' baskets, and chip baskets or punnets for fr.

Barralocus *Dicorynia paraensis*, trop. S.Am., a commercial timber.

Bass Inner fibrous b., esp. of *Tilia*. — **wood**, *Tilia americana* and other spp.; N.Am. a commercial timber, also called 'American lime'.

Bass-broom fibre *See* piassava fibre.

Bassine fibre *Borassus flabellifer*, used for brushes.

Bastard acacia *Robinia pseudoacacia:* — **balm**, *Melittis melissophyllum:* — **box**, *Polygala chamaebuxus:* — **cedar**, *Soymida febrifuga, Chukrasia tabularis:* — **cinnamon**, *Cinnamomum cassia:* — **indigo**, *Amorpha fruticosa:* — **mahogany**, *Eucalyptus botryoides*, Aus.: — **peppermint**, *Tristania suaveolens:* — **teak**, *Butea monosperma:* — **vervain**, *Stachytarpheta*.

Bat pollination In the trop. trees of various kinds are commonly pollinated by bats that are attracted by the nectar in the same way as insects. Such fls.

belong mainly to the families Bignoniaceae, Bombacaceae, Leguminosae, and to a lesser extent Loganiaceae, Cactaceae and Pandanaceae (*Freycinetia*). An interesting fact is that many of the bat-pollinated fls. are evil-smelling, at least to the human nostril, such as spp. of *Kigelia, Parkia, Durio, Oroxylum indicum* and *Adansonia digitata*.

Bat's wing fern *Histiopteris incisa*, trop., cult.

Batai wood *Albizia falcata*, Malaysia, considered for paper making; tree may be the fastest growing trop. tree (9–10 m high, 15 cm diam., in 2 years): used for shade w. tea and other crops.

Bath sponge *See* loofah, *Luffa cylindrica* for vegetable substitute.

Batiputa (Braz.) *Ouratea parviflora*, seeds yield oil, used med.

Bauera Evergreen shrubs, Aus. & N.Z., cult., orn., esp. *B. rubioides*.

Bauhinia Several cult. orn. shrubs, trop. and subtrop., esp. *B. purpurea, B. tomentosa, B. variegata, B. galpinii*. Pods of *B. esculenta* eaten by livestock in S.Af.

Bawchan seed *Psoralea corylifolia*, Ind., aromatic bitter seeds, used med.

Bay berry *Myrica californica* and other spp., frs. source of wax at one time.

Bay laurel *Laurus nobilis*, S.Eur., dried ls. much used for flavouring, esp. with meat and fish dishes in France, inc. the national dish 'bouillabaise'.

Bay rum *Pimenta racemosa* (= *P. acris*), W.I., esst. oil distilled from the ls., formerly distilled with rum.

Bayonet grass *Scirpus paludosus*, N.Am.: — **plant**, *Aciphylla squarrosa*, N.Z., spiny ls.

Bdellium The bdelliums are oleo-gum-resins often resembling myrrh and derived from *Commiphora* spp.

Beach grass *See* marram grass. — **heath**, *Hudsonia tomentosa*, N.Am.: — **pea**, *Lathyrus japonicus:* — **plum**, *Prunus maritima*, N.Am.: — **sunflower**, *Helianthus debilis:* — **wort**, *Batis maritima*, trop.

Beads, seeds used as A number of seeds or frs. mainly of trop. or subtrop. origin, may be used as beads for necklaces, rosaries and orn. purposes. Among them are the following – *Abrus precatorius* (crab's eyes, red and black), *Adenanthera pavonina* (Circassian seed, red), *Afzelia quanzensis* (Rhodesian mahogany, red and black), *Caesalpinia bonduc* (bonduc, grey or white, spherical), *Canna bidentata* (Indian shot, black), *Cardiospermum halicabum* (black w. heart-shaped mark), *Coix lacryma-jobi* (Job's tears, grey and shiny), *Delonix regia* (flamboyant, long, light brown), *Elaeocarpus ganitrus* and other spp. (brown, wrinkled, spherical), *Erythrina* spp. (small, red and black), *Jacquinia armillaris*, Peru, *Leucaena glauca* (flat, brown), *Melia azedarach* (Persian lilac), *Ormosia dasycarpa, Pithecolobium unguis-cati* (black), *Sapindus saponaria* (soapberry, black). Others are olive, peach and styrax (*S. officinalis*) stones or seeds, hazel nuts and *Raphia* fr.

Beak rush *Rhynchospora* spp.

Beans Pulses, including beans, are, like cereals, imp. human and animal food all over the world. Beans are a staple human food in many countries, such as the soya bean in the Orient and 'haricot' beans (*Phaseolus vulgaris*) in C. & S. Am. and Af. In fact these two beans or classes of bean are probably produced in greater quantity throughout the world than any other bean. **Adsuki —**, *Phaseolus vulgaris*, Jap.: **asparagus —**, *Vigna sesquipedalis:* **Bengal —**

Mucuna aterrima: **black** —, *Castanospermum australe:* **Boer** —, *Schotia* spp.:
bog —, *Menyanthes trifoliata:* **broad** —, *Vicia faba:* **Burma** —, *Phaseolus lunatus:* **Bushman** —, *Guibourtia coleosperma,* Rhodesia: **butter** —, *Phaseolus lunatus:* **Calabar** —, *Physostigma venenosum* (poisonous): **Cherokee** —, *Erythrina arborea:* **cluster** —, *Cyamopsis psoraloides:* **donkey eye** —, *Mucuna urens:* **duffin** —, *Phaseolus lunatus:* **dwarf** —, *Phaseolus vulgaris:* **field** —, *Vicia faba* var. *equina:* **Florida velvet** —, *Mucuna deeringiana:* **French** —, *Phaseolus vulgaris:* **gemsbuck** —, *Bauhinia esculenta:* **Goa** —, *Psophocarpus tetragonolobus:* **ground** —, *Kerstingiella geocarpa:* **haricot** —, *Phaseolus vulgaris:* **Hibbert** —, *P. lunatus,* W.I.: **horse** —, *Vicia faba* var. *equina:* **horse-eye** —, *Mucuna sloanei:* **hyacinth** —, *Dolichos lablab* (= *Lablab niger*): **Indian** —, *Catalpa bignonioides,* N.Am.: **jack** —, *Canavalia ensiformis:* **jumbi** —, orn. beans in W.I., e.g. *Abrus, Erythrina, Ormosia, Adenanthera:* **kaffir** —, *Vigna unguiculata:* **kidney** —, *Phaseolus vulgaris:* **kotenashi** —, *P. vulgaris,* Jap.: **lablab** —, *Dolichos lablab:* **Lima** —, *Phaseolus lunatus:* **locust** —, *Ceratonia siliqua,* Eur.: **lucky** —, *Erythrina, Afzelia, Thevetia:* **Mauritius** —, *Mucuna aterrima:* **moth** —, *Phaseolus aconitifolius:* **mung** —, *P.* (= *Vigna*) *mungo:* **navy** —, *P. vulgaris:* **nicker** —, *Caesalpinia bonducella:* **otenashi** —, *Phaseolus vulgaris,* Jap.: **ordeal** —, *Physostigma venenosum:* **overlook** —, *Canavalia ensiformis,* W.I.: **Paternoster** —, *Abrus precatorius:* **pigeon** —, *Vicia faba* var. *minor:* **pole** —, *Phaseolus lunatus:* **potato** —, *Apios americana* (= *A. tuberosa*): **Rangoon** —, *Phaseolus lunatus:* **rice** —, *Phaseolus calcaratus* (= *Vigna calcarata*): **sabre** —, *Canavalia ensiformis:* **scarlet runner** —, *Phaseolus multiflorus:* **sea** —, *Entada gigas* (= *E. scandens*), *Mucuna:* **seaside** —, *Canavalia obtusifolia, C. maritima,* W.I.: **soya** —, *Glycine max:* **St Ignatius'** —, *Strychnos ignatii:* **string** —, *Phaseolus vulgaris:* **sugar** —, *P. lunatus:* **sword** —, *Canavalia ensiformis:* **Tepary** —, *Phaseolus acutifolius:* **tonka** —, *Dipteryx odorata:* **vanilla** —, *Vanilla planifolia:* **velvet** —, *Mucuna deeringiana:* **winged** —, *Psophocarpus tetragonolobus:* **yam** —, *Pachyrhizus erosus:* **yard long** —, *Vigna sesquipedalis* (= *V. unguiculata* var. *sesquipedalis*).

Bear berry *Arctostaphylos uva-ursi,* Eur. & Am., prostrate shrub, ls. med., mucilaginous and astringent: **black** — —, *A. alpina:* **—'s breech,** *Acanthus mollis,* Medit., fol. used as motifs by Greek and Roman sculptors: **—'s ear,** *Primula auricula:* **—'s foot,** *Helleborus foetidus:* **— grass,** *Yucca* spp., *Xerophyllum tenax.*

Beard flower, *Pogonia,* N.Am.: **— grass,** *Polypogon monspeliensis,* Eur., *Bothriochloa,* N.Am.: **—ed orchid,** *Calochilus robertsonii, C. paludosus,* Aus.: **— tongue,** *Penstemon* spp., N.Am.

Beaufortia W.Aus., heath-like shrubs, some w. showy scarlet frs., cult.

Beaumontia *B. grandiflora,* Ind., orn. climber w. large white fls., cult. trop.

Beauty berry *Callicarpa americana,* cult. shrub, violet frs.: **— bush,** *Kolkwitzia amabilis,* China, cult. fl. shrub.

Beaver poison *Cicuta maculata:* **— tail cactus,** *Opuntia basilaris,* SW. U.S.: **— wood,** *Celtis occidentalis.*

Bedstraw *Galium* spp.: **hedge** —, *G. mollugo:* **lady's** —, *G. verum.*

Bee brush *Aloysia* (= *Lippia*) *gratissima.*

BEEFSTEAK ACALYPHA

Beefsteak acalypha *A. wilkesiana* (=*A. tricolor*) ls. splashed with red and crimson on a coppery-green ground colour: widely cult., warm cli.

Bee plants or **bee pasturage** The prevalence of those pls. that afford nectar for the hive bee are, of course, all-imp. for the successful production of honey. Such pls. are commonly termed 'bee pls.', and 'bee forage' or 'bee pasturage' *en masse*. Not all nectar-producing fls. are of value to the honey bee with her relatively short tongue (about 6–7 mm) compared with other nectar-gathering insects such as bumble bees (tongue 10–11 mm), butterflies and moths. The production of nectar by fls. may be greatly influenced by environmental or cli. conditions and esp. soil moisture. These factors vary greatly with different bee pls. and in different countries. Many garden fls., e.g. those made double by cult., are of no value to the honey bee. Some pls. supply only pollen which bees must have for brood rearing.

In Br. and many other temp. countries, the following are numbered among the major sources of nectar to the honey bee – clover (notably white clover *Trifolium repens*, equally important in N.Am. and other countries): lime (*Tilia* spp.): fruit tree blossom (apple, pear, cherry, plum): heather: sainfoin: mustard and charlock: hawthorn: sycamore: blackberry: willow herb: field beans (*Vicia*): dandelion. Honey bees are greatly attracted by many gdn. fls. in their season and the fls. of certain cult. trees and shrubs, although it is seldom that the fls. are present in sufficient abundance to affect the honey yield of the beekeeper. The following are often particularly attractive to honey bees – anchusa: borage: *Buddleja globosa*: catmint: *Caryopteris*: *Cotoneaster* (esp. *C. horizontalis*): crocus: *Echinops*: flowering currant: *Helenium*: *Helianthus* (sunflower): Michaelmas daisy: mint: *Salvia* (notably *S. superba*). Many weeds, e.g. thistles and melilot, are useful producers of nectar for the beekeeper.

In countries where veg. and fl. seed production is undertaken on a large scale (several hectares), as in the U.S., unusual honeys may be obtained, e.g. onion, celery, parsnip etc. The honey from onion flowers has a taste of onion when fresh but this disappears as the honey ripens in the hive [Howes, F. N. (1945) *Plants and beekeeping*, 224 pp.]

Beech *Fagus* spp. The **common** or **Eur. beech**, *F. sylvatica*, is widely grown for timber and as an orn., shade or shelter tree. The straight-grained, even textured wood with excellent bending properties, has a multitude of uses esp. for furniture, domestic woodware and in turnery. The woods of Am. and Jap. beech are similar and are also commercial woods. The seeds of the common beech, beech nuts or beech mast, are ed. and yield a cooking oil, once freely used. In the past beech mast was all-imp. in the feeding of pigs and agric. land or woodland was valued according to the amount of 'pannage for swine' (beech mast and acorns) it was likely to provide. **Af.** —, *Faurea saligna:* **Am.** —, *Fagus grandifolia:* **Antarctic** —, *Nothofagus cunninghamii, N. antarctica:* **Aus.** —, *Nothofagus moorei:* **bitter** —, *Eupatorium villosum*, Jam.: **black** —, *Nothofagus solandri*, N.Z.: **blue** —, *Carpinus caroliniana:* **brown** —, *Cryptocarya glaucescens:* **Cape** —, *Rapanea melanophloeos:* **copper** —, orn., purple leaved form of Eur. beech: **Eur.** —, *Fagus sylvatica:* **— fern**, *Dryopteris:* **Jap.** —, *Fagus crenata* and allied spp.: **mountain** —, *Nothofagus cliffortioides*, N.Z.: **myrtle** —, *Nothofagus cunninghamii:* **negro-head** —, *Notho-*

fagus moorei: **N.Z.** —, *Nothofagus fusca:* — **orchid**, *Townsonia viridis:*
Oriental —, *Fagus orientalis:* **red** —, *Nothofagus fusca, Flindersia,* Aus.:
Roble —, *Nothofagus obliqua:* **silver** —, *Nothofagus menziesii:* **southern** —,
Nothofagus spp.: **Tasm.** —, *Nothofagus cunninghamii:* **white** —, *Gmelina
leichhardtii,* Aus.
Beef apple *Achras* (= *Manilkara*) *zapota,* W.I.: — **suet tree,** *Shepherdia
argentea,* N.Am., buffalo berry, fr. ed.: — **wood,** *Casuarina equisetifolia,
Grevillea striata,* Aus.; *Mimusops,* W.I.; *Pisonia fragrans,* W.I.; *Stenocarpus
salignus, S. sinuatus, Orites excelsa,* Aus.
Beer, native or **African** Usu. made from *Sorghum* or other millets such as
spp. of *Panicum, Pennisetum, Eleusine* etc.; maize and rice may be used and
barley and wheat in Ethiopia. R. crops may also be used, such as sweet
potatoes and cassava in the Congo. A national beverage in Madag. called
'betsa-betsa' is made by fermenting sugar cane juice. Banana and pineapple
are similarly used in parts of trop. Af., also the cashew fr. or 'apple'.
Beetroot or **garden beet** *Beta vulgaris,* considered to be derived from the
wild beet of the seashore (*Beta maritima*): rs. red, much used in salads,
canned: numerous vars., varying much in size and shape of r.; some are
yellow fleshed. *See also* sugar beet.
Beetle grass *Diplachne fusca,* Aus.
Beggar tick *Bidens* spp., esp. *B. pilosa,* a common weed: — **weed,** *Desmodium
tortuosum,* a cover crop in Am. citrus groves: **beggar's lice,** *Cynoglossum* spp.,
Desmodium spp. frs. stick to clothing.
Begonia A large genus, over 1000 spp., many cult. and many cultigens: two
main groups, tuberous and fibrous rooted: fls. of rich colour and beautiful
form, popular bedding and room pls. Some have med. uses in countries of
origin.
Belalie *Acacia stenophylla,* Aus., a fodder tree.
Belamcanda *B. chinensis,* cult., iris-like pl. w. large spotted orange fls.
Belladonna *Atropa belladonna,* Eur., wild and cult. (deadly nightshade):
drug consists of dried aerial parts of pl. taken in fl.: action due to atropine:
much used for colic, allays griping: r. also used.
Bell-apple *Passiflora laurifolia,* W.I., ed. fr.
Bella umbra or **belombra** *Phytolacca dioica,* S.Am., cult. subtrop., a shade
tree.
Bell flower *Campanula:* **Chilean** — —, *Lapageria rosea:* **Chinese** — —,
Platycodon grandiflorus, Abutilon: **climbing** — —, *Littonia modesta,* S.Af.:
ivy-leaved — —, *Wahlenbergia hederacea:* **S.Af.** — —, *Wahlenbergia* spp.:
— **pepper,** *Capsicum:* — **tree,** *Halesia:* — **wort,** *Uvularia* spp., N.Am.
Bells of Ireland *Molucella laevis,* cult. orn. fl.
Bellucia Trop. Am. trees, ed. frs.
Belly-ache bush *Jatropha gossypifolia,* S.Am., cult. orn. shrub, warm cli.
Beloperone *B. guttata,* Mex. cult., a room pl. (shrimp pl.).
Ben, oil of *Moringa oleifera,* cult. trop., oil from seeds, pods ed.
Benedictine Flavour of this liqueur due to angelica, *Angelica archangelica.*
Bengal bean *Mucuna aterrima:* — **cardamom,** *Amomum aromaticum:* —
clock vine, *Thunbergia grandiflora:* — **kino,** *Butea monosperma:* — **quince,**
Aegle marmelos.

BENI SEED

Beni seed *Polygala butyracea*, W.Af., seeds yield ed. oil and stems a fibre (name applied to *Sesamum*): **black — —**, *Hyptis spicigera*, cult., W.Af., seeds yield oil and stem a fibre.

Benjamin, gum- *See* benzoin. **Stinking —**, *Trillium erectum*, N.Am., cult., fls. purple, evil-smelling: **— tree**, *Ficus benjamina*.

Bent grass *Agrostis* spp. Includes many valuable agric. and lawn grasses in temp. and subtemp. cli. **Black —**, *A. gigantea*: **brown —**, *A. canina* var. *montana* (lawns): **colonial —**, *A. tenuis*: **common —**, *A. tenuis*: **creeping —**, *A. stolonifera* (lawns): **silky —**, *Cynosurus*: **velvet —**, *Agrostis canina*. The term bent grass may be used for any wiry grass that grows on a bent or common.

Benzoin Or 'gum-benjamin', a dried resinous balsam obtained from several spp. of *Styrax*, Sumatra benzoin being the best known (from *S. benzoin* and *S. paralleloneurus*): used for inhalations, an ingredient of Friar's balsam. A Medit. sp., *S. officinalis*, was exploited at one time. **Siam —**, *S. tonkinensis*.

Ber (Ind.) *Zizyphus mauritiana*, well-known host tree of the lac insect, fr. ed.

Bere or **bear** Old name for barley, *Hordeum vulgare*.

Bergamot *Monarda didyma*, N.Am., a popular gdn. pl., scented, red fls. **— orange,** *Citrus aurantium* var. *bergamia*, cult., Bergamot oil obtained from rind, used in perfumery, hair oils etc.

Bergenia Cult. orn. peren. usu. w. large thick ls. and fls. in early spring: also called 'Megasea' or Siberian saxifrage. Ls. and rs. of *B. crassifolia* rich in tannin.

Berlinia *B. grandiflora*, W.Af., a commercial timber.

Bermuda grass (Am.) *Cynodon dactylon*, pasture, lawn grass and sand binder in warm cli. (couch in Aus.): **— arrowroot,** *Maranta arundinacea*: **— cedar,** *Juniperus bermudiana*: **— palm,** *Sabal blackburnianum*.

Berry A fleshy fr. containing no hard part except the seeds: the term may be loosely or popularly used to cover other kinds of fleshy frs. Typical or indehiscent berries in *Actaea, Berberis, Ribes, Solanum, Vaccinium, Vitis*: dehiscent in *Akebia, Myristica*: constricted between seeds in *Maerua, Unona* etc. **Bay —,** *Myrica cerifera*: **— bearing bamboo,** *Melocanna bambusoides*, fr. large and ed., unique among bamboos: **bil—,** *Vaccinium myrtillus*: **black—,** *Rubus*: **buffalo —,** *Shepherdia argentea*: **checker—,** *Gaultheria procumbens*: **cloud—,** *Rubus chamaemorus*: **cow—,** *Vaccinium vitis-idaea*: **cran—,** *V. oxycoccus*: **crow—,** *Empetrum nigrum*: **dew—,** *Rubus caesius*: **goose—,** *Ribes grossularia*: **huckle—,** *Gaylussacia*: **partridge —,** *Gaultheria procumbens*: **rasp—,** *Rubus idaeus*: **straw—,** *Fragaria vesca*: **— tree,** *Flacourtia flavescens*, W.Af., fr. ed.: **trimble—,** *Rubus occidentalis*: **whortle—,** *Vaccinium myrtillus*: **yellow —,** *Rhamnus infectoria*.

Berseem clover Or Egyptian clover. *Trifolium alexandrinum*, cult., Medit.

Bertolonia Dwarf herbac. pls., trop. Am., cult. for orn. or marked ls.

Beschorneria Mex. *Agave*-like pls., cult., *B. yuccoides* a hardy sp.

Besom A gdn. broom commonly made of birch or ling; butcher's broom used in Italy; gorse sometimes used.

Betel nut *Areca catechu*, trop. As. The hard seed of this palm cut into slices and rolled in a betel pepper l. (*Piper betle*) and mixed with a little lime is the traditional masticatory of the East. It turns the saliva red.

Betony *Betonica officinalis* (= *Stachys betonica*), Eur., med., used in home remedies.

Betsa-betsa The national drink of Madag., made by fermenting sugar cane juice.

Beverage plants Most of the popular beverages in common use are of veg. origin, such as tea, coffee, cocoa, beer, wines, spirits and fr. juices. They form the basis of large industries in many countries. It would seem that man has, throughout the ages, always sought beverages that are both palatable and stimulating. This has led to the production of beverages of many kinds, both alcoholic and non-alcoholic. Some beverages such as tea, coffee and perhaps certain French wines (champagne) and Scotch whisky have proved so popular that their use has become almost world-wide. Others are of local importance only and not appreciated outside their countries of origin, a taste for them having often to be acquired. In reviewing the many beverages in use it becomes obvious that the more popular ones, such as tea, coffee, cocoa, yerba maté and guarana (Brazil) and drinks of the cola type all contain caffeine or theine or closely related compounds. Undoubtedly it is the presence of this stimulating substance that accounts for their popularity. Much the same may be said about alcoholic drinks.

Among non-alcoholic or 'soft drinks', fr. juices play an important part. Almost any palatable, juicy fr. may be used with or without the addition of sugar or water (tomato juice). The most widely used are the citrus frs., notably orange, lemon, grapefruit or lime. Other popular fr. drinks are those made from pineapple, passion fr. or granadilla, pomegranate (esp. in Medit.), raspberry, blackcurrant and rose-hip, the two last mentioned being valued for their high vitamin content. Throughout the tropics the 'milk' of the fresh coconut is a popular and thirst-quenching drink. As an example of a localized beverage, tubers of a sedge (*Cyperus esculentus*) are made into a drink ('horchata') in Spain.

A large number of pls. afford tea substitutes or herbal teas in different parts of the world, some being considered to have med. properties. The dried fls. of the lime (*Tilia*) are much used in France. In N.Af. 'khat' ls. (*Catha edulis*) are popular and in S.Af. the 'Cape' or bush teas (*Cyclopia*). There are also coffee substitutes which include various seeds and rs. (dandelion root). Chicory and liquorice are also used in beverages.

Alcoholic beverages include wines, beers, spirits and liqueurs. Wine has been used in Medit. as far back as history relates. Somewhat similar to the true wines, made from grapes, are the numerous so-called home-made wines made from a variety of frs., vegs. and wild pls. (elderberry). Cider (apple), perry (pear) and mead (from honey) have been esteemed for centuries in Eur. In the trop. various palm wines are used. Other fermented beverages or beers are prepared in many parts of the world, even by primitive peoples. For this various cereals or other starchy products may be employed. Throughout Af. sorghum or Guinea corn is much used and in Jap. rice. The Mex. drink 'Pulque' is prepared from *Agave* after tapping the flowering stalks. In Braz. the cashew fr. or 'apple' is the source of a popular alcoholic beverage. Fermented sugar cane juice constitutes the national drink of Madag. and is popular in other countries. Among the well known spirits, rum is a product of sugar cane. Brandy is distilled from grape wines and whisky, like beer, is derived from barley. Other cereals and potatoes are used in preparing different kinds of potable spirit, especially on the Continent of Eur., e.g. gin, schnapps and vodka.

Bhaji *Amaranthus* spp., ls. commonly used as spinach in warm cli.

Bhang Or ganja. Common Ind. names for cannabis.

Bharbur grass *Ischaemum angustifolium*, Ind., used for paper, matting and cordage.

Bhendi tree *Thespesia populnea*, cult. orn. tree, trop.

Bhutan pine *Pinus wallichiana*, temp. Himal., cult. orn.

Bible plants A large number of pls. are mentioned in the Bible and several books have been written on the subject. The pls. that occur in the Holy Land today are somewhat different from what they were in the days of Christ. For instance an Am. 'prickly pear' (*Opuntia ficus-indica*) is now one of the commonest pls. Fossil remains and the testimony of early historians show that the date palm was extremely common. Today it exists only as a cult. palm. Likewise the numerous cedar trees (*Cedrus libani*) that clothed the slopes of Mt. Lebanon and other mt. ranges are no longer present, the few remaining trees being carefully fenced in and protected by the Government lest the sp. be completely exterminated in its own homeland.

Nevertheless many of the pls. mentioned in the Bible, wild or cult. are still everyday pls. in the Holy Land, as in other parts of the E.Medit. Among the common food pls. mentioned are wheat, barley, millet, beans, lentils and the onion. Frs. and nuts included the almond, apricot, balanites, carob, date, fig, sycamore fig, grape, mulberry, olive, pistachio nut, pomegranate, walnut and desert date (*Zizyphus*). Spices and flavouring materials mentioned include fenugreek, anise, coriander, saffron, cumin, mint, rue and marjoram. Timbers referred to include acacia (*Acacia nilotica*), box, cedar, cypress, ebony, juniper, manna ash, pine (*Pinus halepensis*), plane, poplar and oak. Cotton and flax are the textile fibres mentioned.

It is of interest to note that some economic pls. now freely cult. in the Medit. or Holy Land receive no mention. Presumably they had not yet been introduced in the days of Christ. Among them are the orange and other citrus frs., the loquat and the peach, all natives of the Orient. [Hepper, F. N. (1973) *Plants of Bible lands*. Moldenke, H. N. & A. L. (1952) *Plants of the Bible*, 328 pp.]

Bicuhyba fat *Virola bicuhyba*, Braz., seeds yield fat used for candles: tree yields a commercial timber, 'bicuiba'.

Bidi A cheap form of Ind. cigarette or cigar, the tobacco being rolled in a leaf, not paper.

Bidi-bidi *Acaena* spp., N.Z., creeping shrubs w. stout rootstocks.

Bifrenaria Trop. S.Am. orchids, cult.

Big tree *Sequoiadendron giganteum*, *see* mammoth tree.

Bigarade Seville or bitter orange, *Citrus aurantium*. — **oil**, prepared from peel and used for flavouring.

Bignonia *Arrabidaea* (=*Bignonia*) *magnifica*, Colombia, cult. orn. climber, purple fls. 9 cm across: other spp. cult.

Bilberry *Vaccinium myrtillus*, Eur., black or purple frs. eaten or cooked in various ways.

Bilimbi *Averrhoa bilimbi*, trop. As., acid frs. ed., mainly stewed.
Billardiera Aus., evergreen climbers, some cult.
Billbergia Bromeliads of C. & S.Am., many v. orn. and cult., warm cli., esp.
  B. nutans.
Billian *Eusideroxylon zwageri*, Borneo, Indon., a commercial timber.
Billion dollar grass *Echinochloa crus-galli*, 'barnyard millet', luscious grass of
  warm cli.
Billy buttons (Aus.) *Craspedia uniflora*, yellow fl. heads over 3 cm across.
Billy-goat weed (Aus.) *Ageratum conyzoides, A. houstonianum*, troublesome
  weeds in many warm countries: have peculiar odour.
Bindweed *Calystegia, Convolvulus, Polygonum*. **Black** —, *Polygonum convol-
  vulus:* **common** or **field** —, *Convolvulus arvensis:* **great** or **hedge** —, *Caly-
  stegia sepium, C. silvatica:* **sea** —, *C. soldanella:* **Spanish** —, *Convolvulus
  tricolor.*
Bintangor *Calophyllum* spp., Malaya, a commercial timber.
Binuang *Octomeles sumatrana*, E.I., a commercial timber.
Biophytum *B. sessile*, E.Af., small herb, rosettes of pinnate ls. sensitive to touch.
Birch *Betula* spp. The **common** Eur. —es (*B. pubescens, B. pendula*) have a
  wide distribution reaching the northern limits of tree growth. The wood is
  freely used in many countries. It is straight-grained, bends and stains readily:
  much used for furniture, esp. chairs and for plywood. Some N.Am. birches
  are similar. **Am. black** —, *Betula lenta:* **brown** —, *B. pubescens:* **canoe** —,
  B. papyrifera: **cherry** —, *B. lenta:* **dwarf** —, *B. nana:* **Karelian** —, name
  for unusually marked Eur. birch wood: **paper** —, *B. papyrifera:* **silver** —,
  B. pendula, grown for orn.: **W.I.** —, *Bursera simaruba:* **yellow** —, *Betula lutea.*
Birch bark The outer b. is very water resistant, hence its use for canoes by
  Amerindians and for roofing shelters. In Eur. birch b. has long been used in
  tanning leather. It contains 10–15 % tannin and yields a pliable, light-coloured
  leather, suited for the 'uppers' of footwear.
Bird cactus *Pedilanthus tithymaloides*, trop., a hedge pl.: — **cherry**, *Prunus
  padus:* — **flower** (Aus.), *Crotalaria cunninghamii:* **golden** — **flower**, *C.
  macrocarpa:* — **pepper**, *Capsicum minimum* or *C. frutescens:* — **plant**, *Heterotoma lobelioides*, Mex., cult., orn. pl.: —'s **eye**, *Veronica* spp.: —'s **eye
  maple**, much favoured, figured furniture wood of the late Victorian period,
  cut from burrs of the Am. sugar maple, these having been caused by buds
  unable to force their way through the bark: the wood was freely used as
  veneer. —'s **eye primrose**, *Primula farinosa:* —'s **nest bromeliad**, *Nidularium* spp., orn. ls., room pls.: —'s **nest fern**, *Asplenium nidus*, trop., cult.:
  —'s **nest orchid**, *Neottia nidus-avis*, Br.: **yellow** —'s **nest**, *Monotropa
  hypopitys:* —'s **foot**, *Ornithopus perpusillus:* —'s **foot trefoil**, *Lotus corniculatus:* —'s **tongue**, *Ornithoglossum.*
Bird-catching trees *Pisonia* spp., trop., many spp. have viscid frs. which
  stick to animals or birds. Small birds may pick up so many that their movements are hampered and they perish: one of the few fruits that will stick to
  feathers.
Bird of Paradise flower *Strelitzia reginae*, S.Af. cult.: — — **shrub**, *Poinciana gilliesii:* — — **tree**, *Strelitzia nicolai.*
Bird lime, plants used for The sticky juice or latex of certain pls. or their

frs. is, or has been, used for making bird lime with which to catch birds. Bark of Eur. holly, *Ilex aquifolium*, and Jap. holly, *Ilex integra*, has been used, as has the bark of another Jap. tree, *Trochodendron aralioides*. In Ind. the green fr. of the marking nut (*Semecarpus*) is pounded and made into bird lime, as is the fruit of the widely cult. Sebesten tree, *Cordia myxa*. In W.Af. the latex of *Euphorbia balsamifera* and *Chrysophyllum africanum* is used. Plants in the following genera are also recorded – *Artocarpus, Clusia, Conopharyngia, Clermontia, Ficus, Loranthus, Sapium* and *Viscum*.

Biriba *Rollinia deliciosa*, Braz. cult., ed. fr.

Birthwort *Aristolochia* spp., esp. *A. clematitis*, Eur. inc. Br.

Biscuit root *Lomatium ambiguum, L. macrocarpum*, N.Am.

Bishop's cap *Mitella* spp., N.Am.: — **weed**, *Aegopodium podagraria*, Br.; *Ptilimnium capillaceum*, N.Am.: — **wort**, *Stachys officinalis*.

Bissabol A name for myrrh.

Bissy nut (Jam.) Cola nut.

Bistort *Polygonum bistorta*.

Bitter almond *Prunus amygdalus* var. *amara*, yields oil: — **aloes**, *Aloe* spp.: — **apple**, *Citrullus colocynthis:* — **ash** or **bark**, *Picrasma antillana* (= *Picraena excelsa*), W.I.: — **bark** (Aus.), *Petalostigma quadriloculare*, a good cabinet timber: — **beech** (Jam.), *Eupatorium villosum* a substitute for hops: — **berry**, *Solanum racemosum*, W.I.: — **bloom**, *Sabatia angularis*, N.Am.: — **bush**, *Purshia tridentata*, N.Am.: — **cress**, *Cardamine bulbosa*, N.Am.: — **cucumber**, *Citrullus colocynthis:* — **nut**, *Carya cordiformis*, N.Am.: — **orange**, *Citrus aurantium* var. *bigaradia:* — **root**, *Lewisia rediviva*, N.Am.: — **sweet**, *Solanum dulcamara:* — **vetch**, *Vicia orobus*, Eur.: — **weed**, *Hymenoxys*, W. U.S., *Helenium tenuifolium*, N.Am., taints milk and butter: — **wood**, *Quassia amara*.

Bkilawan nut Marking nut, *Semecarpus anacardium*.

Black bean *Kennedia nigricans*, Aus.: —**berry**, *Rubus fruticosus* and other spp.: — **bindweed**, *Polygonum convolvulus:* —**boy**, *Xanthorrhoea* spp., Aus.: — **bryony**, *Tamus communis:* — **butt**, *Eucalyptus pilularis*, Aus., an imp. hardwood: —**cap**, *Rubus occidentalis*, N.Am.: — **cherry**, *Eugenia domingensis*, W.I.: —**currant**, *Ribes nigrum:* — **dammar**, *Canarium:* — **eyed Susan**, *Thunbergia alata, Rudbeckia* spp.: — **fella's orange**, *Eustrephus latifolius:* — **gin**, *Kingia australis*, Aus.: — **grass**, *Alopecurus myosuroides:* — **haw**, *Viburnum prunifolium*, N.Am.: —**jack**, *Bidens pilosa*, a weed: — **Jessie**, *Pithecolobium unguis-cati*, W.I.: — **laurel**, *Gordonia lasianthus:* — **locust**, *Robinia pseudoacacia:* — **mulberry**, *Morus nigra:* — **mustard**, *Brassica nigra:* — **nicker tree**, *Sapindus*, W.I.: — **nightshade**, *Solanum nigrum:* — **pepper**, *Piper nigrum:* — **raspberry**, *Rubus occidentalis*, N.Am.: — **root**, *Pterocaulon undulatum*, N.Am.: — **roseau**, *Bactris major*, prickly palm, W.I.: — **rosewood**, *Dalbergia latifolia*, Ind.: — **sloe**, *Prunus umbellata*, N.Am.: — **snakeroot**, *Cimicifuga racemosa*, N.Am.: — **stinkwood**, *Ocotea bullata*, S.Af., a valuable timber.

Blackthorn or sloe, *Prunus spinosa*, Eur. inc. Br., a common wild shrub noted for its white fls. appearing in early spring before the ls., and for its plum-like frs. used for making sloe gin. The wood is hard and tough, used for hay-rake teeth and for the Irish club or shillelagh.

Blackwood Af. —, *Dalbergia melanoxylon:* **Aus.** —, *Acacia melanoxylon:* **Ind.** — *Dalbergia latifolia.* Hard, heavy, close-grained woods taking a good polish and suitable for furniture, cabinet work and musical instruments.

Bladder campion Or catchfly, *Silene vulgaris:* — **fern,** *Cystopteris fragilis, Hymenophyllum:* — **nut,** *Staphylea pinnata,* S.Eur., As. Minor, cult. orn. shrub; *S. triflora,* N.Am.: — **pod,** *Lesquerella gordoni,* SW. U.S., a desert pl.: — **seed,** *Physospermum, Levisticum:* — **senna,** *Colutea arborescens,* Medit., naturalized in Br.: —**wort,** *Utricularia vulgaris* and other spp.

Blaeberry *Vaccinium myrtillus,* Eur. incl. Br., ed. fr., common on moors.

Blakea Trop. Am. trees or shrubs, large showy fls., cult.

Blandfordia Rhizomatous pls. with large handsome fls., E. Aus., cult.

Blanket flower *Gaillardia pulchella,* N.Am., and other spp., cult.: — **leaf,** *Verbascum thapsus,* v. woolly ls.

Blaspheme vine *Smilax laurifolia,* N.Am.

Blazing star *Mentzelia laeviculus,* N.Am.; *Chamaelirium luteum,* N.Am.; *Liatris* spp., N.Am.

Blechnum Cult. orn. ferns, *B. gibbum* a favourite greenhouse subject and described as 'a tree fern in miniature'.

Bleeding heart *Dicentra spectabilis* cult. orn. pl.; *Clerodendron thomsoniae* cult., orn., red and white fls.

Bleedwood tree *Pterocarpus angolensis.*

Bletia Mainly terrestr. orchids, trop. Am., cult.

Bletting The apparent or deliberate over-ripening of fruits, commonly practised w. medlars.

Blimbing *See* bilimbi.

Blinding tree (Sri Lanka) *Excoecaria agallocha,* latex harmful to eyes.

Blinks *Montia fontana,* Eur. incl. Br., pasture pl., ls. used in salads.

Blite *Chenopodium bonus-henricus* or other spp.

Blister bush (S.Af.) *Peucedanum galbanum,* produces dermatitis or bad blistering 40–50 h after contact. — **buttercup,** *Ranunculus scleratus.*

Blood berry *Rivina humilis,* trop. Am., cult., orn. pl. w. scarlet berries: — **flower,** *Haemanthus multiflorus* and other spp.; *Asclepias curassavica,* cult. and a trop. weed, red fls.; *Calothamnus sanguineus,* W.Aus.: — **leaf,** *Iresine* spp., trop. Am., cult., orn. foliage, red or crimson: — **plum,** *Haematostaphis barteri,* W.Af., ed. fr.; Satsuma or Jap. — plum: — **root,** *Sanguinaria canadensis,* dried rhizome red, med., cult.; *Haemodorum coccineum,* Aus.: — **wood,** *Eucalyptus gummifera* and other spp. exuding red sap when bark is cut; *Pterocarpus indicus,* Ind.; *P. angolensis,* trop. Af.; *P. erinaceus,* S.Af.

Blood in the bark *Ceratopetalum succirubrum,* Aus. forest tree.

Blowballs (Am.) *Taraxacum,* the round seed heads.

Blow grass *Agrostis avenacea,* Aus., seed heads break off and are carried by wind.

Blue amaryllis *Griffinia hyacinthina* var. *maxima,* Braz., a favoured pot plant: — **beard,** *Salvia horminum,* cult., coloured bracts: —**bell** (Am.), *Mertensia;* (Aus.), *Wahlenbergia, Sollya;* (English), *Endymion non-scriptus* (= *Scilla nutans);* (Scottish), *Campanula rotundifolia;* (S.Af.), *Wahlenbergia, Gladiolus;* (N.Z.), *Wahlenbergia;* (W.I.), *Barleria:* **Spanish** —, *Endymion hispanicus* cult.: — **bell creeper,** *Sollya fusiformis,* W.Aus., fls. bright blue: — **berry** (Am.),

Vaccinium spp., ed. fr.; (Aus.), *Billardiera longifolia;* (N.Z.), *Daniella intermedia:* — **bonnet**, *Lupinus sericeus* and other spp., N.Am.: — **bottle**, *Centaurea scabiosa:* — **boys**, *Pycnostachys urticifolia:* — **bush**, *Eucalyptus macrocarpa*, Aus., red fls., orn. fol., cult.: — **buttons**, *Scabiosa succisa:* — **cardinal fl.**, *Lobelia siphilitica*, N.Am.: — **China orchid**, *Caladenia gemmata*, Aus.: — **cohosh**, *Caulophyllum thalictroides*, N.Am.: — **curls**, *Trichostema lanatum*, N.Am.: — **devil**, *Eryngium rostratum*, Aus.: — **eyed grass**, *Sisyrinchium* spp., N.Am.: — **eyed Mary**, *Omphalodes verna*, cult., orn. pl.: — **fairies orchid**, *Caladenia deformis*, Aus.: — **flag**, *Iris* spp.: — **toad flax**, *Linaria canadensis:* — **gem**, *Craterostigma plantagineum*, S.Af.: — **grama**, *Bouteloua gracilis:* — **grass**, *Poa* spp., *Festuca* spp.: — **gum**, *Eucalyptus globulus:* — **haze**, *Selago spuria*, S.Af.: — **jacket**, *Tradescantia reflexa:* — **lace flower**, *Trachymene caerulea*, cult. orn. fl.: — **mahoe**, *Hibiscus elatus*, W.I., inner bark used for hat and rope making: — **moor grass**, *Sesleria caerulea:* — **poppy**, *Meconopsis:* — **star**, *Amsonia:* — **stem**, *Andropogon scoparius*, N.Am.: — **tinsel lily**, *Calectasia cyanea:* — **toadflax**, *Linaria canadensis:* — **vine**, *Clitoria ternata*, Ind., cult.

Bluet *Vaccinium* spp., N.Am.; *Houstonia caerulea*, N.Am., cult., orn. fl.: **mountain** —, *Centaurea montana*, N.Am.

Blumea Ls. of several As. spp. used med. locally: *B. balsamifera* yields Ngai camphor.

Blushing bride *Serruria florida*, S.Af., shrub w. beautiful fls.

Blysmus, red *Blysmus rufus*, Br., a salt marsh pl.

Bo tree (Sri Lanka) *Ficus religiosa*, trop. As., much cult., a sacred tree.

Boab (Aus.) = Baobab.

Boat lily *Rhoeo spathacea* (= *R. discolor*), W.I., orn. ls., a room pl.

Bobartia Tough rush-like ls. used for baskets, S.Af.

Bocconia Shrubs, C. & S.Am.; some w. local med. uses.

Boehmeria Some spp. yield a strong bast fibre used locally. *See* ramie.

Boerboon (S.Af.) *Schotia afra:* **weeping** —, *S. brachypetala:* **Boer tea**, *Cyclopia* spp.

Boerhavia *B. diffusa*, a trop. weed.

Bog asphodel *Narthecium ossifragum*, N.Hemisph. inc. Br., iris-like pl., yellow fls.: — **bean**, *Menyanthes trifoliata*, Eur. inc. Br., rhizome med. (tonic): — **myrtle**, *Myrica gale*, Br.: — **pimpernel**, *Anagallis tenella:* — **rhubarb**, *Petasites hybridus:* — **rosemary**, *Andromeda polifolia:* — **rush**, **black**, *Schoenus nigricans:* — **spruce**, *Picea mariana:* — **torches**, *Orontium aquaticum*, N.Am.: — **violet**, *Pinguicula vulgaris*.

Boga medaloa *Tephrosia candida*, trop. As., cover and green manure crop.

Bogota tea *Symplocos theiformis*, Colombia, l. infusion resembles tea.

Bois de Rose oil *Aniba panurensis*, S.Am., esst. oil from wood.

Bois ravine (W.I.) *Calliandra tergemina*, a hedge pl.

Boldo *Peumus boldus*, Chile, ls. med., fr. ed., dye from bark.

Bollea Handsome trop. Am. orchids, cult.

Bolly gum *Litsea* spp., *Beilschmiedia* spp., Aus., forest trees.

Bolobolo fibre *Clappertonia* (= *Honckenya*) *ficifolia*, W.Af., a jute-like bast fibre used locally.

Bomah nut (S.Af.) *Pycnocoma macrophylla*, seeds rich in oil.

Bomarea *B. lehmannii* and other spp., trop. Am., cult. orn. pls., ed. fr.

Bombax *Bombax bounopozense*, W.Af., and allied spp., soft commercial wood, seed floss of several spp. used for stuffing.

Bombay aloe *Agave cantala:* — **ebony**, *Diospyros montana:* — **hemp**, *Crotalaria juncea:* — **mastic**, *Pistacia mutica.*

Bombway, white *Terminalia procera*, Andamans, a commercial timber.

Bonace (W.I.) *Daphnopsis tinifolia*, fibrous inner b. used for cordage.

Bonatea Trop. Af., terrestrial orchids w. remarkable development of rostellum.

Bonavist (W.I.) *Dolichos lablab* (= *Lablab niger*), a much cult. ed. bean.

Bonduc nut *Caesalpinia bonducella*, trop., orn. marble-like seeds used for necklaces and games: a common drift seed.

Boneset *Eupatorium* spp., e.g. *E. perfoliatum*, *E. lancifolium*, N.Am.: **false —**, *Kuhnia*, SW. U.S.

Bonnet orchid *Cryptostylis erecta*, Aus.

Bonsai This is the ancient Jap. art of growing trees and other pls. in small pots or other containers and, by special treatment, inducing them to become exact miniatures of the normal pl., or to become dwarfed or twisted or to acquire an appearance of great age. The special and highly skilled treatment that is applied, which may extend over a considerable period, includes special pruning and training, or maintaining the subjects in a pot-bound condition to discourage too much active growth. This cult has attracted much attention in western countries, especially the U.S.

Bonsai may be commenced with seeds, cuttings, grafts or the training of wild seedlings and there are certain traditional styles which are followed, each with its own Jap. name. *Mame Bonsai*, or miniature Bonsai, is the production of very small subjects in small or even minute containers. Special attention is given to the nature of the potting composts used, to potting and repotting, watering and manuring. Cords and wire, or coiled wire, are much used in training to the desired shapes.

Among the spp. favoured for bonsai in Jap. are several Jap. conifers, such as black pine (*Pinus thunbergii*), white pine (*P. parviflora*), yeddo spruce (*Picea jezoensis*), Hinoki cypress (*Chamaecyparis obtusa*), juniper (*Juniperus rigida*), cedar (*Cryptomeria japonica*), larch (*Larix kaempferi*), golden larch (*Pseudolarix amabilis*), umbrella pine (*Sciadopitys verticillata*), Sargent's juniper (*Juniperus sargentii*). Other favoured trees are zelkova (*Zelkova serrata*), flowering apricot, flowering cherry, crab apple, maple, quince, pomegranate and citrus or citron. Other shrubs or gdn. pls. used in Jap. include cotoneaster, pyracantha, tamarisk, viburnum, jasmine, wisteria, azaleas and chrysanthemums. Among trees of other countries which have been shown by enthusiasts to be suitable for bonsai are N.Am. buckeyes (*Aesculus* spp.), black locust or robinia (*Robinia pseudoacacia*) and some of the ornamental Aus. spp. of *Acacia*.

Boobialla *Myoporum ramulosa*, Aus.

Boombi (Aus.) *Lomandra* (= *Xerotes*) *multiflora*, a rush used in bag making.

Boomerang trigger plant *Stylidium breviscarpum*, W.Aus. *See Stylidium* in Willis.

Boonaree (Aus.) *Heterodendron oleifolium*, red frs. eaten by aborigines.

Boophone or **buphane** S.Af. poisonous pls. (bushman's arrow poison) sometimes cult. for orn.

BOOTLACE TREE

Bootlace tree (W.I.) *Eperua falcata*, trop. Am., long thin fl. stalks: wood much used for shingles.

Borage *Borago officinalis*, Eur., N.Af., a popular old-time culinary herb: much used for flavouring claret cup before the adoption of cucumber: the 'true blue' fls. may be used for garnishing salads (*see Borago* in Willis). Fls attractive to hive bees for nectar.

Borassus palm Or palmyra palm, *Borassus flabellifer*, imp. source of jaggery, economically imp. in India (*see Borassus* in Willis). The Af. borassus palm (*B. aethiopum*) is similar.

Borecole Kale, *Brassica oleracea* var. *acephala*.

Borse (Aus.) *Acacia cana, A. pendula*, useful fodder trees.

Boriti poles *Rhizophora mucronata*, mangrove, used for building, exported from E.Af. to Arabia and Iran on dhows for this purpose.

Borneo camphor *Dryobalanops aromatica*, a large forest tree, camphor used mainly in China: — **ironwood**, *Eusideroxylon zwageri:* — **mahogany**, *Calophyllum inophyllum:* — **rubber**, *Willughbeia edulis, W. firma:* — **tallow**, *Shorea aptera, S. stenocarpa*.

Boronia Aus. evergreen shrubs, fls. often fragrant, cult.: **brown** —, *B. megastigma:* **pale pink** —, *B. floribunda*, highly perfumed and grown commercially in Aus. as a cut fl.: **pink** —, *B. pilosa:* **pinnate** —, *B. pinnata:* **soft** —, *B. mollis; B. purdiana*.

Borzicactus Low cylindrical pls., S.Am., orange or scarlet diurnal fls., cult.

Bossiaea Aus. evergreen shrubs, some cult. for orn.

Boston ivy *Parthenocissus tricuspidata*, a popular creeper, cult.

Bottle-brush *Callistemon* spp. and other pls. w. inflor. resembling a bottle-brush. **Aus.** — —, *Callistemon, Melaleuca:* — — **grass**, *Asperella hystrix*, N.Am. cult. gdn. pl., used in bouquets: **green** — —, *Callistemon viridiflorus*, Aus.: **mountain or Natal** — —, *Greyia sutherlandii*, S.Af.: — — **orchid**, *Dendrobium smilliae:* **red** — —, *Mimetes lyrigera*, S.Af.

Bottle gardens, plants for A revival of interest in this particular form of indoor gardening has directed attention to those pls. that are best suited to it. Many pls. that will not thrive in the dry atmosphere of a centrally heated living room will grow happily in an enclosed or glassed-in space where humidity may build up. The bottle gdn. or glazed 'Wardian case' (fern case) of Victorian times provides such conditions. The glass also provides some protection against fluctuating temperature, draughts, dust and tobacco smoke or other fumes. Any transparent container may serve, such as large flagons (as used for cider), carboys, goldfish bowls, bell jars, sweet jars etc.

With the old-time bottle gdn., skill is needed in introducing the pls. through the restricted opening and successfully planting them, special tools or appliances being generally employed. Ferns, including the difficult filmy ferns, selaginellas, mosses, liverworts, and lichens are favourite subjects. Small or cut-leaved forms of common ivy are often effectively used as are small hot-house plants such as Af. violets (*Saintpaulia*). With flowering pls. it is necessary to remove dead fls. otherwise they become unsightly, being unable to fall to the ground out of sight. Other advantages of the bottle gdn. are that it is generally free from insect pests and can be left for long periods, watering being very rarely required.

Bottle-tree *Brachychiton rupestris,* Aus., peculiar swollen trunk, emergency fodder tree.

Bougainvillea Climbing shrubs, trop. Am., cult., a colourful feature of gdns. throughout the trop. and subtrop.: 3 fls. surrounded by 3 coloured bracts: numerous hybrids and cultivars (over 100) derived from 4 spp. – *B. spectabilis* (common purple —), *B. glabra, B. peruviana* and *B. buttiana.* The common bougainvillea, suitably trained, makes an effective defensive hedge. Some modern vars. change fl. colour from youth to age. Some will not stand hard pruning.

Boulder fern *Dennstaedtia punctiloba,* N.Am.

Bouncing Bet *Saponaria officinalis,* Eur.

Boundary mark (W.I.) *Cordyline terminalis,* cult. trop., ls. may be coloured.

Bouquet garni Small bundle of fresh herbs used for flavouring soup, meat, or other dishes.

Bourbon palm *Latania commersonii,* Mauritius, orn. palm, cult.

Bouvardia Mainly Mex. shrubs, cult., fls. often fragrant and freely produced.

Bow-wood *Maclura pomifera,* N.Am. (*see Maclura* in Willis): **—man's r.,** *Gillenia trifoliata,* N.Am., med.

Bower plant *Pandorea jasminoides,* Aus., cult. trop.

Bowstring hemp *Sansevieria* spp. Many Af. and As. spp. yield a good l. fibre freely used locally. There have been many attempts at commercial exploitation, esp. w. *S. roxburghiana* (Ind.), *S. zeylanica* (Sri Lanka) and *S. cylindrica, S. ehrenbergii* and *S. thrysiflora* of trop. Af. From breeding work in Florida the hybrid 'Florida H-13' (*S. trifasciata* × *S. deserti*) shows promise.

Box or **boxwood Common** or **Eur.,** *Buxus sempervirens,* wild and cult. evergreen shrub or small tree, much used for topiary, hedges and edging. There are dwarf forms. Wood v. firm and close-grained, long favoured for carving and printers' blocks: used for inlaying, furniture, mathematical instruments, chess-men and fancy articles. Other similar commercial boxwoods are – **Cape** or **East London —,** *Buxus macowani:* **Knysna —,** *Gonioma kamassi,* S.Af.: **Maracaibo** or **W.I. —,** *Gossypiospermum* (= *Casearia*) *praecox:* **San Domingo** or **Venezuelan —,** *Phyllostylon brasiliensis.* The names box or boxwood are applied to some other trees or woods, e.g. **Aus. —,** *Eucalyptus:* **brush —,** *Tristania conferta,* Aus.: **Ceylon —,** *Gardenia:* **desert —,** *Eucalyptus microtheca,* Aus.: **— elder,** *Acer negundo:* **Florida —,** *Schaefferia frutescens:* **Jap. —,** *Buxus japonica:* **— myrtle,** *Myrica nagi,* Orient: **Siamese —,** *Gardenia:* **yellow —,** *Planchonella pohlmaniana,* Aus. rain forest tree.

Boxthorn *Lycium* spp., shrubs, usu. spiny. **Af. —,** *L. afrum,* cult., an efficient barrier pl., but can become a nuisance: **Chinese —,** *L. chinensis,* naturalized in Br., common on walls: *L. halimifolium* is similar, a good seaside pl.

Boys and girls *Mercurialis annua,* a common weed.

Boysenberry A hybrid berry fr. resembling loganberry.

Bracatinga *Mimosa bracatinga,* Braz. wild and cult. quick-growing firewood tree.

Bracelet tree *Jacquinia armillaris,* Peru, shining yellow and brown seeds made into bracelets.

Brachyglottis *B. repanda,* N.Z. cult. orn. shrub, large ls., felted below.

Brachystelma Perennials, some spp. have ed. tubers.

BRACKEN FERN

Bracken fern *Pteridium aquilinum*, cosmop. often a bad weed, can destroy good pasture. Rhizomes eaten in some areas. Several vars. *P. esculentum* Aus. and N.Z. is very similar.

Bradley grass (S.Af.) *Cynodon bradleyi* (= *C. incompletus*), a lawn grass; akin to *Cynodon dactylon.*

Bragging Tom (Jam.) Name for a white-fleshed yam w. long tubers.

Brahea Orn. palms, S. U.S., Mex. *B. dulcis* and others w. ed. frs.

Brake Common name for bracken. **Curled rock** —, *Cryptogramma crispa*: — **fern**, *Pteridium aquilinum*: **jungle** —, *Pteris umbrosa*, Aus.: — **root**, *Polypodium vulgare*: **tender** —, *Pteris tremula*, Aus.

Bramble *Rubus* spp., blackberry. **Arctic** —, *R. arcticus*: **Cape** —, *R. rosifolius.*

Brandy bottle *Nuphar luteum*, Eur. inc. Br., yellow water lily: — **bush**, *Grewia flava*, S.Af., fr. eaten by bushmen.

Brasiletto Woods of *Caesalpinia* and *Peltophorum* spp.

Brassia Trop. Am. orchids, cult.

Brassica Many imp. veg. and farm crops belong to this genus such as cabbage, borecole, broccoli, Brussels sprouts, cauliflower, kale, kohlrabi, mustard, rape, swede and turnip. *See* separate headings and *Brassica* in Willis.

Brauna *Melanoxylon brauna*, Braz., strong tough wood, used for wheels, bridges etc.

Bravoa Mex. bulbous pls., *B. geminiflora*, cult., fls. in twos.

Brazil or **Brazilian arrowroot** = Sweet potato, *Ipomoea batatas*: — **cherry**, *Eugenia* spp.: — **copal**, *Hymenaea coubaril*: — '**nut**', *Bertholletia excelsa* this well known dessert nut, used so much in confectionery, esp. nut chocolate, is the product of a large forest tree in the Amazon region: — **nutmeg**, *Cryptocarya*: — **pepper**, *Schinus*: — **pine**, *Araucaria angustifolia*: — **redwood**, *Caesalpinia brasiliensis*, *Brosimum paraense*: — **rosewood**, *Dalbergia nigra* and other spp.: — **satinwood**, *Euxylophora paraensis*: — **tea**, *Ilex paraguariensis*, *Stachytarpheta jamaicensis.*

Brazil wood Name originally applied to woods used as dyes, such as *Caesalpinia sappan* (sappan wood) but later to other woods, notably *Caesalpinia echinata*, a commercial wood, also called bahia, para or pernambuco wood.

Bread and cheese tree *Pithecellobium unguis-cati*, W.I.: **Bread and jam fl.**, *Darwinia fascicularis.*

Bread-fruit *Artocarpus altilis* (= *A. incisa*) South Sea Isles, cult. trop., seedless forms the most esteemed: **Nicobar** — —, *Pandanus leram*, fr. ed.: — **nut**, *Brosimum alicastrum*, trop. Am., seeds ed., roasted: — **root**, *Psoralea esculenta*, N.Am.: — — **sword** (Jam.), the fls. of the bread-fruit tree used to make a crystallized confection.

Briar root *Erica arborea*, Medit., the traditional material for the manufacture of smoking pipes, the name being a corruption of the French 'bruyère' or tree heath. This heath has the peculiarity of forming large woody nodules on the roots and base of the stem, which is the part used in pipe-making after curing and treatment (boiling).

Bridal bouquet *Porana volubilis*, trop., climber w. small white fls. in profusion; *P. paniculata* is similar: — **wreath**, *Petrea kohautiana*, *Francoa sonchifolia*, Chile.

36

Bridelia *B. macrantha*, trop. Af., food pl. of the *Anaphe* silkworm.

Bridewort *Filipendula ulmaria*, meadow-sweet.

Brigalow (Aus.) *Acacia* spp. esp. *A. hypophylla*, dominant in some of the dried areas of E.Aus., hard, heavy wood.

Brimstone tree *Morinda lucida*, W.Af., wood yellow when fresh; *Terminalia ivorensis*, W.Af.

Brinjal *See* egg-fruit.

Bristle fern *Trichomanes radicans:* — **grass**, *Setaria viridis*, Eur., *Trisetum, spicatum*, Aus.: —**tail grass**, *Psilurus incurvus* Medit.: **yellow** — **grass**, *Setaria lutescens*, Medit.: — **scirpus**, *Scirpus setaceus:* — **stem**, *Psathyrotes*, SW. U.S.

Britoa *B.* (=*Campomanesia*) *acida*, Braz., small tree, ed. fr., 'para guava'.

Brittle bush *Encelia farinosa*, SW. U.S., stems exude resin, once used for incense.

Broad bean *Vicia faba*, a widely cult. veg.: smaller seeded forms constitute the field bean or horse bean and the pigeon bean: — **leaf tree**, *Terminalia latifolia*, W.I.: — **path**, *Alternanthera*, W.I.

Broccoli *Brassica oleracea* var. *botrytis*, a much used temp. veg.: **sprouting** —, *B. oleracea* var. *italica*.

Brodiaea Cult. orn. fls. Am., fls. usu. blue in clusters.

Broke back *Dichapetalum toxicarum*, W.Af. shrub, seeds used for poisoning rats, also well water, during tribal warfare.

Broken hearts *Clerodendron thomsoniae*, trop. Af., cult. orn. shrub, fls. red and white.

Brome grass *Bromus* spp. Some spp. are of agric. value, others are better known as weeds. Soft brome, *B. mollis* (=*Serrafalcus mollis*) is the most common sp. in Br. It occurs in many other countries. **False** — —, *Brachypodium:* **Hungarian** — —, *Bromus inermis*, awnless, cult. for fodder.

Bromelia Trop. Am., some spp. cult., orn., others yield fibre used locally, or ed. fr.

Bronze leaf *Rodgersia pinnata*, cult. gdn. pl., several vars.

Brook lime *Veronica beccabunga*, in streams and damp places: — **weed**, *Samolus valerandi*, on wet ground.

Broom *Cytisus, Genista, Spartium* etc. Cult. orn. shrubs. **Common** or **Eur.** —, *Sarothamnus scoparius* (=*Cytisus scoparius*) widely distributed, esp. on heaths (*see Sarothamnus* in Willis): **giant flowered** —, *Carmichaelia williamsii*, N.Z.: **coral** —, *Corallospartium crassicaule*, N.Z.: **Mount Etna** —, *Genista aethnensis:* **native** —, *Viminaria juncea*, Aus.: **pink** —, *Notospartium carmichaeliae*, N.Z.: **purple** —, *Polygala virgata*, S.Af.: **scented** —, *Carmichaelia odorata*, N.Z.: **Spanish** —, *Spartium junceum*, yields fibre: **Tenerife** —, *Cytisus supranubeus:* **white Portuguese** —, *Cytisus albus:* **wild** (**S.Af.**) —, *Lebeckia* spp.

Broom corn *Sorghum bicolor* (= *S. dochna* var. *technicum*), Florence, Venetian or Italian whisk, cult., esp. in Italy, large wiry inflor. used for brooms, brushes etc.: — **palm**, *Thrinax argentea*, W.I., ls. made into brooms: — **plant**, *Sida* spp.: — **rape**, *Orobanche* spp. r. parasites (*see Orobanche* in Willis): — **root**, *Muhlenbergia* (=*Epicampes*) *macroura*, Mex. whisk, a commercial fibre: — **weed**, *Corchorus, Sida*, W.I.: — **wort**, *Scoparia*.

37

Broussa tea *Vaccinium arctostaphylos*, Caucasus.

Browallia Cult. orn. pls., trop. Am., anns. and perens.

Brown beaks *Lyperanthus suaveolens*, Aus., a terrestrial orchid: — **eyed Susan**, *Rudbeckia hirta*, N.Am.: — **George**, a Jamaican confection consisting basically of dry maize, parched, ground and mixed w. sugar and salt: — **Peru bark**, *Cinchona officinalis:* — **pine**, *Podocarpus elatus*, Aus.

Brunfelsia Handsome free-flowering trop. Am. shrubs, widely cult., *B. calycina* is one of the best known, fls. change colour as they age.

Brunsvigia Bulbous S.Af. pls., cult. orn.

Brush box *Tristania conferta*, yields a commercial timber: a cult. orn. tree, quick-growing, a much used street tree in Sydney.

Brussels sprouts *Brassica oleracea* var. *gemmifera*, a widely grown veg. w. well-developed lateral buds resembling a small cabbage – the part eaten.

Bruyère *See* briar root.

Bryony Bastard —, *Cissus* spp., W.I.: **black** —, *Tamus communis*, once used med.: **white** —, *Bryonia dioica*.

Bryophyllum (= *Kalanchoe*) Cult. orn. pls., trop. *B. pinnatum* (= *B. calycinum*), a popular room pl., new pls. develop on fallen ls.

Buaze *Securidaca longipedunculata*, E.Af., fibre used for fishing nets.

Buba (Jam.) A palm l., mat.

Bubinga *Guibourtia demeusei* and other spp., W.Af., a commercial timber, used for furniture in Br.

Buchu *Agathosma betulina* (= *Barosma betulina*), S.Af., a small shrub of the W. Cape province w. aromatic ls.: these, dried, constitute the drug: mainly used for infections of the urinary tract: a popular household remedy from the days of early settlers in S.Af.: — **grass**, *Cymbopogon excavatus*, S.Af.

Buck-bean *See* bog bean: — **berry**, *Vaccinium* spp. N.Am.: — **eye**, *Aesculus* spp. N.Am., cult. orn. trees and shrubs: **Mex. —eye**, *Ungnadia speciosa*.

Bucklandia *B. populnea*, Trop. As., handsome tree w. useful timber.

Buckler fern *Dryopteris* spp. Several in Br., **broad** — —, *D. dilatata:* **crested** — —, *D. cristata:* **hay-scented** — —, *D. aemula:* **narrow** — —, *D. carthusiana:* **rigid** — —, *D. villarsii.*

Buckthorn *Rhamnus* spp. Several yield dyes or a close-grained wood used for special purposes. **Common** or **Eur.** —, *R. cathartica*, berries yield the artist's pigment known as sap green, also a med. syrup. **Alder** —, *Frangula alnus:* **cascara** —, *see* cascara; **Dahurian** —, *R. dahurica*, ls. yield ' Chinese green': **dyer's** —, *R. tinctoria:* **Jap.** —, *R. japonica:* **sea** —, *Hippophae rhamnoides.*

Buckwheat *Fagopyrum esculentum* (= *F. sagittatum*). Thought to have originated in China: starchy seeds much used as human and animal food, esp. in Eur. & N.Am.: husks a packing material: grown as soiling crop: fls. a good nectar source, give dark honey. **Siberian, Tartary** or **Kangra** —, *Fagopyrum tartaricum*, cult. in the Himalayas, is similar. **Bush** —, *Cliftonia monophylla*, N.Am., fls. a source of commercial honey: **climbing false** —, *Polygonum scandens*, N.Am.: **wild** —, *Eriogonum microthecum*, N.Am.

Budding *See* grafting.

Buddleja Cult. orn. fl. shrubs, some med. in countries of origin.

Budge gum *Bursera simaruba*, W.I., partly evaporated juice forms a resinous transparent varnish.

Budgerigar flower *Asclepias syriaca*, cult., the green frs. inverted are made to look like budgerigars (sold to tourists in Switzerland).

Buffalo bean *Astragalus caryocarpus*, N.Am.; *Mucuna*, S.Af.: — **berry,** *Shepherdia argentea:* — **bur,** *Solanum rostratum*, N.Am. an aggressive weed: — **clover,** *Trifolium reflexum*, N.Am.: — **currant,** *Ribes odoratum*, N.Am.: — **gourd,** *Cucurbita digitata*, N.Am.: — **grass,** *Buchloe dactyloides*, N.Am.; *Panicum*, N.Am.; *Setaria sulcata*, S.Af.; *Stenotaphrum secundatum*, S.Af.: — **nut,** *Pyrularia pubera*, N.Am.: — **thorn,** *Zizyphus mucronata*, S.Af.: — **weed,** *Ambrosia trifida*, N.Am.: — **wood,** *Burchellia*, S.Af.

Buffel grass (S.Af.) *Urochloa mosambicensis, Cenchrus pennisetiformis*, a valuable pasture grass in dry trop.

Bugbane *Cimicifuga foetida*, used in Siberia to drive away bugs, cult. orn. fl.: **Am.** —, *C. americana.*

Bugseed *Corispermum*, Eur.

Bugle *Ajuga* spp. **Aus.** —, *A. australis:* **common** or **Eur.** —, *A. reptans:* **garden** —, *A. genevensis*, cult.: **yellow** —, *A. chamaepitys:* — **weed,** *Lycopus virginicus*, N.Am., med.

Bugloss *Echium lycopsis*, Eur. incl. Br.: **viper's** —, *Echium vulgare*, Eur. incl. Br., cult.

Bukchie (Ind.) *Vernonia anthelmintica*, frs. used for skin diseases and repelling moths from woollens.

Bulbine Cult. orn. pls., mainly S.Af., some w. med. uses.

Bulblet fern *Cystopteris* spp., N.Am.

Bulbophyllum Large genus of mainly trop. orchids (900 spp.) some w. curious fls., cult.

Bull bay *Magnolia grandiflora*, E. N. Am. orn. and timber tree, cult.: — **brier,** *Smilax rotundifolia*, N.Am.: — **grass,** *Paspalum boscianum:* — **hoof,** *Passiflora*, W.I.: —**'s horn thorn,** *Acacia cornigera* and other spp., Am.: — **nettle,** *Cnidoscolus stimulosus:* — **oak,** *Casuarina equisetifolia:* — **pine,** *Pinus sabiana.*

Bullace *Prunus insititia*, a small purple plum, usu. preserved. — **grape,** *Vitis rotundifolia*, N.Am.

Bullet or **bully tree** *Manilkara bidentata, Dipholis salicifolia*, W.I.

Bullock brush *Templetonia retusa*, W.Aus.

Bullock's heart = custard apple.

Bulnesia *B. arborea*, S.Am., hard heavy wood, like lignum vitae.

Bulrush *Typha latifolia* (reed-mace); *Scirpus lacustris* (club rush), used for chair bottoms, mats, baskets, hassocks etc.: — **millet,** *Pennisetum typhoides:* **narrow-leaved** —, *Typha angustifolia:* **prairie** —, *Scirpus paludosus*, N.Am.

Bumbo tree *Daniella thurifera*, W.Af., a source of copal.

Bunch berry *Cornus canadensis:* — **flower,** *Melanthium virginicum*, N.Am., cult. orn. fl., *Tofieldia, Zygadenus:* — **grass,** *Andropogon scoparius.*

Bunny orchid (Aus.) *Eriochilus* spp., small terrestrial orchids.

Bunya-bunya pine *Araucaria bidwillii*, Aus. well known tree and commercial timber.

Bupleurum Cult. orn. pls., small yellow fls. in umbels.

Bur or **burr** A hooked fr. or a woody growth on a tree trunk or stem. The latter, when large, are valued for their figure and used as veneers or plywood in furniture and cabinet making, e.g. bur-walnut, bur-maple etc.

Bur bark *Triumfetta*, W.I.: — **clover**, *Medicago* spp.: — **cucumber**, *Sicyos angulatus*, N.Am.: —**dock**, *Arctium lappa* and other spp., fr. characteristic: — **grass**, *Cenchrus* spp., *Tragus australiensis*, Aus.: — **head**, *Echinodorus:* — **marigold**, *Bidens:* **Noogoora** —, *Xanthium:* — **nut**, *Tribulus:* — **oak**, *Quercus macrocarpa:* — **reed**, *Sparganium:* —**weed**, *Medicago*, *Sparganium*, *Xanthium*, esp. *X. strumarium.*

Burgundy pitch Obtained from *Picea abies* (=*P. excelsa*), Eur. spruce, used med. for plasters.

Burmese lacquer tree *Melanorrhoea usitata*, yields on tapping the black varnish used in Burmese lacquer ware. — **rosewood**, *Pterocarpus indicus.*

Burnet *Sanguisorba officinalis* and other spp. **Canadian** —, *S. canadensis:* — **saxifrage**, *Pimpinella major:* — **rose**, *Rosa pimpinellifolia.*

Burning bush *Kochia scoparia*, cult. fol. pl., red or reddish in autumn; *Euonymus americanus*, *E. atropurpureus*, *Dictamnus albus;* in S.Af. *Combretum microphyllum.* **Burn-mouth vine**, *Rhynchosia minima*, W.I.: **burn nose**, *Daphnopsis tinifolia*, W.I., inner bark used for cordage.

Burrawang *Macrozamia spiralis*, E. Aus., palm-like pl., seeds eaten by aborigines.

Bursaria *B. spinosa*, Aus., cult. orn. shrub, fragrant white fls.

Bush butter *Dacryodes edulis*, trop. Af., seeds oily and ed.: — **fruits**, small ed. frs. which grow on bushy or woody pls., currants, gooseberries etc. — **clover**, *Lespedeza:* — **iris**, *Patersonia sericea*, Aus.: — **lawyer**, *Rubus australis*, N.Z.: — **nut**, *Macrozamia*, Aus.: — **pea**, *Pultenaea stipularis:* — **rose**, *Eucalyptus rhodantha*, Aus.: — **tea**, *Cyclopia*, *Aspalathus*, S.Af.: — **willow**, *Combretum apiculatum* and other spp.

Bushman's candle *Sarcocaulon burmannii*, S.Af.: — **poison bush**, *Acokanthera spectabilis*, S.Af.: — **tea**, *Catha edulis.*

Bussu palm *Manicaria saccifera*, Braz., fibrous spathes have many uses.

Busy Lizzie *Impatiens wallerana* (= *I. sultani*, *I. holstii*), trop. Af., popular room pls.

Butcher's broom *Ruscus aculeatus*, Eur., inc. Br., cult. evergreen: once used by butchers to clean their blocks.

Butter-cup *Ranunculus* spp. wild and cult.: **ankalaki** —, *Polygala butyracea*, W.Af.: — **and eggs**, *Linaria vulgaris* (toad flax): — **bur**, *Petasites:* **dika** —, *Irvingia gabonensis*, W.Af.: **goa** —, *Garcinia indica:* **illipe** —, *Shorea*, trop. As.: **kombo** —, *Pycnanthus kombo:* — **nut**, *Juglans cinerea*, N.Am.: **otoba** —, *Myristica otoba*, S.Am.: **shea** —, *Butyrospermum parkii*, W.Af.: — **tree**, *Pentadesma butyracea*, W.Af., seeds yield ed. fat: — **weed**, *Senecio*, N.Am.: — **wort**, *Pinguicula vulgaris*, Eur., As.

Butterfly flowers Some fls. are particularly attractive to butterflies. Among them are spp. of *Buddleja*, *Clerodendron*, *Cuphea*, *Daphne*, *Eupatorium*, *Gentiana*, and *Lonicera.* — **bush**, *Clerodendron myricoides*, E.Af.: — **flower**, *Schizanthus:* — **lily**, *Hedychium coronarium:* — **orchid**, *Oncidium papilio*, *Epidendrum tampense*, N.Am., *Plantanthera:* — **orchis**, *Habenaria:* — **pea**, *Clitoria ternata*, *Centrosema:* — **plant**, *Oncidium papilio:* — **shrub**, *Clerodendron ugandense:* — **tree**, *Bauhinia* spp., *Erblichia odorata*, C.Am., large orange-red petals: — **weed**, *Asclepias tuberosa.*

Button bush *Cephalanthus occidentalis*, E. N. Am. cream fls. in globular heads: — **clover**, *Medicago orbicularis:* — **flower**, *Cotula, Ouratea:* — **mangrove**, *Conocarpus erectus:* — **snakeroot**, *Eryngium yuccifolium, Liatris:* — **weed**, *Spermococe tenuior*, W.I., a common weed; *Diodia; Borreria:* — **wood**, *Platanus occidentalis*, N.Am. a commercial wood, 'Am. plane'; *Laguncularia racemosa; Conocarpus erectus.*

Byblis Aus. insectivorous undershrubs with stalked and sessile glands like *Pinguicula.*

Byrsonima Trop. Am. trees, several yield good furniture woods.

C

Caa-ehe *Stevia rebaudiana*, Paraguay, shrub, ls. w. remarkable sweetening properties.

Caapi *Banisteriopsis caapi*, trop. S.Am., woody vine, cult., ls. used for narcotic beverage causing hallucinations.

Cabbage *Brassica oleracea* var. *capitata*. Widely grown veg., origin uncertain, numerous vars.: red-leaved forms favoured for pickling, savoy cabbage for winter use, special vars. for sauerkraut. **Chinese** —, *Brassica chinensis*, 'pak-choi', *B. pekinensis*, 'pe-tsai': **Isle of Man** —, *Rhynchosinapis monensis:* **Kerguelen** —, *Pringlea antiscorbutica*, Kerguelen Island, has antiscorbutic properties: **Lundy** —, *Rhynchosinapis wrightii:* **musk** —, *Bedfordia salicina*, Aus.: **palm** —, the bud or growing point of several palms is eaten as 'cabbage', e.g. *Euterpe oleracea* trop. Am.; *Roystonea regia*, trop. Am.; *Livistona australis*, Aus.; *Elaeis guineensis*, W.Af.: **Shantung** —, *Brassica chinensis:* — **tree**, *Anthocleista, Vernonia*, W.Af.; *Cussonia spicata* and other spp., S.Af., softwood used for wagon brake blocks; *Cordyline australis*, N.Z., pith ed.; *Andira inermis*, W.I.

Cabelluda *Eugenia tomentosa*, Braz. ed. tree. fr.

Cabreuva oil *Myrocarpus frondosus*, Braz. esst. oil from wood.

Cabuya fibre *Furcraea cabuya*, C.Am., used for cordage and sacking.

Cacao *See* cocoa.

Cachibou *Calathea discolor*, W.I., large ls. used to line baskets.

Cactus The word 'cactus' is popularly applied to almost any plant that is succulent and prickly, such as spp. of *Aloe* and *Agave*. It should of course be restricted to the true cacti or members of the family Cactaceae which are all, with one exception (*Rhipsalis*), indigenous to the New World. The family Cactaceae consists of about 2000 spp., some of which (*Opuntia* spp.) are now widely distributed throughout the world. Cacti vary in size from the size of a thumb to the giant cactus (*Carnegiea*) of tree size and great weight.

The cult. of cacti attracts a large following and is a hobby of many people in many lands. There may be several reasons for this, such as the bizarre form of many of the spp., their attractive fls. and their ability to resist drought. The enthusiastic collector or hobbyist draws his subjects from many different genera in the family, some of the more notable being – *Mammillaria* (there are

200–300 spp., mostly small), *Cereus, Aporocactus, Astrophytum, Echinocactus, Coryphantha, Echinocereus* (large fls. on small pls.), *Echinofossulocactus, Echinopsis* (easily grown from offsets), *Ferocactus, Gymnocalcium, Harrisia* (fls. open at night), *Opuntia. See also* Cactaceae in Willis.

Other cacti are – **barred** —, *Echinocactus wislizenii*, v. large: **brittle** —, *Opuntia fragilis:* **candy** —, *Echinocactus wislizenii:* **Christmas** —, *Schlumbergera truncata:* **cholla** —, *Opuntia:* **cochineal** —, *Opuntia, Nopalea:* **comb hedgehog** —, *Echinocereus pectinatus:* **crab** —, *Schlumbergera truncata:* **dahlia** —, *Wilcoxia:* **deer horn** —, *Peniocereus greggi:* **Devil's** —, *Opuntia:* **fishhook** —, *Echinocactus:* **flapjack** —, *Opuntia engelmannii:* **fuchsia** —, *Schlumbergera:* **giant** —, *Carnegiea gigantea* (=*Cereus giganteus):* **lace** —, *Echinocereus caespitosus:* **living rock** —, *Ariocarpus fissuratus:* **mountain** —, *Pediocactus simpsonii:* **night-flowering** —, *Selenicereus grandiflorus, Acanthocereus:* **nipple** —, *Mammillaria:* **old man** —, *Cephalocereus senilis:* **pencil** —, *Wilcoxia poselgeri:* **pincushion** —, *Mammillaria:* **rainbow** —, *Echinocereus dasyacanthus:* **strawberry** —, *Echinocactus:* **glory of Texas** —, *Echinocactus bicolor:* **torch** —, *Echinocereus viridiflorus:* **Turk's cap** —, *Melocactus communis:* **Turk's head** —, *Echinocereus coccineus.*

Cade, oil of *Juniperus oxycedrus,* Medit., med. uses.

Cadushi (Curaçao) *Cereus repandus,* de-spined young stems eaten.

Caffeine Occurs in various pls., notably coffee, tea, maté, cola, guarana etc. Commercially it may be obtained from tea waste or coffee, or prepared synthetically. It facilitates the performance of muscular work and has med. uses.

Cafta *Catha edulis,* Khat or Arabian tea.

Caihuba Or ucahuba. *Virola surinamensis,* S.Am., and other spp., yield veg. butter.

Cajuado A Braz. beverage made from the cashew 'apple'.

Cajuput oil *Melaleuca cajuputi* (=*M. leucadendron, M. minor),* Malaysia, oil distilled from ls., mainly in the Moluccas: has med. uses, a counter-irritant in ointments and liniments.

Cake or **oil cake** The residue after crushing oil-yielding seeds for oil, much used for feeding livestock esp. cotton seed and linseed oil cake, also as fertilizer.

Calaba *Calophyllum calaba,* Sri Lanka, seeds yield oil: useful timber.

Calabar bean *Physostigma venenosum,* W.Af. seed poisonous, ordeal bean, has med. use, source of physostigmine (eserine).

Calabash gourd *Lagenaria siceraria* (=*L. vulgaris),* cult. trop. and subtrop., *see* gourds. — **nutmeg,** *Monodora myristica,* W.Af., seeds used as a spice: **sweet** —, *Passiflora maliformis,* shell of fr. v. hard, juice grape-like: — **tree,** *Crescentia cujete,* W.I. and trop. Am., frs. like ordinary calabash.

Calabura *Muntingia calabura,* trop. Am., ed. fr.

Caladium Cult. orn. fol. pls., trop. Am., esp. *C. bicolor* w. many vars.

Calalu (W.I.) Name used for tender young ls. of various pls. used as spinach.

Calamander or **coromandel wood** *Diospyros quaesita,* Sri Lanka, greyish-brown wood w. black bands, used for high class furniture (Sheraton and Regency) at one time, now rare.

Calamint *Calamintha* spp. cult. orn. fls., esp. *C. coccinea:* **common —,** *C. ascendens:* **cushion —,** *Clinopodium vulgare:* **wood —,** *Calamintha sylvatica.*

Calamus *See* rattan cane. **— root,** *Acorus calamus,* Eur. inc. Br., rhi. w. strong sweet smell, ls. spread on church floors at one time.

Calandrinia Mainly fleshy and often trailing pls. w. brilliant fls. open only in the sun, cult.

Calathea Orn. fol. pls., trop. Am., cult.

Calceolaria Cult. orn. fls., mainly from the Andes, numerous hybrids ('slipper-fl.').

Calendula *C. officinalis,* S.Eur., common or pot marigold, many vars.: var. *prolifera* has secondary fl. heads from axils of involucral bracts.

Calico bush *Kalmia latifolia,* E. N. Am., cult. orn. shrub: **— flower,** *Aristolochia elegans,* Braz., cult. trop.

Californian allspice *Calycanthus occidentalis,* aromatic shrub, cult.: **— bluebell,** *Nemophila:* **— buckeye,** *Aesculus californica:* **— hyacinth,** *Brodiaea:* **— juniper,** *Juniperus californica:* **— laurel,** *Umbellularia californica:* **— lilac,** *Ceanothus:* **— nutmeg,** *Torreya californica:* **— pepper,** *Schinus molle:* **— poppy,** *Eschscholzia californica,* *Platystemon:* **— redwood,** *Sequoia:* **— rose,** *Rosa californica:* **— tree poppy,** *Romneya.*

Caliphruria Orn. bulbous pls. of the Andes, cult.

Calisaya bark *Cinchona calisaya,* quinine.

Calliandra Cult. orn. fl. shrubs, mainly trop. Am., some w. local med. uses.

Callicarpa *C. japonica* and other spp., cult. orn. shrubs.

Callipsyche Bulbous S.Am. orn. pls., cult.

Callistemon *See* bottlebrush.

Callitris *See* cypress pine.

Calocedrus *C.* (= *Libocedrus*) *decurrens,* N.Am., incense cedar.

Calocephalus Aus. pls., usu. small, sometimes grown for edging.

Calochortus Bulbous Am. pls., many w. beautiful fls., cult.

Calomeria *C.* (= *Humea*) *elegans,* Aus., incense pl., ls. scented, cult., biennial, may cause dermatitis.

Calonyction *See* moonflower.

Calophyllum *C. tacamahaca,* Madag. yields tacamahac resin. *C. inophyllum* yields good timber.

Calotropis *C. gigantea,* Ind., yields a b. fibre and seed floss., fls. candied as a sweetmeat. *C. procera,* trop. Af., also yields fibre.

Caltrops Devil thorn or Maltese Cross, *Tribulus terrestris,* cosmop. weed: **water —,** *Trapa spinosa.*

Calumba Or Colombo root, *Jateorhiza palmata* (= *J. columba*), E.Af., forest climber, dried sliced r. med.: **false —,** *Coscinium fenestratum,* Sri Lanka.

Calvary clover *Medicago echinus,* the twisted pod considered to resemble the Crown of Thorns.

Calythrix Evergreen heath-like shrubs, Aus. and Tasm., cult.

Camash or **Quamash** *Camassia, Zigadenus* spp., N.Am., bulbous pls., some poisonous to livestock: **common —,** *Camassia quamash:* **death —,** *Zigadenus nuttallii:* **white —,** *Z. glaucus.*

Camel's foot *Bauhinia* spp., the ls. of some are the shape and size of a camel's foot-print: **camel-thorn**, *Acacia giraffae*, S.Af.; *Alhagi maurorum*, Egypt to Ind., a desert shrub which secretes a kind of manna.

Cameline oil *Camelina sativa*, Medit., seeds (gold of pleasure) yield oil like rape oil.

Camellia *C. japonica* and other spp. cult. fl. shrubs numerous vars. and hybrids: seeds yield oil esp. *C. sasanqua* (tea seed oil).

Camoensia *C. maxima*, W.Af., creeper, cult., w. large scented white fls.

Camomile *See* chamomile.

Campanula Or bellflower, several spp. are popular gdn. pls., such as Canterbury Bells. In E.Af. *Adenophora liliifolia* is called 'campanula'.

Campeachy wood or **campeche** (W.I.) Names for logwood.

Camphor *Cinnamomum camphora*, Taiwan, an aromatic crystalline substance obtained by distillation of the wood: may also be prepared synthetically: camphor has various med. uses, in liniments and as a counter-irritant in the treatment of fibrositis and neuralgia. **Borneo** or **Sumatra** —, *Dryobalanops aromatica*, a large forest tree of the E.I., the camphor occurs as crystals in the wood and is collected, used mainly in China and Jap.: **E.Af.** — **wood**, *Ocotea usambarensis*, scented, a commercial timber: **Ngai** —, *Blumea balsamifera*, China, ls. distilled, med. uses: — **plant**, *Heterotheca subaxillaris*, N.Am.; *Chrysanthemum balsamita*, cult: — **weed**, *Pluchea camphorata*, N.Am.; *Heterotheca subaxillaris*, N.Am.: — **wood** *Tarchonanthus camphoratus*, trop. S.Af., a fodder tree.

Campion *Silene*, *Lychnis*, *Melandrium*. **Alpine** —, *Lychnis alpina:* **bladder** —, *Silene vulgaris:* **moss** —, *S. acaulis:* **sea** —, *S. maritima*.

Campomanesia *C. aromatica*, W.I., Guyana, tree, w. ed. fr.

Campos (Braz.) Open grass country with patches of trees.

Camwood *Baphia nitida*, W.Af., the heartwood of a small tree once used as a source of a red or red-brown dye and also used for walking sticks, now replaced by *Pterocarpus soyauxii* (African padouk).

Canada balsam *Abies balsamea*, N.Am., the balsam or resin occurs in ducts or 'blisters' in the bark, its high refractive index and little tendency to granulate accounts for its use in optical and microscopic work: now replaced to some extent by synthetic preparations. — **blue grass**, *Poa compressa:* — **garlic**, *Allium canadense:* — **hemlock**, *Tsuga canadensis:* — **lily**, *Lilium canadense:* — **milk vetch**, *Astragalus canadensis:* — **pea**, *Vicia cracca:* — **pitch**, *Abies balsamea:* — **plum**, *Prunus canadensis:* — **rice**, *Zizania aquatica:*—**thistle**, *Cirsium arvense:*—**turpentine**, *Abies balsamea:*—**violet**, *Viola canadensis*.

Canaigre *Rumex hymenosepalus* SE. U.S., Mex., tubers rich in tannin (30–35 % dry wt.), once used in leather tanning.

Cananga oil *Cananga odorata*, trop. As., fragrant fls. distilled for esst. oil, mainly in Indon.

Canarina *C. canariensis*, Canary Isles., orn. climber, cult., fr. ed.

Canarium Name used for commercial timber. **Af.** —, *Canarium schweinfurthii*, W. & E.Af.: **Ind.** —, *C. euphyllum*, Andamans: **Malayan** —, *Canarium* spp.

Canary creeper *Tropaeolum peregrinum*, Peru, creeper w. lemon yellow fls., cult: — **grass**, *Phalaris canariensis*, cult., seeds much used to feed canaries and other cage birds: — **madrone**, *Arbutus canariensis:* — **Island palm**,

Phoenix canariensis: — **pine,** *Pinus canariensis:* **reed** — **grass,** *Phalaris arundinacea,* a useful grazing and hay grass: — **whitewood,** *Liriodendron tulipifera,* U.S., a commercial wood: — **wood,** *Morinda citrifolia.*

Cancer bush *Sutherlandia frutescens,* S.Af., once thought to cure cancer: — **jalap,** *Phytolacca americana:* — **weed,** *Salvia lyrata,* N. Am.: — **wort,** *Veronica.*

Candelabra flower *Boophone disticha* and other spp., S.Af.; *Brunsvigia* spp.: — **tree,** *Araucaria angustifolia; Pandanus* spp., *Euphorbia ingens* and other cactiform euphorbias.

Candied peel The preserved peel of the citron, *Citrus medica* (or other *Citrus*), much used in confectionery, cakes, puddings etc.

Candles *Stackhousia monogyna,* Aus., cult., white fls. in dense racemes: **candle berry,** *Myrica cerifera,* N.Am., wax from fr. used for candles by early settlers; *Aleurites moluccana,* trop. As., oily seeds burn readily: — **nut,** *A. moluccana:* — **plant,** *Dictamnus albus,* cult., gdn. pl.; it is said that the volatile oil secreted by the pl. may be ignited on calm hot days: — **tree,** *Parmentiera cerifera,* cult. trop., candle-like frs.; *Cassia abbreviata,* S.Af.; *Croton gossypiifolia* cult. trop.: — **wood,** *Dodonaea viscosa,* trop. shrub; *Amyris balsamifera,* W.I., trop. Am.

Candy carrot *Athamanta cretensis,* Medit., used for flavouring (liqueurs): — **root,** *Polygala,* N.Am.: —**tuft,** *Iberis amara, I. umbellata* cult., many vars.: — —, **burnt,** *Aethionema saxatile,* Medit., cult.: — **weed,** *Polygala,* N.Am.

Cane A term used commercially for the stems of the smaller bamboos, for rattans (climbing palms) and for reeds or the larger grasses; also used for 1 year old stems of grapes, blackberries, raspberries, loganberries etc. **Bamboo** —, *Bambusa, Phyllostachys, Arundinaria* etc., much used for furniture, screens, gdn. canes etc.: **bush** —, *Costus afer,* W.Af.: **Carolina** —, *Arundinaria macrosperma:* **dumb** —, *Dieffenbachia seguine,* trop. Am.: **giant** —, *Arundinaria gigantea,* N.Am.: **Malacca** —, *Calamus scipionum* and other spp., favoured for walking sticks: — **palm,** *Chrysalidocarpus lutescens,* Madag., cult. orn., good pot palm: **partridge** —, *Rhapis flabelliformis,* China: **rajah** —, *Eugeissona minor,* Borneo, favoured for walking sticks: **rattan** —, *see* rattan: — **reed,** *Costus speciosus,* Ind., cult. orn., large white fls.: **soft** —, *Marantochloa flexuosa,* W.Af., used for mats and baskets: **switch** —, *Arundinaria tecta,* N.Am.: **Tongking** —, *Arundinaria amabilis,* S.China, a superior cane, much in demand, used for split cane fishing rods.

Canella bark *Canella winteriana,* W.I., Florida, small evergreen tree, aromatic b. used as a condiment and med., also called white cinnamon.

Canihua *Chenopodium pallidicaule,* Peru, cult., seeds used as cereal, like quinoa.

Canistel *Pouteria campechianum,* S.Am., ed. tree fr.

Canker berry *Solanum bahamense,* W.I.: — **root,** *Coptis groenlandica,* N.Am.

Canna *C. indica* and other spp., widely cult. orn. pls. in trop. and subtrop., numerous vars.: **edible** —, *Canna edulis,* cult. for starchy ed. tubers used as human or animal food (Queensland arrowroot or 'tous-les-mois').

Cannabis *Cannabis sativa,* Ind. hemp., cult., yields a narcotic resin, the source of the drug. *See Cannabis* in Willis. The same pl. grown in cooler climates yields hemp: fr. or 'seed' a source of oil and a cage bird food.

Cannon-ball tree *Couroupita guianensis*, trop. S.Am., cult., has large brown globular frs. hanging from trunk: curious pink and white fleshy fls.

Cantala fibre Or maguay, *Agave cantala*, C.Am., cult.

Cantaloupe Or musk melon, *Cucumis melo*, long cult., known to ancient Egyptians.

Canterbury bell *Campanula medium*, E.Eur., popular gdn. fl., many vars.

Canton fibre *Musa* sp., Philippines, said to be a cross between Manila hemp and an edible or wild banana: fibre inferior to Manila hemp.

Caparrosa *Neea theifera*, Braz., ls. of this tree or shrub used as tea.

Cape ash *Ekebergia capensis*, S.Af., wood resembles common ash: — **asparagus**, *Aponogeton distachyus:* — **aster**, *Felicia:* — **box**, *Buxus macowani:* — **chestnut**, *Calodendron capense:* — **cowslip**, *Lachenalia aloides:* — **ebony**, *Euclea pseudebenus:* — **figwort**, *Lachenalia aloides:* — **forget-me-not**, *Anchusa:* — **gooseberry**, *Physalis peruviana*, S.Am., cult. for ed. fr., esp. for jam: — **hawthorn**, *Aponogeton:* — **honeysuckle**, *Tecoma capensis:* — **jasmine**, *Gardenia jasminoides:* — **lily**, *Crinum:* — **mahogany**, *Trichilia emetica:* — **pondweed**, *Aponogeton distachyus*, cult.: — **primrose**, *Streptocarpus:* — **shamrock**, *Oxalis:* — **thorn**, *Zizyphus mucronata:* — **weed**, *Lippia nodiflora:* — **willow**, *Salix capensis*, wood used for brake blocks.

Caper *Capparis spinosa*, S.Eur., pickled fl. buds much eaten as flavouring, esp. w. mutton. — **spurge**, *Euphorbia lathyrus:* **bean** —, *Zygophyllum fabago*, S.Eur., N.Af., fl. buds used as substitute for capers.

Capsicum *Capsicum* spp., the source of red or Cayenne pepper (*C. frutescens*) a much used spice, and of chillies. The large fruited, less pungent forms (*C. annuum*) are used as vegetables and as paprika. The drug capsicum consists of the dried ripe frs. of *Capsicum frutescens*, *C. minimum* (bird pepper), the pungent principle (capsicin) being used as a counter-irritant in the treatment of rheumatism, lumbago and neuralgia.

Capulin *Prunus serotina* var. *salicifolia*, C.Am., ed. fr.

Caragana As. shrubs or small trees, some excessively spiny such as *C. spinosa* and *C. jubata*. Pods of *C. arborescens* eaten as a vegetable.

Caraguata fibre *Bromelia serra*, S.Am., used for sacks, Argen.

Caramba or **carambola** *Averrhoa carambola*, trop. As., cult. trop. for rather acid fr., usu. stewed or used for jam or pickles.

Caranday wax *Copernicia australis*, S.Am., a veg. wax resembling carnauba wax is obtained from the young ls.

Carapa or **crab oil** *Carapa guianensis*, *C. procera*, trop. Am., seeds yield oil.

Caraway *Carum carvi*, Eur. a well known spice, drug and esst. oil much used for flavouring: its use goes back to early times: favoured for meat dishes esp. sausages, also used in rye bread, biscuits, cheese and for liqueurs (Kümmel): the oil has med. uses as a carminative. Caraway is much grown in Holland. **Black** —, *Carum bulbocastanum*, Eur., As., starchy tubers eaten.

Cardamom *Elettaria cardamomum*, S.Ind., Sri Lanka. The seeds are a much used spice and have med. uses: have long been used in Ayurvedic med. and were imported into Eur. in Roman times: a popular spice in Arab countries and some Eur. countries, esp. for meat dishes, hamburgers and meat roll. The frs. and seeds of some other spp. of *Elettaria* and *Amomum* or *Aframomum* are also known as cardamom.

CAROLINA ALLSPICE

Cardinal flower *Lobelia cardinalis*, N.Am., *Quamoclit coccinea*, Mex.: **red —**, *Erythrina arborea:* — **wood**, *Brosimum paraense*.

Cardiospermum *C. halicacabum*, a climbing annual of warm cli., black seeds used as beads.

Cardoon *Cynara cardunculus*, S.Eur., a veg., fleshy l. stalks eaten boiled, after blanching.

Carib grass *Eriochloa polystachya*, trop. Am., annual, cult. for cattle fodder.

Caricature plant *Graptophyllum pictum* (= *G. hortense*), trop., small shrub w. curiously variegated or marbled ls., the marbling sometimes resembling faces: some vars. w. reddish ls.

Carissa Thorny shrubs, Old World trop. and subtrop., some w. ed. frs., e.g. *C. grandiflora*, Natal plum, cult., a good hedge pl.; *C. carandas*, Ind.

Carnation *Dianthus caryophyllus*, Eur. inc. Br., widely cult., includes picotee and clove pink, numerous vars., some favoured for commercial cult. under glass.

Carnauba wax *Copernicia cerifera*, Braz., the young ls. of the palm are coated w. wax. These are cut, the wax particles beaten off and melted down: the most imp. veg. wax, used for polishes of many kinds esp. those for leather footwear.

Carnegiea *C. gigantea*, SW. U.S., Mex., the largest of the cacti, has candelabra-like branching.

Carnival bush (S.Af.) *Ochna atropurpurea*.

Carnivorous plants Or insectivorous pls. are widespread throughout the world and probably number between 450 and 500 spp., belonging to 15 different genera. The term carnivorous pls. is regarded as more suitable than insectivorous pls. because some of these pls., especially those that are aquatics, commonly capture forms of animal life other than insects as nutriment.

The largest genus of carnivorous pls. is *Utricularia* (cosmop.) followed by *Drosera* (also cosmop.) and *Nepenthes* (trop. As. and Madag.). Others are *Pinguicula* (N.Hemisph.), *Sarracenia* (N.Am), *Genlisea* (trop. As., Af.), *Heliamphora* (S.Am), *Byblis* (Aus.), *Biovularia* (S.Am., Cuba) and *Polypompholyx* (Aus.): plus the five monotypic genera *Darlingtonia* (Calif.), *Dionaea* (Venus' fly-trap, N.Am), *Aldrovanda* (Eur., As., Aus.), *Drosophyllum* (Eur., N.Af.) and *Cephalotus* (W.Aus). *See* individual genera in Willis.

Carnivorous pls. have been classified according to the nature of their trapping mechanisms. Digestion of the prey when caught may be assisted by the secretion of acids and enzymes by the pl. Various means of luring the prey exist, such as the secretion of nectar (*Nepenthes*) or of special odours (*Sarracenia, Drosophyllum*). Many carnivorous pls. are highly coloured. [Lloyd, F. E. (1942) *The carnivorous plants*, 352 pp.]

Caroa fibre (Braz.) *Neoglaziovia variegata*, used locally for cordage, baskets etc.

Carob *Ceratonia siliqua*, Medit., well known, the sugary pod used as food for man and beast since Biblical times (St John's bread). A gum resembling gum tragacanth, w. similar uses, is prepared from the seeds which may be removed before feeding the pods to sheep or other livestock. A diabetic flour has been prepared from the seeds.

Carolina allspice *Calycanthus floridus*, N.Am., aromatic b. used as spice by Amerindians, wood has camphor-like odour, esp. when dry. — **aster**, *Aster* *carolinianus:* — **cherry laurel,** *Prunus caroliniana:* — **jessamine,** *Gelsemium sempervirens.*

Carotene May be prepared commercially by extraction from lucerne or alfalfa and from carrots and palm oil (*Elaeis*). It is utilized as a precursor of vitamin A: used in colouring agents for margarine, bakery products, pharmaceuticals, also livestock and poultry foods.

Carpenter's weed *Prunella vulgaris,* N.Hemisph., so called because fl. in profile resembles a carpenter's bill-hook: — **square,** *Scrophularia marylandica,* N.Am.

Carpet grass *Axonopus compressus,* cult., a trop. pasture and lawn grass: — **plant,** *Ionopsidium acaule,* Portugal, cult. rock gdn. or house pl.: — **weed** (Am.), *Mollugo verticillata.*

Carpinus *See* hornbeam.

Carpodinus Several spp. have been sources of inferior rubber in trop. Af. in the past.

Carrabeen or **carribin** *Sloanea* and *Geissois* spp., Aus. forest trees.

Carrion flower Name commonly used for *Stapelia,* the fls. (evil-smelling) being attractive to carrion flies, which no doubt assist in pollination. In N.Am. the name has been applied to *Smilax herbacea.* Carrion fls. are also to be found in *Amorphophallus* and in other genera of the Araceae and Asclepiadaceae.

Carrot *Daucus carota,* native of Eur., long cult., widely grown as a veg. and stock food: numerous vars., the larger coarser kinds, often white or yellow, used as animal food.

Cartagena bark *Cinchona cordifolia.*

Cartwheel flower *Heracleum mantegazzianum,* Eur., has an enormous inflor. *See* hogweed, giant.

Caryopteris Cult. orn. shrubs or herbs, the hybrid *C.* × *clandonensis* w. bright blue fls., now widely grown.

Cascalote (Mex.) Divi-divi.

Cascara *Rhamnus purshiana,* W. N.Am., the b. of this small tree yields the imp. drug or aperient cascara or 'cascara sagrada': the bark needs to be stored at least one year before use. The tree has been grown experimentally in other countries, e.g. Br. (Kew) & E.Af.

Cascarilla bark *Croton eluteria,* W.I., aromatic b. used med.: *Cascarilla* spp., trop. S.Am., b. may contain *Cinchona*-ripe alkaloids in small amounts.

Cashew nut *Anacardium occidentale,* this trop. Am. or Braz. nut tree is now widely distributed and cult. in other parts of the trop., esp. Ind., E.Af., W.I. etc. It was disseminated by the early Portuguese explorers. The kernels are in demand for dessert or confectionery. Cashew nut shell liquid (CNSL) is used in the manufacture of brake linings, belting and clutches and plastic resins. The swollen receptacle of the fruit (cashew apple) is ed. and the juice used for beverages.

Cassareep The name given to the concentrated juice of cassava roots, *Manihot esculenta,* rendered harmless by boiling: used locally for culinary purposes: may also be used for certain table sauces.

Cassava Mandioc or mandioca, *Manihot esculenta* (= *M. utilissima*) trop. Am., an imp. r. crop in many trop. countries, esp. Braz. & W.Af. Tapioca is derived from it. *See Manihot* in Willis.

Cassia *Cassia* spp., trop., many spp. cult. as orn. trees and shrubs in warm cli.: most have yellow fls. Some spp. are weeds, e.g. *C. occidentalis*, seeds used as coffee substitute (negro coffee). **Burmese pink** —, *C. renigera:* **foetid** —, *C. tora:* **golden** —, *C. fasciculata:* **horse** —, *C. grandis*, rose coloured fls.: **Java** —, *C. javanica*, fls. pink: **purging** —, *C. fistula*, seeds embedded in a laxative pulp: **red** —, *C. marginata:* **tanner's** —, *Cassia auriculata. See* senna.

Cassia bark or **cassia lignea** *Cinnamomum cassia* (=*C. tamala*) SE.China, tree cult., b. a spice, resembles cinnamon: ls. distilled, yield oil of Cassia.

Cassie Popinac or Opopanac, *Acacia farnesiana*, shrub or small tree widely distributed in warm cli., fls. yield a perfume.

Castanet plant *Crotalaria* spp., seeds rattle in the pod.

Castor oil *Ricinus communis*, a shrub widely distributed in warm cli. and cult., many vars.: oil from seeds a well known purgative and has other med. uses, in ointments and as a vehicle for eye drops. The oil is also used as a lubricant. The seeds are poisonous.

Casuarina Evergreen trees or shrubs, mainly Aus., often called she oaks, sheoakes, or forest oaks: cult., esp. in sandy conditions near the sea where few other trees will grow: wood may be hard and durable.

Catalpa Or bean tree, *Catalpa bignonioides*, N.Am., cult. orn. tree w. bean-like frs.

Cat-berry *Nemopanthus mucronata*, E. N. Am. (mountain holly): — **brier**, *Smilax*, N.Am.: —**gut**, *Tephrosia virginiana:* —**mint**, *Nepeta mussinii*, cult.: —**nip**, *Nepeta cataria.*

Cat plants Some pls. appear to have a strange attraction of fascination for cats, which is not easy to understand. The common gdn. catmint, *Nepeta mussinii*, is one of them, as is the allied catnip, *Nepeta cataria*, a wild pl. in Br. Cats will rub themselves against the pls. or roll on them. Cat thyme, *Teucrium marum*, a rock gdn. pl. from the W.Medit. is similarly attractive and cats may tear it to pieces. Some spp. of *Actinidia* appear to be attractive, esp. *A. kolomikta*, a climbing shrub from the Orient often grown over old tree trunks or against a wall in Br.

Cat's claw *Acacia greggii*, N.Am., a common, though disliked desert pl. w. strong recurved thorns, also called 'devil's claw' or 'tear blanket'. — **ear**, *Hypochoeris*, from the shape of the ls.: — **foot**, *Gnaphalium dioicum*, from its soft fl. heads, *Glechoma hederacea, Antennaria dioica:* — **paw**, *Anigozanthos humilis*, W.Aus.: — **tail**, *Typha, Amaranthus caudatus, Bulbinella*, S.Af.: — **tail grass**, *Koeleria phleoides*, Medit., *Phleum* spp.: — **tail millet**, *Setaria glauca:* — **whin**, *Ulex minor:* — **whiskers**, *Gynandropsis gynandra*, trop., a pot-herb.

Catananche *C. caerulea*, S.Eur. (Cupid's dart), cult. orn. fl. for cutting or as everlasting.

Catasetum Trop. Am. epiphytic orchids, cult., have short stout pseudobulbs.

Catawba tree *Catalpa speciosa*, N.Am., a timber tree.

Catchfly *Silene* spp. a large genus which includes many gdn. pls.

Catechu *See* cutch and gambier.

Caterpillar plant *Scorpiurus* spp. S.Eur., thick tubercled pods.

Cativo *Prioria copaifera*, W.I., C.Am., a commercial timber.

Catjang, kachang or dhal *See* pigeon pea, *Cajanus cajan.*

CATTLEYA

Cattleya Magnificent orchids of trop. Am., some with v. large fls., many hybrids.

Cauliflower *Brassica oleracea* var. *botrytis*. In this well known veg. it is the floral organs or fl. stems that have been artificially modified in the course of cult. The fls. have tended to become abortive and their stalks to gain in substance. These thickened, tender white, floral branchlets constitutes the 'curd' or part eaten. There are numerous vars. early, mid-season, late etc. **Wild** —, *Verticordia brownii*, W.Aus.

Cavendishia Evergreen shrubs native to mts. of trop. Am., some w. showy fls., cult.

Cay-cay fat *Irvingia oliveri*, Indo-China, ed. fat from seeds.

Cayenne pepper *Capsicum frutescens*. Cayenne or red pepper consists of pulverized dried chillies, not to be confused with black or white pepper derived from an entirely different pl. (*Piper nigrum*).

Ceara rubber Or manicoba rubber, *Manihot glaziovii*, S.Am. exploited for rubber at one time and cult. experimentally elsewhere but eclipsed by *Hevea* rubber.

Cedar The true cedars belong to the genus *Cedrus*, such as **Atlas** —, *C. atlantica:* **Cyprus** —, *C. brevifolia:* — **of Lebanon**, *C. libani* and **Ind.** — or **Deodar**, *C. deodara*. Numerous other trees that resemble the true cedars in some way, or w. similar fragrant wood, have acquired the name 'cedar'. Among those that furnish commercial timbers are – **Af. pencil** —, *Juniperus procera*, E.Af.: **Aus.** —, *Toona ciliata* var. *australis:* **Burma** or **Moulmein** —, *Cedrela toona:* **C.Am.** —, *Cedrela mexicana:* **incense** —, *Calocedrus decurrens:* **Port Orford** or **Oregon** —, *Chamaecyparis lawsoniana:* **S.Am.** —, *Cedrela fissilis:* **southern white** —, *Chamaecyparis thyoides*, N.Am.: **Virginian pencil** —, *Juniperus virginiana:* **western red** —, *Thuja plicata*, N.Am., an imp. commercial timber, v. resistant to decay, used for shingles, greenhouses, gdn. frames, boats, beehives etc.: **yellow** —, *Chamaecyparis nootkatensis*, W. N.Am.: **white** —, *Thuja occidentalis*, E. N.Am.

Other trees or timbers that have been termed 'cedar' are – **bastard** —, *Guazuma tomentosa*, trop. Am., cult. avenue trees in trop.; *Soymida febrifuga*, trop. As.: **Bermuda** —, *Juniperus bermudiana*, valued locally: **Chilean** —, *Austrocedrus chilensis:* **cigar box** —, *Cedrela odorata*, W.I.: **Jap.** —, *Cryptomeria japonica:* **Mlanje** —, *Widdringtonia whytei:* **E.Af.** —, *Guarea thompsonii:* **Siberian** —, *Pinus cembra:* **W.I.** —, *Cedrela odorata:* **white** —, *Melia azedarach*.

Cedrat citron.

Cedron *Quassia* (= *Simaruba*) *cedron*, C.Am., bitter seeds and b. med.

Ceiba *C. pentandra*, trop. Af. and *C. occidentalis*, trop. Am., commercial timbers.

Celandine *Ranunculus ficaria*, Eur. inc. Br.: **greater** —, *Chelidonium majus:* **tree** —, *Bocconia frutescens*, C.Am., W.I., shrub, cult.

Celastrus Cult. orn. shrubs and climbers, some w. local med. uses.

Celeriac Or turnip-rooted celery, *Apium graveolens* var. *rapaceum*, a tuberous rooted form of celery, the r. not the l. stalks, being eaten.

Celery *Apium graveolens*, an imp. veg., the blanched l. stalks being eaten raw, in salads, cooked, in soup or for flavouring. Celery seed (dried frs.) also much

50

used in flavouring, 'Celery salt' is a mixture of salt and ground celery seed:
— **wood,** *Polyscias elegans,* Aus. used for inlaying.

Cellophane plant *Echinodorus berteroi,* an aquarium pl. w. thin membraneous ls.

Celmisia Aus. & N.Z. composites, some v. handsome, cult.

Celosia Cult. orn. pls. inc. cock's-comb and *C. argentea* (= *C. plumosa*) w.
its large feathery inflor., red or gold in colour. Some spp. are pot-herbs in
countries of origin.

Celsia Cult. orn. fls., warm cli., favoured as pot pls., esp. *C. arcturus.*

Celtis, African *Celtis soyauxii,* trop. Af., a commercial timber.

Centipede grass *Eremochloa ophuroides,* trop. As., used for erosion control.

Centrosema *C. pubescens,* Malaya and *C. plumieri,* S.Am., trop. cover crops.

Century plant *Agave americana,* C.Am., a large pl. w. prickly ls., widely
distributed in warm cli. and often used as a defensive hedge. Fls. after many
years, according to conditions, produce an enormous tree-like inflor. In Mex.
this may be tapped, yielding much sap, which is fermented to produce
'pulque', a national beverage.

Cephalotaxus Cult. orn. shrubs, E.As., allied to yews (called plum yew).

Ceratopteris *C. thalictroides,* trop. and subtrop., an aquatic fern, floating
fleshy fronds eaten as a veg.

Cerbera *C. manghas* (= *C. odollam*) small coastal tree, Old World trop., fr.
common in ocean drift.

Ceratotheca *C. sesamoides,* W.Af., yields an oil seed like sesamum: is also a
pot-herb.

Cereus Several spp. cult., large fls. and large fleshy, usu. ed. frs.

Ceriman *Monstera deliciosa,* trop. Am., cult., house pl., ed. fr.

Ceropegia Cult. climbers w. distinctive waxy frs.: fls. may be insect traps.
See *Ceropegia* in Willis.

Cestrum C. & S.Am. shrubs, some v. orn., widely cult. in warm cli., e.g.
C. nocturnum (lady of the night), *C. diurnum* (day jasmine), *C. purpureum,*
C. parqui.

Ceterach Ferns related to *Asplenium,* cult.

Cevadilla Or Sabadilla, *Schoenocaulon officinale,* C.Am., seeds med. and used
in veterinary medicine.

Ceylon gooseberry *Doryalis hebecarpa* (= *Aberia gardneri*), small spiny tree
or shrub, purple to black velvety frs. 2.0–2.5 cm across, mainly used for
preserves: — **olive,** *Elaeocarpus serratus:* — **willow,** *Ficus benjamina.*

Chaff flower *Alternanthera, Achyranthes.*

Chaguar fibre *Bromelia serra,* Argen., used by Matico Amerindians.

Chahomilia *Salvia cypria,* Cyprus, used for a beverage.

Chain fern *Woodwardia, Lorinseria,* N.Am.

Chairmaker's rush *Scirpus americanus,* N.Am.

Chalice vine *Solandra guttata,* W.I., large showy cream fls., cult.

Chalk glands Pores on the margins of some ls. that secrete water, like nectaries.
If the water contains lime, a lime deposit may be left on the leaf surface:
seen well in some saxifrages (*Saxifraga*), & Plumbaginaceae.

Chamaecyparis Conifers related to *Cupressus,* several widely cult. w. many
vars., e.g. *C. lawsoniana* (Lawson's cypress), *C. nootkatensis* (Nootka cypress),
C. obtusa (Hinoki).

CHAMAEDOREA

Chamaedorea Small reedy palms, trop. Am., some cult. orn.: young fl. clusters of some Mex. spp. eaten as veg.

Chamomile *Anthemis nobilis*, Eur., wild and cult.: the double flowered form or 'Roman' chamomile is cult. for med. purposes, the fl. heads being steam distilled for the oil used in hair dyes, hair washes and shampoos, esp. for blonde hair. The single fl. chamomile or 'German' chamomile is considered inferior. **Corn** —, *A. arvensis:* **scentless** —, *Tripleurospermum maritimum:* **stinking** —, *Anthemis cotula:* **sweet** —, *Chamaemelum nobile:* **wild** —, *Matricaria recutita* (= *M. chamomilla):* **yellow** or **golden** —, *Anthemis tinctoria*, fls. yield a yellow dye.

Chamomile lawns Popular in Elizabethan times, due no doubt to the scent given off by the foliage when walked on. They were easy to establish and maintain, pls. spreading quickly and resisting dry spells (better than grass). Gdn. seats of chamomile were also popular. Chamomile lawns still exist in some Royal parks in England and at Buckingham Palace. It is considered that Sir Francis Drake played his famous game of bowls on a chamomile lawn.

Champak or **champ** *Michelia champaca*, Ind., esst. oil from fls., useful timber.

Chan *Shorea* spp., Thailand, commercial timber.

Chanar *Gourliea decorticans*, Argen. ed. pods, cattle food.

Chandelier tree *Pandanus candelabrum*, trop. Af.

Chaparral or **chamisal** The xerophytic scrub on the hills of California, includes spp. of *Adenostoma, Arctostaphylos, Baccharis, Ceanothus, Eriodictyum, Garrya, Rhus*, dwarf oaks, currants, buckeye, roses etc. : — **yucca**, *Hesperoyucca whipplei*, SW. U.S.

Chaplash *Artocarpus chaplasha*, Ind. & Burma, useful golden brown wood.

Charas Cannabis.

Charcoal Charcoal is perhaps the most ancient of manufactured pl. products, dating from prehistoric times. It is widely used as fuel, particularly in countries without coal. It is employed for some special purposes and in the metallurgical and chemical industries. Both hardwoods and softwoods are used. Scrapwood unsuited for commercial timber is often profitably utilized for charcoal. Kilns vary from the small primitive earth kiln to large brick-built ceramic kilns. Med. charcoal may be prepared from sawdust, peat, cellulose and coconut shells and may be chemically purified. It is valued for its absorptive power and may be given by mouth in the treatment of poisoning.

Chards There are two kinds – (1) blanched summer shoots of the globe artichoke and (2) the young flowering shoots of salsify. Swiss chard or leaf beet is a form of beet (*Beta*).

Charlock *Sinapis arvensis*, a bad cornfield weed but easier to control w. modern herbicides: **jointed** —, *Raphanus raphanistrum* (wild radish).

Chaste tree Or monk's pepper, *Vitex agnus-castus*, Medit., cult. for orn. fls. violet: there is a white var.: regarded as a symbol of chastity from ancient times. **Chinese** — —, *Vitex incisa*.

Chats Small or undersized potatoes, usually fed to livestock in Br.

Chaulmoogra oil *Hydnocarpus wightianus*, SW.Ind., *Taraktogenos kurzii*, Burma: seed oil at one time used in treatment of leprosy.

Chawstick (W.I.) *See* chewstick.

Chay root (Ind.) *Oldenlandia umbellata*, r. yields a red dye.

Chayote *Sechium edule*, cult. warm cli., popular veg. w. single large seed.

Chayotilla (Mex.) *Hanburia mexicana*, climber w. explosive fr.

Cheat or **chess** *Bromus secalinus*, Eur., a common weed grass ('ryebrome' in Br.).

Checkerberry (Am.) *Gaultheria procumbens*, ls. yield wintergreen oil and are used as tea.

Cheese weed *Hymenoclea salsola*, SW. U.S., crushed foliage has cheesy odour.

— **wood**, *Pittosporum undulatum*, E.Aus., a light-coloured, close-textured wood: turnery and parquet floorings: **cheeses**, *Malva neglecta*, Eur. inc. Br., fr. has a fancied resemblance to a flat cheese; name used for other spp. of *Malva* and for *Lavatera*.

Cheilanthes Distinctive ferns, some v. decorative, cult.

Chelone N.Am. perens., showy fls., cult.

Chena Shifting cultivation, as practised in trop. Af. and other trop. countries: the forest is burned or destroyed to grow crops for a short period.

Chenelle plant *Acalypha hispida*, cult. orn. fl.

Cherimoyer or **cherimolia** *Annona cherimolia*, C.Am., ed. fr. like custard apple, widely cult. in warm cli.: **wild** —, *A. longifolia*, Mex.

Cherokee bean *Erythrina arborea*, N.Am.: — **rose**, *Rosa laevigata*.

Cherry The cult. cherries are derived from two wild Eur. cherries, both of which occur in Br., *Prunus avium* having given rise to the sweet or dessert cherries and *Prunus cerasus* to the culinary or cooking cherries (morellos or amarelles). Numerous vars. of both kinds are cult., the Romans having brought some kinds to Br. Light soils overlying chalk provide the best conditions for commercial cherry production in Br.

Cherry wood is in demand for various purposes. **Eur. cherry** (*P. avium*) and **Am.** or **black** — (*Prunus serotina*) are both commercial woods. Cherry wood is used in high-class cabinet making, for musical instruments and as a substitute for antique mahogany in repairing furniture. Cherry wood tobacco pipes are made from the wood of the **St Lucie** — (*P. mahaleb*). The wood may be buried for a time to intensify aroma. **Af.** —, *Mimusops heckelii*, W.Af., a commercial timber: **Barbados** —, *Malpighia glabra*, ed. cherry-like fr., a hedge pl.: **bastard** —, *Ehretia*, W.I.: **bird** —, *Prunus padus*: **broad-leaved** —, *Cordia macrophylla*: **Cayenne** —, *Eugenia uniflora*: **choke** —, *Prunus virginiana*: **clammy** —, *Cordia collococca*, W.I.: **Cornelian** —, *Cornus mas*: **Jam.** —, *Muntingia calabura*: — **laurel**, *Prunus laurocerasus*: **Liberian** —, *Saccoglottis gabonensis*, fr. not ed.: **Mex. ground** —, *Physalis ixocarpa*: **morello** —, *Prunus cerasus*: **native** — (Aus.), *Exocarpus cupressiformis*: **Peruvian** —, *Physalis peruviana* ('Cape gooseberry'): — **pie**, *Heliotropium peruvianum* orn. pl.: **Pitanga** —, *Eugenia uniflora*: — **plum**, *Prunus cerasifera*: **prairie** —, *P. gracilis*: **Rocky Mt.** —, *P. melanocarpa*: **rum** —, *P. serotina*: **sand** —, *P. depressa*, N.Am.: **sour** —, *Syzygium* spp., Aus.: **Surinam** —, *Eugenia uniflora*: **sweet** —, *Prunus avium*: **W.I.** —, *Malpighia punicifolia*: **wild** —, *Prunus avium*, Eur., walking sticks: **winter** —, *Physalis alkekengi*, orn. frs.

Chervil, garden *Anthriscus cerefolium*, S.Eur., an ann. culinary herb resembling parsley: ls. w. delicate aniseed flavour favoured for seasoning and salad. Chopped w. parsley and chives may be used as 'fines herbes' in omelettes, sauces and salad dressings. For winter use ls. are cut and quickly dried to preserve green colour. Chervil is claimed to bring out the flavour of other herbs. **Bur** —, *Anthriscus caucalis:* **turnip rooted** —, *Chaerophyllum bulbosum:* **wild** —, *Chaerophyllum temulentum.*

Chess *See* cheat, *Bromus secalinus.*

Chestnut *Castanea sativa.* The **common Eur.**, **Spanish** or **sweet chestnut,** cult. and naturalized in Br., has been an imp. food for man and beast in many southern Eur. countries from the earliest times. Unlike most other nuts it is essentially a carbohydrate or starchy food and not rich in fat or protein. Flour or meal, much used in cooking, esp. in Italy, is prepared from it. Oriental and Am. chestnuts (*Castanea*) are basically similar to the Eur. chestnut or 'English chestnut' as it may be called in Am.

Chestnut wood superficially resembles oak and may be used as a substitute for it, esp. in chair making and for coffins. Coppice poles are used for split-pale fencing. The wood of old trees is rich in tannin and has been used for the preparation of tannin extract in S.Eur. **Am.** —, *Castanea dentata,* trees largely destroyed by disease: **Aus.** —, *Castanospermum australe,* cult. orn. or street tree, subtrop.: **Cape** —, *Calodendron capense,* cult. orn. fl. tree: **Chinese** —, *Castanea mollissima:* **dwarf** —, *C. pumila,* N.Am.: **horse** —, *Aesculus hippocastanum:* **Jap.** —, *Castanea crenata:* **golden** —, *Chrysolepis chrysophylla:* **low veld** —, *Sterculia murex,* S.Af.: **Moreton Bay** —, *Castanospermum australe:* **palm** —, *Guilielma australe,* trop. Am., fr. ed.: **Polynesian** —, *Inocarpus edulis:* **Rhodesian** —, *Baikiaea plurijuga,* useful timber: **Tahiti** —, *Inocarpus edulis:* — **vine,** *Tetrastigma voinierianum,* orn. ls., a room pl.: **Virginia** —, *Castanea pumila:* **water** —, *Trapa natans, T. bispinosa, Eleocharis tuberosa,* ed. tubers: **wild** —, *Castanopsis,* N.Am.; *Pachira,* trop. Am.; *Brabejum stellatifolium,* S.Af.

Chewing gum *See* chicle, *Achras zapota. See* Jelutong.

Chewstick In many countries, esp. throughout Af. and the Ind. sub-continent, the twigs or rs., of certain pls., are chewed by local inhabitants or used like a toothbrush as a means of cleaning the teeth. In W.Af. the so-called yoroba chewstick, *Vernonia amygdalina,* is highly esteemed for the purpose and is sold on the markets. The twig or r. may be used, the b. having been first removed by scorching. The rs. of *Calotropis procera* are also much used in many parts of Af. Other W.Af. chewsticks are derived from spp. of *Acacia, Anogeissus, Bandeiraea, Baphia, Marsdenia, Sorindeia* and *Teclea.* In the W.I. a popular chewstick, or 'chawstick' as it may be termed there, is *Gouania lupuloides,* a woody vine. When used a soap-like froth develops in the mouth, considered beneficial. A stick about the thickness of the little finger is selected and first chewed at the end to soften it and then applied like a toothbrush. Among common trop. trees or shrubs sometimes used are the gdn. hibiscus (*Hibiscus rosa-sinensis*), the tamarind, mango, and neem tree. *Salvadora persica* and *Acacia farnesiana* (Cassie) are also used.

Chia seeds *Salvia chia,* Mex., *S. columbariae,* W. N.Am. used for beverages.

Chian turpentine *Pistacia terebinthus,* Medit. an oleo-resin w. med. uses.

Chibasa *Juncus* sp., Colombia, aromatic rs. used for skin affections.

Chibou *Bursera gummifera,* trop. Am., W.I., elemi resin.

Chica *Arrabidaea chica,* S.Am., red pigment from ls. an Amerindian body paint.

Chick pea *Cicer arietinum*, Medit., imp. pulse crop in many warm countries. numerous vars., some large-seeded.

Chicken claws *Salicornia europaea*, occurs in muddy tidal marshes: — **grape**, *Vitis vulpina*, *V. cordifolia*, N.Am.

Chickrassy *Chukrasia tabularis*, Ind. & Burma, a commercial timber.

Chickweed *Stellaria*, *Cerastium*. **Common** —, *Stellaria media:* **daisy-leaved** —, *Mollugo nudicaulis*, W.Af.: **greater** —, *Stellaria neglecta:* **mouse-eared** —, *Cerastium* spp.: — **wintergreen**, *Trientalis europaea*.

Chicle *Achras zapota* (= *Manilkara zapotilla*), C.Am., coagulated latex used as a base for chewing gum.

Chicory *Cichorium intybus*, Eur., dried rs., roasted and ground, are mixed with coffee: blanched ls. used as winter salad (witloof chicory).

Chiku (Malaya) *Achras zapota*, trop. dessert fr., much esteemed.

Chilean crocus *Tecophilaea cyanocrocus*, blue fls., cult., several vars.: — **jasmin**, *Mandevilla laxa:* — **nut**, *Gevuina avellana:* — **pepper tree**, *Schinus latifolius:* — **pine**, *Araucaria araucana:* — **wineberry**, *Aristotelia maqui:* — **wine palm**, *Jubaea chilensis* (= *J. spectabilis*), the sap of the palm is thickened by boiling for 'miel de palma' or palm honey, or made into wine.

Chillies *Capsicum frutescens* or other spp., the frs. are a much used spice and seasoning agent in all warm countries, dried and ground they constitute red or Cayenne pepper.

Chilte *Cnidoscolus elasticus* and other spp., Mex., resinous latex yields a gutta-percha-like material, prepared commercially.

Chimney bellflower *Campanula pyramidalis*, Eur., a popular gdn. pl.

Chimonanthus *C. praecox*, China, cult. orn. shrub, w. scented early fls.

China or **chinese** — **anise**, *Illicium verum:* — **apple**, *Pyrus prunifolia:* — **artichoke**, *Stachys sieboldii:* — **aster**, *Callistephus sinensis*, gdn. aster: — **bell fl.**, *Platycodon grandiflorus:* — **briar**, *Smilax bona-nox:* — **box**, *Murraya exotica*, trop., hedge pl.: — **berry**, *Melia azedarach:* — **cabbage**, *Brassica chinensis*, *B. pekinensis:* — **chestnut**, *Castanea mollissima:* — **coir**, *Trachycarpus fortunei* (= *T. excelsus*): — **crab apple**, *Pyrus hupehensis:* — **date**, *Diospyros kaki* (persimmon): — **date plum**, *Zizyphus jujuba:* — **banana**, *Musa nana* 'Dwarf Cavendish': — **ephedra**, *Ephedra sinica:* — **fir**, *Cunninghamia sinensis:* — **flower** (S.Af.), *Adenandra* spp.: — **grass**, *see* ramie: — **galls**, large insect galls forming on *Rhus semialata*, rich in tannin, used med.: — **gooseberry**, *Actinidia chinensis*, ed. fr., cult. for export in N.Z.: — **hat pl.**, *Holmskioldia speciosa:* — **haw**, *Photinia serrulata:* — **hemlock**, *Tsuga sinensis:* — **horse chestnut**, *Aesculus chinensis:* — **jujube**, *Zizyphus jujuba:* — **jute**, *Abutilon avicennae:* — **lantern**, *Physalis alkekengi*, *P. franchetti*, inflated red calyx: — **matrimony vine**, *Lycium chinense:* — **olive**, *Canarium* spp.: — **parasol**, *Firmiana simplex:* — **pink**, *Dianthus chinensis:* — **plum**, *Prunus salicina:* — **pumpkin**, *Benincasa hispida:* — **quince**, *Cydonia sinensis:* — **raisin**, *Hovenia dulcis*, ed. fr.: — **red bud**, *Cercis chinensis:* — **root**, *Smilax china*, med., once used for gout: — **sacred lily**, *Narcissus tazetta:* — **tree**, *Melia azedarach:* — **white pine**, *Pinus armandii:* — **wood oil**, *see* tung oil: — **wythe** (W.I.), *Smilax balbisiana:* — **yew**, *Taxus chinensis*.

Ching quat (Hong Kong) *See* water spinach.

Chingma (Manchuria) *Abutilon* sp., a textile fibre.

Chincherinchee *Ornithogalum thyrsoides*, S.Af., handsome white fls., exported in cool storage from Cape to Eur. during winter: name indicative of sound made by stems rubbing together in wind.

Chinquapin *Chrysolepis* (=*Castanopsis*) *chrysophylla*, ed. nuts, once much used by Amerindians: *Castanea pumila*, N.Am.

Chionanthus Cult. orn. shrubs, *C. virginicus*, N.Am. and *C. retusus*, China.

Chir pine *Pinus roxburghii* (=*P. longifolia*), Ind., tapped for rosin and turpentine, wood much used, esp. for railway sleepers.

Chiretta *Swertia chirata*, Himal., med., bitter tonic: **Jap.** —, *S. chinensis*.

Chironja A Puerto Rican hybrid citrus fr. of great promise: believed to be a cross between orange and grapefruit.

Chittam *Cotinus americanus* (=*Rhus cotinoides*), SE. U.S. shrub cult. for autumn fol., wood yields an orange dye.

Chittagong wood *Chukrasia tabularis*, Ind., valued for furniture, fls. yield red or yellow dye.

Chive *Allium schoenoprasum*, Eur., a much used culinary herb w. onion-tasting ls., the part usu. used: favoured for soups, salads, egg and cheese dishes: freely cult. in N.Eur.: **Chinese** —, *Allium tuberosum*.

Chlorophyll The green colouring matter found in pls.: has various uses: imp. commercial sources are lucerne (alfalfa) and nettles: used in medicine and by confectioners: green icing sugar is coloured with it: used in popular deodorants and in toothpaste.

Chlorophytum *C. comosum*, S.Af., form w. variegated ls. a popular house pl.

Cho-cho *Sechium edule*, trop. Am. gourd-like veg. widely cult. in warm cli., also called chou-chou, sou-sou or chayote.

Choisya *C. ternata*, Mex. cult. orn. evergreen shrub.

Chokeberry *Aronia* spp., N.Am., attractive fl. shrubs, cult.: **black** —, *A. melanocarpa*: **purple** —, *A. floribunda*: **red** —, *A. arbutifolia*.

Cholla (Am.) Any cylindrical stemmed member of the genus *Opuntia*: — **gum**, *Opuntia fulgida* and other spp., may be used like gum arabic.

Chondrorhyncha Trop. C. & S.Am. orchids, cult.

Chonemorpha *C. macrophylla* (=*C. fragrans*), Ind., cult. orn. woody climber w. large white fls. 7–8 cm across: b. yields fibre.

Chorizema Orn. shrubs, Aus., *C. ilicifolium* w. orange pea-like fls.

Chowlee (Ind.) *Vigna unguiculata*, cowpea.

Christmas bell Name applied to various pls. that flower at about Christmas time in different parts of the world, e.g. *Blandfordia nobilis*, *B. punicea*, Aus.; *Sandersonia aurantiaca*, Natal, S.Af.: — **berry**, *Chironia baccifera*, S.Af.: *Lycium carolinianum*, Rhacoma, Am.: — **bush**, *Prostanthera lasianthos*, *Ceratopetalum gummiferum*, Aus.; *Pavetta bowkeri* S.Af.; *Alchornia cordifolia*, W.Af.; *Eupatorium odoratum*, W.I.: — **cactus**, *Schlumbergera truncata* (*Zygocactus*), a popular room pl., free flowering: — **cholla**, *Opuntia leptocaulis*, scarlet fls., long spines (darning needle cactus, Am.): — **fern**, *Polystichum acrostichoides*, Am.: — **flower**, *Euphorbia pulcherrima*: — **gambol**, *Ipomoea sidifolia*, W.I.: — **hope**, *Tecoma stans*, W.I.

Christmas rose *Helleborus niger*, Eur. cult. orn., fls. in mid-winter: perianth turns green after fertilization: petals are modified into funnel-shaped nectaries. An extract of the dark rhizomes has been used med.

Christmas tree The common Christmas tree of Eur. inc. Br., is the Eur. spruce, *Picea abies*, often cult. for the purpose. Many other coniferous trees may be used, esp. pines in N.Am. Where conifers are not readily obtainable other trees may be used. Trees known as 'Christmas tree' include – *Nuytsia floribunda*, W.Aus., *Metrosideros excelsa*, N.Z., *Dichrostachys cinerea*, S.Af.

Christopher, herb *Actaea spicata*, N.Hemisph., fr. poisonous.

Christophine (W.I.) *Sechium edule, see* chayote.

Christ's thorn *Paliurus spina-christi*, S.Eur., W.As. Christ's Crown of Thorns may have been made from this pl., or from *Zizyphus spina-christi* which is more common around Jerusalem. The name has been used for other pls., e.g. *Euphorbia milii*, *Carissa carandas*, a prickly hedge pl. — **eye**, *Inula oculis-christi*, E.Eur.

Chrysalidocarpus *C. lutescens*, Madag., cult. orn. palm, forms tufts of stems.

Chrysanthemum Gdn. and greenhouse chrysanthemums cover a wide range in habit and in size, shape and colour of the blooms, with both the ann. and peren. kinds. In the Orient they have been intensively cult. for centuries and there is uncertainty as to the wild sp. from which they have been evolved.

Chufa *See* tiger nut, *Cyperus esculentus*.

Chuglam, white *Terminalia bialata*, Andamans, a commercial timber.

Chumprak *Tarrietia cochinchinensis*, Thailand, a commercial timber.

Chupandilla (Mex.) *Cyrtocarpa procera*, a little known tree fr.

Churnwood *Citronella moorei*, Aus. rain forest tree.

Churras or **charas** (Ind.) Cannabis.

Chusan palm *Trachycarpus fortunei* (= *T. excelsus*), China, a hardy palm cult. outdoors in Br.

Chysis Trop. Am. orchids w. showy bright coloured fls.

Cibotium Trop. tree ferns, some cult.: used for stuffing pillows in Hawaii.

Cicely, sweet *Myrrhis odorata*, Eur., fragrant herb, wild and cult., used in home remedies and for flavouring; also called gdn. myrrh.

Cider tree *Eucalyptus gunnii*, Tasm., cult., hardy in Br.

Cidra Or sidra. *Cucurbita ficifolia*, E.As., cult., in Spain fr. made into a kind of marmalade (Malabar gourd).

Cigar flower *Cuphea ignea*, Mex., popular greenhouse pl., scarlet fls. tipped black and white: — **tree**, *Catalpa speciosa*, N.Am., cigar-like frs.

Cinchona Or quinine, *Cinchona* spp., Andes, several spp. yield b. rich in quinine and similar alkaloids, e.g. *C. officinalis*, *C. succirubra*, *C. calisaya*, *C. ledgeriana*. A cult. hybrid tree (*C. ledgeriana* × *C. succirubra*) accounts for much commercial b. In the treatment of malaria quinine has been largely displaced by other synthetic drugs. *See Cinchona* in Willis.

Cineraria Cult. orn. fls. w. varied and vivid colours: many vars.: derived from spp. native to the Canary Isles.

Cinnamon *Cinnamomum zeylanicum*, Sri Lanka, cult. trop., grown as coppice: b. in form of quills: a much-used spice: oil also used in flavouring: has med. uses. In the U.S. *Cinnamomum cassia* (Cassia bark) is also called cinnamon: the b. is coarser and thicker. — **bells**, *Gastrodia sesamoides*, Aus.: **Chinese** —, *Cinnamomum cassia*: — **fern**, *Osmunda cinnamomea*, N.Am.: **mt.** —, *Cinnamodendron corticosum*, Jam.: — **rose**, *Rosa cinnamomea*, Euras.: **Saigon** —,

3-2

Cinnamomum loureirii: — **vine**, *Apios americana* (=*A. tuberosa*), N.Am.: **white** —, *Canella winteriana*, W.I.: **wild** —, *Litsea zeylanica*, As.

Cinquefoil *Potentilla reptans* and other spp., some are gdn. pls.

Circassian tree *Adenanthera pavonina*, trop., orn. red seeds used as beads.

Cirrhaea Trop. Am. orchids w. curiously contorted fls., some cult.

Cirrhopetalum Trop. orchids, mainly Old World, cult., some w. showy or bizarre fls.

Cirsium Thistles, some are troublesome weeds, e.g. *C. arvense.*

Cissus Cult. orn. climbers, *C. adenopodus* and *C. discolor* good roof climbers.

Cistus Cult. sun-loving shrubs, Medit., many hybrids.

Citrange A hybrid citrus fr. (*Citrus sinensis* × *Poncirus trifoliata*). — **quat**, a hybrid between sweet orange and kumquat.

Citron *Citrus medica*, cult. warm cli., peel candied, candied peel being much used in confectionery and cooking: several vars. inc. **fingered** — and **Jewish** —, or etrog used at the Feast of Tabernacles.

Citronella oil *Cymbopogon nardus*, *C. winterianus*, trop. grass oil, cult., used in perfumery, esp. for soap.

Citrus fruits *Citrus* spp. Widely cult., trop. and subtrop., imp. commercially, the better known being – citron, grapefruit, kumquat, lemon, sweet lemon, lime, mandarin or tangerine, pummelo, Bergamot orange, Seville or sour orange, sweet orange (*see* separate headings). — **hybrids.** There are several hybrid citrus frs. Some have arisen naturally, others produced by pl. breeders, e.g. bitter × sweet orange calarin, chironja, citrange, citrangequat, orangelo, oranguma, ortanique, satsumelo, siamor, sopomaldin, tangelo, tangelolo, tangerona, ugli. Most of these are grown in private gdns. or by enthusiasts and are not commercially grown, possible exceptions being chironja, citrange, ortanique, tangelo and ugli.

Clammy weed *Polanisia trachysperma*, N.Am.

Clanwilliam cedar (S.Af.) *Widdringtonia juniperoides*, hard fragrant wood.

Clappertonia *C.* (=*Honckenya*) *ficifolia*, W.Af. yields a jute-like fibre.

Clarkia *C. elegans*, W. N.Am., popular ann. w. showy fls., many vars.

Clary sage *Salvia sclarea*, Eur., cult. orn. gdn. pl., once used for culinary purposes in place of ordinary sage: oil distilled from it used for flavouring, also med.

Classification, plant Various systems of classification of the flowering pls. have been put forward by different authors over the years, dating from the purely artificial systems based on habit (herbaceous, woody etc.) of the early herbalists. The system of Linnaeus, based on floral or sexual characters, marked a great advance. As knowledge of pls. increased other systems were put forward, the two best known being those of Bentham and Hooker and of Engler. The former, presented in the well known book *Genera Plantarum*, gave good family and genera descriptions of all the flowering pls. known at the time. It was widely used in Britain and Commonwealth countries. In Eur. & N.Am. the system of Engler and Prantl (*Syllabus der Pflanzenfamilien*) became the standard system. Later systems of classification included those of Hutchinson and Takhtajan. [Jeffrey, C. (1968) *An introduction to plant taxonomy*, 128 pp.]

Clearing nut *Strychnos potatorum*, Ind., the seed rubbed inside earthenware water vessels causes muddy water to clear.

Clearweed *Pilea* spp. N.Am.

Cleavers Or goose grass, *Galium aparine*.

Clematis Mainly climbing shrubs, many cult., and some, esp. hybrid forms, have v. showy fls.: **bush** —, *Clematopsis stanleyi*, S.Af.; *Clematis paniculata*, N.Z.: **leafless** —, *C. afoliata*, N.Z., scented.

Clementine A form or variety of tangerine.

Cleome Pls. with showy spider-like fls., cult.

Clerodendron Shrubs and woody climbers, many orn. and cult. in warm cli., e.g. *C. thomsoniae*, *C. fallax*, *C. squamatum*.

Clethra Cult. orn. shrubs, Am., As., fragrant fls.

Clianthus Cult. orn. fls., Aus., *C. dampieri* (glory pea) red and black fls., for greenhouse culture grafted on *Colutea arborescens; C. puniceus*, 'parrot's bill', brilliant red fls.

Clidemia Trop. Am. hairy shrubs, cult., *C. vittata* valued for fol.

Cliff brake *Pellaea*, Am.

Climbing bell *Littonia modesta*, S.Af., like *Gloriosa:* — **fern**, *Lygodium*, trop.

Climbing plants Generally most abundant in trop. forests where they may grow large and woody (lianes or lianas). There are four main groups of climbers – (1) **twiners**, plants whose tips nutate in search of support (*Apios, Apocynaceae, Araliaceae, Aristolochiaceae, Basella, Bauhinia, Bignoniaceae, Bowiea, Calystegia, Camoensia, Ceropegia, Cassythea, Combretaceae, Connaraceae, Convolvulaceae, Cuscuta* w. sensitive stems like tendrils, *Cynanchium, Dipladenia, Freycinetia, Gnetum, Hoya, Ipomoea, Jasminum, Lardizabalaceae, Loasaceae, Lonicera, Lygodium, Malpighiaceae, Menispermaceae, Phaseolus, Phytocrene, Plumbago, Polygonum, Rhodochiton, Ruscus, Schizandra, Solanum, Tamus, Thunbergia, Wisteria*): (2) **climbers with sensitive organs**, usually tendrils, which may be modified *stems* (*Antigonon, Landolphia, Passiflora, Vitis*), ls. (*Bignoniaceae, Cobaea, Corydalis, Leguminosae* e.g. *Lathyrus & Vicia, Mutisia*), *sensitive hooks*, which clasp and become woody (*Ancistrocladus, Artabotrys, Bauhinia, Gouania, Hugonia, Landolphia, Paullinia, Strychnos, Uncaria, Unona, Uvaria*); *sensitive ls.* occur in *Gloriosa, Littonia* etc., *petioles* in *Clematis, Dalbergia, Fumaria, Hablitzia, Maurandya, Rhodochiton, ˙Tropaeolum, midrib* in *Nepenthes, lateral branches* in *Hippocratea, Machaerium, Salacia, Securidaca, Uvaria:* (3) **hook climbers**, sprawling and catching by hooks (*Caesalpinia, Calamus, Capparis, Combretaceae, Desmoncus, Dipladenia, Galium, Hugonia, Lycium, Pereskia, Plectocomia, Smilax, Ventilago*): (4) **root climbers** with special negatively heliotropic adventitious rs. that adhere to the support (*Araceae, Araliaceae, Begonia, Bignoniaceae, Clusia, Ficus, Hedera, Hoya, Kendrickia, Norantea, Piper, Rhus, Salacia, Sapindaceae, Tecoma*).

Climbers are often of anatomical interest, presenting many abnormal features, especially the trop. lianas.

Clitandra Rubber has been obtained from various spp., trop. Af.

Clitoria *C. ternata*, Ind., cult. trop., twiner w. handsome blue fls. (butterfly pea).

Clivia S.Af. bulbous pls. w. large umbels of reddish fls., house pls., esp. *C. miniata* and *C. gardenii*.

CLOAKFERN

Cloakfern *Cheilanthes* spp. (=*Notholaena*).

Clock vine, Bengal *Thunbergia grandiflora*, Ind., large climber w. blue fls. 7–8 cm long.

Clotbur *Xanthium spinosum*, Medit. and other spp.

Cloudberry *Rubus chamaemorus*, N.Eur. inc. Br., orange fr. ed.

Clove gilliflower or **pink** *Dianthus caryophyllus*, *D. plumarius.*

Clover The many wild and cult. clovers are all-imp. as fodder pls. Most, but not all, belong to the genus *Trifolium.* **Alsike** —, *Trifolium hybridum:* **Alyce** —, *Alysicarpus vaginalis:* **Aztec** —, *Trifolium amabile:* **ball** —, *T. nigrescens:* **beach** —, *T. fimbriatum:* **Berseem** —, *see* Egyptian —: **bighead** —, *T. macrocephalum:* **Bokhara** —, *see* sweet —: **buffalo** —, *T. reflexum:* **bur** —, *Medicago hispida:* **bush** —, *Lespedeza:* **button** —, *Medicago orbiculata:* **Calvary** —, *Medicago echinus:* **carnation** —, *Trifolium incarnatum:* **cluster** —, *T. glomeratum:* **crimson** —, *T. incarnatum:* **Dutch** —, *T. repens:* **Egyptian** —, *T. alexandrinum:* **holy** —, *Onobrychis viciifolia:* **hop** —, *Trifolium procumbens:* **Hungarian** —, *T. pannonicum:* **Italian** —, *T. incarnatum:* **Jap.** —, *Lespedeza striata:* **Korean** —, *Lespedeza stipulacea:* **owl** —, *Orthocarpus tenuifolius:* **Persian** —, *Trifolium resupinatum:* **pin** —, *Erodium: cicutarium:* **prairie** —, *Petalostemum* spp.: **purple** —, *Trifolium purpureum:* **rabbit or hare's foot** —, *T. arvense:* **red** —, *T. incarnatum:* **reversed** — *T. resupinatum:* **running** —, *T. stoloniferum:* **shield** —, *T. clypeatum:* **snail** —, *Medicago scutellata:* **sorrel** —, *Oxalis:* **star** —, *Trifolium stellatum:* **strawberry** —, *T. fragiferum:* **subterranean** —, *T. subterraneum:* **Swedish** —, *see* Alsike —: **sweet** —, **white**, *Medicago alba;* **yellow**, *M. officinalis:* **— trefoil**, *Trifolium:* **— tree**, *Goodia latifolia:* **Uganda** —, *Trifolium johnstonii:* **yellow** —, *Medicago lupulina:* **zigzag** —, *Trifolium medium. See Trifolium* in Willis.

Clovenberry (W.I.) *Samyda serrulata.*

Cloves *Eugenia caryophyllus* (=*E. caryophyllata: E. aromatica: Syzygium caryophyllus*). This well known spice consists of the dried fl. buds, the E.I. tree being cult. mainly in Zanzibar, Madag. and Malaysia: much used for flavouring pickles, puddings, cakes etc.: the oil has analgesic properties, used in toothache remedies and to flavour dentifrices: much used to flavour cigarette tobacco in Indon. ('bretek' cigarettes). **Braz.** —, *Dicypellium caryophyllatum:* **Madag.** —, *Ravensara aromatica*, whole pl. aromatic: **W.I.** —, *Pimenta acris* (name wrongly applied to allspice): **wild** —, *Psammisia urichiana*, W.I. orn. vine w. fls. like those of clove.

Clown's mustard *Iberis amara* (candytuft), Eur. seeds sometimes used med. or as mustard substitute.

Clubrush *Scirpus lacustris* and other spp.

Cluster bean *See* guar.

Clytostoma Trop. Am. climbers w. showy fls. cult. (called bignonia).

Coach-whip *Fouquieria splendens*, N. or C.Am. thorny shrub, used for hedges.

Coachwood *Ceratopetalum apetalum*, Aus., a commercial timber.

Coast-blite *Chenopodium rubrum*, Eur. **— dairy bush**, *Olearia ramulosa*, Aus.

Cobaea *C. scandens*, trop. Am. orn. vine (cup and saucer vine) grows fast, 6–8 m in a year. *See Cobaea* in Willis.

Cobnut *See* hazel nut: **Jam.** —, *Omphalea triandra*, ed. kernel.

Coca *Erythroxylum coca*, Peru, Bolivia, cult. elsewhere, the dried ls. are the source of cocaine and traditionally have been a stimulant and masticatory w. Amerindians. Cocaine is the oldest local anaesthetic, now used mainly in ophthalmic, ear, nose and throat surgery: used w. morphia for relief of pain in advanced malignant disease. **Truxillo** —, *E. truxillense.*

Cocaine *See* coca.

Cocculus indicus Name for frs. of *Anamirta cocculus*, Ind., used med. and for poisoning fish.

Coccus wood *Brya ebenus*, trop. Am., a handsome dark wood much used for door handles at one time.

Cochineal A colouring matter formerly obtained from the dried female cochineal insect (*Dactylopius coccus*) which fed on *Opuntia* or *Nopalea:* dye now produced synthetically: the Canary Isles. industry was based on cult. of *Nopalea cochenillifera* (thornless).

Cochlioda S.Am. orchids, cult., brilliant hues, many hybrids.

Cochliostema *C. odoratissima*, trop. S.Am., cult., scented showy fls.

Cocillana bark *Guarea rusbyi*, Bolivia, med., used for coughs.

Cock tree *Erythrina glauca*, Guyana.

Cockatoo orchid *Caleana major*, Aus.

Cockle bur *Xanthium strumarium*, *Lychnis:* **corn** —, *Agrostemma githago:* **cow** —, *Saponaria vaccaria:* **sticky** —, *Silene noctiflora:* **white** —, *Silene alba.*

Cockrico bush *Cassia bacillaris*, W.I., Trinidad.

Cockroach grass *Zebrina pendula*, W.I.: — **plant**, *Haplophyton crooksii*, extract of ls. plus molasses kills cockroaches.

Cockscomb *Celosia cristata* (=*C. argentea*) cult. orn. pl., a monstrosity in which fasciation of the fls. and inflor. occurs; young fl. heads may be eaten as spinach.

Cocksfoot grass *Dactylis glomerata* an imp. pasture grass in all temp. countries, inc. N.Z., ('orchard grass' in U.S.).

Cockspur *Pisonia aculeata*, W.I.: — **grass**, *Echinochloa crus-galli*, cosmop.: — **thorn**, *Acacia eburnea*, Sri Lanka.

Coco or **coco-yam** *Calocasia esculenta* (=*C. antiquorum*) a widely cult. r. crop in trop., esp. W.Af.: — **de mer**, *see* double coconut: — **grass**, *Cyperus rotundus* (nut grass), a bad trop. weed: — **plum**, *Chrysobalanus icaco*, W.I., cult., ed. fr. usu. stewed.

Cocoa *Theobroma cacao*, S.Am., imp. trop. crop., two main types 'Forastero' and 'Criollo', but there are numerous vars. or cultivars, descended also from *Theobroma leiocarpa* or other spp. Trees range from 5–20 m in height. Fls. and frs. on trunk and branches, each fr. or 'pod' w. 20–50 seeds or cocoa beans embedded in pulp which is fermented. Main use of beans is for chocolate. Shells have been used in heat and sound insulation.

Cocobolo *Dalbergia retusa* and allied spp., S.Am., a commercial timber.

Coconut *Cocos nucifera*, trop. One of the world's most imp. economic pls. There is doubt as to its country of origin. The nuts may be spread by ocean currents. Both tall and dwarf vars. are grown. The two most imp. commercial products are copra for oil and coir or coconut fibre. In some parts of the trop. the coconut palm supplies almost all the needs of the inhabitants. *See Cocos*

COCOS WOOD

in Willis. Large quantities of coconut oil are used in soap and margarine manufacture. Desiccated coconut is prepared from fresh nuts in the trop.

Cocos wood *Brya ebenus*, W.I. a highly orn. wood used for special purposes, esp. musical instruments (flutes).

Codlin A green or golden apple, without stripes, usu. early and a good 'cooker'.

—'s and cream, *Epilobium hirsutum*.

Codonopsis As. pls. w. bell-shaped fls., often malodorous, cult.

Coelogyne Trop. As. orchids, cult.

Coffee *Coffea* spp. The roasted and ground seeds of certain spp., mainly **Arabian** —, *C. arabica* and **robusta**, *C. robusta*, *C. canephora*, and to a lesser extent **Liberian**, *C. liberica*, **Sierra Leone**, *C. stenophylla* and **Senoussi**, *C. excelsa*. There are many cultivars and hybrids of these, esp. of Arabian coffee, which is usu. grown at the higher altitudes in the trop. Some kinds of coffee are 'low country' coffees. Cult. and production methods vary with different countries. *See Coffea* in Willis. **Abeokuta** —, *Coffea liberica*: **— bush**, *Clerodendron aculeatum*, trop.: **Congo** —, *Coffea robusta*: **Inhambane** —, *C. racemosa*: **Kentucky** —, *Gymnocladus canadensis*: **Maragogipe** —, *Coffea arabica*: **Mocha** —, *C. arabica*: **Mussaenda** —, *Gaertnera canadensis*, seeds a coffee substitute: **negro** —, *Cassia occidentalis*: **Quillou** —, *C. canephora*: **— weed**, *Cassia occidentalis*: **wild** —, *Triosteum perfoliatum*.

Coffin nail *Anacardium occidentale*: **— tree**, *Machilus nanmu*, planted round temples in China, good timber.

Cogwood *Zizyphus chloroxylon*, W.I., hard tough wood: *Ceanothus*.

Cohosh, blue *Caulophyllum thalictroides*, N.Am., med.: **black** —, *Cimicifuga racemosa*, med.; *Actaea spicata* med., used for muscular rheumatism.

Cohune nut *Attalea cohune*, C. & S.Am., kernel oil like coconut oil, palm w. many other uses.

Coigue *Nothofagus dombeyi*, Chile, a commercial timber.

Coir Fibre prepared from the husk of the coconut, *Cocos nucifera*. It has numerous uses such as mattress fibre, matting, door mats, brushes, cord and rope: **— dust**, or coconut peat, a by-product, is used for mulching, soil conditioning, rooting and as a seed germinating medium.

Cola Or Kola. *Cola* spp. W.Af. The seeds of some 'Kola nuts' are the popular masticatory of W.Af. and are a stimulant for the weary labourer or traveller. They contain caffeine and are normally chewed in the fresh state before they dry and harden when they have a somewhat bitter taste. A liking for them has to be acquired. There is a large trade in the nuts in W.Af. for they enter into the social life of the people. For export they are dried; used in the manufacture of some popular beverages. **Abata** —, *Cola acuminata*: **Bamenda** —, *Cola anomala*: **Gbanja** —, *Cola nitida*: **Owé** —, *Cola verticillata*: **'bitter'** —, *Garcinia kola*, *Carapa procera*, (not real cola). [Russell, T. A. (1955) The kola of Nigeria and the Cameroons. *Trop. Agric., Trin.* **32** (3), 210–40.]

Colchicine *See* colchicum.

Colchicum *C. autumnale*, C. & S.Eur., inc. Br., meadow saffron, the dried corm and seed is the source of the alkaloid colchicine (0.25–4.0%) used med. (for relief of pain) and in pl. breeding work to induce chromosome doubling.

Coleus Cult. orn. pls. w. colourful ls., mainly derived from *C. blumei,* Java: *C. rotundifolius,* trop., is grown for its ed. tubers: aromatic ls. of *C. aromaticus* (Spanish thyme) used in place of thyme in W.I.

Colic root *Aletris farinosa,* N.Am., med.

Collards (Am.) A form of cabbage, *Brassica oleracea* var. *acephala.*

Collecting, plant The following notes may assist those wishing to collect pls. in an intelligent manner or to make a reference collection or herbarium of their own. Fresh pls. cannot of course be maintained in their natural state for long. The use of a preserving fluid in glass or plastic containers retains the original structure very well and is commonly used for orchid fls. and some fleshy pls. but for general use it is expensive and cumbersome, or storage presents a problem. Dried or herbarium specimens have the advantage of cheapness and easy or inexpensive storage.

How and what to collect. The collector should bear in mind that scraps of pls. should be avoided for they are seldom of any value and are anathema to the herbarium botanist concerned with the accurate identification of pls. He needs good or complete material for his work showing all parts of the pl. if possible. The minimum amount collected of each sp. should be sufficient to fill a standard herbarium sheet, which measures 41.6 × 26.8 cm (or 40.6 × 29.4 cm in the U.S.). With herbaceous pls. the whole pl., including the underground part, should be collected. A number of pls. to show the range in size is much better than a single one. When the pl. is longer than a herbarium sheet the upper part may be bent back when pressed and thus accommodated on a single herbarium sheet, or a pl. may be cut up to occupy more than one sheet. With very small pls. a number sufficient to fill a herbarium sheet should be collected. In the case of larger pls., such as trees and shrubs, small branches or leafy twigs, sufficient to fill a sheet, should be obtained. Collecting should not be carried out in wet weather if it can be avoided.

Material should always be collected in fl. or fr. It may be necessary to return later for the fruiting material. Flowering and fruiting material should be obtained from the same pl. if possible. In the case of unisexual pls. care must be taken to see that both male and female fls. are represented. Frs. that are large or large fl. heads such as those of thistles or proteas may be cut in two longitudinally to facilitate pressing and later mounting. Some pls. require special treatment. Many heath-like pls. and many conifers retain their ls. better on drying if they are first momentarily dipped in boiling water. This treatment is also beneficial with many succulent pls. in that it causes them to dry out more quickly and prevents their continuing to grow in the drying press, when fls. may change to useless rudimentary frs. Small floating aquatic pls. may be best collected by slipping a sheet of paper under them in the water, then lifting the sheet with specimens attached and placing it in the drying press. Covering with butter muslin prevents drying paper sticking. Where specimens, particularly carpological specimens, such as cycad or large conifer cones, are too large or woody for mounting on a herbarium sheet, they are commonly dried off as they are and placed in boxes. If prone to disintegrate they may be encased in nylon or plastic netting. Some large fleshy specimens are best preserved in fluid. Alcohol (70%) is commonly used, or a preserving fluid made up as follows – commercial formalin 5 parts, glacial acetic acid 5 parts and 50 % alcohol 90 parts.

Collecting equipment. This need not be elaborate or expensive. Many good collections have been made using home made equipment and newspaper in place of botanical drying paper. In the field one may either use a light portable collecting press or port-folio consisting basically of sheets of paper held tightly together by two pieces of stiff cardboard or plywood, these held together by two light webbing straps. A carrying handle may be devised. Alternatively a vasculum or stout polythene bag may be used to carry the specimens, in a fresh state, until the return to base is made and the specimens put into a drying press. Very small specimens may be put into air-tight tins to avoid crushing by larger specimens. Flimsy or delicate fls. are often placed between sheets of thin paper, tissue paper or polyethylene film before placing them in a press and are retained in them.

Other equipment needed is a good pocket knife or secateurs, trowel or suitable implement for obtaining underground parts, field labels, notebook and possibly camera and sketch book. In collecting in the trop. other equipment, such as a cutlass or hatchet, for cutting lianas may be needed. Every specimen is labelled with the collector's number and for this purpose jewellers' tag-labels, available in many sizes with string already attached, are very convenient.

Importance of field notes. The making of field notes, on the spot, where the collection is made, is very important. It is not safe to trust to memory to make the notes at a later time. The object of a herbarium specimen and the notes that go with it is to record as far as possible the exact appearance of the pl. For this reason any characters not revealed by a dried specimen and likely to be lost or altered in drying, such as fl. and fr. colour, scent, duration of flowering, insect visitors, nature or colour of underground organs etc., should be carefully recorded. The name of the collector, his or her number of the specimen, the locality and date are among the first essentials to be recorded. Every collector should maintain a single series of collection numbers throughout his lifetime. All duplicates of course bear the same number and this is important if the collector's specimens are to be lodged in any of the large herbaria. Other matters which should be recorded by the collector, especially if collecting in little known areas, relate to altitude and habitat, whether the pl. grows in shade or full sun, terrain, nature of soil, rainfall or other relevant climatic data. Any economic uses the plant may have, whether grazed by animals, should be recorded, also local names if any. Caution needs to be taken over this as names can be invented merely to oblige!

Pressing and drying specimens. The important point to remember in pressing and drying specimens is that they should be dried off quickly. This obviates the likelihood of mould developing and ensures a good colour, at any rate with most specimens. The specimens brought back from the field should be gone over and placed between dry absorbent paper in drying presses which are similar to collecting presses or portfolios. If the presses can be placed in hot sunshine and the papers changed daily (after sunset) for a few days, the wet papers being spread out to dry, drying soon takes place. Of course collection may be followed by wet or dull weather when this method becomes impossible. The modern tendency is to dry specimens with artificial heat, electrical driers being very satisfactory. Where these are used the presses are stood vertically or on edge over the source of heat and each press contains a number of ventilators, i.e. corrugated cardboard or aluminium sheet which allows the warm dry air to pass through the press and effect the drying rapidly. Care must be taken not to overdry specimens otherwise they become too brittle and inclined to break when handled.

A simple specimen drier can be made at home by using a wooden box of suitable dimensions to take several drying presses arranged vertically, with ventilators, and resting on wire mesh or wire netting. Underneath a source of heat from a small electric heater or a series of light bulbs is installed, the box being slightly raised from the floor to allow ventilation. Where electricity is not available a small safe oil heater such as a car radiator heater, or two or three of them, may be considered.

In transferring specimens from the collecting press (polythene bag or vasculum) to the drying press the opportunity should be taken to straighten out or arrange the specimens as they are to appear on the mounting sheet. At least one leaf should be arranged with the undersurface facing upwards so that characters there such as veining, nature of hairs etc. may be readily seen without having to detach a leaf. When the pls. are dry and packed between sheets of paper in bundles it is a wise precaution to sprinkle an insect repellent amongst them, naphthalene or *para*dichlorobenzene being commonly used. Many insects, especially beetles, are liable to attack dried plants or herbarium specimens.

Mounting specimens. Before being placed in a herbarium cabinet it is necessary or desirable that specimens be mounted. This consists of attaching the specimen firmly to a stout sheet of paper (mounting paper) or cardboard in the case of heavy specimens. Specimens may be fixed to the paper by means of an adhesive or they may be strapped or stitched depending on the nature of the specimen. The adhesive needs to be strong and not prone to crack or flake off Fish glue has been much used but synthetic or plastic adhesives are now favoured in many herbaria. Specially tough paper (with a high rag content) is used for mounting. The size of mounting sheets is usually about 41.6 × 26.8 cm but there is variation between different national herbaria. A mounting label needs to be attached to each sheet on which the correct name of the plant is written and information from the collector's label. [Arnold Arboretum, U.S. (1968) Notes on making an herbarium: suggestions for collecting particular kinds of plants. *Arnoldia,* **28** (8–9), 69–111.]

Colletia S.Am. shrubs, some v. spiny, cult.

Colliguaji bark *Colliguaja odorifera,* Chile, b. used as soap.

Collinsia N. & C.Am. annuals resembling *Pentstemon,* cult., esp. *C. bicolor,* many vars.

Colocasia *See* cocoyam.

Colocynth *Citrullus colocynthis,* S.Eur., N.Af., Middle East, wild and cult., fr. med. dried pulp used, a strong purgative.

Colomba root *See* calumba.

Colombian berry *Rubus macrocarpus*, fr. large ed., 5 cm long: — **blueberry,** *Vaccinium floribundum:* — **mahogany,** *Cariniana pyriformis*, good timber.

Colophony The crude, oleo-resin tapped from pine trees is distilled yielding rosin or colophony and turpentine.

Colorado grass *Panicum texanum*, a fodder grass.

Coltsfoot *Tussilago farfara*, Eur. inc. Br., a weed, ls. med. and used in herbal tobacco and cigarettes.

Columba root *See* calumba.

Columbine Can. —, *Aquilegia canadensis:* **Eur.** or **garden** —, *A. vulgaris*, many cultivars.

Columbus grass *Sorghum* × *almum* (hybrid), good fodder grass under dry conditions.

Columnea Shrubs or climbers, trop. Am., showy fls., cult. some are popular room pls.: *C. gloriosa* has striking red and yellow tubular fls. 7–8 cm long.

Colutea Cult. orn. shrubs, Medit. & As., esp. *C. arborescens* (bladder senna).

Colvillea *C. racemosa* ('Colville's Glory'), Madag., cult. orn. tree. w. orange-red fls.

Colza The seed and oil of 'colzas' are similar to those of rape. **Eur.** —, *Brassica campestris* var. *oleifera:* **Ind.** —, *Brassica campestris* var. *sarson.*

Comb fringe grass *Dactyloctenium aegyptiacum*, trop. Af., seeds eaten.

Comfrey *Symphytum* spp., cult. for fodder or orn., med. *S. officinale*, fls. whitish or pinkish purple, rs. med.: **blue** —, *S. peregrinum:* **rough Russian** or **prickly** —, *S. asperum*, Caucasus, cult. for fodder: **tuberous** —, *S. tuberosum*, fls. yellowish: **Turkish** —, *S. orientale* fls. white.

Comino *Aniba perutilis*, Colombia, a handsome furniture wood.

Commelina Herbaceous perens., several cult. gdn. pls., e.g. *C. coelestis, C. tuberosa:* some are weeds, e.g. *C. nudiflora.*

Commiphora *C. berryi* and other spp. used as live fences in dry trop. Af. and Ind.

Common dragon *Dracunculus vulgaris*, S.Eur. cult.

Comparettia Elegant Andean orchids, cult.

Compass plant *Silphium laciniatum*, N.Am., pilot weed or polar pl., in the young state the ls. often present their edges north and south: *Lactuca scariola.*

Condiments *See* spices, flavouring.

Condor vine or **condurango** *Marsdenia condurango*, Colombia, white bell-shaped fls. b. med., reputed cure for snake bite.

Conessi Or kurchi, *Holarrhena antidysenterica*, Ind., ls. b. and seeds med., wood favoured for carving and turnery.

Cone flower *Rudbeckia, Ratibida, Echinacea:* — **head,** *Strobilanthes.*

Confetti-tree *Maytenus senegalensis.*

Congea *C. tomentosa*, cult. trop., scandent shrub, persistent mauve bracts.

Congo copal *Guibourtia* (=*Copaifera*) *demeusii:* — **jute,** *Urena lobata:* — **wood** (Am.), *Lovoa klaineana:* — **stick,** trade name for chestnut walking stick (coppice).

Conifers The term conifer is used in a broad sense for any member of the Coniferae, the most imp. class of gymnosperms. Economically conifers are of very great importance for they furnish the bulk of the world's timber (softwoods) and pulpwood for paper, including as they do the pines, larches, spruces, hemlocks, cypresses and junipers. *See Coniferae* in Willis. **Dwarf —.** Numerous dwarf or slow-growing conifers of various genera showing a wide range of mutations, dwarf, weeping, prostrate, colour forms etc. are in cultivation. [Welch, H. J. (1968) *Dwarf conifers*, 334 pp.]

Conophor nut *Tetracarpidium conophorum*, W.Af., kernel ed. and yields a drying oil, conophor oil.

Consumption weed *Baccharis halimifolia*, N.Am.

Contrayerva (W.I.) *Aristolochia odoratissima*, roots med. (vermifuge). **— root,** *Dorstenia contrajerva*, med. and used for flavouring cigarettes.

Convolvulus Cult. orn. climbers, many w. large colourful, trumpet-shaped fls.: some are bad weeds, e.g. bindweed *C. arvensis*. **N.Z. —,** *Calystegia turguriorum:* **sand —** (N.Z.), *Calystegia soldanella*. The name Convolvulus is also applied to some spp. of *Ipomoea*, e.g. *I. purpureum* (blue fls.), *I. cardinalis* (red fls.), *I. rubrocaerulea* (morning glory) and *I. setosa* (mauve fls.). *I. triloba* (pink convolvulus) is a bad cane weed in Queensland.

Convulsion root *Monotropa uniflora*, N.Am., r. med., sedative.

Conyza *C. canadensis*, a weed in Br.

Cooba *Acacia salicina*, Aus.

Cooktown orchid *Dendrobium bigibbum*, Aus., probably Australia's most handsome and best known orchid: the floral emblem of Queensland: a variable sp.

Coolabar or **coolibar** *Eucalyptus microtheca*, Aus.

Coolwort *Tiarella cordifolia*, N.Am., med., urinary complaints.

Copaiba balsam *Copaifera langsdorfii*, Braz. oleo-resin, med.

Copal The name copal used for a large group of hard resins much used in varnish and paint manufacture at one time. The most imp. of these commercially was Congo copal, dug up from the ground, believed to be derived from *Guibourtia* (=*Copaifera*) *demeusii*. Other W.Af. fossilized resins, probably from *Copaifera* spp., were also exploited. E.Af. fossilized copal was from *Trachylobium verrucosum*. S.Am. copals were from *Hymenaea coubaril* or other spp. of *Hymenaea*. The source of E.I. or Manila copal was *Agathis alba*. N.Z. Kauri 'gum' or copal was from the Kauri pine, *Agathis australis*.

Copalchi bark *Croton niveus*, Mex. b. med., substitute for cascarilla b.

Copper cups *Pileanthus peduncularis*, Aus.: **— leaf,** *Acalypha wilkesiana* and other spp., cult., colourful ls.

Coppice Wood cut over at intervals for poles, gdn. stakes or fencing (cleft chestnut). In Br. hazel and sweet chestnut mainly grown: cut at intervals of about 5, 7 or 14 years according to requirements.

Copra Dried flesh or endosperm of the coconut, imp. commercially for coconut oil (*see* coconut).

Coprosma Aus. and N.Z. shrubs, some cult.

Coquilla palm *Attalea funifera*, Braz. the thick-shelled nut takes a high polish and is used for carving and fancy articles: l. bases furnish piassava fibre (*see* piassava).

Coquito palm *Jubaea chilensis* (=*J. spectabilis*) yields 'palm honey', cult. for orn.

Coracan *Eleusine coracana*, finger millet, cult. trop., esp. Ind.

Coral bead *Cocculus carolinus*, N.Am.: — **bean**, *Erythrina* spp.: — **bell**, *Heuchera sanguinea:* — **berry**, *Symphoricarpus orbiculatus*, N.Am.: — **bush**, *Russelia juncea*, *Templetonia retusa*, Aus.: — **creeper**, *Antigonon leptopus*, *Kennedia* spp., Aus.: — **fern**, *Gleichenia* spp.: — **flower**, *Erythrina* spp.: — **heath**, *Epacris microphylla*, Aus.: — **honeysuckle**, *Lonicera sempervirens*, N.Am.: — **pea**, *Kennedia* spp., W.Aus., cult.: — **peony**, *Paeonia corallina:* — **plant**, *Jatropha multifida:* — **root**, *Corallorhiza odontorhiza*, N.Am., leafless parasite, scented rs. med. (sedative); *Cardamine bulbifera:* — **sumac**, *Metopium toxiferum:* — **tree**, *Erythrina* spp.: — **vine**, *Antigonon leptopus*, cult. warm cli.

Corallila or **coronilla** *Porana paniculata*, Ind., shrubby climber, small white fls.

Corallita *Antigonon leptopus*, popular trop. climber, coral fls.

Cord grass *Spartina maritima*, Eur. inc. Br. in tidal salt marshes: **Am.** — —, *S. pectinata* and other spp.: **smooth** — —, *S. alterniflora*, N.Am. introduced to Br.: **Townsend's** — —, *S. × townsendii*, *S. anglica*, used for stabilizing and raising level of maritime mud flats.

Cordeauxia *See* yeheb.

Cordia Several S.Am. spp. yield valuable timbers: some are trop. orn. trees and shrubs, e.g. **scarlet** —, *C. sebestina*, widely cult.

Cordyla *C. africana*, W.Af., has a yellow ed. fr. (bush mango).

Cordyline *C. terminalis* and other spp., cult. fol. pls.: house pls. w. colourful ls.: Maoris used ls. for garments and ate roots.

Coreopsis Cult. gdn. fls., many vars., largely derived from *C. drummondii* or *C. tinctoria*.

Coriander *Coriandrum sativum*, Medit., an ancient spice or flavouring, referred to in Biblical times: the aromatic frs. or 'seeds' are used in many food preparations and in curry powder and liqueurs: considered to give an Oriental flavour to casseroles and sauces: has med. uses, added to purgatives to prevent griping.

Coriaria *C. japonica* and other spp., cult. orn. shrubs, petals change colour.

Cork, commercial The cork oak (*Quercus suber*) is native to W. Medit. and cork is known to have been used since pre-Christian times. Countries producing commercial cork include Portugal, Spain, France, Italy, Algeria, Morocco and Tunisia. There are over two million hectares of cork forest. The trees, which occur naturally as they have done for hundreds of years, are now conserved and protected from fire, their main enemy, by removing inflammable undergrowth. Trees live for 150–200 years. When the trees are large enough, about 20 years of age or 25 cm in diam., the corky b. is stripped off. Thereafter stripping is done every 8–10 years. The pieces of b. are first boiled to soften or remove tannin and other impurities, scraped to remove the hard or crusty outer b., and then pressed flat, dried and baled for export.

The b. or cork varies in thickness and in quality. Its ultimate use depends upon this. Some of the best quality cork is to be seen in stoppers for champagne bottles. The poorest grades and scraps are commonly used for granulated cork, which w. appropriate modern adhesives or resins is so extensively used for cork insulation boards, mats, gaskets, and for numerous other purposes. An age-old use for cork is as floats for fishing nets. Other uses are with sporting goods, inner soles of footwear, cork tiles, a constituent of linoleum, polishing wheels for glass and ceramics, deep freeze cabinets and cork-tipped cigarettes. Cork has many unique properties and is still widely used in spite of the advent of plastics and other man-made materials. Among these properties are compressibility, resiliency, low thermal conductivity, sound and vibration absorbency, high coefficient of friction (non-slip, hence value for bottle corks or stoppers), lightness and imperviousness to water and other liquids (valued for life belts). The use of cork for stoppers greatly increased with the invention of the modern glass bottle in 1760. [Cooke, G. B. (1961) *Cork and the cork tree*, 121 pp.]

Cork tree *Ochroma pyramidale*, trop. Am., balsa tree, cult.; *Phellodendron* spp., China, corky b.; *Millingtonia hortensis*, Ind. orn. fl. tree w. cracked corky b.

Corkwood *Duboisia myoporoides*, Aus., soft, used for carving; *Entelea arborescens*, N.Z., wood half the weight of cork, used for fishing floats; *Hakea* spp. Aus.; *Leitneria floridana*, N.Am. soft light wood used for fishing floats; *Musanga smithii*, W.Af. v. light wood; *Myrianthus arboreus*, Congo, wood lighter than cork.

Corn Term used for the kind of grain commonly used in a country: in Br. usu. refers to wheat and in Am. to maize. — **bind**, = bindweed: — **bluebottle** *Cyanus segetum* (=*Centaurea cyanus*): — **cockle**, *Agrostemma githago*, a bad weed, seed poisonous: Ind. —, *Zea mays:* — **marigold**, *Chrysanthemum segetum:* **pop**—, *Zea mays* subsp. *praecox*, grains burst or pop when exposed to heat: — **poppy**, *Papaver rhoeas:* — **salad**, *Valerianella locusta:* — **salad, Italian**, *Valerianella eriocarpa:* **squirrel** or **turkey** —, *Dicentra canadensis:* — **violet**, *Legousia* (=*Specularia*) *hybrida*.

Cornel Or dogwood, *Cornus* spp., hard, close-grained wood.

Cornelian cherry *Cornus mas*, Eur. cult. orn. small tree valued for its early yellow fls., hard dense wood: fr. ed., preserved.

Cornflour Maize product much used in cooking.

Cornflower *Cyanus segetum* (=*Centaurea cyanus*), Eur., popular ann., grown for market in Br., many colour forms. Am. —, *Plectocephalus americanus* (=*Centaurea americana*).

Cornish moneywort *Sibthorpia europaea*.

Corokia Cult. orn. shrubs, N.Z., *C. cotoneaster* is the hardiest in Br.

Coromandel wood Or ebony, *Diospyros quaesita*, *D. hirsuta* Ind.

Corozo palm *Corozo oleifera* (=*Elaeis melanocarpa*), Braz.; *Acrocomia* spp. W.I.; *Phytelephas macrocarpa*, C.Am.

Corpse plant *Monotropa uniflora*, N.Am. a saprophyte w. flesh-pink waxy stems.

Correa Aus. evergreen orn. shrubs, cult. esp. *C. speciosa*.

Cortusa *C. matthioli* and other spp., Eur. to Jap., mt. woodland pls. allied to *Primula*, orn., cult.

Coryanthes Trop. Am. orchids w. strange or bizarre fls., cult. *See* Willis.

Corydalis Several spp. are gdn. pls., tubers of some spp. eaten: **climbing** —, *C. claviculata:* **purple** —, *C. solida:* — **yellow**, *C. lutea*.

Corylopsis Early spring fl. shrubs, NE.As., cult., fls. appear before ls.

Corylus *See* hazel.

Coryphantha Round or cylindrical cacti, cult.

Cosahuico *Sideroxylon tempisque*, Mex., a little known tree fr.

Cosmos or **cosmea** *Cosmos* spp. cult. orn. fls., esp. *C. tripinnatus*, Mex., may be a weed.

Costmary Or alecost, *Pyrethrum* (=*Chrysanthemum*) *balsamita*, S.Eur., an old-time herb, the strong smelling camphoraceous ls. were once put into ale.

Costus root Or kuth, *Saussurea lappa*, Kashmir, rs. aromatic, costus of the Ancients, med. in E.

Coto bark *Nectandra coto*, Bolivia, b. med.: **false** — —, *Ocotea pseudocoto*.

Cotoneaster Cult. orn. shrubs, mainly Chinese, some make good shelter belts.

Cotton *Gossypium* spp. There are four main spp. of cotton from which the present-day commercial kinds of cotton are considered to have been derived, two from the New World, *G. barbadense* and *G. hirsutum*, and two from the Old World, *G. herbaceum* and *G. arboreum*. The main commercial kinds of cotton are American upland, sea island, Egyptian and Asiatic. There are many vars. in each. The length of the seed hairs of cotton or lint is imp. in determining its value. Cotton is one of the oldest textile fibres and was used for cloth in India at least 2500 years ago. It probably reached Egypt from India.

Cotton seed, after removal of the lint, is an imp. source of edible oil, much used for margarine and as a substitute for olive oil. The remaining seed-cake is a good stock food. — **bush**, *Pimelea nivea*, Aus.: **devil's** —, *Abroma augusta*, Malaysia: — **grass**, *Eriophorum* spp., seed hairs once used for stuffing pillows; *Imperata cylindrica*, a bad weed: — **gum**, *Nyssa aquatica*, N.Am.: **kidney** —, *Gossypium brasiliense*, seeds stuck together in the shape of a kidney: **lavender** —, *Santolina chamaecyparissus*, ls. white-felted: — **linters**, the short hairs or fuzz on cotton seed, used for paper: **sea island** —, *Gossypium barbadense*, a v. superior cotton: — **sedge**, *Eriophorum angustifolium*, N.Am.: **silk** —, *Ceiba pentandra*: — **thistle**, *Onopordon*: **tree** —, *Gossypium arboreum; Ochroma*: — **weed**, *Froelichia* spp., N.Am.: — **wild** —, *Cochlospermum, Asclepias*: — **wood**, *Populus* spp., esp. *P. deltoides, P. trichocarpa*, N.Am. commercial timbers; *Cassinia* spp. N.Z.

Cotula Cult. orn. pls., often used for edging.

Cotyledon Genus of succulent leaved S.Af. plants, cult.

Couch grass Or twitch, *Agropyron repens*, a bad cosmop. weed quickly spreading: in Aus. *Cynodon dactylon* may be called couch.

Cough root Or Indian balsam, *Leptotaenia multifida*, N.Am.

Coula *Coula edulis*, W.Af. a commercial timber, mahogany substitute: seeds ed., fresh or cooked (Gabon nut).

Coumarin A sweet-scented substance, the smell of newly mown hay, notably the grass *Anthoxanthum odoratum*: also characteristic of some seeds, e.g Tonka bean, *Dipteryx odorata*.

Counter wood *Chlorophora excelsa*, W.Af. Iroko, used for counter tops.

Courbaril *Hymenaea coubaril*, trop. Am., a commercial timber.

Cous root Or biscuit root, *Lomatium ambiguum*, N.Am., and other spp., once an imp. food of the Amerindian.

Cousin mahoe *Urena lobata*, W.I., yields aramina fibre or 'Congo jute'.

Cover plants and crops Pls. that are able to form a ground cover to suppress weeds, or to protect the soil have attracted increased attention with the growing shortage or expense of hand labour for weeding in so many parts of the world. With gdns. or orchards in temp. cli. ground covers are now widely used in shrubberies, areas under trees or out of the way places. Among the ground cover pls. used in this way in Br. are ivy (*Hedera*) esp. the large leaved Irish ivy and ferns, also spp. of *Alstroemeria, Bergenia, Dicentra, Erica, Gaultheria, Geranium, Hosta, Hydrangea, Hypericum, Lamium* (notably the variegated dead-nettle), *Lysimachia, Pachysandra, Teucrium, Tradescantia* and many more.

In the trop. and subtrop. cover crops are cult. to protect the soil from erosion by torrential rain as well as for weed control. This applies esp. with many of the trop. plantation crops such as rubber, tea, coffee, oil palm, citrus etc. Cover crops may also serve to enrich the soil by providing humus and in the case of leguminous cover crops in some climates supplying nitrogen. Leguminous cover pls. used in warm cli. in this way include spp. in the following genera – *Alysicarpus, Crotalaria, Cassia, Centrosema, Desmodium* (notably *D. tortuosum*), *Dolichos, Indigofera* (esp. *I. endacaphylla*), *Leucaena, Medicago, Melilotus, Pueraria, Tephrosia* (esp. *T. candida*), *Teramnus, Trifolium, Trigonella, Vicia* and *Vigna*. [Fish, M. (1970) *Ground cover plants*, 144 pp.]

Cow bane *Cicuta virosa*, poisonous: — **berry**, *Vaccinium vitis-idaea:* — **cockle**, *Saponaria vaccaria:* — **cress**, *Lepidium campestris:* — **foot**, *Peperomia pellucida*, cosmop. trop. weed: — **itch**, *Mucuna pruriens*, cosmop. in trop., stinging hairs on pods cause intense irritation: — **parsnip**, *Heracleum sphondylium:* — **pea**, *Vigna unguiculata* (= *V. sinensis, V. catjang*) trop. and subtrop. pulse crop, much variation in shape, size and colour of seeds: —**slip**, *Primula veris: Am.* —**slip**, *Dodecatheon* spp. cult., resemble cyclamen: **Cape** —**slip**, *Lachenalia* spp.: —**slip orchid**, *Caladenia flava*, Aus.: **Virginian** —**slip**, *Mertensia virginica*, cult.: — **tree**, *Brosimum utile*, S.Am., latex resembles cow's milk: — **wheat**, *Melampyrum pratense* and other spp.

Cowrie pine *See* kauri.

Coyal palm *Acrocomia mexicana*, Mex., frs. ed., cooked, and source of oil.

Crab apple or **crab** *Malus* spp., many spp. and hybrids cult. as orn. fl. trees, frs. may be used for jelly: **Iowa** —, *M. ioensis:* **Siberian** —, *M. baccata:* **Oregon** —, *M. rivularis:* **Sikkim** —, *M. sikkimensis:* **wild** —, *M. sylvestris* (= *Pyrus malus*), Br.

Crab's eyes Or jequerity seeds, *Abrus precatorius*, trop., orn. red and black seeds: — **grass**, *Digitaria sanguinalis* and other spp.; *Panicum; Eleusine:* — **oil**, *Carapa guianensis*, trop. Am., seed oil, med.: — **wood**, *Carapa guianensis*, a mahogany substitute.

Crackerberry *Cornus canadensis*, N.Am.

Crambe Some spp. cult. as foliage pls., large ls.

Cramp bark *Viburnum opulus*, Eur., (Guelder rose) bark med.

Cranberry *Vaccinium oxycoccus*, Eur. mt. bogs, fr. used for preserves. **Am.** —, *V. macrocarpum:* **mt.** —, *V. erythrocarpum*, N.Am.: **small** —, *V. microcarpum.*

Crane flower *Strelitzia reginae* S.Af., widely cult. several vars.

Cranesbill *Geranium* spp. Many cult. as gdn. pls., some are good smother pls.

for weeds. **bloody** —, *G. sanguineum:* **meadow** —, *G. pratense:* **shining** —, *G. lucidum:* **tuberous** —, *G. tuberosum.*

Crape fern *Todea barbara,* S.Hemisph., cult. orn. fern: — **myrtle,** *Lagerstroemia indica,* China, orn. fl. tree, widely cult. subtrop.

Crassula A large genus of succulents, mainly S.Af., cult.

Crataego-Mespilus Hybrids between hawthorn and medlar, cult.

Crataegus Cult. orn. trees or shrubs, frs. of some spp. ed. or used for preserves, *See* hawthorn.

Crattock *Ficus glomerata,* Ind., a shade or avenue tree.

Craw-craw plant *Cassia alata,* W.Af., l. juice used for skin complaints.

Crazyweed *Oxytropis* spp., N.Am., harmful to livestock.

Cream cups *Platystemon californicus,* N.Am., cult. also called Calif. poppy; *Oenothera spinulosa.*

Creeper Canary — *Tropaeolum peregrinum:* **creeping Jenny,** *Lysimachia nummularia:* — **lily,** *Gloriosa:* — **partridge berry,** *Gaultheria procumbens:* — **sailor,** *Saxifraga stolonifera:* — **snowberry,** *Chiogenes hispidula,* N.Am.: — **soft grass,** *Holcus mollis,* a weed: — **trumpet,** *Tecoma:* **Virginia creeper,** *Parthenocissus tricuspidata.*

Creole tea *Sauvagesia erecta,* W.I.

Creosote bush *Larrea tridentata* (=*L. mexicana*), W. N.Am., Mex., common desert pl., resinous twigs used med. by Amerindians: binds drifting sand.

Crepe flower *Lagerstroemia indica,* China, orn. fl. tree widely cult. subtrop.: — **jasmin,** *Tabernaemontana coronaria,* W.I., fragrant fls.

Cress *Lepidium sativum,* Persia, **common** or **garden** —, an ann. of v. rapid growth, usu. eaten in the early seedling or cotyledon stage w. mustard at a similar stage and sown 4 days later. **Aus.** or **golden cress** is a dwarf yellowish variety, '**curled**' and '**broad-leaved**' are others. **Am.** —, *Barbarea praecox:* **bitter** —, *Cardamine pennsylvanica:* **Braz.** —, *Spilanthes oleracea:* **cow** —, *Lepidium campestre:* **creeping yellow** —, *Rorippa sylvestris:* **hoary** —, *Cardaria draba:* **Ind.** —, *Tropaeolum majus:* **land** —, *Barbarea praecox:* **meadow** —, *Cardamine pratensis:* **Para** —, *Spilanthes oleracea:* **penny** —, *Thlaspi arvense:* **rock** —, *Arabis:* **shepherd's** —, *Teesdalia nudicaulis:* **spring** —, *Cardamine bulbosa:* **violet** —, *Ionopsidium acaule:* **water**—, *Nasturtium officinale,* extensively cult. in watercress beds: **winter** —, *Barbarea praecox,* cult.

Crested dog's tail grass *Cynosurus cristatus,* Br., mats and baskets have been made from it: — **hair grass,** *Koeleria cristata,* Br.

Crinum Handsome bulbous pls., large fls., cult., popular room pls.

Crin vegetal Or vegetable hair, fibre from Eur. fan palm, *Chamaerops humilis,* used in upholstery.

Crocosmia S.Af. cormous pls., cult., dried fls. placed in water smell of saffron.

Crocus Hardy cormous pls. valued for early spring flowering: gdn. forms derived largely from *C. aureus,* E.Eur. and *C. vernus,* Eur. *See Crocus* in Willis. **Autumn** —, *Colchicum autumnale* (*see Colchicum*): **Chilean** —, *Tecophilaea cyanocrocus,* cult.; *Zephyranthes* spp.: **purple** —, *Crocus purpureus:* **saffron** —, *C. sativus:* **warren** —, *Romulea columnae.*

Crookneck *Cucurbita pepo, C. moschata,* cult. ed. gourd, several vars.

Crosnes Name for Chinese artichoke, *Stachys sieboldii*.

Crossandra Evergreen fl. shrubs, mainly trop. Af., cult.

Cross of Jerusalem *Lychnis chalcedonica*, from the resemblance of the petals to the arms of a Maltese or Jerusalem cross: — vine, *Doxantha* (= *Bignonia*) *capreolata*, N.Am., stem cut in cross section shows a Greek cross: —wort, *Galium cruciatum*, cross-placed ls.; *Crucianella*, ls. crosswise; *Lysimachia*.

Crotalaria Some spp. cult. for orn. in warm cli. *See* sun hemp.

Croton *Codiaeum variegatum* and other spp. Most of the orn. shrubs with colourful or unusual ls. cult. in warm cli. and called croton are botanically *Codiaeum* not *Croton*.

Croton oil *Croton tiglium*, trop., cult., the seed oil is a powerful purgative.

Crowa fibre Name for pineapple fibre.

Crow berry *Empetrum nigrum*, frs. are food for moorland birds, Br. **Portuguese** — —, *Corema album*, cult. orn. shrub: **purple** — —, *Empetrum atropurpureum*, N.Am.: — **apple**, *Owenia venosa*, Aus.: — **ash**, *Flindersia australis:* — **foot**, *Ranunculus*, *Dactyloctenium*, N.Am.: — **poison**, *Zygadenus densus*, N.Am.

Crown bark *Cinchona officinalis:* — **beard**, *Verbesina* spp., N. Am.: — **gum**, *Achras chicle*, Mex., a product resembling true chicle: — **Imperial**, *Fritillaria imperialis*, a handsome and well known gdn. pl.: — **of Thorns**, *Paliurus spina-christi*, *Zizyphus spina-christi*, *Euphorbia milii*.

Crucifer Any member of the Cruciferae (cabbage family).

Crucifix orchid *Epidendrum*, Aus.: — **thorn**, *Koeberlinia spinosa*, S. U.S., Mex., an excessively thorny shrub w. ls. reduced to scales.

Crunch weed *Brassica kaber*, N.Am.

Cryptadenia Orn. pls., S.Af. (Cape region), cult.

Cryptanthus Trop. Am. bromeliads w. orn. fol., house pls., easily grown.

Cryptomeria *C. japonica*, China, Jap., imp. timber tree, cult. for orn., many vars.

Cryptophoranthus Cult. trop. Am. orchids, 'window orchids', sepals with apertures.

Cuba bast *Hibiscus elatus*, bark fibre used for rope and cordage. — **hemp**, *Furcraea cubensis:* — **lily**, *Scilla peruviana*.

Cubebs *Piper cubeba*, Malaysia, the dried unripe frs. used in Eur. as a spice in the 17th century; mainly used med. in the Orient. **Af.** —, *Piper clusii:* **Guinea** —, *P. guineense*.

Cuckold *Bidens* spp. N.Am., often occur as weeds.

Cuckoo flower *Cardamine pratensis*, common in moist meadows in Br.; *Lychnis flos-cuculi* (in Am.): — **pint**, *Arum maculatum*, starchy tubers have been used for ed. purposes (Portland arrowroot) after treatment.

Cucumber *Cucumis sativus*, N.Ind., usu. eaten raw, in salads, or pickled, esp. the smaller kinds or gherkins. Cucumbers may be successfully prepared by a good cook and eaten as a veg.: numerous vars. for different conditions: some are parthenocarpic, frs. develop without pollination (hothouse kinds). **Af.** —, *Momordica charantia:* **bur** —, *Sicyos angulatus*, N.Am.: **horned** —, *Cucumis metuliferus:* **lemon-scented** —, *Cucumis melo* var. *dudaim:* — **orchid**, *Dendrobium cucumerinum*, Aus.: **prickly** —, *Echinocystis lobata*, N.Am.: — **root**, *Medeola virginiana*, N.Am., r. tastes like cucumber:

CUDJOE WOOD

squirting —, *Ecballium elaterium*, when ripe, seeds are squirted out, fr. is med., source of drug elaterium: — **tree**, *Averrhoa bilimbi*, trop.; *Magnolia acuminata*, N.Am., frs. cucumber-like: **wild** —, *Acanthosicyos naudinianus*, S.Af.

Cudjoe wood *Jacquinia keyensis*, S. U.S.

Cudweed *Gnaphalium* spp., *Filago* spp.

Cuica resin *Cercidium spinosum*, Colombia, a poor resin.

Culilawan *Cinnamomum culilawan*, E.I., b. used as spice and yields oil.

Culver's root *Veronica virginica*, N.Am., has emetic and cathartic properties.

Cumin seed *Cuminum cyminum*. The frs. or 'seeds' are an old spice known to the early Eur. civilizations: uses similar to caraway by which it has been largely replaced: an ingredient of curry powders and used to flavour cheese, cakes, bread, soups and liqueurs: used in veterinary medicine.

Cumquat *See* kumquat.

Cunila oil *C. origanoides* (=*C. mariana*), E. U.S. ls. med. and used as tea by early settlers.

Cunninghamia Chinese conifers, fragrant durable timber, cult. for orn.

Cunure *Calliandra guildingii*, W.I., orn. tree w. red, brush-like stamens, cult.

Cup flower *Nierembergia gracilis*, Argen., good pot pl.: — **grass**, *Eriochloa*: — **of gold**, *Solandra grandiflora*, trop. orn. climber w. yellow fls.: — **plant**, *Silphium perfoliatum*, N.Am.: — **and saucer plant**, *Holmskioldia sanguinea* Ind. shrub w. showy fls.

Cupania *C. glabra*, W.I., and other trop. Am. spp. yield good timber.

Cuphea *C. ignea*, Mex., 'cigar fl.' and other spp. cult., some as house pls.

Cupid's darts *Catananche caerulea*, S.Eur., cult. gdn. pl., narrow ls.: — **flower**, *Ipomoea quamoclit*, a weed in Queensland sugar cane: — **paintbrush**, *Emilia javanica*, trop., a weed.

Cuprea bark *Remijia pedunculata*, *R. purdieana*, Colombia, b. contains quinine and cupreine.

Cupressocyparis *C. leylandii*, a valuable quick-growing hybrid conifer, esp. for screening: cross between *Cupressus macrocarpa* and *Chamaecyparis nootkatensis*.

Cupressus *See* cypress.

Curare Originally a S.Am. arrow poison, prepared chiefly from b. of *Strychnos toxifera* and *Chondodendron tomentosum*, now of great importance med. in anaesthesia and as a curative and palliative drug. [Bryn Thomas, K. (1964) *Curare: its history and usage*, 144 pp., Lippincott.]

Curculigo *C. latifolia*, Indon. l. fibre used for fish-nets, rope and clothing.

Curled rock brake *Cryptogramma crispa*, a small Arctic fern.

Curlewberry *Empetrum nigrum*, Eur., inc. Br., a moorland pl.

Curly grass *Schizaea pusilla*, N.Am. small fern w. twisted wiry fronds: — **greens**, *Brassica oleracea* var., a veg.: — **heads**, *Clematis ochroleuca*: — **palm**, *Howeia belmoreana*, Lord Howe Island, seeds exported for raising orn. pot pls.

Currant *Vitis vinifera* var. *corinthiaca* or *apyrena*, the ordinary or grape currant, used dried so extensively for bakery products, home cooking etc. Greece has long been an imp. producer. The name has been applied to other small berried frs. such as the black and red currant. **Alpine** —, *Ribes alpinum*, Br.: **Am. black**—, *R. americanum*: **Aus.** —, *Leucopogon* spp.: **black**—,

Ribes nigrum, widely used fr. and imp. vitamin source, many vars.: **buffalo** —, *R. odoratum, R. aureum,* N.Am.: **bush** —, *Clidemia* spp., *Miconia* spp., W.I.: **flowering** —, *Ribes sanguineum,* many vars.: **golden** —, *R. aureum:* **Missouri** —, *R. aureum:* **native** —, *Leptomeria* spp., Aus., used in preserves: **prickly** —, *Ribes lacustre,* N.Am.: **red** —, *R. rubrum,* a widely used fr. (jelly): **skunk** —, *R. glandulosum,* N.Am.: — **tree,** *Jacquinia,* W.I.: **white** —, a variety of red currant: **wild** —, *Rhus* spp., S.Af.

Curry leaf *Murraya koenigii,* Ind., small tree, ls. much used for flavouring curries.

Curua palm *Attalea spectabilis,* Braz., ls. much used for thatching, seeds yield oil.

Cuscus root *See* vetiver.

Cush-cush *Dioscorea trifida,* W.I., a palatable and popular yam.

Cushion calamint *Calamintha clinopodium,* Eur. inc. Br.: — **pink,** *Silene acaulis* Eur. inc. Br., cult., many vars.: — **plant,** *Pygmaea* (= *Veronica*) *pulvinaris,* N.Z.: — **spurge,** *Euphorbia epithymoides,* cult. orn. pl.

Cusparia bark *See* angostura bark.

Custard apple *Annona reticulata,* trop. Am., a widely cult. and popular dessert fr. in warm cli., also called bullock's heart: **wild** — —, *A. chrysophylla* (= *A. senegalensis*), W.Af., fr. ed.: — **orchid,** *Thelymitra villosa,* Aus.

Cutch Or catechu, the solid tanning or dyeing extract prepared from the heartwood of *Acacia catechu* in Ind. and Burma. It has been a trade product from the earliest times. Other uses are for treating fishing nets and sails to give them long life and as a masticatory with betel. The name cutch has also been used for mangrove b. extracts and pale catechu from gambier.

Cut-grass Grasses or sedges liable to cut the hands: *Leersia oryzoides,* Eur. inc. Br.; *Leersia* spp., N.Am.; *Scleria* spp., W.I.; *Carex* spp. N.Z.

Cyathea Trop. and subtrop. tree ferns, some cult. for orn.: starchy pith used as food in some countries (includes '*Alsophila*').

Cycad General term applied to members of the order Cycadales, palm-like pls. that occur in trop. and subtrop., esp. spp. of *Cycas* and *Encephalartos.* The pith of some spp., being starchy, may be used as food by local inhabitants, e.g. *Cycas circinalis* (trop. As.) and *C. revoluta* (Japan). *See also Cycadales* in Willis.

Cycas Pith and seed of some spp. used as food. Ls. of *C. revoluta* imported to Br. from Jap. for decoration.

Cyclamen The common cyclamen is one of the most colourful and outstanding of greenhouse or room pls.: has the advantage of winter flowering: the numerous vars. are derived largely from *C. persicum,* E.Medit.

Cycnoches Trop. Am. orchids w. large fls., resembling *Catasetum,* cult.

Cymbidium As. and Aus. orchids, cult., many hybrids.

Cynoglossum Widely distributed herbaceous pls., some orn. and cult.

Cyperus *See* sedge.

Cyphel *Cherleria sedoides* on Scottish mts.

Cyp or **cypre** *Cordia alliodora,* W.I., a light brown, much used wood.

Cypress The true cypresses (*Cupressus* and *Chamaecyparis*) are evergreen trees widely distributed throughout the warm temp. regions of the N.Hemisph. Many are cult. for timber, for orn. and for hedges and shelter belts. Trees of various other genera may be termed 'cypress'. **Af.** —, *Widdringtonia:* **Arizona** —, *Cupressus arizonica:* **bald** —, *Taxodium distichum:* **black** —, *Callitris endlicheri* (=*C. calcarata*), Aus.: **Calif.** —, *Cupressus goveniana:* **Chinese deciduous** —, *Glyptostrobus lineatus:* **Chinese weeping** —, *Cupressus funebris:* **Formosan** —, *C. formosensis:* **funereal** —, *C. funebris:* **Himalayan** —, *C. torulosa:* **Hinoki** —, *Chamaecyparis obtusa,* Jap.: **Italian** —, *Cupressus sempervirens:* **Jap.** —, *Chamaecyparis obtusa:* **Lawson's** —, *Chamaecyparis lawsoniana;* SW. U.S. widely cult. for orn. and timber, used for shingles, sleepers and posts, also called Port Orford cedar: **Medit.** —, *Cupressus sempervirens:* **Mex.** —, *C. lusitanica, Taxodium:* **Monterey** —, *C. macrocarpa,* cult. in many countries inc. Br.: **Patagonian** —, *Fitzroya patagonica:* — **pine,** *Callitris* spp., Aus.: **Port Jackson** —, *Callitris rhomboidea:* — **root,** *Cyperus longus,* scented: **Sawarra** —, *Chamaecyparis pisifera:* — **shrub,** *Lawsonia inermis* (=*L. alba*): **southern** —, *Taxodium distichum,* S. U.S.: **summer** —, *Kochia:* **swamp** —, *Taxodium distichum:* **weeping** —, *Cupressus funebris:* **yellow** —, *Chamaecyparis nootkatensis,* W. N.Am.

Cypripedium A large genus of orchids of special horticultural and botanical interest (2 fertile stamens instead of 1): many cult.

Cyrtanthus Af. bulbous pls. related to *Crinum,* some cult.

Cyrtomium Cult. orn. ferns, SE.As., *C. falcatum* is regarded as one of the toughest greenhouse ferns and tolerates poor light conditions.

Cyrtopodium Trop. Am. orchids, some w. large panicles of showy fls., cult.

Cyrtosperma *C. edule* and other spp., Polynesia, large starchy rootstocks, eaten cooked.

Cyrtostachys Stately Malaysian palms, cult.

Cytisus Many cult. orn. shrubs, numerous hybrids. In the common broom (*Cytisus* or *Sarothamnus scoparius*) the fl. has an interesting explosive mechanism in regard to pollination. The style is long and there are two lengths of stamens. When an insect alights on the fl. (there is no nectar) the keel begins to split from the base towards the tip, and presently the pollen of the short stamens is shot out upon the lower surface of the visitor; immediately afterwards, the split having reach the tip, the other pollen and the style spring violently out and strike the insect on the back. As the stigma touches first there is thus a chance of a cross, if the insect bears any pollen. Afterwards the style bends right round and the stigma occupies a position just above the short stamens so that another chance of cross fertilization is afforded if other insects visit the fl. (in most exploding fls. there is only one chance).

+ *Laburnocytisus* (=*Cytisus*) *adami* is a curious graft-hybrid between *Cytisus purpureus* and *Laburnum anagyroides,* the latter being used as the stock.

D

Daboecia Heath-like orn. shrubs, cult., esp. vars. of *D. cantabrica.*

Dadap *Erythrina lithosperma,* shade tree for trop. crops.

Daddy long legs orchid (Aus.) *Caladenia pumila.*

Daemonorops Slender spiny palms, trop. As.: *D. draco* is the source of the resin dragon's blood.

Daffodil *Narcissus pseudonarcissus*, Eur., cult. orn. fl., many vars. **Spanish —**, *N. hispanicus:* **Tenby —**, *N. obvallaris:* **wild —**, *N. pseudonarcissus.*

Dagame (Cuba) *Calycophyllum candidissimum,* wood once used for bows.

Dagga (S.Af.) *Cannabis sativa* (Indian hemp or cannabis): **wild —**, *Leonotis leonurus,* S.Af.

Dagger fern *Polystichum acrostichoides,* N.Am.: **— orchid** (Aus.), *Dendrobium pugioniforme:* **— plant** (W.I.), *Yucca* spp.

Dahl (Ind.) Name for pigeon pea, *Cajanus cajan.*

Dahlia *Dahlia rosea* (= *D. variabilis*) and other spp. (Mex.) and hybrids, cult. orn. fls.: introduced to Eur. in 1789 as a food plant, tubers being eaten in Mex. The numerous modern cultivars are divided hort. into about a dozen groups, cactus, collarette, pom-pom etc.

Dahoma *Piptadenia africana,* W. & E.Af., a trade timber.

Dahoon *Ilex cassine,* N.Am.

Daikon *See* radish, Japanese.

Daily dew (N.Am.) *Drosera* spp.

Dais *Dais* spp., cult. orn. pls. *D. glaucens* yields b. fibre (Madag.).

Daisy The **common — or** Eur. (*Bellis perennis*) is a widespread weed, esp. on lawns, but there are many cult. orn. forms. A general term for any member of the family *Compositae.* **Alpine —**, (Aus.), *Brachycome scapigera:* **Aus. —**, *Vittadinia:* **Barberton —**, *Gerbera jamesonii:* **bog —** (N.Z.), *Celmisia graminifolia:* **burr —** (Aus.), *Calotis:* **bush —** (Aus.), *Olearia tomentosa:* **carmel —** (Medit.), *Scabiosa prolifera:* **Clanwilliam —** (S.Af.), *Euryops athanasiae:* **crown —** (Medit.), *Chrysanthemum coronarium:* **cut-leaf —** (Aus.), *Brachycome multifida:* **dog —**, *Leucanthemum vulgare:* **everlasting —** (Aus.), *Helichrysum, Helipterum, Waitzia* etc.: **globe —** *Globularia:* **golden paper —** (Aus.), *Helichrysum bracteatum:* **kingfisher —** (Aus.), *Felicia bergeriana:* **lazy —** (N.Am), *Aphanostephus skirrobasis:* **Livingstone —**, *Dorotheanthus bellidiformis:* **mat —** (N.Z.), *Raoulia:* **Michaelmas —**, *Aster:* **mountain —** (N.Z.), *Celmisia:* **Namaqualand —**, *Dimorphotheca sinuata:* **ox-eye,** *Leucanthemum vulgare:* **paper —** (Aus.), *Helipterum:* **purple —** (Aus.), *Brachycome:* **rock tree —** (N.Z.), *Brachystegia insignis:* **Rottnest Island —**, *Trachymene caerulea:* **shasta —**, *Leucanthemum maximum:* **snow —** (Aus.), *Celmisia longifolia:* **Swan river —** (Aus.), *Brachycome iberidifolia:* **— tree** (N.Z.), *Olearia;* (S.Am.), *Montanoa:* **true blue —** (S.Af.) *Charieis heterophylla.*

Dalbergia Several spp. yield 'rosewood'. **False —**, *Afromosia laxiflora,* W.Af., hard heavy timber resembling lignum vitae.

Dalechampia *D. roezliana,* C.Am., cult. orn. fl., pink bracts.

Dalli *Virola surinamensis* and allied spp., a trade timber.

Dallis grass *Paspalum dilatatum,* valued for fodder, warm cli.

Dalmatian insect powder *Pyrethrum cinerariifolium.*

Damaceen (W.I.) *Chrysophyllum argenteum,* cult., ed. fr.

Damask *Hesperis matronalis,* cult. orn. fl.: **— rose,** *Rosa damascena:* **— violet,** *Hesperis matronalis.*

Dame's violet *See* damask violet.

DAMIANA

Damiana *Turnera diffusa*, C.Am., ls. med.

Dammar The many dammar resins are derived, by tapping, from forest trees belonging to several genera of the family Dipterocarpaceae in the Malayan region, the name dammar (or damar) being derived from a Malay word signifying any resin, or a torch made from resin. The dammars were valued mainly for the manufacture of pale-coloured spirit varnishes. One of the most esteemed forms of dammar, pale in colour, is 'dammar penak', from *Balanocarpus heimii*, an important timber tree in Malaya, known as 'chengal'. Other dammars are from spp. of *Hopea, Shorea, Vatica, Canarium* etc.

Dampiera, blue (Aus.) *Dampiera stricta*, orn. fl.

Damsel (W.I.) *Phyllanthus distichus* (= *P. acidus*) fr. ed. (preserves).

Damson or **damson plum** *Prunus insititia*, long cult. in Damascus, favoured for preserves, many vars. **Butter** — (W.I.), *Simaruba amara* root b. med.: — **plum** (W.I.), *Chrysophyllum*: **W.Af.** —, *Sorindeia juglandifolia*, fr. ed.

Dancing girls Or opera girls. *Mantisia saltaloria*, Ind., cult., strange fls.: — **orchid**, *Calendenia discoidea*.

Dandelion *Taraxacum officinale*, cosmop. weed, ls. and r. med., r. roasted as a coffee substitute, blanched ls. (of special vars.) used as salad: **dwarf** —, *Krigia*, N.Am.: **false** —, *Agoseris glauca* N.Am.: **Russian** —, *Taraxacum koksaghyz* (a rubber plant).

Danewort *Sambucus ebulus*, Eur. incl. Br.

Dangleberry (N.Am.) *Gaylussacia frondosa*, bright blue ed. fr.

Danta (W.Af.) *Nesogordonia papaverifera*, a trade timber.

Danube grass The common reed, *Phragmites australis*.

Daphne Cult. orn. shrubs, mostly small w. scented fls.: some spp. have local med. use: the fibrous b. of *Daphne cannabina* is the source of Nepal or Bhutia paper: **native** — (Aus.), *Eriostemon myoporoides*, *Pittosporum undulatum*.

Dari *Sorghum*, giant millet or Guinea corn.

Darnel *Lolium temulentum*, a Medit. grass, often a weed.

Darning needle cactus *See* Christmas cactus.

Darwinia Evergreen Aus. shrubs, cult., orn.: **lemon-scented** —, *D. citriodora*, W.Aus., cult.

Dasheen *Colocasia esculentum* (= *C. antiquorum*), starchy tubers ed., much cult. in trop.

Dassie bos (S.Af.) *Salvia rugosa*.

Dasylirion Cult. orn. pls., some spp. with their drooping ls. favoured for subtrop. bedding: ls. of *D. simplex* used for basket making in Mex.

Date The well known dried fruit of the date palm (*Phoenix dactylifera*), cult. in warm dry cli., esp. Asia Minor and Egypt: numerous vars. **Canary — palm**, *Phoenix canariensis*: **central Af.** —, *Balanites*: **Chinese** —, = jujube: **desert** —, *Balanites*: **dwarf** —, *Phoenix reclinata*: **Indian** —, = tamarind: **Trebizond** —, *Elaeagnus angustifolia*: **wild** —, *Phoenix reclinata*.

Date-plum *Diospyros lotus*, Medit., fr. ed., eaten fresh, dried or bletted: (Am.), *D. virginiana*: (Orient), *D. kaki*.

Datisca *D. cannabina*, W.As., cult. orn. fol. pl.: yields a dye.

Dattock (W.Af.) *Detarium senegalense*, mealy pods ed., good timber.

Datura Some spp. cult., large orn. fls., e.g. *D. sanguinea* and *D. arborea* (angel's trumpet): some spp. contain narcotics, used by Amerindians in S.Am., and dacoits in Ind. and SE.As.

Dau *Dipterocarpus alatus, D. costatus,* Indoch., trade timber.

Davallia Cult. orn. ferns, old world trop.

David's root (W.I.) *Chiococca* spp. *C. alba* used for snake bite: — **harp** or **Solomon's seal,** *Polygonatum multiflorum,* rhiz. med.

Davidia *D. involucrata,* China, cult. orn. tree, fl. heads subtended by two large white or creamy bracts (30 cm long, 15 cm wide): has been called 'handkerchief tree'.

Davidson's plum *Davidsonia pruriens,* Queensland, fr. ed. (preserves).

Daviesia Aus. evergreen shrubs, cult.

Dawa-dawa (W.Af.) *Parkia filicoidea,* mealy pod-pulp ed.

Dawn flower *Ipomoea learii,* cult. orn. climber, trop.: — **cypress** or **redwood,** *Metasequoia glyptostroboides.*

Day flower *Commelina* spp.: — **lily,** *Hemerocallis* spp.

Dead finish (Aus.) *Acacia tetragonophylla,* a fodder tree. — **nettle,** *Lamium:* variegated dead nettle a popular gdn. pl. or 'smother pl.': **red — —,** *L. purpureum:* **white — —,** *L. album.*

Dead sea apple *Quercus lusitanica,* Medit.

Deadly dwale (W.I.) *Acnistus:* — **nightshade,** *see* belladonna.

Deal, yellow or **red** *Pinus sylvestris:* **white —,** *Picea abies.*

Death-camus *Zygadenus* spp. N.Am., some spp. cause much poisoning of livestock, e.g. *Z. gramineus, Z. venenosus, Z. paniculatus.*

Decaisnea *D. insignis,* Himal., cult. orn. shrub w. yellow ed. fr.: *D. fargesii,* W. China, blue fr. (3–6 cm long), resembles a caterpillar.

Deciduous cypress *Taxodium distichum.*

Deer berry (Am.) *Vaccinium stamineum:* — **bush,** *Ceanothus integerrimus,* Calif.: — **flowers** or **grass,** *Rhexia* spp., Am.; *Trichorphorum caespitosum,* Am.: — **meat** (W.I.), *Centropogon surinamensis:* — **tongue,** *Saxifraga erosa, Trilisia odoratissima, Liatris odoratissimum,* Am., ls. used for flavouring tobacco.

Degami (W.I.) *Calycophyllum candidissimum,* a trade timber, strong, elastic.

Deleb palm (W.Af.) *Borassus aethiopum.*

Delphinium Cult. orn. fls.: popular gdn. pls.

Dendé (Braz.) *Elaeis guineensis,* Af. oil-palm, cult.

Dendrobium Many orn. cult. spp., some used med. locally.

Dennstaedtia Cult. orn. ferns, pantrop.

Deodar *Cedrus deodara,* Himal., imp. timber and orn. tree.

Dermatitis, plants causing As most gardeners are aware, certain cultivated or ornamental pls. may cause dermatitis or skin affections w. people who handle them. Some persons may be severely affected while others handling the same pls. in the same way may be quite unaffected. A Chinese primula (*Primula obconica*) is the most common cause of plant dermatitis in Britain. It was introduced in 1880 and rapidly became popular as a greenhouse or room pl. because of its beauty, long flowering and ability to thrive under a wide range of conditions and to withstand neglect. *Calomeria* (= *Humea*) *elegans* is another greenhouse pl. liable to cause dermatitis. Some vars. of gdn.

chrysanthemum can cause dermatitis. This may be severe w. susceptible persons concerned with picking and packing the blooms for market. Asparagus has been held to be responsible for dermatitis w. market gardeners in France, also the globe artichoke. Celery and hops are also claimed to affect some individuals. The giant hogweed may also cause skin affections under certain conditions (*see* hogweed).

In N.Am. the juice of poison ivy (*Rhus toxicodendron*) and other spp. of *Rhus* such as *R. radicans* and *R. vernix* (poison sumac) may cause severe dermatitis or blistering in contact w. the skin. Another member of the same family (*Anacardiaceae*) *Metopium toxiferum* (poison wood) in N.Am. acts similarly. In S.Af. the blister bush, *Peucedanum galbanum*, may cause severe blistering of the skin as many who ascend Table Mt. and accidentally brush against the plant will know. The acrid juice in the pericarp or shell of the cashew nut (*Anacardium occidentale*) will also cause severe blistering. Some timbers are known to be capable of causing dermatitis or skin affections among woodworkers dealing with them, if they happen to be allergic.

Derris root The rs. of some spp. of *Derris*, woody climbers or climbing shrubs, notably *D. elliptica* and *D. malaccensis*, have long been used to stupefy or capture fish in the Malayan region. Subsequently they came to be extensively used as insecticides, derris powder having become a household word.

Desert Aster *Aster tephrodes*, U.S.: — **broom**, *Baccharis* spp.: — **catalpa**, *Chilopsis linearis*, U.S.: — **dandelion**, *Rafinesquia neomexicana*, U.S.: — **date**, *Balanites aegyptica:* — **hyacinth**, *Brodiaea pulchella*, U.S.: — **ironwood**, *Olneya tesota*, U.S.: — **lemon**, *Atalanta glauca:* — **lily**, *Hesperocallis undulata*, U.S.: — **mallow**, *Sphaeralcea angustifolia*, U.S.: — **marigold**, *Baileya multiradiata*, U.S.: — **mariposa**, *Calochortus kennedyi*, U.S.: — **milkweed**, *Asclepias subulata*, U.S.: — **mistletoe**, *Phoradendron californicum*, U.S.: — **pea**, *Clianthus formosus*, Aus.: — **pepperweed**, *Lepidium fremontii:* — **phlox**, *Sphaeralcea angustifolia*, U.S.: — **poppy**, *Eschscholtzia glyptosperma*, U.S.: — **pride**, *Erimophila mackinlayi*, W.Aus.: — **ramona**, *Salvia carnosa*, N.Am.: — **rose**, *Adenium obesum*, E.Af.; *Hermannia stricta*, S.Af.: — **sage**, *Salvia carnosa*, N.Am.: — **seepweed**, *Suaeda suffrutescens*, U.S.: — **senna**, *Cassia bauhinioides*, U.S.: — **sunflower**, *Geraea canescens*, U.S.: — **tree tobacco**, *Nicotiana glauca*, U.S.: — **trumpet**, *Eriogonum inflatum*, U.S.: — **willow**, *Chilopsis linearis*.

Desmazeria *D. sicula*, cult. annual orn. grass.

Desmodium Some spp. are trop. cover crops or green manures.

Detarium *D. senegalense*, nearly pod ed.

Deutzia Cult. orn. shrubs, As., allied to *Philadelphus*.

Devil, blue *Echium vulgare*, *Eichhornia crassipes:* — **in the bush**, *Nigella:* —'s **apple**, *Mandragora officinalis:* —'s **bit**, *Chamaelirium*, *Succisa:* —'s **boots**, *Sarracenia purpurea:* —'s **claw**, *Martynia*, U.S.: —'s **club**, *Oplopanax horridus*, U.S.: —'s **coach whip**, *Stachytarpheta jamaicensis (indica)*, W.I., trop weed: —'s **cotton**, *Abroma augusta:* —'s **darning needle**, *Clematis virginiana*, U.S.: —'s **ear**, *Enterolobium cyclocarpum*, W.I.: —'s **fig**, *Argemone*, U.S.: —'s **grandmother**, *Elephantopus tomentosus*, U.S.: — **grass**, *Cynodon dactylon*, N.Am.: —'s **guts**, *Cuscuta*, *Equisetum:* — **herb**, *Plumbago scandens:* —'s **ivy**, *Scindapsus aureus:* — **leaf**, *Urtica spatulata:* —'s **paintbrush**,

Hieracium aurantiacum, N.Am.: — **pepper**, *Rauvolfia:* —'s **pins**, *Hovea pungens*, W.Aus.: —'s **plague**, *Daucus carota*, N.Am.: —'s **potato**, *Echites umbellata*, N.Am.: — **thorn**, *Tribulus terrestris*, *Emex australis*, *Pretrea* (=*Dicerocaryum*): — **tree**, *Erythrina corallodendron* W.I.: —'s **trumpet**, *Datura stramonium:* —'s **walking stick**, *Aralia spinosa:* — **wood**, *Osmanthus*.

Dewberry *Rubus caesius*, ed. fr., Eur. inc. Br.: **northern** —, *R. flagellaris*, N.Am.: **southern** —, *R. trivialis*, S. U.S.: **golden dew-drop**, *Duranta repens*, N.Am.: **dew-flower**, *Drosanthemum* S.Af.: **dew-thread**, *Drosera filiformis*.

Dhak tree (Ind.) *Butea monosperma* (=*B. frondosa*), yields Bengal kino (med.) yellow dye from fls. and cordage fibre from inner bark.

Dhal *See* lentil, pigeon pea.

Dhawa (Ind.) *Anogeissus latifolia*, ls. used in tanning.

Dhob or **dhub grass** *Cynodon dactylon*, much used for lawns, world wide.

Dhoby's tree *See* porcupine tree.

Dhup *Canarium euphyllum:* **red** —, *Parishia insignis*, both trade timbers from Andamans.

Dhupa fat *Vateria indica*, oil or fat from kernels.

Diabetes plant *Solanum sanitwongsei*, a reputed cure, Thailand.

Dianthus Cult. orn. pls., noted for fragrance and brilliance of fls.: the genus includes carnations, pinks and Sweet William.

Dichaea Elegant trop. Am. orchids, cult.

Dichopogon Cult. orn. pls., anthers w. appendages.

Dichorisandra Trop Am. perens., cult., orn. fls. and fol.

Dicksonia Handsome tree ferns of Aus. & N.Z., cult.

Dieffenbachia Cult. trop. orn. fol. pls.

Dierama Af. grass-like pls. w. pretty pendulous fls., cult.

Diervillea *D. florida*, and other spp., cult. orn. shrubs.

Digger's speedwell (Aus.) *Veronica perfoliata*.

Digitalis or **foxglove** *Digitalis purpurea*, Eur., As., wild and cult., orn. gdn. pl., ls. source of digitalin, a much used drug, acting mainly on the cardio-vascular system increasing irritability of cardiac muscle and producing more forceful palpitations: **Austrian** —, *D. lanata*, Eur., cult., med., source of digoxin and other alkaloids.

Dika (W.Af.) *Irvingia gabonensis* (=*I. barteri*), kernels contain oil or fat (Dika butter): — **bread**, the mashed kernels eaten. *See* Gaboon chocolate.

Dikamali (Ind.) *Gardenia lucida; G. gummifera*, yield a resinous gum, keeps flies away.

Dildo (W.I.) *Cephalocereus* (=*Cereus*) *swartzii:* **flat-hand** —, *Opuntia dillenii*.

Dill *Anethum graveolens*, S.Eur., cult., ann. herb. w. flavour like fennel, favoured for fish and in pickling (gherkins). Seeds and ls. (dill-leaf or dill-weed) used in soups and salads. Dill water is a well known remedy for flatulence in infants: **Ind.** or **E.I.** —, *Anethum sowa*.

Dillenia *D. indica*, orn. tree, cult. trop., large fls. (15 cm diam.) used for curries and for jelly.

Dilodendron Single S.Am. sp., kernels yield oil.

Dimorphotheca Star of the veld: S.Af. sun-loving gdn. pls., many hybrids and new strains developed.

DINGAAN'S APRICOT

Dingaan's apricot (S.Af.) *Doryalis caffra*, fr. ed. (Kei apple).

Dingleberry (N.Am) *Vaccinium erythrocarpum*, fr. ed., used for jellies.

Dionysia As. mt. pls., mainly tufted, cult., rock or alpine gdn.

Dioon *D. edule*, Mex., starchy seeds ed.

Dioscorea *See* yam.

Diosgenin Cortisone and related compounds, a group of efficient modern drugs, have been prepared from diosgenin, which occurs in wild yams (*Dioscorea*) and some other pls. Diosgenin has been used successfully as a starter in the production of cortisone, progesterone, and related sex hormones.

Dipelta *D. ventricosa*, *D. floribunda*, China, cult, orn. shrubs.

Dipladenia Cult. orn. fls., trop. and subtrop., mainly climbers.

Diplazium Cult. orn. ferns, related to *Asplenium*.

Diplectria Cult. orn. fls., Malaysia.

Disa Terrestrial orchids, mainly Af., cult., *D. uniflora* is a well known Cape sp. w. a single large red fl.

Dish cloth gourd *Luffa cylindrica*, trop., cult.

Dispersal of seeds This may be *occasional*, as by floating trees, ice, tornadoes, mud on birds' feet, etc. or *regular* by methods which may be classified under four headings – (i) **Wind**, which may carry *directly* the spores of ferns etc., seeds of *Pyrola*, *Orchidaceae*, some *Caryophyllaceae* etc. and by censer-mechanisms, (*Campanula*, *Delphinium*, *Iridaceae*, *Liliaceae*, *Papaver*); winged seeds (*Bignoniaceae*, *Bromeliaceae*, *Casuarina*, *Millingtonia*, *Pinus*, *Zanonia*); winged fruits (*Abroma*, *Aceraceae*, *Bignoniaceae*, *Carpinus*, *Dipterocarpaceae*, *Fraxinus*, *Liriodendron*, *Malpighiaceae*, *Ptelea*, *Pterocarpus*, *Rumex*, *Serjania*, *Terminalia*, *Ulmus*, *Ventilago*); hairs forming a parachute mechanism (seeds of *Apocynaceae*, *Asclepiadaceae*, *Epilobium*, *Gossypium*, *Salix* etc. and fruits of *Anemone*, *Clematis*, *Compositae*, *Eriophorum*, *Typha*, *Valerianaceae* etc.). (ii) **Animals**, carried as 'inside passengers' are the seeds of ed. frs. and as 'outside passengers' hooked frs. of various kinds (*Asperula*, *Bidens*, *Blumenbachia*, *Cenchrus*, *Circaea*, *Daucus*, *Galium*, *Geum*, *Harpagophytum*, *Martynia*, *Medicago*, *Tragoceros*, *Triumfetta*, *Xanthium* etc.), also glandular fruits or seeds (*Allionia*, *Boerhavia*, *Pisonia*, *Plumbago*, *Sigesbeckia* etc.). (iii) **Water** (*Cerbera*, *Cocos*, *Crinum*, *Nuphar*, *Nymphaea*, *Potamogeton* etc.), some seeds and fruits are known to be carried long distances by ocean currents without losing viability (*see* drift seeds). (iv) **Explosive mechanisms**, as a rule this method of seed dispersal involves relatively short distances (*Alstroemeria*, *Balsaminaceae*, *Biophytum*, *Buxus*, *Cardamine*, *Cyclanthera*, *Dorstenia*, *Ecballium*, *Eschscholtzia*, *Geranium*, *Hura*, *Impatiens*, *Ricinus*, *Ulex*, *Viola* etc.).

Modern intercontinental air travel may well be responsible for unusual seed dispersal, as seeds are easily carried in the mud on passengers' footwear or on the tyre of landing wheels of aircraft, the mud having dropped on to the runway from wheels of servicing vehicles. Noxious weed seeds could be dispersed in this way.

Dis or **diss grass** *Ampelodesma tenax*, S.Eur., Af., grows w. esparto, used for paper and locally for fish nets, ropes etc.

Dissotis Usu. hairy pls. with showy fls., Af., cult.

Dita bark *Alstonia scholaris*, Old World trop., med.

Ditch grass (N.Am.) *Ruppia maritima. See Ruppia* in Willis.

Dittander *Lepidium latifolium,* Medit., As., med. herb, once used as salad.

Dittany *Origanum dictamnus,* Medit., formerly med. (vulnerary): **Am.** —, *Cunila mariana:* **bastard** —, *Dictamnus albus:* **false** —, *Ballota acetabulosa,* Medit., seed vessel used as a floating wick in olive oil lamps.

Divi-divi A commercial and local name in Colombia and Venezuela for a small tree, *Caesalpinia coriaria,* or its dry pods, which have long been a commercial tanning material. The dry pods contain an abnormally high percentage of tannin (40–45 %). Divi-divi yields a light-coloured leather, its tanning properties resembling those of sumac.

Djave (W.Af.) *Baillonella toxisperma* (= *Mimusops djave*), ed. fat from seeds: yields a commercial timber, 'moabi'.

Dock roots *Rumex* spp., common weeds, the rs. are relatively rich in tannin and may be used med. as astringents (*see* canaigre). **Alpine** —, *R. alpinus:* **broad-leaved** —, *R. obtusifolius:* **curled** —, *R. crispus:* **patience** —, *R. patienta:* **prairie** —, *Silphium terebinthaceum:* **sharp** —, *Rumex conglomeratus:* **swamp** —, *R. verticillatus:* **Spanish rhubarb** —, *R. abyssinica:* **water** —, *R. hydrolapathum.*

Dockmackie *Viburnum acerifolium,* E. N.Am., inner bark med.

Doctor's gum (W.I.) *Metopium, Symphonia,* resin med. or made into a kind of glue, used in joinery.

Dodder Or scald, *Cuscuta* spp., leafless and rootless total parasites, without chlorophyll: can be serious pests with some crop pls. such as lucerne, clover and flax. *See Cuscuta* in Willis.

Dodecatheon N.Am. and As. herbs w. reflexed corolla, like *Cyclamen,* cult.

Dodo cloth (W.Af.) *Conopharyngia pachysiphon,* b. fibre.

Dodonaea *D. viscosa,* useful windbreak or defensive hedge in dry trop. and subtrop.

Dog apple *Asimina reticulata,* N.Am.: — **bane,** *Apocynum* spp.: **climbing** — **bane,** *Trachelospermum difforme,* N.Am.: — **berry,** *Ribes cynosbati; Pyrus americana,* N.Am.: — **daisy,** *Leucanthemum vulgare:* — **fennel,** *Eupatorium capillifolium,* N.Am.: — **liver,** *Kalanchoe crenata,* W.Af.: — **nettle,** *Urtica urens:* — **plum,** *Ekebergia capensis:* — **rose,** *Rosa canina:* — **senna,** *Cassia obovata:* —**'s tail grass,** *Cynosurus cristatus,* Eur., valued pasture and fodder grass: —**'s tongue,** *Erigonum tomentosum:* — **tooth violet,** *Erythronium dens-canis,* cult.: — **violet,** *Viola canina:* — **weed,** *Dyssodia,* SW. U.S.: — **willow,** *Capparis flexuosa,* W.I.

Dogwood *Cornus* spp., cult. orn. shrubs, wood of some spp. favoured for shuttles and bobbins. **Am.** —, *C. florida,* E. N.Am., *C. nuttalii,* N.Am.: **Aus.** —, *Jacksonia scoparia, Ceratopetalum* spp.: **blood twig** —, *Cornus sanguinea:* **Eur.** —, *C. mas:* **flowering** —, *C. florida:* **Jam.** —, *Piscidia erythrina:* **pagoda** —, *Cornus alternifolia:* **poison** —, *Rhus vernix,* N.Am.: **swamp** —, *Cornus sanguinea:* **Tasm.** —, *Bedfordia salicina.*

Doll's eyes *Actaea pachypoda,* N.Am.

Dolly Varden *Tephrosia virginiana,* N.Am.

Domatia tree *Endiandra discolor,* Aus. rain forest tree.

Domba oil (Ind.) *Calophyllum inophyllum,* seed oil med. and illuminant.

DONKEY EYE

Donkey eye (W.I.) *Mucuna sloanea:* — **orchid**, *Diuris longifolia*, Aus.: —'s **tail**, N.Am., *Sedum morganianum.*

Doob grass (Ind.) *Cynodon dactylon*, a lawn grass.

Doodia Orn. ferns, trop. As., Aus., cult.

Doom bark *Erythrophleum suaveolens* (=*E. guineense*), W.Af., ordeal poison.

Doon (Ceylon) *Doona* spp. timber and resin.

Doorweed *Polygonum aviculare*, a common weed.

Doronicum N. Temp. orn. spring fls., cult.

Doronoki (Jap.) *Populus suaveolens, P. maximowicii*, used for matches.

Doryopteris Ferns, w. small fronds, some w. bulbils (*D. pedata*), cult.

Double claw *Martynia proboscidea.*

Double coconut Or coco de mer, *Lodoicea maldivica* (=*L. callipyge, L. seychellarum*), Seychelle Isles., the large fr. or seed of this palm, one of the largest known, takes 10 years to develop. It was found by mariners floating in the sea long before the palm was known and was thought to originate in the sea. It was highly valued by eastern potentates, the nut showing some resemblance to the female human form.

Double tails orchid *Diuris* spp. Aus.

Douglas fir *Pseudotsuga menziesii* (=*P. douglasii, P. taxifolia*), W.Can. & U.S.; may reach 100 m and 3 m diam., fast growing, widely cult., many cultivars; timber extensively used, esp. in building. Related spp. are – **Colorado** or **blue** — —, *P. glauca:* **Forrest's** — —, *P. forrestii:* **Jap.** — —, *P. japonica:* **larged coned** — —, *P. macrocarpa.*

Doum palm Or gingerbread tree, *Hyphaene thebaica*, N.Af., exceptional among palms in having branched stems; fleshy fibrous part of fr. considered to resemble gingerbread in colour and taste; the hard kernels have been used as veg. ivory (buttons, etc.).

Dove flower or **orchid** *Peristeria alata*, W.I.: — **tree**, *Davidia involucrata*, China, cult., large white bracts: — **weed**, W.I., *Euphorbia prostrata.*

Down tree Name for balsa wood tree.

Downy myrtle *Rhodomyrtus tomentosa*, cult. orn. shrub w. red ed. frs., subtrop.

Doxantha *D. capreolata*, trop. Am., cult. orn climber.

Draba Some spp. of this large N.temp. genus cult., mainly as rock gdn. pls.

Dracaena Some spp. used as hedges or boundary pls., trop.

Dracocephalum Several spp. cult., hardy border pls. w. showy fls.

Dracontium *D. gigas*, S.Am., a remarkable giant aroid, cult. trop.

Dragon arum *Dracunculus vulgaris* (=*Arum dracunculus*) Eur., *Arisaema* spp., N.Am.: — **claw**, *Corallorhiza odontorhiza:* **false** — **head**, *Dracocephalum* spp.: — **plant**, *Dracaena sanderiana*, orn. ls. a room pl.: — **root**, *Arisaema dracontium*, cult. orn. pl.: — **tree**, *Dracaena draco, Harungana madagascariensis*, yellow sap turns red.

Dragon's blood Name has been applied to several red or reddish resins of different botanical origins. That known to the Ancients from Socotra was from *Dracaena cinnabari*, that from the Canary Isles from *Dracaena draco*. Later E.Ind. or Malayan dragon's blood was derived from several spp. of *Daemonorops*, climbing jungle palms. Has been used med. in various ways and for varnishes.

Drakensberg silky grass *Beckeropsis uniseta* (=*Panicum unisetum*) S.Af.

Drambuie A Scottish liqueur, made from whisky and heather honey (*Calluna vulgaris*).

Drift seeds The seeds of some pls., reaching the sea through rivers and streams, may be carried long distances by ocean currents and be washed ashore amongst driftwood etc, in another country or continent. It has been shown that some seeds of this kind retain their ability to germinate in spite of the long sea journey, such as the hard seeds of certain spp. of *Mucuna, Dioclea, Caesalpinia* etc. One of the most common drift seeds is the so-called 'sea-bean', *Entada gigas* (=*E. scandens*), sometimes washed up on the coasts of Br. having been brought by the Gulf Stream from the Am. trop.

Drimiopsis Cult. orn. fls., S. and trop. Af.

Drinks, plants affording *See* beverage pls.

Dropseed grass (N.Am.) *Sporobolus, Muhlenbergia.*

Dropwort *Filipendula vulgaris:* **water —,** *Oenanthe crocata,* v. poisonous.

Drosophyllum *D. lusitanicum* N.Af., insectivorous pl. *See Drosophyllum* in Willis.

Druggists' bark *Cinchona* or quinine.

Drugs, vegetable Many noteworthy drugs of veg. origin have been superseded to a greater or lesser degree by synthetic drugs. A good example is quinine, for so long the established drug for use against malaria but no longer extensively used for that purpose. The following include the veg. drugs among fl. pls. and ferns that are used in western med. – **Trop.** or **subtrop.** aloes, capsicum, cardamom, cascara, castor oil, cinnamon, cloves, coca or cocaine, ginger, ipecacuanha, nutmeg, nux-vomica or strychnine, pepper, quinine or cinchona, senna. **Temp.** aconite, althaea, belladonna, chamomile, caraway, colchicum, coriander, fennel, foxglove or digitalis, gentian, henbane or hyoscyamus, lavender, liquorice, lobelia, male fern, peppermint, poppy (opium alkaloids), rhubarb, stramonium, valerian.

Other med. products of veg. origin often used by herbalists include – alder bark, alkannet root, ammoniacum, almond oil, anise, araroba, arnica, asafoetida, benzoin, betel nut, bitter orange peel, black haw bark, bloodroot, buchu, Burgundy pitch, cajuput oil, Calabar bean, calendula, calumba, camphor, canella bark, cannabis, cascarilla, *Cassia fistula*, catechu, cherry bark, cinnamon, cloves, cocoa butter, colocynth, colophony, copaiba balsam, cubebs, cumin, dandelion root, dill, dragon's blood, elaterium, elecampane, eucalyptus oil, euonymus bark, euphorbium, fennel, figs, galbanum, galls (Turkey), gamboge, gelsemium, gentian root, ginger, grindelia, guaiacum, Guinea grains, gum arabic, hellebore, hydrastis, isphaghul seed, jaborandi, jalap, juniper berries, krameria, lemon peel, linseed, liquorice, manna (ash), musk root, oil of cade, oil of lemon, olive oil, orris, peanut oil, Peru balsam, pimento, podophyllum, psyllium, quince seed, rauvolfia, rose hips, rosemary, saffron, sandalwood, sandarac, santonica, sarsaparilla, savin tops, scammony root, senega root, sepentary, slippery elm bark, snakeroot, squill, star anise, stavesacre, strophanthus, styrax, tolu balsam, tonka bean, tragacanth, turmeric, turpentine, uva ursi, vanilla, willow bark, witch hazel, wood tar, wormseed (Am.).

Drumsticks *Moringa oleifera*, trop., unripe pods a curry vegetable: *Isopogon anemonifolius*, Aus.

DRYANDRA

Dryandra Aus. evergreen shrubs w. orange or yellow fls., cult.

Dryopteris Large genus of ferns, many spp., cult.

Duboisia *Duboisia* spp. The ls. of at least three spp. of these Aus. shrubs (*D. myoporoides, D. leichardtii* and *D. hopwoodii*) contain alkaloids of interest or value med. as substitutes for *Atropa* and *Hyoscycamus*, the former being used in childbirth and psychiatry and the latter as a source of atropine for dilating the pupil.

Duca *Clusia duca*, Colombia, resin burned as incense.

Duck acorn *Nelumbo lutea*, N.Am.: — **eyes**, *Cotula* spp. S.Af.: — **grass**, N.Am., *Eriocaulon aquaticum:* — **meat** = duckweed: — **orchid**, Aus., *Caleana* spp.: — **plant**, *Sutherlandia frutescens* S.Af. (the inflated seed pods placed in a bowl of water look like ducks): — **potato**, N.Am., *Sagittaria latifolia.*

Duckweed *Lemna minor* and other spp., floating aquatic pls., much eaten by wild fowl: **gibbous** —, *L. gibba:* **great** —, *L. polyrhiza:* **ivy** —, *L. trisulca.*

Duffin bean *Phaseolus lunatus.*

Dumb cane (W.I.) *Dieffenbachia seguine* and other spp.: said to render speechless anyone chewing the stem, once used to torture slaves. Some spp. w. orn. ls. are popular room pls., esp. *D. picta.*

Dumori butter (W.Af.) *Mimusops heckelii*, ed. fat from kernels.

Dune plants There are many pls. that are characteristic of sand dunes whether in warm or temp. countries, the latter including spp. of *Ammophila, Carex, Elymus, Hippophae.* In warm cli. spp. of *Canavalia, Ipomoea, Cassia, Tephrosia* and *Tamarix* often help to bind the sand.

Dunkal *Commiphora drake-brockmanii*, gum-resin used by Somalis for poisoning leopards.

Dunks (W.I.) *Zizyphus mauritiana* (= *Z. jujuba*) naturalized in W.I.

Duppy needles (W.I.) *Bidens pilosa*, a common weed.

Duranta *Duranta* spp., trop. Am., W.I., cult. orn. shrubs.

Durban grass *Dactyloctenium australe*, stoloniferous, a useful sand binder.

Durian *Durio zibethinus*, a well known ed. fr. of the Malayan region noted for its pleasant aromatic taste but disagreeable smell.

Durma mats (Ind.) Made from *Phragmites australis* (= *P. communis*).

Duroia S.Am. myrmecophilous (ant) pls. See *Duroia* in Willis.

Durra *Sorghum* or Guinea corn.

Dusty miller *Artemisia stellerana, A. cineraria.*

Dutch clover *Trifolium repens*, seed once imported to Br. from Holland: — **grass** (W.I.), *Eleusine indica*, cosmop. weed.

Dutchman's breeches *Dicentra spectabilis*, China, cult. orn. fl.: — **pipe**, *Aristolochia sipho, A. durior*, orn. fl. climbers, warm cli.

Dwarf bamboo *Arundinaria vagans:* — **fan palm**, *Chamaerops humilis:* — **gorse**, *Ulex gallii:* — **mallow**, *Malva neglecta:* — **palmetto**, *Sabal minor:* — **pansy**, *Viola kitaibeliana:* — **sedge**, *Carex humilis:* — **spurge**, *Euphorbia exigua.*

Dyers' chamomile *Anthemis tinctoria:* — **greenweed**, *Genista tinctoria*, fls. yield a yellow dye, when mixed with woad gives a fine green – Kendal green. See *Genista* in Willis. — **mulberry**, *Morinda tinctoria:* — **rocket**, *Reseda luteola:* — **weld**, *Reseda luteola*, yields a yellow dye.

Dyes, vegetable Veg. dyes have been used by mankind in different parts of the world as far back as history relates, in both the Old and the New Worlds. There are in fact hundreds of different pls. that are known to have been used for dyeing or which are still used by primitive peoples or in the more remote regions. It would seem that the people of bygone days, especially in the Middle Ages, were just as fond of gaily coloured clothes as are their modern counterparts.

In Eur. countries the indigenous dye pls. such as madder, woad, saffron, weld and the lichen dyes were of great importance in ancient and medieval days. In the 17th and 18th Century, however, they began to be displaced by imported dye woods such as logwood, 'Brazil' wood and fustic. In the 19th Century they received a further blow with the development of chemical or synthetic dyes with the result that, with one or two exceptions, veg. dyes became of little importance commercially. The new man-made dyes, largely coal tar derivatives, possessed many overwhelming advantages. They were cheaper, quicker to use, faster (less liable to lose colour) than the older pl. dyes. Other well known veg. dyes such as indigo, and also cochineal, were soon displaced by these chemicals. The use of a few natural dyes for special purposes, such as annatto for colouring cheese and butter, did, however, persist. Henna is still used, esp. in Eastern countries.

In the dyeing of cloth and clothing the use of veg. dyes has persisted to some extent as a cottage industry in the more remote areas, as in the Highlands of Scotland, where they are employed in dyeing woollens. There are those who argue that 'no chemical dye has that lustre, that underglow of rich colour, that delicious aromatic smell, that soft light and shadow that give so much pleasure to the eye' (Thurston).

An essential process in dyeing with veg. dyes, which are water soluble, is to render them insoluble, otherwise they would run. This is done by the use of mordants which are salts of various metals, esp. iron, aluminium, chromium and tin, w. which the dye forms an insoluble compound.

The following are among the better known veg. dyes – annatto, camwood, cutch, dyer's greenweed, fustic, gambier, gamboge, henna, indigo, lichens (various), logwood, madder, some mosses, quercitron, safflower, saffron, sappan wood, Turkey red, turmeric, weld, woad. [Thurston, V. (1936) *The use of vegetable dyes*, 51 pp.]

Dysentery herb (S.Af.) *Monsonia biflora* and other spp.

E

Eagle fern *Pteridium aquilinum:* — **wood** or **aloe wood**, *Aquilaria agallocha*, Ind., Burma, fragrant resinous wood (partly decomposed) used in oriental medicine and perfumery, also as incense.

Eardrop vine *Brunnichia cirrhosa*, S. U.S.

Earth almond *Cyperus esculentus*, tubers ed.: — **chestnut**, *Lathyrus tuberosus:* — **nut**, *Arachis hypogaea*, *Conopodium majus* r. ed. when roasted, *Carum* *bulbocastanum:* — **star**, *Cryptanthus* spp. orn. fol., room pls.: — **smoke**, *Fumaria officinalis*, Eur.

East African cardomom *Aframomum mala:* — — **sandalwood**, *Osyris tenuifolia*.

East Indian arrowroot *Curcuma angustifolia*, *Tacca leontopetaloides:* — — **hemp**, *Crotalaria juncea:* — — **rosewood**, *Dalbergia latifolia:* — — **screw tree**, *Helicteres isora:* — — **walnut**, *Albizia lebbeck*.

Easter cactus *Rhipsalidopsis rosea*, *Schlumbergera gaertneri*, cult.: — **flower**, *Securidaca diversifolia*, W.I.: — **lily**, *Hippeastrum puniceum*, *Lilium longifolium*.

Eau de créole (W.I.) *Mammea americana*.

Ebony Various woods with dark heartwood have been termed 'ebony', but the name is normally regarded as covering all spp. of *Diospyros* with predominantly black heartwood. Ebony which shows mottling is known as coromandel or calamander wood. The country of origin is commonly used as a prefix. The commercially imp. ebonies include – **Ceylon** or **E.I.** —, *Diospyros ebenum:* **Af.** —, *D. crassiflora*, *D. piscatoria* and other spp.: **Macassar** —, *D. celebica*, Celebes. Other ebonies or so-called ebonies include – **black** —, *Euclea psuedebenus:* **Gaboon** —, *Diospyros dendo*, *D. crassiflora*, *D. viridicans:* **Lagos** —, *D. mespiliformis:* **Maracaibo** —, *Caesalpinia granadillo:* **Mozambique** —, *Dalbergia melanoxylon:* **W.I.** —, *Brya ebenus* (coccus wood).

Ebony heart *Elaeocarpus bancroftii*, Aus. rain forest tree.

Ebor or **eboe tree** *Dipteryx oleifera*, S.Am., seeds resemble tonka beans.

Eccremocarpus *E. scaber*, cult. orn. climber in warm cli.

Echeveria Cult. orn. fls., ls. in rosettes, often colourful and covered in fine hairs, like plush.

Echinocactus Orn. cacti, often cult.

Echinocereus Cacti, often cult., spiny frs., may be ed.

Echinops *E. ritro* and other spp. cult. orn. pls. (globe thistles). *E. strigosus* once used for making tinder in Spain.

Echites Trop. Am. orn. climbing shrubs, cult.

Echium Many cult. orn. spp., ls. often scabrid or coarsely hairy.

Economic botany In the broad sense the study of pls. from the point of view of their usefulness to man. These pls. are often divided into main groups or categories such as pls. yielding – (1) *foods*, which include cereals, r. crops, pulses, frs. and nuts, vegs., foods and fodder for domestic animals etc.; (2) *fibres*, including pls. used for man-made fibres, fibre board and paper; (3) *timbers*; (4) *drugs* and med. products; (5) *rubber* or rubber-like products; (6) *oils*, fixed and ess.; (7) *gums* and *resins*; (8) *tannins*; (9) *dyes*; (10) *miscellaneous products* such as tobacco, insecticides, waxes, cork, veg. ivory, pls. used for ornament etc. [Uphof, J. C. T. (1959) *Dictionary of economic plants*, 591 pp.]

Eddoes (W.I.) *Colocasia esculenta* (= *C. antiquorum*) or dasheen, a common trop. r. crop.

Edelweiss *Leontopodium alpinum*, the well known pl. of the Alps. fl. heads covered w. white cottony down: **N.Z.** —, (N. Island), *Leucogenes leontopodium:* **N.Z.** — (S. Island), *Leucogenes grandiceps*.

Eel grass or **grass-wrack** *Zostera marina*, used as packing material, for stuffing mattresses, sound insulation etc.; called 'alva marina'.

Efwatakala grass *Melinis minutiflora,* trop. Af., strong smelling, alleged to repel insects, also called molasses grass.

Egg fruit Or brinjal, *Solanum melongena,* a common veg. in trop. and subtrop., also called gdn. egg or egg pl., many vars., frs. being white, black, purple, violet or yellow.

Eggs and bacon (Aus.) *Aotus, Dillwynia, Phyllota* and *Pultenaea:* (Br.), *Lotus corniculatus.*

Egyptian barley *Hordeum trifurcatum:* — **bean,** *Nelumbo nucifera:* — **clover,** *Trifolium alexandrinum:* — **grass,** *Dactyloctenium aegyptiacum,* fodder and grain, also a weed: — **lettuce,** *Lactuca scariola:* — **lotus,** *Nymphaea lotus:* — **paper reed,** *Cyperus payprus,* papyrus.

Ejow palm *See* gomuti.

Ekki *Lophira alata* (*= L. procera*), W.Af., a commercial timber.

Elaeagnus Cult. trees and shrubs, often scaly or spiny, orn. fol. and fragrant fls., frs. sometimes ed., used for jelly.

Elaeocarpus *E. ganitrus* and other spp., trop. As., hard tubercled seeds used as beads, for necklaces, rosaries etc.

Elaterium *See* cucumber, squirting.

Elbow orchid (Aus.) *Spiculaea* spp.

Elder or **elderberry** *Sambucus nigra,* Eur. & W.As., shrub or small tree, common, quick-growing, often a weed: fls. med., frs. used for wine, pith of young stems for section cutting (microscope) and optical work: wood of old trees v. hard and close grained, has been used as a substitute for box. **Am** —, *S. canadensis:* **blue** —, *S. glauca,* W. N.Am.: **box** —, *Acer negundo:* **dwarf** —, *Sambucus ebulus:* **poison** —, *Rhus vernix:* **red-berried** —, *Sambucus pubens, S. racemosa:* **stinking** —, *S. pubens:* **W.I.** —, *S. intermedia:* **white** — (S.Af.), *Nuxia floribunda.*

Elecampane *Inula helenium.* Eur. & As., r. med.

Elemi Name applied to a number of resinous products or oleo-resins, often aromatic and of soft consistency, that have appeared in commerce from time to time and derived from several different botanical sources (mostly Burseraceae). They have been used to some extent in varnishes, in printers' inks and med. (ointments). **Af.** —, *Canarium schweinfurthii:* **Am.** —, *Bursera gummifera:* **Braz.** —, *Protium heptaphyllum:* **Carana** —, *Protium carana:* **E.Af.** —, *Boswellia frereana:* **Manila** —, *Canarium luzonicum:* **Mex.** —, *Bursera jorullensis:* **W.I.** —, *Dacryodes hexandra:* **Yucatan** —, *Amyris plumieri.*

Eleocharis *E. tuberosa,* E.As., ed. tubers.

Elephant's apple *Feronia elephantum,* trop. As., fr. ed. and yields a gum: — **climber,** *Argyreia speciosa,* trop., cult., large ls. and fls.: — **ear,** *Begonia* spp., *Elaphoglossum* spp., ferns: — **food** (S.Af.), *Portulacaria afra:* — **foot,** *Dioscorea* (*= Testudinaria*) *elephantipes, Elephantopus scaber,* Am.: — **foot yam,** *Amorphophallus campanulatus:* — **grass** *Pennisetum purpureum,* cult., trop. fodder: — **tusks,** *Martynia:* — **wood** (S.Af.), *Bolusanthus speciosus.*

Elkhorn fern *Platycerium* spp.: **elk nut** (N.Am.), *Pyrularia pubera,* oily ed. nut.

Elm *Ulmus* spp. **Common** or **English elm,** *U. procera* (*= U. campestris*), C. & S.Eur. incl. Br., cult. for orn. and timber: wood has numerous uses being durable outdoors, weather boards, paving blocks, Windsor-chair seats, coffins etc. Other elms which may be commercially imp. for timber include the following – **Dutch** —, *U.* × *hollandica* var. *hollandica:* **Flemish** or **French** —, *U. carpinifolia:* **hickory** — (Am.), *U. thomasii:* **Jap.** —, *U. laciniata, U. davidiana* (= *U. japonica):* **mountain** — (Am.), *U. glabra:* **red** — (Am.), *U. procera:* **rock** —, (Am.), *U. thomasii:* **Scotch** —, *U. glabra:* **soft** — (Am.), *U. americana.*

Other so-called elms are – **Af. false** —, *Celtis integrifolia:* **rock** —, *Chlorophora excelsa:* **water** —, *Planera aquatica:* **W.I.** —, *Guazuma ulmifolia.*

Elsholtzia *E. stauntonii,* China, cult. orn. fl. shrub.

Embelia *E. ribes,* trop. As., seeds med. and used to adulterate black pepper.

Emblic myrobalan *Phyllanthus emblica,* trop. As., fr. ed., ls., bark and young frs. used in tanning and dyeing.

Embothrium *E. coccineum,* S.Am., orn., shrub, crimson fls., cult.

Embuia Or imbuya *Phoebe porosa,* Braz., a handsome furniture wood.

Emmer or **emmer wheat** *Triticum dicoccum,* cult. more in ancient times.

Emu apple *Owenia acidula,* Aus.: — **bush,** *Eremophila oppositifolia,* Aus.: **spotted** — **bush,** *E. maculata,* Aus., orange fls., cult.

Enamel orchid *Elythranthera emarginata,* W.Aus.

Encephalartos Cycads of trop. and S.Af., cult. orn. pls., meal prepared from pith by natives.

Endive *Cichorium endivia,* widely cult., ls. a much used salad.

Eng *Dipterocarpus tuberculatus,* Indoch., a commercial timber.

Engelhardtia *E. spicata,* Ind., wood used for building and carving.

Enkianthus *E. japonicus,* N. & E.As., and other spp., cult. orn. shrubs.

Ensete Or inset, *Ensete ventricosum* (= *Musa ensete),* Ethiopia, cult. for food and fibre, stem pulp, tuber and young shoots cooked and eaten.

Enzymes Living pl. cells contain substances called enzymes which enable them to carry out certain functions more readily. Their role may in fact be likened to that of a catalyst in chemical reactions. Enzymes form a large part of the protein in an active cell. They increase in amount as the cell grows. They are easily destroyed or inactivated, being very sensitive to temperature, acidity etc. The enzyme *pepsin* hydrolyses proteins from dead insects in pitcher pls. The enzyme *papain* obtained from the latex of the green papaw fr. (*Carica papaya*) is a commercial product. Enzymes have many industrial uses, ranging from the conversion of starch to fermentable sugars, through proofing of beer to bating of hides for leather. There are 100–150 well known or fairly well known enzymes, classified in the following main categories – (1) Hydrolases, (2) phosphorylases, (3) desmolases, (4) hydrases and (5) oxidoreductases.

Epacris Several cult. orn. fl. shrubs, Aus. & N.Z.

Ephedrine *Ephedra* spp., several spp. may yield the drug, e.g. *E. sinica, E. equisitina* and *E. distachya;* employed in nasal sprays to relieve symptoms of acute nasal catarrh, sinusitis, asthmatic attacks etc.

Epidendrum A laige genus of trop. orchids, many cult.

Epigaea Evergreen creeping shrubs, cult.

Epimedium Several spp. cult., esp. in rock gdns.

Epiphyllum Trop. Am. cacti: many cult., some as room pls., showy fls.

Episcia Trop. Am. perens., many w. showy fls. or fol., cult.

Eranthemum Trop., SE.As., cult. orn. fol. pls.

Eremurus Stately perens. w. tall scapes, W. & C. China, cult.

Eria Large genus of trop. As. epiphytic orchids, many cult.

Erigeron Aster-like pls. favoured for the fl. border or rock gdn.

Erimado *Ricinodendron heudelotii*, trop. Af., a trade timber.

Eriochloa Trop. and subtrop. fodder grasses.

Eriope *E. crassipes*, Braz., has explosive pollination mechanism, like gorse.

Eriopsis Trop. Am., cult. orn. fls.

Eriospermum S.Af. cult. orn. fls.

Eryngo *Eryngium campestre*, *E. creticum*, Medit., bluish-green thistle-like pls., rosettes of young ls. used as veg.

Erysimum Some spp. cult., showy border or rock gdn. pls.

Erysipelas plant *Heliotropium indicum*, Ind., heliotrope, trop.

Erythrina Several spp. cult. orn. trees or shrubs in warm cli.: colourful red fls. and black and red seeds (lucky beans): some grown for shade or support.

Escallonia *E. macrantha* and other spp., orn. fl. shrubs, hedges.

Eschscholtzia W. N.Am., cult. orn. fls., esp. *E. californica* (Calif. poppy), numerous cultivars in brilliant colours.

Esenbeckia B. of some Braz. spp. which may be used like angostura b.

Esparto grass *Stipa tenacissima*, W.Medit., used in high class paper manufacture and locally for baskets and other articles.

Espercet (Spanish) *Hedysarum coronarium*, fodder pl.

Essential oils *See* oils.

Ettercap, crested *Pogonia ophioglossoides*, N.Am.

Ethulia *E. conyzoides*, Old World trop., a weed, esp. in rice.

Etrog Or Jewish citron, *Citrus medica*, a small citron used by the Jews during the Feast of the Tabernacles.

Eucalypts *Eucalyptus* spp., the dominant feature of Aus. vegetation (about 500 spp.). They vary in size from dwarf shrubs, such as the mallees, to v. large forest trees, among the world's tallest. A notable feature is the small size of the seeds, some being exceedingly minute. Many eucalypts occur naturally in the hot and arid regions of Aus., under conditions where few other evergreen trees would grow. Such trees have proved valuable in afforestation or in ornamental planting in other parts of the world with similar climatic conditions and low rainfall. This has applied in all the other continents. Many eucalypts are quick-growing and are prized for their timber, which varies greatly with different spp.

Several spp. of *Eucalyptus* are exploited in Aus., or in other countries where they have been cult., for the oil yielded by steam distillation of ls. and twigs. The oil is used in perfumery and for med. purposes. With some eucalypts the fls. yield nectar freely and are the main basis of commercial honey production in Australia. The wood of wandoo (*Eucalyptus wandoo*) of W.Aus. was at one time a commercial source of tannin extract for leather tanning.

Eucalyptus bark The b. of some spp. of *Eucalyptus* is rich in tannin and was used in earlier days in Aus. for tanning leather, esp. mallet b. (*E. astringens*). It was also exported to Eur. This b. is one of the world's richest tan bs. with 40–50 % tannin. Mugga or red iron b. (*E. sideroxylon*) was also exploited. [Howes, F. N. (1953) *Vegetable tanning materials*, 325 pp.]

EUCALYPTUS OIL

Eucalyptus oil Obtained by steam distillation of the ls. and twigs of various spp. of *Eucalyptus*, esp. *E. globulus*, the blue gum, in Spain (from cult. trees). A lemon-scented oil is obtained from *E. citriodora*. Other spp. exploited include *E. dives* in Aus. Eucalyptus oil has various med. uses; it may be inhaled to relieve symptoms of bronchitis and sinusitis, used in pastilles to relieve colds or applied externally as a counter-irritant.

Eucharis S.Am. bulbous pls., cult., esp. *E. grandiflora*, large white scented fls., known as Amazon or Eucharist lily.

Eucommia *E. ulmoides*, a Chinese tree, sometimes cult., remarkable for producing caoutchouc or rubber in the ls. in a solid form (not latex): the strands of rubber may be seen on breaking a l.: has been considered for commercial rubber production (unsuccessfully).

Eucryphia Cult. orn shrubs, *E. glutinosa* the best known.

Eugenia Many spp. in this large trop. and subtrop. genus have ed. frs., such as the rose-apple and Braz. cherry.

Eulalia *Miscanthus sinensis*, E.As., an orn. grass.

Eulophiella Madag. orchids, cult., showy fls.

Euonymus Orn. trees and shrubs, cult., showy frs. and orn. autumn fol.

Eupatorium Many cult. orn. spp., mostly Am.

Euphorbia A large widespread genus, many spp. cult. for orn., some are weeds: tree or cactiform spp. yield latex freely, this contains rubber but much resin as an impurity: some spp. used as hedge or barrier pls.

Euphorbium Or gum euphorbium. The dried latex of *Euphorbia resinifera* (Morocco): has been used as a vesicant in veterinary practice.

Euryale *E. ferox*, a large aquatic, trop. As., young stalks, rs. and seeds eaten.

Evening primrose *Oenothera biennis:* — **star** (Am.), *Cooperia* spp. — **trumpet flower**, *Gelsemium sempervirens*.

Evergreen beech *Nothofagus:* — **laburnum**, *Piptanthus laburnifolius*, As.: — **oak**, *Quercus ilex*, S.Eur.

Everlasting Many fls. that do not suffer from loss of colour and shape on drying are known by this name. Several are commonly cult., seed being available from seedsmen. The fls. are cut when at their best and hung upside down to dry. They will last several years and are very popular w. the French who call them *immortelles*, and use them freely for wreaths. They are coloured vars. of *Helichrysum bracteatum*. One of the best everlastings to grow in Br. is considered to be *Helipterum manglesii; H. roseum* is also good. Among genera with everlastings are – *Achyrachaena, Acroclinium, Antennaria, Ammobium* esp. *A. alatum, Anaphalis* esp. *A. margaritacea, Gnaphalium, Helichrysum, Helipterum, Limonium, Rhodanthe, Syngonanthus, Waitzia, Xeranthemum.* **Cape** —, *Helichrysum vestitum:* **common** —, *Gnaphalium polycephalum:* **Crete** —, *Helichrysum vestitum:* — **grass** (Am.), *Eriochloa:* **mountain** —, *Antennaria dioica:* **pearly** —, *Anaphalis margaritacea:* — **pink** (S.Af.), *Phoenocoma prolifera:* **Swan River** —, *Rhodanthe.*

Exacum *E. affine*, cult. orn., violet-like fls., a pot pl.

Exile tree *Thevetia neriifolia* (= *T. peruviana*), cult., orn., trop.

Exochorda *E. racemosa* (= *E. grandiflora*), N.China and other spp., cult. orn. fl. shrubs.

Eye bane *Euphorbia maculata*, N.Am.: — **bright**, *Euphrasia officinalis*, Eur., ls. formerly used as an eye-wash. *See Euphrasia* in Willis: —**lash orchid** (C.Am.), *Epidendrum ciliare*, fls. white, lip much fringed.

F

Fabiana S.Am. cult. orn. shrubs.

Faham tea *Angraecum fragrans*, fragrant ls. used as tea.

Fair maids of France (or **Kent**) *Ranunculus aconitifolius*, an old gdn. favourite.

Fairy duster (N.Am.) *Calliandra eriophylla:* — **flax**, *Linum catharticum:* — **wand**, *Chamaelirium luteum*, N.Am., cult, orn. pl.

Falling stars (S.Af.) *Crocosmia aurea* var. *maculata*.

False acacia *Robinia pseudoacacia*, N.Am., cult. orn. and timber tree: — **aloe**, *Agave virginiana:* — **asphodel** (Am.), *Tofieldia:* — **beech drops** (Am.), *Monotropa* (= *Hypopitys multiflora:* — **boneset** (Am.), *Kuhnia:* — **box** (W.I.), *Schaefferia frutescens:* — **brome grass**, *Brachypodium:* — **buckthorn** (Am.), *Bumelia:* — **buckwheat**, *Polygonum:* — **buffalo grass**, *Munroa*, N.Am.: — **bugbane**, *Trautvetteria:* — **dandelion** (N.Am.), *Pyrrhopappus:* — **dittany**, *Dictamnus albus:* — **dragon head** (Am.), *Physostegia:* — **flax**, *Camelina:* — **foxglove** (Am.), *Gerardia:* — **garlic** (Am.), *Nothoscordum:* — **goat's beard**, *Astilbe:* — **gromwell** (Am.), *Onosmodium:* — **hellebore**, *Veratrum:* — **hemp**, *Datisca:* — **indigo**, *Amorpha*, *Baptisia:* — **jalap**, *Mirabilis jalapa:* — **jasmine**, *Gelsemium sempervirens:* — **lettuce**, *Cicerbita* (= *Mulgedium*): — **lily of the valley**, *Maianthemum:* — **loosestrife**, *Ludwigia:* — **mallow**, *Malvastrum*, *Sphaeralcea:* — **mermaid** (Am.), *Floerkea:* — **mistletoe** (Am.), *Phoradendron:* — **mitrewort**, *Tiarella:* — **nettle**, *Boehmeria:* — **oatgrass**, *Arrhenatherum:* — **pennyroyal**, *Isanthus:* — **pimpernel**, *Lindernia:* — **Solomon's seal**, *Smilacina:* — **spikenard**, *Smilacina:* — **spiraea**, *Sorbaria:* — **tarragon**, *Artemisia dracunculus*.

Fame flower (Am.) *Talinum:* **orange** — —, *Talinum aurantiacum*.

Fan flower *Scaevola ramosissima*, Aus.: — **fern**, *Sticherus* spp., Aus.: — **palm**, *Chamaerops*, *Corypha*, *Sabal*, *Thrinax* etc.: **dwarf** — **palm**, *Chamaerops humilis*, the only palm occurring naturally in Eur., fibre from l. bases used in upholstery (veg. horsehair): — **weed**, *Thlaspi arvense*, Eur.

Fancy or **florist's fern** *Dryopteris spinulosa*, cult., many vars.

Farkleberry (Am.) *Vaccinium arboreum*, fr. black, inedible.

Fat-hen *Chenopodium album*, seeds relished by birds and poultry: — **pork** (W.I.), *Chrysobalanus icaco*.

Fat, vegetable *See* oils.

Fatsia *F. japonica*, cult. orn. evergreen fol. pl., much favoured in Victorian and Edwardian eras.

Feaberry Old name for gooseberry (*Ribes grossularia*).

Feather bells (Am.) *Stenanthium gramineum:* — **climber**, *Acridocarpus natalitius:* — **fleece**, *Stenanthium gramineum:* — **flower** (Aus.), *Verticordia* spp.: **pink** — **flower**, *Verticordia picta:* — **fleece** (Am.), *Stenanthium* *gramineum:* — **foil** (Am.), *Hottonia inflata:* — **grass**, *Stipa* spp., *Leptochloa* spp.; in *S. pennata,* the awn is long and feather-like curling up in dry weather, uncurling when moist, and this helps drive the fr. into the soil: — **head** (S.Af.), *Phylica pubescens:* — **top**, *Calamovilfa.*

Feijoa *Feijoa* (= *Acca*) *sellowiana,* S.Am. small tree, cult. for ed. fr. in subtrop., much esteemed (pineapple guava).

Felicia *F. bergeriana,* S.Af., handsome fls. (kingfisher daisy).

Felwort *Gentianella campestris* and other spp., N.Am.

Fennel *Foeniculum vulgare,* a well known and much used spice or flavouring, esp. in Medit. and eastern countries: much used w. fish also for garnishing, and, after blanching, for salads: oil distilled from seeds used med. (carminative) and for certain liqueurs. **Bitter** —, *Foeniculum vulgare:* **dog** —, *Eupatorium capillifolium:* **Florence** — or **finocchio,** *Foeniculum dulce,* l. bases eaten: — **flower,** *Nigella sativa,* used for flavouring: **giant** —, *Ferula communis:* **hog's** — , *Peucedanum officinale:* **sweet** —, *Foeniculum vulgare:* **water** —, *Oenanthe aquatica* (poisonous).

Fenugreek *Trigonella foenum-graecum.* Medit. long cult. for fodder or food: seeds mucilaginous, popular where meatless diets are customary, used med. and as a tonic, also in veterinary practice (condition powders): a source of diosgenin (steroid drugs).

Ferns Ferns are, in many ways, an interesting group of pls. Like other large groups that are widely distributed they show a considerable range of diversity of form and size varying from large tree forms, over 15 m high, to small ferns only a few cms in height. A large number of ferns are shade and moisture loving pls. but on the other hand many are xerophytic or alpine and show the usual characteristics of such pls., such as reduced l. surface, thick cuticle, hairiness, incurving of ls. etc. and even succulence in one or two instances (e.g. *Drymoglossum carnosum* of N.Ind. and *Cyclophorus adnascens* of trop. As. and Polynesia). The tree ferns and many others have water-storage tissue in the stem. Many ferns are epiphytic, esp. in the trop., although some may be found growing this way in the cli. of Br.

Ferns occur in profusion in the trop., where some are climbers reaching to the tops of trees. They extend even to the Arctic and Antarctic regions. Some are extremely widely distributed, occurring in all or several different continents. Examples are the ubiquitous bracken, the royal fern (*Osmunda regalis*) and the maidenhair fern. A reason for this wide distribution may be the small size of fern spores and their extreme lightness, making it possible for them to be carried long distances by wind or even light air currents. Certain characteristics are linked with the ferns of certain regions. For instance those native to N.Am. are largely deciduous, whereas most of the ferns of Jap. are evergreen, the fronds having a glossy appearance and thick texture. It has been stated that the native Br. ferns have a tendency to produce crested or depauperate forms. A few exotic ferns have produced crested forms after having been cult. in Br. The tree ferns indigenous in the New World generally have thin stems or trunks whereas those in Aus. & N.Z. have comparatively thick trunks.

The ferns, like the orchids, which are a group of pls. of approximately similar size, have relatively few economic uses. The drug male fern, the dried rhizome and l. bases of *Dryopteris filix-mas*, has long been used as a valuable taenicide. Osmunda fibre is much used and held in high esteem in the pot cult. of orchids. In Br. supplies have been obtained largely from Jap. and the U.S. Ferns are extensively grown as orn. pls., often in ferneries and several are well suited for and popular as house or room pls. Many tree ferns are extremely graceful and pleasing to the eye, those of N.Z. and Aus. being notable in this respect. A craze for fern cultivation developed in Br. during Victorian times and wild ferns were ruthlessly dug up everywhere, often by dealers, threatening them with extinction in some instances. It is thought that the development of the Wardian case at about this time and modified as a 'fern case' for indoor cult., affording as it does constant humidity, may have sparked off the craze [Allen, D. E. (1969) *The victorian fern craze, a history of pteridomania*, 21 pp.]

Under cultivation some ferns are easy to grow and will tolerate the usually unfavourable conditions (excessive dryness, smoke or fumes of a living room), while other ferns are extremely difficult to grow and will tax the capabilities of the most skilful gardener. Different spp. are of course selected for indoor and outdoor ferneries. Some are esp. favourable for hanging baskets. Many trop. ferns have to be treated as stove pls., requiring constantly high temperature and humidity.

Beech fern, *Dryopteris:* **birds' nest** —, *Asplenium nidus:* **bladder** —, *Cystopteris:* **bristle** —, *Trichomanes:* **cinnamon** —, *Osmunda cinnamomea:* **eagle** —, *Pteridium aquilinum* (bracken): **elkhorn** —, *Platycerium:* **filmy** —, *Hymenophyllum,* etc.: — **grass,** *Catapodium:* **hard** —, *Lomaria:* **hart's tongue** —, *Phyllitis:* **holly** —, *Polystichum:* **lady** —, *Athyrium:* **leaf tree,** *Filicium decipiens,* trop. As., fern-like fol.: **maidenhair** —, *Adiantum:* **male shield** —, *Aspidium:* **marsh** —, *Acrostichum:* **northern** —, *Blechnum:* **oak** —, *Dryopteris:* **ostrich** —, *Onoclea:* **parsley** —, *Cryptogramma:* **prickly shield** —, *Polystichum:* **royal** —, *Osmunda:* **shield** —, *Dryopteris:* **staghorn** —, *Platycerium:* — **tree,** *Filicium decipiens, Jacaranda:* **walking** —, *Camptosorus.*

Ferocactus Usu. large Am. cacti w. cylindrical stems and large spines, some cult.

Ferraria Cult. orn. bulbous pls. w. spotted fls., Af.

Fescue *Festuca, Vulpia.* The genus *Festuca* contains several valuable pasture and lawn grasses. **Bearded** —, *Vulpia ambigua:* **Chewing's** —, *Festuca rubra,* lawns: **green leaf** — (N.Am.), *F. viridula:* **meadow** —, *F. pratensis,* pasture and hay: **rat's tail** —, *F. myuros:* **red** or **creeping** —, *F. rubra,* pasture and lawns: **rigid** —, *Catapodium rigidum:* **rough** —, *Festuca scabrella:* **sheep's** —, *F. ovina,* pasture and lawns: **squirrel tail** —, *Vulpia bromoides:* **tall** —, *Festuca arundinacea.*

Feterita (Sudan) *Sorghum caudatum* var. *feterita,* imp. food grain.

Fetter bush (Am.) *Leucothoe catesbaei,* spreading evergreen shrub.

Fever bark *Croton* sp., *Alstonia constricta:* — **bush,** *Lindera,* N.Am. *Garrya elliptica,* N.Am., *Ilex verticillata,* N.Am.: — **few,** *Chrysanthemum parthenium,* cult. orn., med.: — **leaf,** *Ocimum viride,* W.Af.: — **plant,** *Ocimum viride,* W.Af.: — **tree,** *Acacia xanthophloea,* S.Af., *Anthocleista zambesiaca,* S.Af.; *Pinckneya pubens,* N.Am., *Eucalyptus globulus,* Aus., *Zanthoxylum capense,* S.Af.: — **wort,** *Triosteum perfoliatum,* N.Am.

FIBRES

Fibres, vegetable The fibres from some 2000 spp. of pls. have been recorded as used by man in different parts of the world but only a fraction of these (30 to 40) could rank as commercial fibres. Over the years the trade in veg. fibres has seen many changes, particularly w. the advent of artificial or manmade fibres. The following are numbered among the commercial veg. fibres – abaca or Manila hemp, aramina (*Urena lobata*), broom root, cabuya (*Furcraea*), cantala, caroa, China jute (*Abutilon*), chingma, coir, cotton, crin végétal, esparto, fique (*Furcraea*), flax, guaxima, hemp, henequin, ixtli (Jaumave and Tula), jute, kapok, kenaf or Deccan hemp (*Hibiscus cannabinus*), kittool, Mauritius hemp, N.Z. flax or phormium, palmetto, palma fibre, piassava, pineapple, raffia, ramie, roselle, sisal, Spanish moss, sunn hemp. [Kirby, R. H. (1963) *Vegetable fibres*, 464 pp.]

Ficus A large genus, mainly of trees and shrubs, trop. and subtrop.: many cult. as shade or street trees and as room pls.: some yield bark cloth or have a rubbery latex once exploited: frs. of some are ed. (*see* fig) much eaten by birds, bats and monkeys.

Fiddle dock *Rumex pulcher:* — **heads**, *Osmunda cinnamomea*, N.Am.: **fiddle neck** (Am.), *Amsinckia intermedia*, a common desert pl.

Fiddle wood *Citharexylum quadrangulare*, trop. Am., *Petitia domingensis*, W.I., good furniture wood, name a corruption of *bois-fidèle:* **black** — —, *Vitex divaricata*, W.I., an excellent, much used timber: **five leaf** — —, *Vitex capitata:* **white** — —, *Citharexylum fruticosum*, *C. spinosum*.

Field botanists, notes for *See* collecting.

Field madder *Sherardia arvensis*, common in S.England.

Fig *Ficus carica*, the common fig of age-old cult. in Medit. countries, widely grown elsewhere for fresh and dried fr.: a large trade exists in dried figs: an extract of them is med., a demulcent and mild laxative ('syrup of figs'); numerous cultivars. **Blue** — (Aus.), *Elaeocarpus grandis*, seeds used as beads: **bush** —, *Ficus capensis*, S.Af.: **Hottentot** —, *Carpobrotus* (= *Mesembryanthemum*) *edulis:* **Indian** —, *Opuntia* spp.: **India rubber** —, *Ficus elastica*, a popular room pl.: **mulberry** —, *F. sycomorus:* **sour** —, *Carpobrotus edulis:* **sycamore** —, *F. sycomorus*, frs. eaten in Egypt.

The frs. of numerous other wild figs are eaten, or are the food of birds, bats and monkeys. Some figs are popular room pls. such as *F. lyrata* (**fiddle leaf** —) and *F. pumila* (**climbing** —).

Figwort *Scrophularia nodosa* and other spp., fls. secrete nectar freely, much visited by bees. **Cape** —, *Phygelia capensis*, cult. orn. pl.: **water** —, *Scrophularia auriculata* (= *S. aquatica*).

Filbert *See* hazelnut.

Filmy ferns *Hymenophyllum* spp., *Trichomanes* spp. mainly trop. moisture loving ferns characterized by their thin, semi-transparent fronds or ls. only one cell thick and without stomata. In Br. the Killarney fern (*Trichomanes radicans*) and Tunbridge filmy fern (*Hymenophyllum tunbridgense*) occur.

Fimbristylis *F. globulosa*, a sedge cult. in As. for mats and baskets.

Finger Aralia *Dizygotheca elegantissima*, New Caledonia, orn. ls., cult., a room pl.: — **leaf morning glory**, *Ipomoea digitata:* — **grass**, *Digitaria* spp.: **Abyssinian** — —, *D. abyssinica*, used in erosion control: **hairy** — —, *D. sanguinalis:* **Richmond** — —, *D. diversinervis*, a good grazing grass: **W.Af.**

— —, *D. debilis*, *D. horizontalis:* **woolly** — —, *D. pentzii*, S.Af., good pasture grass: — **root**, *Uvaria chamae*, W.Af. carpels in finger-like clusters, r. med.

Fingrigo (W.I.) *Pisonia aculeata*, trop., used as a live fence.

Finocchio *See* fennel.

Fiorin *Agrostis stolonifera*, a widespread grass, often a weed.

Fique *Furcraea* spp. Some supply commercial fibre in trop. Am., e.g. *F. gigantea* (Mauritius hemp) and *F. macrophylla*.

Fir *Abies* spp., the name fir, usually with a prefix, is used for numerous spp. of *Abies* and sometimes for other conifers. Among the spp. that yield timber of commercial importance are the following – **Alpine** —, *Abies lasiocarpa*, W. N.Am.: **amabilis** —, *A. amabilis*, W. N.Am.: **balsam** —, *A. balsamea*, W. N.Am.: **Douglas** — (*see* separate heading): **grand** —, *A. grandis*, W. N.Am.: **noble** —, *A. procera* (=*A. nobilis*), W. N.Am.: **silver** — or **whitewood**, *A. alba* (=*A. pectinata*), Eur., cult. in Br., the tallest Eur. tree, much used for building, constructional work and telegraph poles.

Other well known firs include – **Calif. red** —, *A. magnifica:* **Caucasian** —, *A. nordmanniana:* **Greek** —, *A. cephalonica*, used for ship building in ancient Greece: **Himal.** —, *A. pindrow:* **Jap.** —, *A. firma:* **lowland** —, *A. grandis*, N.Am.: **Norway** —, *Pinus sylvestris:* **white** or **giant** —, *Abies grandis*, tree reaches about 100 m *See also Abies* in Willis.

Fire ball *Haemanthus multiflorus*, *H. cinnabarinus*, E.Af., brilliant scarlet fls.: — **bush**, *Pyracantha coccinea*, *Haemelia patens*, N.Am.; *Keraudrenia*, Aus.: **Chilean** — **bush**, *Embothrium coccineum:* —**cracker plant**, *Russelia juncea*, Mex., much cult.: — **pink**, *Silene virginica:* —**sticks**, *Philippia mafiensis*, E.Af.; *Maytenus senegalensis*, W.Af.; *Asclepias fruticosa; Brachylaena elliptica*, S.Af.: —**thorn**, *Pyracantha coccinea*, Medit. orn. shrub and hedge, cult.: — **tree**, *Nuytsia floribunda*, Aus., brilliant orange yellow fls.: — **pink**, *Silene virginica:* —**weed**, *Epilobium* spp., *Erechtites* spp., N.Am. —**wheel tree**, *Stenocarpus sinuatus*, Aus.

First of May *Saxifraga granulata*, Eur., cult. orn. fl., many vars.

Fish berries *Anamirta cocculus*, frs. once used to poison fish: — **bone fern**, *Nephrolepis cordifolia*, Aus.: — **fuddle tree** (W.I.), *Piscidia erythrina:* — **poison bean**, *Tephrosia vogelii*, trop. Af.

Fish-poison plants In many parts of the world primitive peoples have, from time immemorial, made use of pls. for poisoning or stupefying fish to facilitate their capture. Because of the wholesale destruction of fish of all sizes that is involved the practice has been discouraged or made illegal in many countries and has fallen into disuse. The poisoned or intoxicated fish float to the surface. Used as human food they seem to be in no way harmful. Some of the pls. used have proved to be of value in other ways, e.g. as insecticides, such as *Derris* from the Malayan region and *Lonchocarpus* from S.Am. Derris root (*Derris elliptica* and other spp.) is perhaps the best known of the Asiatic fish poisons, especially in Malaysia where it is known as *tuba*. The fr. of *Anamirta cocculus* is another well known fish poison in tropical As. In some As. countries the frs. of certain spp. of *Diospyros* and a number of Euphorbiaceous pls. are also used.

In Af. certain spp. of *Tephrosia* are among the widely used fish poisons, esp. *T. vogelii* in trop. Af. and *T. macropoda* in southern Af. The latter was commonly used by the Zulus at one time, the fleshy, irregularly shaped rs. being simply mashed between stones at the side of the pool it was decided to treat. Very soon fish would float to the surface. *Mundulea suberosa* is another leguminous pl. that has been much used in parts of Af. for fish poisoning, as have some spp. of *Adenium*.

A large number of fish-poison pls. have been recorded from S.Am., some having been long cult. by the inhabitants for the purpose. Two of the best known and most potent are the forest lianas black and white 'haiari' of Guyana (*Lonchocarpus* spp.) and 'cube' of Peru (*Lonchocarpus nicou*). *Clibadium sylvestre* ('conami' of Guyana) is also widely used as are several sapindaceous pls.

In Aus. the aborigines use pls. for poisoning and catching fish. Often the pls. used are rich in tannin or saponin and it is these substances rather than alkaloidal poisons which affect the fish. A few pls. have been used in Eur., especially S.Eur., for poisoning fish, notably spp. of *Verbascum* (frs. and seeds). The rs. of *Oenanthe crocata* are also said to have been used.

Fish tanks, plants for *See* aquarium plants.

Fitsroot *Astragalus glycyphyllos*, Eur., As., inc. Br.

Fittonia Cult. orn. fol. pls. from S.Am., some w. beautifully marked ls., house pls.

Fitweed *Eryngium foetidum*, W.I.

Fitzroya *F. cupressoides* (= *F. patagonica*), Chile, large evergreen conifer with straight-grained, reddish-brown wood, favoured for shingles.

Five corners (Aus.) *Styphelia adscendens, S. tubiflora:* — **fingers** (Am.), *Potentilla reptans: Syngonium auritum:* — **minute grass** (Aus.), *Tripogon loliiformis.*

Fixed oils *See* oils.

Flag, blue *Iris versicolor*, N.Am.: **cat-tail** —, *I. versicolor*, N.Am.: — **lily, purple,** *Acorus calamus:* **poison** —, *I. versicolor:* — **root,** *Acorus calamus:* **sweet** —, *Acorus calamus:* **yellow** —, *Iris pseudacorus:* **water** —, *Iris pseudacorus.*

Flakes Horticultural term used for a group of carnations w. pure ground colour marked w. splashes of another colour.

Flamboyant *Delonix* (= *Poinciana*) *regia*, Madag., a widely cult. orn. fl. tree, trop. and subtrop., dense red fls., large woody pods, a much used street tree but vigorous rs. inclined to lift or dislodge paving. **Yellow** —, *Peltophorum* (= *Poinciana*) *roxburgii*, Ind., cult. orn. trop. tree w. yellow fls.

Flame flower *Ixora coccinea*, cult. trop., many vars.; *Kniphofia aloides:* — **of the forest** (W.Af.), *Mussaenda erythrophylla*, (Ind.) *Butea monosperma* (= *B. frondosa*): — **tree** (Aus.), *Brachychiton acerifolium*; (W.Aus.), *Nuytsia floribunda;* (trop.Af.), *Spathodea campanulata* (= *S. nilotica*); (W.I.), *Delonix regia:* — **vine,** *Bignonia* (= *Pyrostegia*) *ignea*, Braz., cult.: **Mex.** — —, *Senecio* (= *Pseudogynoxys*) *confusus.*

Flames (S.Af.) *Homoglossum merianella*, bright red tubular fls.

Flamingo flower *Anthurium andreanum*, Colombia; *A. scherzerianum*, Costa Rica; red or scarlet spathe, cult. trop. many vars, some used as room pls.

Flannel flower *Actinotus helianthi*, Aus., a densely woolly pl.: — **plant,** *Verbascum thapsus:* — **weed,** *Chrysopsis longii*, N.Am.

Flat crown *Albizia* spp., trop. and S.Af., some w. hard heavy wood: — **pea** (Aus.), *Platylobium obtusangulum*, of prostrate growth.

Flavouring, plants used for All over the world pls. of many kinds are used for flavouring or improving the taste of food or beverages. Some are quite local in their use. Many of the best known have been used in western countries from v. early times. A large proportion of them originated in countries bordering on the Medit., many being freely used in classical times. Examples are bay leaves, capers, coriander, cumin, dill, fennel, fenugreek, mustard seed, rosemary, saffron and sage. Horse radish and caraway may have originated in more northerly parts of Eur.

The following are other well known flavouring pls. or herbs – angelica, alexanders, almond, anise, balm, basil (bush and sweet), borage, burnet, chervil, chives, cicely, garlic, horehound, hyssop, lovage, marjoram, mint, onion, parsley, peppermint, poppy seed, purslane, rosemary, rue, sage, samphire, savory, sesame seed, shallot, tansy, tarragon, thyme and vanilla. Some esst. oils, such as those of lemon, cloves, nutmeg and cinnamon are freely used in flavouring in domestic cookery and by the food manufacturer. *See also* essential oils and spices. There are some flavouring pls. whose uses are mainly restricted to certain specific purposes such as hops for flavouring beer, juniper berries for gin, borage for claret-cup, and some tobacco flavourings.

Flax *Linum usitatissimum*. The flax pl., widely cult., yields both fibre from the stem, the source of linen, and also linseed oil and press-cake (animal food) from the seed. Flax is one of the oldest textile fibres used by man for clothes. It was well known to the Ancient Egyptians. Mummies were wrapped in linen cloth. **Blue** — S.Eur., *Linum narbonense;* (S.Af.), *Heliophila longifolia:* **bush** — (N.Z.), *Astelia nervosa:* **fairy** —, *L. catharticum:* **false** —, *Camelina sativa:* **flowering** —, *L. grandiflorum:* **N.Z.** —, *Phormium tenax*, cult. for orn. and commercial fibre: **pink** —, *Linum pubescens:* **spurge** —, *Daphne gnidium:* **Rocky Mountain** —, *Linum perenne* var. *lewisii:* — **weed**, *Linaria vulgaris*.

Fleabane *Pulicaria, Erigeron, Pluchea*. **Common** —, *Pulicaria dysenterica:* **Alpine** —, *Erigeron borealis:* **blue** —, *E. acer:* **daisy** —, *E. annuus*, a common weed: **marsh** —, *Pluchea sericea:* **salt marsh** —, *P. purpurascens*, N.Am.: **stinking** —, *P. foetida*, N.Am.: **W.I.** —, *Vernonia arborescens*.

Flea seed *Plantago psyllium. See* psyllium.

Fleawort *Senecio integrifolius:* **marsh** —, *S. palustris*.

Fleur-de-lis *Iris florentina* (source of orris root) is believed to have been the model for the fleur-de-lis of heraldry.

Flint-bark *Diospyros sanza-minika*, trop. Af.

Flixweed *Descurainia* (= *Sisymbrium*) *sophia*, waste places.

Floating bur-reed *Sparganium angustifolium*, Eur.: — **foxtail grass,** *Alopecurus geniculatus:* — **heart,** *Nymphoides*, N.Am.: — **manna grass,** *Glyceria borealis:* — **marshwort,** *Apium inundatum:* — **mud-rush,** *Scirpus fluitans:* — **pondweed,** *Potamogeton natans:* — **rice,** *Oryza glaberrima*, W.Af., seeds eaten: — **sweet grass,** *Glyceria fluitans:* — **water plantain,** *Luronium* (= *Alisma*) *natans*, Eur.

Flopper *Kalanchoe daigremontiana* (= *Bryophyllum daigremontianum*) cult. propagated by ls.

Floral emblems *See* national flowers.

FLORENCE WHISK

Florence whisk *See* broom corn.

Flores de palo *See* wood flowers.

Florida arrowroot *Zamia integrifolia:* — **iris,** *Iris savannarum:* — **flame azalea,** *Rhododendron austrinum:* — **moss,** *Tillandsia usneoides:* — **tree orchid,** *Ionopsis utricularioides:* — **velvet bean,** *Mucuna deeringiana:* — **woodbox,** *Schaefferia frutescens.*

Florist's fern *Dryopteris spinulosa,* N.Hemisph., many vars.

Floss flower *Ageratum houstonianum,* Mex., cult., gdn. ann.

Flower fence *Caesalpinia pulcherrima* orn. fl. shrub, warm cli: — **of a day,** *Hemerocallis* spp., *Tradescantia virginiana:* — **of Jove,** *Lychnis flos-jovis,* cult. orn. fl.

Flowering apricot *Prunus mume, P. armeniaca* and hybrids: — **ash,** *Fraxinus ornus:* — **box,** *Vaccinium vitis-idaea:* — **cherry,** *Prunus serrulata* and hybrids: — **currant,** *Ribes sanguineum:* — **cypress,** *Tamarix* spp.: — **dogwood,** *Cornus florida:* — **grass** (S.Af.), *Dierama* spp.: — **quince,** *Chaenomeles* spp.: — **rush,** *Butomus umbellatus:* — **spurge,** *Euphorbia corollata:* — **straw** (Am.), *Lygodesmia aphylla.*

Flowers as food Some fls. are eaten, usu. cooked in some way, esp. in the trop. The fleshy calyx of roselle (*Hibiscus sabdariffa*) and its uses are well known. The large fls. of *Sesbania grandiflora* are much relished in the East when boiled or fried. *Yucca* fls. are eaten in C.Am. after frying. Fls. of many of the Cucurbitaceous vegs., squash, pumpkin etc. may be eaten. Other fls. consumed as human food include – *Abutilon esculentum,* Braz.: *Madhuca* spp., Ind.: *Bombax,* Burma: *Ipomoea,* Sri Lanka: *Lilium,* China: *Rivea,* Sri Lanka. The cauliflower, broccoli, globe artichoke, capers and cloves afford instances where the fl. bud, not the open flower, is consumed.

Fluellen *Kickxia spuria* and *K.* (= *Linaria*) *elatine,* Eur. inc. Br.: an old name for common speedwell (*Veronica officinalis*).

Fly away grass *Eragrostis scabra,* N.Am.: — **catcher,** *Bejaria racemosa,* S.Am.: — **catcher plant** (Aus.), *Cephalotus follicularis,* ls. form pitchers: — **flowers,** many fls. often evil smelling, attract flies, such as in *Araceae* (*Arum, Amorphophallus*), *Asarum, Cobaea, Compositae, Crassulaceae, Cynanchum, Hedera, Helicodiceros, Paris, Stapelia, Umbelliferae, Veronica* and many more: **fly honeysuckle,** *Lonicera xylosteum:* — **orchid,** *Ophrys insectifera:* — **poison,** *Amianthium muscitoxicum,* N.Am., bulb poisonous: **shoo — —,** *Nicandra physalodes:* — **trap** (Am.), *Apocynum androsaemifolium:* **Venus' — trap,** *Dionaea muscipula:* —**ing duck orchid,** *Caleana major, C. nigrita.*

Foam flower *Tiarella cordifolia,* gdn. pl., spreads rapidly from stolons.

Fodder *See* grass.

Foetid horehound *Ballota nigra.*

Fog fruit *Lippia lanceolata,* N.Am.

Foliage plants Pls. cult. for their orn. fol.: often these are vars. or forms w. variegated fol., particularly in regard to house or room pls. (cf.) and pls. used for bedding such as orn. grasses. Ferns are popular fol. pls. in both warm and temp. countries. Palms are widely grown as fol. pls., both as pot pls. and mature pls. in trop. Many fol. pls, belong to the following genera – *Bambusa* and other bamboos, *Begonia, Caladium, Codiaeum* (crotons), *Coleus,* many conifers, *Cordyline, Cortaderia, Dracaena, Eryngium, Eucalyptus, Ficus,*

Fittonia, Grevillea, Gunnera, Gesneria, Monstera, Panax, Phormium, Ravenala, Rheum, Rhus, Smilax and many trees.

Food of the Gods *Ferula* spp., the dried exudation from tapped rs. is a Persian condiment.

Fool's parsley *Aethusa cynapium*, poisonous, a weed.

Forbidden fruit (W.I.) Pomelo or shaddock.

Forget-me-not *Myosotis* spp. wild and cult. gdn. fls. **Am.** —, *Eritrichium* spp.: **Cape** —, *Anchusa capensis:* **Chatham Island** —, *Myosotidium nobile*, a strange and bizarre fl.: **Chinese** —, *Cynoglossum amabile:* **common** —, *Myosotis arvensis:* **early** —, *M. ramosissima:* **garden** —, *M. sylvatica:* **marsh** —, *M. secunda:* **water** —, *M. scorpioides, M. caespitosa.*

Formio S.Am. name for N.Z. flax, *Phormium tenax.*

Forsythia Eur. and As. cult. orn. early fl. shrubs.

Fothergilla N.Am. shrubs, cult., bottle brush type of fl., no petals.

Fountain plant *Amaranthus salicifolius; Russelia juncea*, cult. trop.: **— tree,** *Spathodea campanulata*, water secreted by the fls.

Four corners (S.Af.) *Grewia occidentalis*, a common shrub.

Four-o'clock plant *Mirabilis jalapa*, fls. open late afternoon.

Fowl blue grass (Am.) *Poa palustris:* **— meadow grass** (Am.), *Glyceria striata*, Poa.

Fox-bane *Aconitum vulparia*, Alps.: **— berry**, *Vaccinium vitis-idaea.*

Fox-glove *Digitalis purpurea, see* digitalis: **Canary — —**, *Isoplexis canariensis:* **false — —**, *Gerardia virginica:* **Mex. — —**, *Tetranema mexicana:* **mountain — —**, *Ourisia macrocarpa*, N.Z.: **— wild** — (S.Af.), *Ceratotheca triloba:* **— grape** (Am.), *Vitis labrusca:* **— nut**, *Euryale ferox*, cult. aquatic, ed.: **— tail grass**, *Alopecurus pratensis*, cult., pasture; *Setaria glauca*, a weed: **bristly — —**, *Setaria barbata*, W.Af.: **giant — —**, *Setaria magna*, N.Am.: **green — —**, *Setaria viridis*, N.Am.: **— — lily**, *Eremurus:* **marsh — —**, *Alopecurus geniculatus*, N.Am.: **— — millet**, *Setaria italica*, widely cult.: **— — pine**, *Pinus aristata.*

Fragile fern *Cystopteris fragilis*, N.Am.

Fragrant cliff fern *Dryopteris fragrans*, N.Am.: *Microsorium scandens.*

Francoa *F. ramosa*, Chile, cult. orn. fl. (wedding fl.).

Frangipani *Plumeria rubra*, trop. Am. (temple tree), orn. fl. tree or shrub, widely cult. trop. and subtrop., fls. scented, many colour forms: **white** —, *P. alba:* **Queensland** —, *Hymenosporum flavum.*

Frankincense The use of this well known aromatic oleo-gum-resin, also called 'gum olibanum' and often referred to in the Scriptures along with myrrh, goes back to very early times. It is still used today in incense. The Ancient Egyptians made use of it in embalming. It is obtained from *Boswellia carteri, B. frereana* and possibly other spp. of *Boswellia* in NE.Af. and Arabia. **— fern**, *Mohria caffrorum*, E. and S.Af. cult. fern w. scented ls.: **— pine**, *Pinus taeda*, N.Am.

Fraxinus *See* ash.

Freesia Cult. orn. cormous pls., S.Af., fls. highly scented, yellow, pink or purple, improved strains have larger fls. and longer stems.

Freijo *Cordia goeldiana*, Braz., a commercial timber.

Fremontia Cult. orn. shrubs, fls. w. coloured calyx and no petals.

French bean *Phaseolus vulgaris*, a widely cult. veg.: — **berries,** *see* yellow, berries: — **cotton,** *Calotropis procera:* — **honeysuckle,** *Hedysarum coronarium,* red fls. cult.: — **jujube,** *Zizyphus jujuba:* — **lavender,** *Lavandula stoechas:*— **marigold,** *Tagetes patula:* — **mulberry,** *Callicarpa americana:* — **plum,** =prune: — **rose,** *Rosa gallica:* — **rye grass,** *Arrhenatherum avenaceum:* — **tamarisk,** *Tamarix gallica:* — **turpentine,** *Pinus pinaster:* — **weed,** *Thlaspi arvensis, Commelina:* — **whisk,** *Chrysopogon gryllus,* a brush fibre: — **willow** (W.I.), *Thevetia peruviana.*

Freycinetia Old World woody pls. allied to *Pandanus,* l. fibre used.

Friar's balsam Med. tincture largely composed of gum benzoin: — **cowl,** *Arisaema vulgare,* Medit., has a hooded spathe.

Friendship plant *See* aluminium plant.

Frijoles *Phaseolus vulgaris,* French or dwarf bean, widely cult.

Fringe lily *Thysanotus tuberosus* and other spp., Aus.: — **myrtle,** *Myrtus* spp.: **small-leaved** — —, *Myrtus microphyllus,* Aus.: — **tree,** *Chionanthus virginicus,* N.Am., cult. shrub: **Chinese** — —, *C. retusa.*

Fritillary *Fritillaria* spp., spring fl. bulbous pls.: **snake's head** —, *F. meleagris.*

Fro-fro (W.I.) *Hernandia sonora,* tall forest tree w. soft wood.

Frog-bit *Hydrocharis morsus-ranae,* a floating aquatic: **Am.** — —, *Limnobium spongia,* often used in aquaria: — **orchis,** *Coeloglossum viride:* — **plant,** *Sedum telephium* (= *S. purpureum*).

Frost flower *Aster* spp., N.Am.: — **grape,** *Vitis vulpina, V. cordifolia,* N.Am.: — **weed,** *Helianthemum canadense.*

Fruits, edible or **dessert** Throughout the world frs. of many kinds are imp. in the human diet. Some, such as the banana, are extensively cult. and shipped in large quantities to other lands. Others are dried or form the basis of large canning industries, such as the peach and the pineapple, the two most extensively canned frs. in terms of tonnage. Some frs., such as the mangosteen and durian may be grown almost entirely for local use. The development of air transport has resulted in various trop. frs., such as the mango and avocado, appearing regularly on the markets of temp. countries.

The following are numbered among the more imp. or better known frs. Some, such as the guava, various citrus frs., the pineapple and the banana are cult. in both trop. and subtrop. Several temp. frs. may be grown successfully at the higher altitudes in trop. *Temp. or subtrop. frs.* – apple, apricot, avocado, banana, blackberry, blueberry, Boysenberry, Cape gooseberry, cherimoyer, cherry, Chinese gooseberry, citron, cranberry, damson, dewberry, feijoa, fig, gooseberry, granadilla, grape, greengage, guava, kumquat, lemon, litchi, loganberry, loquat, medlar, melon, mirabelle (cherry plum), mountain papaw, mulberry, Natal plum, nectarine, olive, orange, papaw, peach, pear, persimmon, pineapple, plum, pomegranate, prune, quince, raspberry, raison, Seville orange, strawberry, sultana, tree tomato, watermelon, wineberry. *Trop. frs.,* akee, avocado, bacury, banana, bilimbi, bread-fruit, bullock's heart, capulin, citrange, coconut, cocoplum, custard apple, date, durian, genipa, giant granadilla, grapefruit, guava, hog plum, ilama, imbu, jaboticaba, jak, jambolan, jambi, jujube, kumquat, langsat, lemon, lime, limequat, litchi, mammey apple, mandarin, mango, mangosteen, monstera or ceriman, Natal plum, orange, ortanique, Otaheite apple, Otaheite gooseberry, passion fr., papaw, pineapple, pulasan, rambutan, red mombin, rose apple, sapodilla, sapote, Seville orange, soursop, star apple, Surinam cherry, sweet granadilla, sweet lemon, sweet lime, sweet-sop, tangelo, tangerine, tuna or prickly pear, wampee, white sapote, yellow mombin, zalacca.

Fruit salad tree *Feijoa sellowiana*, Braz.

Fuchsia Small trees, shrubs or herbaceous pls., C. & S.Am., N.Z., many cult., numerous cultivars. **Calif.** —, *Zauschneria californica*, cult., several vars.; *Ribes speciosum:* **Jam.** —, *Lisianthus longifolius*, cult. shrub: **native** — (Aus.), *Correa reflexa*, *Epacris longifolia*, *Grevillea wilsonii*, *Eremophila maculata:* **tree** —, (N.Z.), *Fuchsia arborescens*, *F. excorticata:* — **tree** (S.Af.), *Halleria lucida*, *Schotia brachypetala*.

Fuel value of woods The fuel value of different woods varies considerably. In general hard, heavy, close-grained woods are superior to soft, light or coarse-textured woods. They burn with greater heat and last longer on a fire. Among the common trees of Br. there are several that yield first class firewood, which gives out much heat, such as oak, beech, yew, hornbeam, hawthorn and the much grown Am. *Robinia* or False Acacia. Many others, although not in the front rank, are very satisfactory as fuel. These include birch, elm, pine, maple, sycamore, plane and sweet chestnut. A few woods are regarded as poor or of little value for fuel, because of dull burning, production of little heat or a tendency to throw out sparks. Among these are numbered lime, alder, larch and some poplars and willows. W. most woods well-seasoned heartwood is superior to sapwood as fuel. Decayed or partly decayed wood is usu. inferior.

Fufu (Ghana) Name for any mashed or pounded starchy food, e.g. yam, cocoyam, plantain etc.

Fuller's herb *Saponaria officinalis* or soapwort. — **teasel**, *Dipsacus fullonum*, the hooked bracts used by wollen cloth manufacturers to give a 'nap' to cloth.

Fumitory *Fumaria* spp.: **common** —, *F. officinalis:* **climbing** —, *Adlumia fungosa*, N.Am., biennial climber: **few-flowered** —, *F. media:* **ramping** or **white** — (Medit.), *F. capreolata*.

Fundi (W.Af.) *Digitaria exilis*, seeds eaten, a millet.

Funereal cypress *Cupressus funebris*, China, drooping branches, useful timber.

Funtumia *See* rubber plants.

Furze *See* gorse.

Fustic *Chlorophora tinctoria*, W.I., trop. Am., a commercial timber, yields the yellow dye fustic: **young** —, *Cotinus coggygria* orn. tree ('smoke tree'), S.Eur. & As.

G

Gaboon chocolate *Irvingia gabonensis*, W.Af., the paste made from the kernels (*see* dika): — **ebony**, *Diospyros dendo* and other spp.: — **mahogany**, *Aucoumea klaineana*, kernels ed. raw or cooked, wood much used.

Gaillardia Cult., orn. fls., N.Am., many cultivars.

Galangal *Alpinia officinarum*, China, rhizome a drug or spice, used more formerly: **greater** —, *A. galanga*, Siamese ginger.

GALAPEE TREE

Galapee tree (W.I.) *Sciadophyllum brownii*, cult., orn. fol.
Galax *G. urceolata*, E. N.Am., cult., ground cover, ls. used by florists.
Galba (W.I.) *Calophyllum antillarum*, strong durable wood.
Galbanum *Ferula galbaniflua* and other spp., As. Minor, med. resin.
Gale, sweet *Myrica gale*, Eur. inc. Br., cult. shrub, ls. used to flavour beer:
    fern-leaved —, *Comptonia aspleniifolia*, N.Am., fragrant, orn. fol., cult.
Galeandra Trop. Am. orchids, terrestrial and epiphytic, cult.
Galingale *Cyperus longus*, tall sedge, cult., esp. by water.
Galinsoga *G. parviflora*, Mex., a widespread weed, as is *G. ciliata*.
Gall Pharmaceutical term for insect galls collected from twigs of the oak,
    Quercus infectoria, in As. Minor. They have a high tannin content and are
    used after grinding as an astringent in ointments and suppositories, esp. for
    piles. — **berry**, *Ilex glabra*, N.Am.: — **of the earth**, *Gentiana quinqueflora*.
Galls, plant Pl. galls are numerous and widespread in the veg. kingdom. They
    may occur on all classes of pls., even freshwater and marine algae. Only a
    few are of economic importance, such as those that have been used in the
    tanning of leather. Many galls are rich in tannin. Galls may arise through
    various agencies, notably parasitic attack of some sort due to bacteria, fungi,
    eelworms, mites or insects, the last being the most common cause. The
    parasite merely provides the stimulus for the localized growth constituting
    the gall, this being derived wholly from the tissues of the host pl. Gall forma-
    tion is invariably to be associated with the reproductive phase of the parasite.
    Five groups of insects are of importance in causing galls in plants, viz.
    Hemiptera–Homoptera (bugs), Hymenoptera (wasps etc.), Diptera (two-winged
    flies), Coleoptera (beetles including weevils) and Lepidoptera (moths). Like
    other parasites, gall causers tend to show a measure of specificity for certain
    hosts. Other insects or organisms may subsequently occupy some pl. galls,
    esp. the larger ones.
    The more imp. pl. galls economically are the oak galls (*Quercus*), Chinese
    galls (*Rhus*), tamarisk galls (*Tamarix*) and *Pistacia* galls (*see* separate headings).
    The use of some galls in leather tanning or for med. purposes goes back to
    very early times, but they are used less now than formerly, having given way
    to other tanning materials. They were used by the Ancient Egyptians and
    in Greece and As. Minor as early as the 4th or 5th Century B.C. At the time
    of the Crusades oak galls were imported from As. Minor. Jap. and Chinese
    galls (*Rhus semialata*) were common in Eur. in the 15th Century.
Gally or **garlic pear** *Crateva tapia*, W.I.
Galphimia *Galphimia* spp. (= *Thryallis*), trop. Am. shrubs, cult., fls. yellow.
Galtonia Handsome S.Af. bulbous pls., cult.
Gama grass (Am.) *Tripsacum dactyloides*.
Gamalote (W.I.) *Setaria* spp. troublesome weed grasses.
Gambia pods *Acacia arabica*, trop. Af., used in local tanning.
Gambian tea bush *Lippia adoensis*, W.Af., ls. used like tea, med.
Gambier Or pale-cutch, *Uncaria gambier*, Malayan scandent shrub (bushy
    under cult.), a solid tannin extract is prepared from ls.; used locally for dyeing
    and preserving sails, nets, clothing etc.: also used as a masticatory w. betel
    and med. as an astringent, w. other ingredients, in the symptomatic treatment
    of diarrhoea.

Gamboge This dye, so familiar to the artist, is obtained from the gum-resin procured by tapping a Siamese tree, *Garcinia hanburyi*. The yellow robes of Buddhist priests are sometimes dyed w. it.

Gamelotte fibre (Mex.) *Fimbristylis spadicea*, a paper stock (esparto mulata).

Gander grass (N.Am.). *Polygonum* spp.

Ganja (Ind.) *Cannabis sativa*.

Garbanzo bean (Mex.) *Cicer arietinum*, see chick pea.

Garcinia Frs. of many spp., ed., e.g. mangosteen.

Gardener's garters *See* ribbon grass.

Gardenia Trop. and subtrop. shrubs, cult., orn. fls., often perfumed, some spp. used in local dyeing: **brilliant —**, *G. lucida*, Ind. orn. tree, large white fls.

Gari W.Af. food made from cassava.

Garigue Type of veg. in dry areas of the Medit. characterized by low shrubs, many of which are of a woolly nature. Some have colourful fls. in the spring. Several well known herbs occur naturally such as hyssop, lavender, rosemary, sage, savory and thyme, also garlic and numerous tuberous pls.

Garland flower *Hedychium* spp., trop. and subtrop., cult.; *Daphne cneorum*, Eur. v. fragrant fls.: **— chrysanthemum**, *Chrysanthemum coronarium*.

Garlic *Allium sativum*, a much used culinary herb or flavouring agent, esp. in Medit. where it has been grown for centuries. Its popularity in many other countries has increased with the advent of dehydrated, powdered or granulated garlic and 'garlic salt'. **bear's —**, *Allium ursinum:* **Can. —**, *A. canadense:* **crow —**, *A. vineale:* **false —**, *A. vineale*, *Nothoscordum*, N.Am.: **field —**, *A. oleraceum:* **giant —**, *A. scorodoprasum:* **great-headed —**, *A. ampeloprasum:* **hedge —**, *Alliaria petiolata* (*Sisymbrium alliaria*): **Levant —**, *Allium ampeloprasum:* **Naples —**, *A. neapolitanum:* **— pear**, *Crateva* spp., W.I.: **rose —**, *A. roseum* Medit. cult.: **round-headed —**, *A. ampeloprasum:* **— shrub**, *Pseudocalymma alliaceum*, W.I.: **striped —**, *A. cuthbertii*, N.Am.: **triangular-stalked —**, *A. triquetrum:* **wild —** (N.Am.), *A. canadense; A. ursinum:* **wood —**, *A. ursinum*.

Garnet berry *Ribes rubrum*, cult. for fr. in Scandinavia.

Garrya *G. elliptica*, SW. U.S., evergreen shrub w. showy silvery tassels, cult.

Gasparillo (W.I.) *Esenbeckia* spp., bark of some spp. med.

Gasteria S.Af. aloe-like pls., cult.

Gastrolobium W.Aus., evergreen shrubs, cult.

Gaub tree *Diospyros embryopteris*, Ind. cult. trop., good shade tree, young, fol. blood-red, yellow frs. (4 cm diam.) w. sticky pulp, used for caulking boats.

Gaultheria Attractive evergreen shrubs, cult.

Gaura *G. lindheimeri*, *G. coccinea*, Am., cult. orn. fls., good for cutting.

Gauze tree *See* lace bark, *Lagetta*.

Gay feather *Liatris* spp., N.Am., showy fl. pls., feathery inflor., cult.

Gazania Cult. orn. pls., S.Af., large showy fl. heads, many cultivars.

Gean *Prunus avium*, Eur., wild cherry, double fl. form cult.

Gebang palm *Corypha utan*, trop. As., the large ls. have many uses.

Gedu nohor *Entandrophragma angolense*, trop. Af., a commercial timber.

Geebung (Aus.) *Persoonia* spp., evergreen shrubs or small trees.

Geelgranaat (S.Af.) *Rhigozum obovatum* ('yellow pomegranate').

Geiger tree *Cordia sebestena*, S. U.S. & W.I., cult. trop. orn. tree, large scarlet fls.

Geissorhiza S.Af. & Madag., cult. orn. fls., resemble *Ixia*.

Geissospermum *G. laeve*, trop. Braz., b. med.

Gelsemium *G. sempervirens* SE. U.S., r. and rhizome med., depresses central nervous system and is mainly used in the treatment of trigeminal neuralgia, and migraine.

Gem of the Rio Grande *Senecio confusus* (= *Pseudogynoxys berlandieri*), cult., climber w. showy red fls.

Gemsbuck bean (S.Af.) *Bauhinia esculenta*, pods eaten by livestock.

Genepi Rich green liqueur of the absinthe type, flavoured with *Artemisia*, *A. glacialis* and *A. mitellina* collected in the Alps for its manufacture.

Genip or **genipa** *Genipa americana*, W.I., used for live fence posts; wood durable: fr. ed., size of an orange, must be kept before use: popular in Braz. and Puerto Rico; used for beverage 'genipapo'.

Genista Popular name for *Cytisus canariensis*, the Canary broom, often cult. The genus *Genista* contains many cult. shrubs.

Gentian *Gentiana* spp. A large genus (about 400 spp.) chiefly alpine, many spp. of hort. value, fls. mainly blue (white in N.Z.). The drug gentian root is derived from *G. lutea*, Eur. It is a bitter and stomachic, used to improve the appetite. **Aus. —**, *G.* (= *Gentianella*) *diemensis*: **bitter —**, *G. lutea*: **Cherokee —**, *G. cherokeensis*, N.Am.: — **clematis**, *Clematis gentianoides*: **closed —**, *G. andrewsii*, N.Am.: **field —**, *G. campestre*, Eur. tonic and stomachic: **marsh —**, *G. pneumonanthe*: **small —**, *G. nivalis*: **spring —**, *G. verna*.

Georgia bark tree *Pinckneya pubens*, b. contains cinchona-like alkaloid.

Geranium *Pelargonium* spp., mainly S.Af., cult. gdn. geraniums are *Pelargonium* spp., numerous cultivars (fancy and ivy-leaved, show geraniums etc.): some spp., notably *P. odoratissimum*, cult. for esst. oil distilled from ls. and shoots, used in perfumery and soap making. Geranium oil is also a trade name for some grass oils, e.g. *Cymbopogon martinii* ('palma rosa') w. similar uses. **Chatham Island —**, *Geranium traversii*, N.Z.: **jungle —**, *Ixora coccinea*: **wild —**, *Ambrosia hispida*.

Gerbera *G. jamesonii*, Barberton daisy, cult., many cultivars.

Germander *Teucrium* spp. **Am. —**, *T. canadense*: **common —**, *T. chamaedrys*, Eur. inc. Br., old-time med. herb: **cut-leaved —**, *T. botrys*: **Medit. —**, *T. polium*, cult. shrub: **tree —**, *T. fruticans* S.Eur.: **wall —**, *T. chamaedrys*: **water —**, *T. scordium*: **wood —**, *T. scorodonia*.

Germination The first stage in the development of a seed. In commercial seed-testing the first appearance of the radicle is regarded as germination The presence of moisture, oxygen and a certain temperature is necessary for germination. There are three main stages (1) absorption of water, (2) solution or reserve foods in the seed (by enzymes) so that they may be moved where needed, (3) the expansion of the embryo, rupture of the testa and appearance of the radicle. With different seeds there is a wide difference in the time taken for germination to commence after sowing. Some seeds may germinate in a day or two, while others take months or even years. A Brazil nut or seed (*Bertholettia excelsa*) may take up to two years to germinate, even with constant high temperature and humidity. With hard-shelled seeds such as those of many Leguminosae scarification is practised to hasten germination i.e. a portion of the testa is filed away or cracked to allow moisture to enter. In field planting of the black wattle (for tan bark) it is common practice for the seed to be treated with boiling water before sowing to improve germination. Where there have been fires to burn brush wood, germination of naturally sown seed is stimulated.

Germiston grass (S.Af.) *Cynodon dactylon.*

Gero (Nigeria) *Pennisetum americanum* (= *P. typhoides*), Hausa name: bulrush or pearl millet.

Geronggong *Cratoxylum arborescens*, Malaya, a medium-weight hardwood.

Gesneria Trop. Am. pls., mainly tuberous rooted, showy fls., cult.

Geum Cult. orn. fls., old favourites for cutting, many cultivars.

Gevuina *G. avellana*, evergreen tree, Chile, ed. nut.

Gherkin *Cucumis sativa*, a small var. of cucumber used mainly for pickling. **W.I.** —, *C. anguria.*

Ghost gum tree (Aus.) *Eucalyptus papuana:* — **plant**, *Monotropa uniflora*, N.Am.: — **tree**, *Davidia involucrata*, China, cult. orn.

Giant bamboo *Dendrocalamus giganteus* (*see* bamboo): — **cactus** or **saguaro**, *Carnegiea gigantea*, the largest Am. cactus reaching 12 tons in weight and believed to live 200 years, a desert pl.: — **fennel**, *Ferula communis:* — **fescue**, *Festuca gigantea:* — **garlic**, *Allium scorodoprasum:* — **grandilla**, *Passiflora quadrangularis:* — **hogweed**, *Heracleum mantegazzianum:* — **hyssop**, *Agastache anethiodora*, N.Am.: — **nettle**, *Laportea gigas:* — **reed**, *Arundo donax:* — **wild rye**, *Elymus condensatus.*

Gibraltar mint Name for pennyroyal.

Gidgee (Aus.) *Acacia homalophylla, A. cambagei.*

Gifdoorn (S.Af.) *Sarcocaulon patersonii*, a waxy inflammable substance is secreted in b. When the woody core has decayed tubes of wax remain. These burn like a torch w. a pleasant odour.

Gigantochloa Tall woody bamboos used for building in Indo-Malaysia, cult., young shoots used as veg.

Gilia Popular bedding and edging pls., mainly from W. N.Am.

Gill (Am.) *Glechoma hederacea:* or gill-over-the-ground.

Gillenia Cult. orn. fls., N.Am., easily grown, need moist peaty soil.

Gilliesia Bulbous pls. w. greenish fls., cult., Chile.

Gilliflower Old name for wallflowers, stocks and carnations.

Gimlet *Eucalyptus salubris*, W.Aus.

Gin Characteristic flavour due to juniper berries (*Juniperus communis*).

Gingelly (Ind.) Name for sesamum (*Sesamum indicum*).

Ginger *Zingiber officinale*, As., widely cult., trop., not known in wild state: the dried rhizomes are the well known spice: fresh rhizomes ('green ginger') preserved in sugar syrup, candied or crystallized: much used in confectionery, cake making, beverages etc. (ginger beer): also has med. uses as an aromatic stimulant and carminative: pungency due to gingerol, a yellowish oily substance. —**bread palm**, *Hyphaene thebaica:* —**bread plum**, *Parinari macrophylla:* **Cassumar** —, *Zingiber cassumar:* **coated** —, trade term for ginger rhizome w. rind still attached: — **grass**, *Cymbopogon martinii:* **Jap.** —, *Zingiber mioga:* — **lily**, *Hedychium* spp.: **Mioga** —, *Zingiber mioga:* **scraped**

GINKGO

—, trade term for ginger rhizome w. rind removed: **Siamese** —, *Alpinia galanga* (galangal): **wild** —, *Asarum canadense* and other spp.

Ginkgo *See* maidenhair tree.

Ginseng *Panax ginseng* (=*Aralia ginseng*), Manchuria, Korea, rs. used in Chinese homeopathic med. or as aphrodisiac: **Am.** —, *Panax* (=*Aralia*) *quinquefolia*, N.Am., wild and cult.

Gipsy-wort *Lycopus europaeus*, Eur., common in Br.

Girardinia Nettle-like pls. w. stinging hairs, trop., fibre from stems.

Giri-giri (W.Af.) *Sphenostylis stenocarpa*, cult., r. crop: *Cyperus rotundus*.

Givotia *G. rottleriformis* Ind., soft light wood, used for carving figures, toys etc.

Gladdon *Iris foetidissima*, stinking iris.

Glade fern *Athyrium pycnocarpon*, N.Am.

Gladiolus Cult. orn. fls., mainly from S.Af., also Medit. & As., many cultivars.

Glasswort *Salicornia* spp.: *Salsola* spp.

Glastonbury thorn *Crataegus monogyna* var. *biflora*, fls. in winter, ls. early.

Glaucidium *G. palmatum*, Jap., showy peren. w. large fls., mauve sepals, no petals.

Gleditsia Trees, trop. and subtrop., usu. w. stout thorns, defensive hedges.

Gleichenia Cult. orn. ferns, mainly creeping, fronds rarely unbranched.

Gliricidia *G. sepium* (=*G. maculata*), S.Am., widely cult. as shade for plantation crops, esp. cocoa, grows quickly from cuttings ('Madre de cacao').

Globe amaranth *Gomphrena globosa*, a gdn. pl. (bachelor's buttons) white, red or purple globular fl. heads: **— artichoke**, *Cynara scolymus*, veg., fleshy young fl. heads eaten: **— daisy**, *Globularia* spp. rock gdn. pl.: **— flower**, *Trollius* spp.: **— mallow**, *Sphaeralcea* spp.: **— thistle**, *Echinops*, gdn. pls., orn. fl. heads.

Globularia Mainly Medit. pls. favoured for dry areas and rock gdns.

Gloriosa *Gloriosa* spp. (Gloriosa or Glory lily), cult. orn. climbers w. large lily-like fls.: the national fl. or floral emblem of Rhodesia: *G. superba* and *G. rothschildiana* are noteworthy spp.

Glory-bower *Clerodendron fragrans*, China, cult. orn. shrub: **— bush**, *Tibouchina semidecandra*, Braz., cult. orn. shrub w. large velvety purple fls.: **— of the snow**, *Chionodoxa* spp.: **— of the sun**, *Leucocoryne ixiodes*, Chile, cult. bulbous pl.: **— pea**, *Clianthus dampieri*, Aus., cult., showy scarlet and black fls.

Gloxinia *Sinningia speciosa* and other spp., cult. orn. fls., esp. for the greenhouse, numerous hybrids and cultivars.

Gluta *G. renghas*, Malaya, useful red timber, latex blisters.

Glyceria *G. maxima* (=*G. aquatica*) and other spp., valued for pasture.

Glycosmis Indomal., frs. ed., citrus relatives.

Gmelina *G. hystrix*, cult. spiny climber (Philippines) with the habit of a *Bougainvillea*, and yellow bell-shaped fls.

Gnetum *G. gnemon*, Malaysia, tree, wild and cult. w. ed. starchy seeds.

Goa bean *Psophocarpus tetragonolobus*, cult., trop., pod and tuberous r. eaten: **— butter**, *Garcinia indica*: **— ipecacuanha**, *Naregamia alata*: **— powder**, *Andira araroba*, Braz., obtained from cavities in the tree, used med. (for ringworm).

Goat's beard *Tragopogon pratensis*, Eur., cosmop. weed; *Aruncus dioicus*, N.Am.; *Diuris* spp., Aus.; *Vinca rosea*, trop. weed: **— foot**, *Oxalis* spp.:

— **grass**, *Aegilops cylindrica*, Eur., a weed: — **nut**, *Simmondsia chinensis* (= *S. californica*): — **rose**, *Vinca rosea* (= *Lochnera rosea*), periwinkle: — **rue**, *Galega officinalis*, once grown for fodder; *Tephrosia virginiana*, N.Am.: — **thorn**, *Astragalus* spp., As. Minor: — **weed**, *Ageratum conyzoides*, pantrop. weed; *Capraria* spp., *Stemodia* spp., W.I.: — **willow**, *Salix caprea*.

Gobbo *Hibiscus* (= *Abelmoschus*) *esculentus*, okra.

Godetia Cult. orn. fls., Am. origin, favoured for bedding and cutting.

Gold band lily *Lilium auratum*, cult.: — **cup**, *Ranunculus bulbosus:* — **dust**, *Alyssum* spp.: — **fern**, *Pityrogramma* spp., cult., gold or silver powder on lower surface of fronds: — **fields**, *Blaeria chrysostoma*, SW. U.S., a colourful ann.: —**fussia**, *Strobilanthes anisophylla*, cult. orn. shrub, warm cli.: —**ilocks**, *Ranunculus auricomus; Crinitaria* spp.: — **in green**, *Chrysogonum virginianum*, E. U.S.: — **mohur tree**, *Delonix regia* (flamboyant): — **of pleasure**, *Camelina sativa*, wild and cult. seed fed to cage birds: — **thread** or **vegetable** —, *Coptis trifolia*, thread-like rs. med.

Golden apple *Spondias cytherea*, W.I., cult., ed. fr.: — **aster**, *Chrysopsis* spp., N.Am.: — **ball**, *Trollius; Allamanda:* — **bell**, *Forsythia:* — **canna**, *Canna flaccida:* — **cassia**, *Cassia fasciculata:* — **chain**, *Laburnum anagyroides:* — **chamomile**, *Anthemis tinctoria:* — **chinquipin** (Am.), *Chrysolepis* (=*Castanopsis*) *chrysophylla:* — **club** (Am.), *Orontium aquaticum:* — **crest** (Am.), *Lophiola aurea:* — **currant**, *Ribes aureum:* — **dewdrop**, *Duranta repens:* — **dock**, *Rumex maritima:* — **drop**, *Onosma frutescens:* — **eye**, *Viguiera multiflora*, N.Am.: —**feather**, *Pyrethrum parthenium*, cult.: — **glory pea** (Aus.), *Gompholobium:* — **gram**, *Phaseolus aureus:* — **guinea flower** (Aus.), *Hibbertia procumbens:* — **lily**, *Lycoris:* — **pine**, *Pseudolarix:* — **ragwort**, *Senecio aureus:* — **rain**, *Laburnum anagyroides*, common laburnum; *Koelreutera paniculata*, N.Am., cult.: — **rod**, *Solidago* spp.: — **samphire**, *Inula crithmoides:* — **seal**, *Hydrastis canadensis*, N.Am. med. tonic: — **shower**, *Bignonia* (=*Pyrostegia*) *venusta*, Braz. showy climber, cult., subtrop.; *Cassia fistula:* — **star**, *Chrysogonum virginianum*, N.Am.: — **tassels** (S.Af.), *Calpurnea aurea:* — **thistle**, *Scolymus hispanicus:* — **thread**, *Coptis trifolia:* — **top** (Am.), *Lamarckia aurea:* — **trumpet**, *Allamanda cathartica:* — **tuft** (W.I.), *Pterocaulon:* — **vine**, *Stigmaphyllon ciliatum:* — **wattle**, *Acacia pycnantha;* —**weed**, *Haplopappus nuttallii*, N.Am.: —**willow**, *Salix vitellina*.

Gombo Or gumbo, name for okra, *Hibiscus* (= *Abelmoschus*) *esculentus*.

Gompholobium Aus. evergreen shrubs, many w. handsome fls., cult.

Gomphrena *G. globosa* (bachelor's buttons) cult., fl. heads used as everlastings.

Gomuti palm *Arenga pinnata* (= *A. saccharifera*), trop. As. cult., source of palm sugar, jaggery, palm wine or toddy: l. sheaths yield brush fibre and the trunk a kind of sago.

Gonagra *See* canaigre.

Gonçalo alves *Astronium fraxinifolium*, *A. peroba*, Braz., a commercial timber, also called zebrawood, locust wood, kingwood or tigerwood.

Good King Henry *Chenopodium bonus-henricus*, cult., ls. may be used as spinach and young shoots as asparagus.

Good luck tree *Thevetia peruviana* (= *T. neriifolia*) cult. trop.

Gooseberry *Ribes grossularia*, a widely cult. temp. fr., bushes reaching a

GOOSEFOOT

great age, many vars., much used for jams and preserves. **Am.** —, *Ribes cynosbati:* **Barbados** —, *Pereskia aculeata:* **Cape** —, *Physalis peruviana,* S.Am., widely cult., subtrop. esp. for jam: **Ceylon** —, *Doryalis hebecarpa,* purple acid frs.: **Chinese** —, *Actinidia chinensis,* cult. commercially in N.Z., exported to Br.: **dwarf Cape** —, *Physalis pubescens,* S.Am.: **Fuchsia-flowered** —, *Ribes speciosum,* W. N.Am.: **Indian** —, *Phyllanthus emblica* (= *Emblica officinalis):* **Otaheite** —, *Phyllanthus acidus,* acid fr. used for preserves: **pinewood** —, *Ribes pinetorum,* N.Am.: **prickly** —, *R. cynosbati,* N.Am.: **smooth** —, *R. oxycanthoides:* **tomato** —, *Physalis peruviana:* **W.I.** —, *Pereskia aculeata, Heterotrichum* spp.: **wild** — (Am.), *Ribes cynosbati.*

Goosefoot *Chenopodium* spp.: *Syngonium vellozianum,* trop. Am., orn. ls., a room pl.

Goosegrass *Galium aparine,* fr. w. hooked prickles: *Eleusine* spp.

Gopher apple *Chrysobalanus oblongifolius.*

Gordonia Trop. and subtrop. trees and shrubs, some cult., orn. fls.

Gorli seed (W.Af.) *Caloncoba* (= *Oncoba) echinata,* oil med.

Gorse *Ulex europaeus* (whin or furze) a v. common spiny evergreen shrub, often a weed (as in N.Z.), cult. as cover for game. **Dwarf** —, *U. minor, U. gallii:* **Irish** —, *U. europaeus* var. *strictus,* has upright growth: **needle** —, *Genista anglica:* **yellow** — (S.Af.), *Aspalathus sarcodes.*

Gourd tree *Adansonia gregorii,* Aus.; *Crescentia cujete,* trop. Am., cult.

Gourds The fleshy frs. of many Cucurbitaceae, often v. large, are used as human food, for feeding livestock and in other ways. The flesh and the nutritious oily seeds have been used by man as food from the earliest times. Some non-edible gourds have med. properties and some are poisonous. Other gourds, the calabash gourds, in the dried state, are believed to be among the first containers used by primitive man. They are still used as water containers in many parts of the world, esp. Af., and for such commodities as milk and native beer. They are also used in making musical instruments of various kinds. The shapes of some gourds are considered to have influenced early pottery design. Sometimes the outer surfaces of gourds used for domestic purposes were elaborately carved. Small calabash gourds may be used as snuff-boxes and in the manufacture of tobacco pipes which were fashionable at one time.

The more important ed. gourds include pumpkins, squashes, marrows, melons and cucumbers, in innumerable forms and vars. Some are cult. in both warm and temp. cli., others are restricted to trop. Pumpkins and water melons are perhaps the most extensively used gourds in the feeding of livestock, some of the coarser pumpkins used for this purpose reaching an enormous size and weight (40 to 90 kg). The snake gourd, *Trichosanthes cucumeria* (= *T. anguina),* of ancient cultivation in eastern As., has a long narrow fr., eaten only when young. The wax gourd, *Benincasa hispida* (= *B. cerifera),* which is covered with a whitish waxy bloom is another much used oriental gourd. In W.Af. the seeds of *Citrullus lanatus* (= *C. vulgaris)* called 'neem', 'egussi' or 'agushi', and in E.Af. the seeds of *Telfairia pedata* (oyster nut) are esteemed for food. Seeds of other gourds are often eaten out-of-hand, after roasting, like peanuts. The sou-sou, chayote or Christophine, *Sechium edule,* with its large single seed, is widely grown as a veg. in warm cli., as is the so-called Malabar melon or Siamese gourd, *Cucurbita ficifolia.*

Among gourds used med. are colocynth, *Citrullus colocynthis*, the squirting cucumber, *Ecballium elaterium* of the Medit. and of ancient cult., and the trop. balsam pear *Momordica charantia*. The towel gourd or luffa, *Luffa cylindrica* is cult. in warm cli. for the tough fibrous network of the fr., used as a bath sponge and in other ways.

Orn. gourds are often grown purely for decoration, e.g. to take the place of fls. during winter months in temp. cli. These include representatives of both the genera *Cucurbita* and *Lagenaria*, some being highly coloured such as the Turk's cap gourd. Other well known orn. gourds are those called crown of thorns, crown gourd, apple gourd, orange gourd, Jap. nest egg-and-spoon gourd.

The numerous squashes in cult. are divided into two main groups, summer and winter squashes, the latter being capable of being stored for some time and often grown for winter use. The numerous vars. available may be derived from *Cucurbita maxima*, *C. pepo*, *C. mixta* or *C. moschata*. Melons are derived from *Cucumis melo* and watermelons from *Citrullus lanatus*.

The trop. Am. calabash tree, *Crescentia cujete*, produces a large, hard-shelled, spherical fr. used for domestic purposes like the calabash gourds. [Organ, J. (1963) *Gourds*, 189 pp.; Whitaker, T. M. & Davis, G. N. (1962) *Cucurbits*, 250 pp.]

Gousbloom (S.Af.) Various spp. of *Arctotis* and *Dimorphotheca*.

Goutweed *See* ground elder.

Gouty stem tree (Aus.) *Adansonia gregorii*.

Govenia Trop. Am. terrestrial orchids, some cult.

Governor plum *Flacourtia indica* (=*F. ramontchi*), frs. make good jelly: cult. trop., also called Ind. plum or ramontchi.

Gowan Scottish word for various yellow fls., buttercup, dandelion etc.

Graft hybrid Or chimera, a plant that has developed from both the stock and the scion after grafting, being subsequently propagated vegetatively. They are not common, two of the best examples being + *Laburnocytisus* and + *Crataegomespilus*.

Grafting The operation of uniting a shoot, or portion of a shoot (termed the scion) with the stem of a rooted pl. (termed the stock) so that the shoot will continue to grow, eventually showing its own characteristics in regard to fl., fr. or l. It is widely practised with economic and orn. pls. of all kinds, esp. fr. trees. The art of grafting is very old and there are several forms of it. It was practised by the Greeks and Romans and by the Chinese at an early period.

Bud-grafting or 'budding' as it is more commonly called, is a modification of ordinary grafting in that a bud only is used in place of the branch or scion to incorporate on the stock. Budding is of comparatively recent origin and was not known to the Ancients.

Gram, black *Phaseolus* (=*Vigna*) *mungo*, trop. pulse crop: **golden —**, *P. aureus* (=*P. radiatus*, *Vigna radiata*): **green —**, *P. aureus*: **horse —**, *Dolichos biflorus*.

Grama or **gramma** (Am.) *Bouteloua* spp., includes several imp. pasture grasses.

Grammatophyllum Large handsome epiphytic orchids, trop. As., *G. speciosum* (=*G. giganteum*) has fls. 13–15 cm across and inflor. 1.5 m long, probably the world's largest orchid.

GRANADILLA

Granadilla *Passiflora* spp., ed. frs.: **giant** —, *P. quadrangularis* cult. trop.: **purple** —, *P. edulis*, widely cult. in subtrop., pulp and juice canned, greatly favoured for fruit salad, a yellow fruited var. much grown in Hawaii: **sweet** —, *P. ligularis*.

Granadillo *Dalbergia granadillo*, Mex., a valued and highly orn. cabinet wood: name also used for *Brya ebenus* and *Caesalpinia* spp.

Grandsir-greybeard (Am.) *Chionanthus virginicus*, beautiful fringe-like inflor.

Granite gooseberry *Ribes* (= *Grossularia*) *curvata*, N.Am.

Grape *Vitis vinifera*. The grape is known to have been cult. for over 4000 years. The fresh and dried frs. have always been much esteemed, as have beverages prepared from grapes, such as the numerous wines, brandies, liqueurs and some non-alcoholic drinks. Endless vars. of grape have been evolved over the centuries, special vars. being grown in different grape-growing regions and for special wines. Sultanas, raisins, and currants are examples of the dried fruit, currants being derived from a small-fruited grape, *Vitis vinifera* var. *corinthiaca*, long grown extensively in Greece. Various by-products are obtained from the pressing of grapes. **Amurland** —, *Vitis amurensis:* **bush** —, *V. acerifolia*, N.Am.: **blue** —, *V. bicolor*, N.Am.: **bullace** —, *V. rotundifolia*, N.Am.: **Calif.** —, *V. californica:* **Cañon** —, *V. arizonica:* **cat** —, *V. palmata*, N.Am.: **chicken** —, *V. vulpina*, N.Am.: **dune** —, *V. riparia*, N.Am.: **fig-leaved** —, *V. thunbergii:* **fox** —, **northern**, *V. labruska*, N.Am., many cultivars: **fox** —, **southern**, *V. rotundifolia*, N.Am.: **frost** —, *V. cordifolia*, N.Am.: **— hyacinth**, *Muscari* spp.: **— ivy**, *Rhoicissus rhomboidea*, a room pl.: **jasmine** —, *Ervatamia coronaria:* **June** —, *Vitis riparia*, N.Am.: **mountain** —, *V. monticola*, N.Am.: **mustang** —, *V. candicans:* **Oregon** —, *Mahonia aquifolium:* **parsley leaved** —, *V. vinifera:* **pigeon** —, *V. cinerea*, N.Am.: **possum** —, *Cissus* spp.: **riverside** —, *Vitis vulpina*, N.Am.: **Rocky Mt.** —, *Mahonia aquifolium:* **sand** —, *Vitis rupestris*, N.Am.: **sea-side** —, *Coccoloba uvifera:* **skunk** —, *Vitis labruska*, N.Am.: **summer** —, *V. aestivalis*, N.Am.: **tree** —, *Coccoloba uvifera:* **wild** — (S.Af.), *Cissus* spp.

Grapefruit *Citrus paradisi*, widely cult. trop. and subtrop., extensively canned (grapefruit segments) and used for juice. Frs. may be in bunches like grapes, hence the name.

Grapple plant *Harpagophytum procumbens*, S.Af., the fiercely armed frs. may cause grazing animals to become lame or starve through having their jaws locked together by them. See *Harpagophytum* in Willis.

Graptophyllum Some spp. cult. for their orn. ls. (*see* caricature plant).

Grass Any sp. of the family Gramineae may be termed a grass. Certain pls. in other families that are grass-like in appearance have also acquired the name. The true grasses constitute one of the most imp. families among fl. pls. and are of inestimable value to man, being the main food of his domestic animals. There are about 9000 spp. of grasses, classified under some 660 genera and spread all over the world. They vary in size from a few centimetres in height to over thirty metres in the tallest bamboos. *See* Gramineae in Willis.

The cereals such as wheat, oats, barley and rye in the cooler regions and rice, maize, sorghum and other millets in trop. and subtrop. provide enormous quantities of food for man and beast. These cereals are today sources of numerous other manufactured products such as proteins, starch, wax, oils, plastics, adhesives and cosmetics. Sugar cane, also a grass, is the world's main source of sugar, while the refuse from milling (bagasse) is much used for fibre board and certain classes of paper. Both cereals and sugar cane are imp. sources of alcohol. Other grasses, notably esparto, are used for high grade paper. Bamboos are also used for paper pulp and have innumerable uses in the Orient, especially for house building and furniture (*see* bamboo). Many grasses are used for thatching, others have local uses for such articles as mats, baskets, brooms, brushes, rope, straw hats, etc. Some grasses are much used in combating soil erosion and in sand dune fixation or in mud binding. Many different grasses w. the ability to stand frequent cutting are used for lawns, playing fields and aerodromes. Others such as pampas grass and some with striped or colourful ls. are favoured for orn. purposes.

From natural and man-made pastures grazing animals obtain their main food in the form of fresh grass or hay. Young grass may be cut for silage or artificially dried and made into meal, cakes or pellets for animal feeding. The main pasture grasses in Br. and in some other temp. countries are – peren. rye grass, timothy, cocksfoot and Italian rye grass. Other valuable agricultural grasses are meadow fescue, crested dog's tail, foxtail, rough meadowgrass, black bent, tall fescue and tall or false oat grass. In warm cli. fodder grasses are usually of taller and coarser growth. They include the following – Guinea grass, Para grass, Rhodes grass, Napier fodder or elephant grass, Guatemala grass, Kikuyu grass, Pangola grass, teff, carpet grass, Bermuda grass, Natal grass, buffalo grass and boer manna (S.Af.). [Hubbard, C. E. (1954) *Grasses*, 428 pp.]

Grass cakes Young grass, artificially dried, is made into biscuits, cakes or nuts for animal feed; superior nutritionally to hay.

Grass of Parnassus *Parnassia fimbriata*, N.Am.: *P. palustris*, Br.

Grass oils Citronella, lemon grass, ginger grass, palma rosa, vetivert.

Grass tree *Xanthorrhoea* spp., Aus., in places a feature of the vegetation: some estimated to grow only 30 cm in 100 years. See *Xanthorrhoea*, also Willis. N.Z. ― ―, *Dracophyllum* spp.

Grass widows *Sisyrinchium inflatum*, N.Am.

Grasswrack *Zostera marina* and other spp., dried ls. used for packing and for stuffing mattresses (ulva or alva marina).

Gratiola *G. officinalis*, formerly med., other spp. cult. orn. pls.

Gravel root *Eupatorium purpureum*, N.Am., med. (urinary complaints).

Greasewood *Sarcobatus*, W. N.Am., a shrub.

Greater celandine *Chelidonium majus*, yellow latex once used for skin ailments.

Greek valerian (Am.) *Polemonium caeruleum*, med., astringent.

Green briar (Am.) *Smilax* spp.: ― **dragon**, *Arisaema dracontium; Berlandiera*, N.Am.: ― **gram**, *Phaseolus aureus:* ― **hood orchid** (Aus.), *Pterostylis curta:* **Kendal** ―, *Genista tinctoria:* ― **milkweed**, *Asclepias viridifolia*, N.Am.: ― **orchis**, *Habenaria* spp.: ― **sapote**, *Calocarpum viride*, trop. Am. ed. fr.: ― **weed, dyer's** *Genista tinctoria:* ― **withe** (W.I.), *Vanilla claviculata:* ― **wood**, *Sarcobatus vermicula.*

Greenheart *Ocotea rodiaei* (=*Nectandra rodiaei*) Guyana, an imp. commercial timber, much used in marine construction, piles etc. **Af.** —, *Cylicodiscus gabunensis*, W.Af., logs exported: **Surinam** —, *Tecoma leucoxylon*.

Greenovia *G. aurea*, Canary Isles., cult. orn. pl., resembles *Sempervivum*.

Greens Word used in Br. for almost any *Brassica* ls. used as a veg. – kale, young cabbage etc.: **sour** —, *Rumex venosus*, N.Am.

Greigia Cult. orn. pls. w. pineapple-like ls., S.Am.

Grevillea *G. robusta*, Aus., a widely cult. tree, yields a commercial timber: used as shade for trop. crops: grown for orn. and as a pot pl., fern-like foliage.

Grewia B. fibre of several spp. used locally in trop. and subtrop.

Greybeard grass *Amphipogon strictus*, Aus.

Grigri palm (W.I.) *Martinezia caryotifolia*, ed. fr.

Groats or **grits** Oats deprived of the husk.

Gromwell *Lithospermum officinale, Mertensia maritima:* **blue** —, *Lithospermum purpurocaeruleum:* **corn** —, *L. arvense:* **false** —, *Onosmodium* spp., N.Am.

Gros michel Variety of banana grown in Jam. and elsewhere catering for temp. markets, being well adapted for shipping.

Ground berry (Aus.) *Astroloma* spp.: — **cherry** (Am.), *Physalis* spp. — **cover pls.**, *see* cover plants and crops: — **elder**, *Aegopodium podagraria*, a pernicious weed w. creeping rhizomes: a form w. variegated ls. cult. — **ivy**, *Glechoma hederacea* (= *Nepeta glechoma); —* **pine**, *Ajuga chamaepitys*.

Groundnut *Arachis hypogaea*, an imp, oil seed widely grown in trop. and subtrop: oil resembles olive oil: often employed as a substitute for it, used in cooking and for margarine. The press-cake is good animal feed. Ed. starchy tubers of *Apios tuberosa* called groundnut in N.Am. **Bambarra** —, *Voandzeia subterranea*, seeds a trop. Af. pulse, wide range in colour and size of seeds.

Groundsel *Senecio vulgaris*, a cosmop. and troublesome ann. weed: **climbing** —, *S. macroglossus:* **sticky** —, *S. viscosus:* **tree** —, *Baccharis halimifolia*, N.Am., *Senecio* subgenus *Dendrosenecio*, E.Af.: **wood** —, *Senecio sylvaticus*.

Grugru palm (W.I.) *Acrocomia* spp., *Astrocaryum* spp., kernels yield oil.

Grumichama (Braz.) *Eugenia dombeyi*, wild and cult., ed. fr.

Guabiroba (Braz.) *Campomanesia guaviroba*, shrub, cult., ed. fr.

Guaco *Clibadium* spp. cult. as fish poison in Braz.

Guaiacum Resinous product obtained from the heartwood of *Guaiacum officinale* and *G. sanctum* (lignum vitae), in the Caribbean reg.: has various med. uses.

Guaje (Mex.) *Leucaena esculenta*, seeds ed., pods sold in markets.

Guako *Mikania guaco*, S.Am., med.

Guapi Or cocillana b., *Guarea rusbyi*, Bolivia, med.

Guar *Cyamopsis psoraloides*, Cluster bean, imp. pulse crop in Ind.

Guarana *Paullinia cupana*, Braz., a stimulating beverage (4–5 % caffeine) made from seeds, the main beverage of Matto Grosso, shrub wild and cult.

Guarea *Guarea cedrata* and *G. thompsonii*, W.Af. a trade timber, used for furniture.

Guatemala grass *Tripsacum laxum*, an imp. trop. fodder grass.

Guava *Psidium guajava*, ed. fr., widely cult. trop. and subtrop., often naturalized, many vars. much used for jelly or canned. 'Guava cheese' a kind of thick jam is popular in W.I.: some vars. have high vitamin value: **black** —, *Guettarda argentea*, W.I.: **Cattley** —, *Psidium cattleyanum:* **Chilean** —,

Myrtus ugni: **Costa Rican** —, *Psidium friedrichsthalianum:* **monkey** —, *Diospyros mespiliformis:* **mountain** —, *Psidium montanum:* **para** —, *Campomanesia* spp.: **pineapple** —, *Psidium cattleyanum:* **purple** —, *P. cattleyanum:* **Spanish** —, *Catesbaea spinosa*, spiny orn. shrub, cult. trop.: **strawberry** —, *Psidium cattleyanum*, widely cult., warm cli., small purple fr. much esteemed.

Guaxima S.Am. name for fibre of *Urena lobata* (Congo jute).

Guayacan *Caesalpinia melanocarpa*, pods used for tanning in Argen.

Guayota *Gonolobus edulis*, Costa Rica, fr. ed.

Guayule *Parthenium argentatum*, Mex. & SW. U.S., shrub, of interest as an emergency rubber pl. (in wartime).

Guayusa *Ilex guayusa*, Peru, ls. used for a stimulating beverage.

Gubgub (W.I.) *Vigna unguiculata*, a var. of cowpea favoured as a veg.

Guelder rose *Viburnum opulus*, shrub, cult., orn. frs., several vars.

Guere palm *Astrocaryum* sp., Colombia, oil from kernels.

Guernsey lily *Nerine sarniensis.*, cult. orn. pl., fls. pale salmon.

Gevuina nut *Gevuina avellana*, Chile, evergreen tree, ed. nut.

Guiac wood *Bulnesia sarmientii*, Argen., esst. oil distilled from wood.

Guinea corn *Sorghum* spp.: — **grains**, *Aframomum melegueta*, also known as Grains of Paradise or Melegueta pepper; a much used spice in W.Af.; imported to Eur. at an early date: used in veterinary med. and flavouring cordials: — **flower**, *Fritillaria meleagris*, gdn. fl.: — **grass**, *Panicum maximum*, an imp. trop. fodder grass: — **hen weed**, *Petiveria alliacea:* — **peach**, *Nauclea latifolia* (= *Sarcocephalus esculentus*): — **pepper**, *Xylopia aethiopica:* — **yam**, *Dioscorea cayennensis.*

Guisaro *Psidium molle*, C.Am., shrub, cult., acid fr. used for jelly.

Guitar plant (Aus). *Lomatia tinctoria*, low shrub, spreads by suckers.

Gul mohur Ind. name for flamboyant, *Delonix regia.*

Gulugulu (S.Af.) *Strychnos gerrardii*, orange-like frs., pulp ed.

Gumbo *Hibiscus* (= *Abelmoschus*) *esculentus*, okra.

Gums Gums are widely distributed throughout the veg. kingdom. They are produced more freely or abundantly in some families than in others but comparatively few are of commercial importance. Their uses by man go back to very early times, their adhesive properties being made use of for paints and pigments. Primitive peoples have used them for food in many parts of the world and the main use of commercial gums, such as gum arabic, is still for food, i.e. in the confectionery trades. The more imp. veg. gums (true gums) are – **gum arabic**, *Acacia senegal* and other *Acacia* gums from the Sudan and other parts of Af.: **gum tragacanth**, *Astragalus* spp. from As. Minor: **Ind. tragacanth** or **karaya gum**, *Sterculia urens* and *Cochlospermum gossypium*, also called katira gum, from the Ind.: **carob seed gum**, *Ceratonia siliqua*. The three last mentioned are gums of the gum tragacanth class and are used as substitutes for gum tragacanth.

Gum arabic is collected from the wild Acacia trees in the dry season and often sun-bleached before export, the paler the colour the higher its value is likely to be. Good solubility and viscosity are imp. Large quantities are used in the manufacture of lozenges and gum sweets. It is also much used as an adhesive, in med. preparations and for inks and water colours.

Gum tragacanth and similar gums constitute an entirely different class of gum. They swell with water, forming a gel, and are much used in the textile industry. Their med. use is mainly for suspending soluble powders in fluid mixtures and as an ingredient in toilet preparations such as skin and dental creams, shaving lotions etc. Other uses are in food preparations of various kinds such as chutneys and pickles and as a smoothing agent in ice cream. Some pl. exudations which are not true gums or soluble or miscible with water have been referred to as gums, such as kauri gum, gum copal, gum accroides etc., which are resins and gum balata, chicle gum (chewing gum), which are of a rubbery nature. **Aden gum,** *Acacia* spp.: **almond** —, *Prunus amygdalus:* **Amrad** —, *Acacia arabica:* **Amritsar** —, *Acacia modesta:* **anami** —, *Hymenaea courbaril:* **angico** —, *Piptadenia rigida:* **Ashanti** —, *Terminalia* spp.: **Barbary** —, *Acacia* spp.: **brea** —, *Caesalpinia praecox:* **Cape** —, *Acacia karroo:* **cashew** —, *Anacardium occidentale:* **cherry** —, *Prunus cerasus:* **E.Af.** —, *Acacia drepanolobium:* **E.I.** —, *Acacia* spp.: **gatty** —, *Anogeissus latifolia:* **hog** —, =tragacanth: **karee** —, *Saccopetalum tomentosum:* **khair** —, *Acacia catechu:* **kolhol** —, *Acacia senegal:* **locust** —, =carob seed gum: **mesquite** —, *Prosopis juliflora:* **Mogador** —, *Acacia* spp.: **Morocco** —, *Acacia* spp.: — **resin,** a mixture of gum and resin: **sweet** —, *Liquidambar styraciflua:* **tahl** —, *Acacia seyal:* **Tartar** —, *Sterculia cinerea.* [Howes, F. N. (1949) *Vegetable gums and resins,* 188 pp.]

Gundabluey (Aus.) *Acacia victoriae,* emergency fodder shrub or tree.

Gungurru *Eucalyptus caesia,* W.Aus.

Gunnera *G. chilensis, G. manicata,* majestic herbaceous pls. w. large rhubarb-like ls. 1–2 m across, w. stalks 2 m or more long, usu. cult. near water. Young peeled l. stalks of *G. chilensis* eaten as a veg. in Chile.

Gunny Ind. name for jute (*Corchorus*).

Gunpowder plant *See* artillery plant.

Gurjun *Dipterocarpus* spp. imp. commercial timber. **Andaman** or **Ind.** —, *D. alatus, D. grandiflorus, D. pilosus:* **Burma** —, *D. alatus, D. turbinatus:* **Ceylon** —, *D. zeylanicus:* — **oil,** obtained from the trunks of some spp.

Gutta percha *Palaquium* spp. a rubber-like substance which softens w. heat, used in dentistry: *P. gutta,* Malaysia, formerly the main source, but other spp. now exploited.

Guzmania Trop. Am. & W.I., bromeliads resembling *Tillandsia,* cult.

Gymea lily *Doryanthes excelsa,* Aus., tall w. large scarlet fls.

Gymnocalycium S.Am. cacti, low spiny pls. w. relatively large fls., cult.

Gynandropsis *G. speciosa,* showy ann., warm cli. violet fls., seeds and ls. eaten: *G. gynandra* a trop. weed.

Gynerium Reed grasses: favour damp situations, fl. stalks of *G. sagittatum* used as arrow shafts by Amerindians in Braz.

Gynura *G. cernua,* W.Af., mucilaginous ls. used in soups.

Gypsophila Cult. orn. fls. for cutting and foam-like effect. Rs. of some spp. rich in saponin, used as soap substitutes.

Gypsywort *Lycopus europaeus,* r. used to stain face brown; fls. attractive to hive bees for nectar: — **weed,** *Veronica officinalis.*

H

Habana oat grass *Themeda quadrivalvis*, trop. weed, esp. in sugar cane, Aus.

Habenaria Many cult. orchids, often fragrant, in this large genus (600 spp.).

Hackberry (Am.) *Celtis occidentalis*, N.Am., yields a commercial timber: **Eur.** —, *C. australis*, wood resembles elm: other spp. of *Celtis* yield useful timber.

Hackmatack (Am.) *Larix laricina: Populus balsamifera.*

Haemanthus Bulbous S.Af. pls. w. a striking, dense inflor., cult.

Haematoxylin Dye obtained from logwood (*Haematoxylum campechianum*).

Hagberry or **hegberry** *Prunus padus: P. avium.*

Haiari *Lonchocarpus* spp., Guyana, lianas used as fish poisons.

Hair grass Various grasses have acquired the name, notably spp. of *Aira, Corynephorus, Deschampsia, Koeleria, Muhlenbergia* and *Trisetum.*

Hakea Evergreen shrubs and trees, Aus., some cult., some have become troublesome (prickly) weeds in other countries. **Dagger** —, *H. teretifolia:* **bottle brush** —, *H. bucculenta.*

Haldu *Adina cordifolia*, Ind., trop. Aus., an imp. commercial timber.

Halimium Shrubs resembling *Cistus*, cult.

Halimodendron *H. argenteum*, As., prickly deciduous shrub, cult.

Hallucinations, plants causing A large number of pls. are known to be capable of causing hallucinations when taken internally, or used as a snuff or smoked. One authority considers there are about a dozen such spp. native to the Old World, the best known being the opium poppy and cannabis or Ind. hemp and 80–100 in the New World, where they are used mainly in religious or ceremonial rites by primitive native peoples. The kind of hallucination varies w. different pls., being commonly visual but may also relate to hearing or taste. Two of the best known hallucinogenetic pls. or groups of pls. of the Americas are the stemless cactus peyote or peyotyl of Mex., *Lophophora* (= *Echinocactus*) *williamsii* and the dendroid or shrubby Daturas or Brugmansias of the Andes, including *Methysticodendron amesianum*. The dried crowns of peyotyl (termed mescal buttons) are chewed during religious ceremonies causing remarkable visions and feelings of well-being. They have been used in Mex. since pre-Colombian times. *Methysticodendron amesianum* is a small tree of the Colombian Andes, an infusion of the ls. being used in witchcraft, mainly by the Ingano and Kamsa Amerindians causing a trance that may last a long time or be prolonged even into days. This and some allied spp. may be cult. by the natives. In Mex. and Siberia certain mushrooms or fungi are known to have been used as hallucinogens for centuries. In C. & S.Am. some spp. of *Banisteriopsis* (notably *B. caapi;* 'caapi' or 'yage') are much used, also some spp. of *Ipomoea, Sophora, Anadenanthera* and *Virola.* [Schultes, R. E. (1970). The New World Indians and their hallucinogenic plants. *Morris Arb. Bull.* **21** (1), 3–14.]

Hamamelis *Hamamelis* spp. (witch hazel), As., N.Am., cult. orn. shrubs:

H. mollis, w. conspicuous yellow fls. in early spring, is widely grown. The drug hamamelis consists of the dried ls. of *H. virginiana,* N.Am. It has astringent properties, used in preparations for treatment of haemorrhoids. Hamamelis water is used in eye lotions and the treatment of sprains and bruises.

Hamelia Handsome free-fl. shrubs, trop. and subtrop. Am., cult.

Hamilla (Sri Lanka) *Berrya cordifolia* (=*B. amonilla*), trincomali wood.

Hamiltonia *H. suaveolens,* Ind., cult. orn. shrub, fragrant white fls.

Hand plant *Cheirostemon platanoides,* Mex. evergreen tree, 5 red stamens like a hand.

Handkerchief tree *See Davidia.*

Harbinger of spring *Erigenia bulbosa,* E. N.Am.

Hard fern *Blechnum spicant, Lomaria:* — **grass,** *Parapholis, Pholiurus:* — **head,** *Xyris pallescens,* N.Am.: — **heads,** *Centaurea nigra:* — **pear** (S.Af.), *Strychnos henningsii.*

Hardback *Spiraea tomentosa,* N.Am., brown felted shoots, rose fls.

Hardenbergia Evergreen climbers, Aus., orn. fls., cult.

Harding grass *Phalaris tuberosa,* cult., warm cli.

Hardwood The term 'hardwood' is commonly used for all timbers that are not coniferous or members of the Coniferae (pine, spruce, larch etc.) regardless of whether they are hard or soft in the physical sense. The principal hardwoods in Br. are – alder, ash, beech, birch, cherry, elm, hornbeam, horse-chestnut, lime, oak, plane, poplar, sweet-chestnut, sycamore, walnut and willow. Hardwoods of minor importance include the following, some being little used or not available in any quantity – apple, blackthorn, box, elder, hawthorn, hazel, holly, maple, mulberry, pear, robinia or false acacia, rowan, service, spindle and white-beam. The imported hardwoods used in the timber trades are numerous. *See* timber.

Hare-bell (England) *Campanula rotundifolia;* (Aus. & N.Z.), *Wahlenbergia* spp. (S.Af.), *Dierama* spp.: — **'s ear,** *Bupleurum* spp.: — **'s foot fern,** *Davallia canariensis:* — **orchid,** *Caladenia* spp., Aus. & N.Z.: — **'s tail grass,** *Lagurus ovatus,* Eur., fl. heads dried for winter decoration; *Eriophorum vaginatum:* (Am.), *E. spissum.*

Haricot bean *Phaseolus vulgaris* widely cult. for green pods and ripe seeds.

Harlequin flower *Sparaxis* spp. S.Af., showy fls., cult.

Harmal *Peganum harmala,* N.Af. to Ind. yields the dye 'turkey red'.

Harrisia *H. martinii,* S.Am., cult. gdn. pl. now a weed in S.Af. cane fields.

Hart's tongue fern *Phyllitis scolopendrium,* N.Hemisph., cult.

Harvest lice (Am.) *Agrimonia* spp.

Hashish *Cannabis* or Ind. hemp; *Cannabis sativa.*

Hatchet cactus *Pelecyphora aselliformis,* Mex., bears hatchet-shaped tubercles.

Hatpins *Eriocaulon decangulare,* N.Am.

Hat-plant, Chinese *Holmskioldia sanguinea,* cult. showy fls. shaped like a Chinese hat.

Hats, plants used for Many different pls. are made into hats in different parts of the world esp. the trop., where palm ls. are widely used, e.g. spp. of *Corypha, Heterospathe, Hyphaene, Phoenix, Raphia, Sabal* etc. Other pls. or pl. products freely used are bamboo, cereal straws, numerous grasses,

Hibiscus and other b. fibres, maize cob husks, *Pandanus* ls., sedges, sola pith or *Aeschynomene,* wood shavings, notably *Cupressus* in Japan, *Tacca* spp. (*see also* Panama hat plant).

Hatstand tree *Rheedia lateriflora,* W.I., branches in whorls, a young tree felled makes a useful hatstand.

Haulm The stems of such pls. as beans, peas, potatoes etc.

Hausa potato *Solenostemon rotundifolius* (= *Coleus dysentericus*), W.Af., tubers ed., cult.

Haustoria The r.-like suckers of parasites.

Hautbois or **hautboy** *Fragaria moschata,* a strawberry, wild. and cult.

Haw *Crataegus monogyna* and other spp., name used for fr. or pl. **Black —,** *C. douglasii,* W. N.Am., fr. ed., used for jellies; *Viburnum prunifolium,* E. N.Am.: **Chinese —,** *Crataegus pentagyna,* China, frs. esteemed: **southern black —,** *Viburnum prunifolium.*

Hawaiian goosefoot *Chenopodium oahuense,* a pot-herb: **— persimmon,** *Diospyros ferrea:* **— raspberry,** *Rubus hawaiiensis:* **— strawberry,** *Fragaria chiloensis:* **— tree fern,** *Cibotium glaucum* pith ed.

Hawk's beard *Crepis* spp. many are common meadow pls. **— bit,** *Leontodon* spp.: **— weed,** *Hieracium* spp.

Haworthia Small S.Af. succulent. pls., cult.

Hawthorn *Crataegus* spp. **Common —,** *C. monogyna* (= *C. oxyacantha*), also called May, quick or whitethorn, Eur. inc. Br., wild and cult., many cultivars w. pink, red or 'double' fls.; a common farm hedge: wood strong and tough, resembles apple or pear and has similar uses, a substitute for box: **Chinese —,** *Photinia serrulata:* **Mex. —,** *C. pubescens,* cult., fr. used as animal food: **river —,** *C. rivularis,* N.Am.: **two-styled —,** *C. laevigata* (= *C. oxyacanthoides*), very similar to common hawthorn.

Hay-fever, plants causing The pollen of a large number of different pls. may be responsible for causing hay-fever among those people who happen to be allergic. It is usually the pollen from wind-pollinated pls., which is light and readily suspended in the atmosphere, that causes the trouble. Very frequently, as in Br. and many other countries, it is grass pollen that is the main cause of hay-fever. As the fl. period of most grasses in temp. cli. is of short duration, taking place in early summer, the period when hay-fever is rife is fortunately also of relatively short duration. A grass considered to be largely responsible for hay-fever in Br. is cocksfoot (*Dactylis glomerata*) which is widespread and common, often cult. This grass is present in most parts of Eur. In the U.S. it is called orchard grass. Timothy grass (*Phleum pratense*), also widely grown for grazing and hay in Br., is another common cause of hay-fever.

In other countries, particularly the U.S., a large number of other pls. are responsible for hay-fever. In some areas there are in fact spring-fl., summer-fl. and autumn-fl. pls. all capable of causing hay-fever. In such regions the possible hay-fever season is of course considerably extended. Many conifers, notably pines and junipers, may cause the complaint. They produce pollen in considerable quantity. The Texas mountain cedar (*Juniperus mexicana* or *J. sabinoides*) is a good example. Some of the worst hay-fever pls. in N.Am. are spp. of *Artemisia* (mug-worts, sage bushes and wormwoods).

HAYRATTLE

They are wind-pollinated and copious pollen shedders. The ragweeds (*Ambrosia* spp.) are other imp. sources of hay-fever. Poplars and the Am. spp. of ash (*Fraxinus* spp.), which are wind-pollinated, both produce pollen freely and may be responsible for hay-fever. In the U.S. there are recognized 'hay-fever resorts' where there is absence of troublesome pollen and where sufferers may escape from the cause of their complaint. These may be in mountain areas or seaside resorts where the prevailing wind is from the sea. A sea voyage is also a means of escape. [Wodehouse, R. P. (1945) *Hay-fever plants*, 245 pp.]

Hayrattle *Rhinanthus minor:* **greater —**, *R. serotinus.*

Hayscented fern *Dennstaedtia* spp. N.Am.

Hazel *Corylus* spp. The **common —**, *C. avellana*, Eur., As., is a shrub or small tree, wild and cult., esp. for coppice (on a 14–17-year rotation) being much used for hurdles and light poles: wood fine-grained, may be used for small turned articles: nuts small but ed. The hazel nut group of nuts constitute the 'small nuts' of the nut trade and include 'Barcelona' and 'Turkish' nuts derived from cult. cobnuts and filberts. The best known cult. var. in Br. is the 'Kentish cob' (really a filbert as the husk covers the nut). These nuts are popular dessert nuts and much used in confectionery, esp. nut-chocolate. **Am. —**, *Corylus americana:* **Aus. —**, *Chorilaena:* **beaked —**, *Corylus rostrata*, N.Am.: **Chinese —**, *C. chinensis:* **Himalayan —**, *C. ferox:* **Jap. —**, *C. heterophylla:* **Tibetan —**, *C. tibetica.* **— wood**, *Symplocos stawellii*, Aus.

Headache tree *Premna integrifolia*, Malaysia, handsome veined wood. **— weed** (W.I.), *Hedyosmum nutans.*

Heal-all (Am.) *Prunella vulgaris:* **healing bush** (S.Af.), *Lobostemon fruticosus.*

Heart-berry *Aristotelia peduncularis*, Aus. evergreen shrub w. orn. frs. **—'s ease**, *Viola tricolor*, *Polygonum persicaria:* **— leaf** (Am.), *Asarum* spp.: **— pea** or **— seed**, *Cardiospermum grandiflorum* (= *C. halicacabum*), trop. Af., used as a veg.

Heath *Erica* spp., name also used for some spp. in other genera. **Antigua —**, *Russelia juncea:* **Aus. —**, *Epacris impressa:* **berry —**, *Erica baccans:* **bridal —**, *Erica bowieana:* **ciliate —**, *E. ciliaris:* **climbing —**, *Prionotes cerinthoides*, Aus., Tasm.: **Cornish —**, *Erica vagans:* **cross-leaved —**, *E. tetralix:* **fine-leaved —**, *E. cinerea:* **fuchsia —**, *Epacris longifolia*, Aus.: **— grass**, *Sieglingia decumbens:* **Irish —**, *Erica erigena:* **jasmine —**, *Erica jasminiflora*, S.Af.: **long-leaved —**, *E. longifolia*, S.Af.: **mountain —**, *Archeria* spp., N.Z.; *Phyllodoce empetriformis*, N.Am.: **prickly —**, *Pernettya* spp.: **Scotch —**, *Erica cinerea:* **sea —**, *Frankenia* spp.: **St Dabeoc's —**, *Daboecia cantabrica:* **sticky-leaved —**, *Erica glandulosa*, S.Af.: **— tea tree**, *Leptospermum:* **Swellendam —**, *Erica walkeria*, S.Af.: **tree —**, *E. arborea*, Eur., rs. used for briar pipes.

Heather Common or **Scottish —**, *Calluna vulgaris*, see ling: **beach —**, *Hudsonia ericoides*, N.Am.: **Himal. —**, *Cassiope fastigiata:* **purple —**, *Erica cinerea.*

Heaven, tree of *Ailanthus altissima*, China, cult. orn. tree, quick-growing, fls. have disagreeable odour, liable to taint honey.

Hechtia Prickly-leaved bromeliaceous pls., C.Am., some cult.

Hedge bedstraw *Galium mollugo:* **— bindweed**, *Calystegia sepium:* **— hyssop,**

Gratiola: — **mustard**, *Sisymbrium officinale:* — **nettle**, *Stachys:* — **parsley**, *Torilis.*

Hedgehog broom *Erinacea pungens*, S.Eur.: — **cactus**, *Echinocereus engelmannii*, SW. U.S.: — **grass**, *Cenchrus*, N.Am.; *Echinopogon cheelii*, Aus.: — **holly**, *Ilex aquifolium* var. *ferox:* — **thistle**, *Echinocactus.*

Hedges, defensive *See* barrier plants.

Hedychium Large As. pls., reed-like stems, orn. fol., scented fls., cult.

Helenium Popular gdn. pls., N. & C.Am., numerous cultivars.

Helianthus Many spp. cult. as gdn. pls. (sunflowers), of Am. origin.

Helichrysum Several spp. of this Old World genus cult. as gdn. pls., fls. of some used as everlastings.

Heliophila S.Af. anns. or subshrubs, colourful fls. cult. (Cape stock).

Heliopsis *H. helianthoides*, *H. scabra*, N.Am., cult. orn. fls.

Heliotrope Gdn. heliotropes cult. for their fragrant fls. are derived from *Heliotropium peruvianum* and *H. corymbosum*, Peru: **Wild** — (S.Af.), *Vernonia* spp.: **winter** —, *Petasites fragrans.*

Helipterum Cult. orn. fls., S.Af., Aus., fl. heads used as everlastings.

Hellebore *Helleborus* spp., Eur., As., cult. gdn. pls., fl. early in the year, many cultivars and hybrids. **Black** —, *H. niger*, the Christmas rose: **false** —, *Veratrum* spp.: **green** —, *Helleborus viridis:* **stinking** —, *H. foetidus:* **white** —, *Veratrum album*, Eur. rhizomes med.

Helleborine *Epipactis* spp., *Cephalanthera* spp., several wild in Br.

Hellfetter *Smilax tamnoides*, N.Am.

Helmet-flower *Scutellaria* spp. (skull cap): — **orchid** (Aus.), *Corybas.*

Helonias *H. bullata*, E. N.Am., cult., a bog gdn. pl.

Heltrot Am. name for Eur. hogweed (*Heracleum sphondylium*).

Helxine *Soleirolia soleirolii* (=*Helxine*), Sardinia, cult., naturalized in Br., may be a weed in lawns, small creeping w. minute fls. ('mind-your-own-business').

Hemerocallis Temp. E.As., several spp. cult. ('day lilies') fls. last a day. Dried fls. of *H. fulva* a condiment in Jap.

Hemionitis Unusual trop. ferns, well suited for indoor cult. or fern cases.

Hemlock *Conium maculatum*, a notoriously poisonous pl. to man and animals: —**parsley**, *Conioselinum chinense:* **water** —, *Cicuta virosa*, extremely poisonous to man and all classes of livestock.

Hemlock bark For a long period, dating from early colonial days, hemlock b. from the Am. eastern hemlock or hemlock spruce (*Tsuga canadensis*) was the most imp. tanning material of the U.S. & Can. It was responsible for the characteristic red colour of Am. leather of last century. B. and later tanning extract were exported to Eur. B. of the western hemlock (*Tsuga heterophylla*) was used to some extent.

Hemlock spruce *Tsuga* spp. Coniferous trees of N.Am. & E.As., the two most imp. or best known being *T. canadensis* and *T. heterophylla*. **Eastern hemlock**, *T. canadensis*, timber is inferior to western hemlock: **Carolina** —, *T. caroliniana:* **Chinese** —, *T. chinensis:* **Formosan** —, *T. formosana:* **Himal.** —, *T. dumosa* (=*T. brunoniana*): **Jap.** —, *T. sieboldii:* **western** —, *T. heterophylla*, W.Can. & U.S., a commercial timber much used for construction work, building, joinery, flooring etc.

HEMP

Hemp *Cannabis sativa*, **true** or **common** —, cult. for fibre, esp. in Russia and Italy: fibre much used for cordage and rope, esp. rose for marine use: the seed is a source of oil and much used as food for cage birds: the pl. is also a source of a drug (*see* cannabis). **Ambari** —, *Hibiscus cannabinus:* **Bahama** —, *Agave sisalana:* **Bombay** —, *Crotalaria juncea:* **bowstring** —, *Sansevieria:* **Chinese** —, *Abutilon avicennae:* **Cuban** —, *Furcraea cubensis:* **Deccan** —, *Hibiscus cannabinus:* **Ind.** —, *Cannabis sativa:* **Madras** —, *Crotalaria juncea:* **Manila** —, *Musa textilis:* **Mauritius** —, *Furcraea gigantea:* — **nettle**, *Galeopsis:* **N.Z.** —, *Phormium tenax:* **Queensland** —, *Sida rhombifolia:* **Russian** —, *Cannabis sativa:* **sisal** —, *Agave sisalana:* **sann** or **sunn** —, *Crotalaria juncea:* — **weed** (Am.), *Mikania scandens.*

Hen-and-chickens *Sempervivum soboliferum* (houseleek): abnormal forms of the common daisy (*Bellis perennis*) and calendula (*Calendula officinalis*) in which branches from the head bear secondary heads.

Hen-bane *Hyocyamus niger*, Eur. inc. Br., cult., source of a drug (hyoscyamus) consisting of the dried ls. and fl. tops: action similar to belladonna: used to counteract the griping action of purgatives and to relieve spasm in the urinary tract.

Henbit *Lamium amplexicaule.* Eur. inc. Br., in cult. ground.

Henequin *Agave fourcroydes*, C.Am., a fibre resembling sisal. **Salvador** —, *A. letonae.*

Henna *Lawsonia inermis* (= *L. alba*) powdered ls. form the oriental dye or cosmetic: much used as a hair dye.

Heraldry, plants in Eur. heraldry is considered to have started in the first half of the 12th Century. At that time a new type of helmet, which covered the entire face, was adopted instead of one covering only the top of the head and nose (as depicted on the Bayeux tapestry). This made it difficult to distinguish friend from foe in battle, so distinguishing marks or crests were adopted on shields and on the body (coat-of-arms, put on over the mail). Fls., frs., ls. or whole pls. along with other natural objects were incorporated. The terminology of heraldry was derived largely from French, 'azure' – blue, 'vert' – green etc. Some special terms were used for pls., 'proper' signified in natural colour, 'blasted' – destitute of leaves, 'fleuri' – bearing fls. etc. One writer lists over 200 spp. of common Eur. pls., wild and cult., that have been used in heraldry. These include common trees (e.g. alder, ash, elm, fir, hawthorn, holly, maple, mulberry, oak, yew), wild and cult. fls. (e.g. daisy, carnation, cornflower, crocus, heather, lily, marigold, rose), fruits (e.g. apple, bramble, cherry, chestnut (nut), fig, olive, pear, pomegranate, quince), vegs. (e.g. bean, cabbage, garlic, leek, turnip). In addition, many other pls. such as clover, wheat and barley and some weeds (nettles) were used.

Herb Any non-woody vascular pl. w. no woody part above ground: a pl. used in seasoning food (culinary herb) or med.: **pot-herb**, one that is boiled and eaten, like spinach. — **Christopher**, *Actaea spicata:* — **of-Grace**, *Ruta graveolens:* — **Paris**, *Paris quadrifolia:* — **patience**, *Rumex patientia*, once grown and used as spinach: — **Robert**, *Geranium robertianum:* **willow** —, *Epilobium* spp.

Herbal Old book w. descriptions of pls. and their uses, esp. med. uses. — **tobacco**, *see* smoking mixtures.

Herbarium A building in which dried or preserved plant specimens are kept: a collection of pressed and dried pls. arranged in systematic order to facilitate reference. A herbarium is usu. concerned with the pls. of a particular area or country, but some large national herbaria such as those at Kew and the British Museum are concerned with pls. on a world basis. Some herbaria may be intended to deal only with special groups of pls., such as weeds, or different groups of cult. pls. such as cereals, where numerous vars. or cultivars exist. The aim of a herbarium is to accumulate in one place all possible information about the pls. with which it is concerned and herbarium sheets should bear detailed notes supplied by the collector (*see* collecting). A herbarium is in effect a repository of information and can be a research tool of great value in botanical work.

Herbary Old English for that part of the gdn. devoted to herbs.

Herbicides Numerous chemicals are in use for controlling the weeds of arable land and cult. crops; new chemicals, more efficient or selective than their predecessors, are constantly being produced. The choice of the best herbicide in any given situation or for any particular crop depends upon many considerations, but is determined mainly by the tolerance of the crop and the susceptibility of the weeds that are present. This applies equally to the crops of temp. cli. and to those of the trop., such as plantation crops.

With non-selective methods of control, i.e. for areas such as paths, railway tracks and roadsides, the more drastic herbicides are used, used as arsenical weedkillers, sodium chlorate, petroleum oils, coal tar acids and others. With growing crops choice is restricted to the selective herbicides, the total number of which is now v. large, including both pre- and post-emergence herbicides.

Herbs, culinary and medicinal The use of culinary herbs varies greatly from one country to another. In Br. only a few are favoured in the average household, these being mint, parsley, sage, thyme and perhaps horse radish, and of course onion or onion-like flavourings. On the continent of Eur. many other herbs are freely used, esp. in France, Italy and Spain, herbs such as fennel, garlic, marjoram, basil, savory, tarragon and sorrel. Aromatic seeds include anise or aniseed, caraway, coriander, dill and fennel. The following are among the better known med. herbs used in western countries – belladonna, chamomile, fox-glove or digitalis, hen-bane or hyoscyamus, lobelia, psyllium, stramonium, valerian. There are many more. *See also* drugs.

Hercules club *Aralia spinosa*, N.Am.: *Zanthoxylum clava-herculis*, N.Am.

Herd's grass (Am.) *Phleum pratense* (Timothy grass).

Herniary Or rupture wort, *Herniaria glabra*.

Heterotoma *H. l·belioides*, Mex. cult. orn. fl.

Heterotrichum Orn. shrubs, trop. Am., showy fls., cult., some w. ed. fr.

Heuchera *H. sanguinea*, Mex., N.Am. (coral bells) cult., many vars.

Hiba *Thuja dolabrata*, Jap., imp. forest tree.

Hibbertia Attractive shrubs, mainly Aus., cult., *H. volubilis* w. large yellow fls.

Hibiscus Many spp. cult. for showy fls., esp. *H. rosa-sinensis*, ('shoe fl.') w. many vars., *H. mutabilis*, *H. schizopetalus* in trop. and *H. syriacus* in colder cli. Many yield bast fibre, used locally. *See* roselle, okra, kenaf. **Bladder —,** *H. trionum*, ann. calyx inflated in fr.: **lilac —,** *H. huegelii* (Aus.): **Norfolk Island —,** *Lagunaria patersonii*.

HICKORY

Hickory *Carya* spp., N.Am. Some half dozen spp. are exploited for their timber in N.Am., mainly E.Can. & E. U.S. Hickory wood is noted for its strength and toughness, much favoured for tool handles. Several spp. yield ed. nuts, the more imp. nut-yielding spp. being – **shag bark** —, *C. ovata:* **big shell-bark** —, *C. laciniosa:* **white** —, **butternut** or **pignut**, *C. glabra:* **red** — or **mocker nut**, *C. tomentosa.* Hickory nuts are noted for their good keeping qualities. Normally the hard thick shells are a drawback but thin-shelled forms have been brought to light. **Aus.** —, *Acacia implexa.*

Hierochloe Grasses notable for smell of newly mown hay, esp. *H. odorata.*

High water grass (Am.) *Spartina patens:* — — **shrub**, *Iva* spp.

Hill gooseberry *Rhodomyrtus tomentosa*, trop. As., ed. fr.

Hippeastrum Bulbous pls., S.Am. large handsome fls., many cultivars.

Hippo, wild (Am.) *Euphorbia corollata.*

Hippocrepis Some are useful fodder pls. in Medit.

Hiptage *H. madablota*, trop. As., cult. shrub, large fragrant showy fls.

Hobble bush *Viburnum alnifolium*, N.Am.

Hodgsonia *H. macrocarpa*, trop. As., large vine, seeds yield ed. oil.

Hoffmannia Handsome trop. Am. fol. pls., colourful ls., cult., esp. *H. roezlii.*

Hog-brake *Pteridium aquilinum* (bracken); *Ambrosia artemisiifolia:* — **fennel**, *Peucedanum officinale:* — **gum**, *Metopium toxiferum*, S. U.S., resin a drastic purgative; *Moronobea*, W.I.: — **nut**, *Carya glabra* N.Am.: — **peanut**, *Amphicarpa monoica*, N.Am.: — **plum**, *Spondias mombria* (= *S. lutea*) and other spp., trop.; *Ximenia americana; Symphonia globulifera; Prunus umbellata:* — **weed**, *Boerhavia* spp., trop. weeds and pot-herbs; *Heracleum sphondylium;* *Ambrosia* spp.: — **weed, giant**, *Heracleum mantegazzianum*, cult. orn. pl. w. enormous ls. and inflor., can cause dermatitis: —**wort**, *Euphorbia capitata.*

Hoheria N.Z. shrubs w. white fls., cult., esp. *H. lyallii*, several vars.

Holly *Ilex aquifolium.* **Common** or **Eur.** —, tree or shrub, wild or cult., esp. for hedges, many orn. forms, variegated ls. coloured frs. etc. Fls. a useful source of nectar to the hive bee. Wood white, hard, heavy and fine-grained, used for inlay work or stained black as an ebony substitute. **Am.** —, *Ilex opaca:* **Calif.** —, *Photinia arbutifolia*, *Rhamnus ilicifolius:* **Dahoon** —, *Ilex cassine:* — **fern**, *Cyrtomium falcatum*, *Polystichum lonchitis:* — **flame-pea**, *Chorizema ilicifolium*, W.Aus., ls. like holly, fls. orange and red: — **grape**, *Mahonia repens*, N.Am.: **Himalayan** —, *Ilex dipyrena:* **Jap.** —, *I. crenata:* **longstalk** —, *I. pedunculatus:* **mountain** —, *Nemopanthus canadensis:* **native** —, *Oxylobium ilicifolium*, Aus.: — **oak**, *Quercus ilex:* — **pea**, *Jacksonia floribunda:* — **rose**, *Turnera*, W.I.: **sea** —, *Eryngium maritimum:* **smooth** —, *Hedycarya angustifolia*, Aus.: **W.I.** —, *Turnera ulmifolia.*

Hollyhock *Althaea rosea*, a popular gdn. pl. and traditional feature of the English cottage gdn.: popularity lessened by increase in hollyhock rust. **Aus.** —, *Lavatera plebeia:* **Mountain** —, *Iliamna rivularis*, W. N.Am.

Holmskioldia *H. sanguinea*, Ind., cult. orn. evergreen shrub.

Holodiscus N.Am. shrubs allied to *Spiraea*, cult.

Holy basil *Ocimum sanctum:* — **clover**, *Onobrychis viciifolia:* —**flax**, *Santolina* spp., S.Eur.: — **grass**, *Hierochloe odorata*, sweet scented, once strewn on church floors: — **herb**, *Verbena officinalis:* — **thistle**, *Silybum marianum.*

Homeria S.Af. pls. w. tunicated bulbs, some poisonous, showy enduring fls.
Hominy (Am.) Hulled and crushed maize or food prepared from it.
Honesty *Lunaria annua*, cult. orn. pl., often grown for the dried frs. which are flat w. a silvery septum, favoured for winter decoration.
Honewort *Trinia glauca*, occurs in W. of England.
Honey The nectar of fls. is sometimes erroneously referred to as honey. It does not become honey until it has been prepared or 'ripened' into honey in the hive by the bees (moisture driven off by fanning of the wings etc.). — **flower**, *Melianthus major*, S.Af., used in subtrop. bedding; *Lambertia formosa* W.Aus.: — **guide**, marks or lines on fls. considered to show the way to the nectar, e.g. *Myosotis:* — **locust**, *Gleditsia* spp.: — **palm**, *Jubaea chilensis* (=*J. spectabilis*), Chile, cult.: — **Richea**, *R. scoparia*, Tasm.: **sham** —, *Lopezia, Parnassia* (*see Lopezia* in Willis): — **wort**, *Cerinthe major*.
Honeydew The name given to the excretion of certain insects that feed by sucking the sap or juice of pls., such as pl. lice or aphids, scale insects and leaf-hoppers. The insect absorbs nourishment from the juice and voids the rest, which when fresh is a colourless sugary fluid. Honey bees frequently collect it, especially if flower nectar is scarce. This may cause honey to be dark, as honeydew often becomes blackish due to the presence of fungi or sooty moulds. In some trop. countries honeydew collected by bees from sugar cane may also spoil honey. In Br. honeydew becomes most noticeable in hot dry weather, particularly on lime trees (*Tilia*) when it may drop on pavements making them slippery, or on parked cars. Other trees that may show honeydew are oak, sycamore, beech, elm, ash, chestnut, hawthorn, fr. trees and various conifers. In parts of Eur. honey from conifer honeydew ('Tannenhonig') is much esteemed. It is thick and dark with a characteristic strong flavour.
Honeysuckle *Lonicera* spp., many spp. are cult. orn. climbers or shrubs, inc. the common Eur. honeysuckle or woodbine, *L. periclymenum.* **Af.** —, *Halleria lucida:* **Aus.** —, *Lambertia multiflora:* **bush** —, *Lonicera standishii* and other spp.: **Cape** —, *Tecomaria capensis:* **Chinese** —, *Lonicera tragophylla:* **coral** —, *L. sempervirens:* **fly** —, *L. xylosteum:* **French** —, *Hedysarum coronarium:* **Jam.** —, *Passiflora laurifolia:* **Jap.** —, *Lonicera nitida*, a popular hedge pl.: **Maori** or **N.Z.** —, *Knightia excelsa:* **swamp** —, *Rhododendron viscosum*, E. N.Am.: — **tree**, *Turraea floribunda*, S.Af.: **trumpet** —, *Lonicera sempervirens.*
Hoop ash *Celtis occidentalis* var. *crassifolia:* — **pine**, *Araucaria cunninghamii*, E.Aus., strong durable wood: — **tree** (W.I.), *Melia:* — **withe** (W.I.), *Colubrina, Rivina.*
Hop *Humulus lupulus*, Eur. wild and cult. for fr. widely used in brewing. — **bush**, *Dodonaea viscosa:* frs. used as hops by early settlers in Aus.: — **clover**, *Medicago lupulina:* **Spanish** —, *Origanum* sp.: — **tree**, *Ptelea trifoliata*, N.Am., cult. orn. tree: — **weed**, *Salvia occidentalis*, W.I.; *Hyptis suaveolens*, W.I.: **wild** —, *Flemingia strobilifera*, W.I.; *Leonotis nepetifolia*, W.I. [Burgess, A. H. (1964) *Hops: botany, cultivation, utilization*, 300 pp.]
Hop-hornbeam *Ostrya* spp. **Am.** —, *O. virginiana*, also called ironwood, a commercial timber, hard, heavy and strong, favoured for tool handles. **Eur.** —, *O. carpinifolia*, also a commercial timber and similar, used in turnery, once used for axles of vehicles.

125

HORA

Hora *Dipterocarpus zeylanicus*, Sri Lanka, a commercial timber.

Horehound *Marrubium vulgare*, or white horehound, med. herb.: **black —**, *Ballota nigra*, often a weed, has been used as a substitute for white horehound: **foetid —**, *Ballota nigra:* **water —**, *Lycopus europaeus.*

Hornbeam *Carpinus betulus*, Eur., As.Minor, wild and cult. tree, esp. cvs. *columnaris* and *fastigiata*. Wood v. hard, favoured for wood-screws, pulleys, tool handles, mallets, skittles etc. **Am. —**, *C. caroliniana:* **heart-leaf —**, *C. cordata:* **Jap. —**, *C. japonica:* **Oriental —**, *C. orientalis.*

Horn-nut *Trapa natans*, fr. horned: **— of-plenty**, *Fedia cornucopiae*, Medit., cult. orn. pl.: **— wort**, *Ceratophyllum demersum* and *C. submersum*, cosmop. water pls., used in aquaria, old ls. translucent and horny.

Horned orchid *Orthoceras strictum*, Aus.: **— poppy, yellow**, *Glaucium flavum:* **— poppy, violet**, *Roemeria hybrida*, Eur. inc. Br.: **— pondweed**, *Zannichellia palustris:* **— rampion**, *Phyteuma.*

Horse balm *Collinsonia canadensis*, E. N.Am., yellow fls.: **— bean**, *Vicia faba*, widely used in animal feeding: **— cassia**, *Cassia grandis*, cult. trop., v. large pods; *Cassia polyphylla.*

Horse chestnut *Aesculus hippocastanum*, E.Eur., **common** or **Eur. — —**, widely cult., esp. in Br. as an orn. or avenue tree, wood soft, light, of even texture, no distinctive heartwood, favoured for dairy utensils, fancy articles etc. **Am. — —**, *Aesculus* spp. (commonly called 'buckeye'): **Chinese — —**, *A. chinensis:* **Himal. — —**, *A. indica*, becoming a popular orn. tree, late fls.: **Jap. — —**, *A. turbinata*, yields a commercial timber: **red flowered — —**, *A. carnea.*

Horse fly weed *Baptisia tinctoria*, E. N.Am.: **— gentian**, *Triosteum perfoliatum*, E. N.Am.: **— gram**, *Dolichos biflorus* (= *D. uniflorus*) trop. pulse and green manure crop, esp. in Ind.: **— hair, vegetable**, *Chamaerops humilis*, Medit.: **— mint** (N.Am.), *Monarda fistulosa;* (Br.), *Mentha longifolia:* **— nettle**, *Solanum carolinense*, N.Am.

Horse radish *Armoracia rusticana* (= *Cochlearia armoriaca*), Eur. wild (naturalized) or cult., r. a popular condiment, used esp. w. roast beef and w. oysters, much used in pickling: dehydrated horse radish is popular. **— — tree**, *Moringa oleifera* (= *M. pterygosperma*), cult. trop., r. smells like horse radish: **— seed bush**, *Dodonaea viscosa*, trop.: **— shoe vetch**, *Hippocrepis comosa:* **— sugar**, *Symplocos tinctoria*, N.Am., ls. sweetish: **— tail**, *Equisetum* spp., troublesome weeds (*see Equisetum* in Willis).

Hosta Handsome perens., E.As., large tubular fls., cult.

Hot water plant *Achimenes grandiflora*, Mex., crimson or purplish fls., cult.

Hottentot bean *Schotia* spp., S.Af.: **— bedding**, *Helichrysum auriculatum:* **— bread**, *Dioscorea* (= *Testudinaria*) *elephantipes:* **— cherry**, *Maurocenia capensis:* **— fig**, *Carpobrotus edulis* (= *Mesembryanthemum edule*) widely cult., naturalized in Medit.: **— head**, *Standeria eriopus:* **— poison bush**, *Acokanthera spectabilis*, *A. venenata:* **— tea**, *Helichrysum serpyllifolium.*

Houlletia S.Am. orchids, large fls., some fragrant, cult.

Hound's tongue *Cynoglossum officinale*, wild in Br., also cult. orn. pl.

Houseleek *Sempervivum tectorum*, cult. orn. pl., planted on cottage roofs to keep slates in position: **— lime**, *Sparmannia africana*, cult. orn. fl. pl.

House plants The increased interest in growing orn. pls. in the house or the home has been attributed to various factors such as the extended use of central heating and improved lighting (strip lighting). Central heating has meant a more uniform temperature, day and night, and the abolition of the cold night period, when fires went out, so harmful to many pls. from warm cli. It has also meant few open coal or gas fires with fumes so harmful to pl. life.

A large number of decor. pls. are now grown as house pls., some for their fls., others for their attractive foliage. Some house pls. are of easy culture and will thrive under a wide range of conditions or even tolerate bad management or neglect, like the *Aspidistra* of Victorian times and music hall fame. Other house pls. are more selective in their requirements and need care and attention if they are to thrive. In the natural state many house pls. are part of the ground flora of trop. forests where they do not have direct sunlight and are in effect essentially shade-loving pls. The dry atmosphere of most living rooms does not suit some house pls. but the following will tolerate it – cacti and succulents w. spp. of *Aechmea, Billbergia, Chlorophytum, Clivia, Ficus, Grevillea, Pilea, Sanseverinia, Vriesia* and *Zebrina*. Where gas fires are used fl. house pls., except *Impatiens* and *Billbergia*, should be avoided.

House pls. grown for their decor. fol. include the following – aluminium plant (*Pilea*), bird's nest bromeliad (*Nidularium*), boat lily (*Rhoeo spathacea*), Canary Island ivy (*Hedera canariensis*), chestnut vine (*Tetrastigma*), cordyline, devil's ivy (*Scindapsus*), dragon plant (*Dracaena*), dumb cane (*Dieffenbachia*), earth star (*Cryptanthus*), fan plant (*Begonia*), fig (*Ficus* spp.), finger aralia (*Dizygotheca*), goosefoot (*Syngonium*), grape ivy (*Rhoicissus*), iron cross plant (*Begonia masoniana*), ivy, maidenhair fern, mind-your-own-business (helxine), monstera, mother-of-thousands (*Saxifraga sarmentosa*), pepper elder (*Peperomia*), philodendron, pick-a-back (*Tolmiea menziesii*), prayer plant (*Maranta*), sansevieria, setcreasea, spider plant (*Chlorophytum; Atalia*), umbrella plant (*Cyperus alternifolius*), wandering jew (*Tradescantia; Zebrina*), weeping fig (*Ficus benjamina*).

Fl. house or room pls. include – Af. violet (*Saintpaulia*), begonia, billbergia, busy Lizzie (*Impatiens wallerana*), clivia, columnea, cupid's bower (*Achimenes*), cyclamen, dwarf orange, flamingo flower (*Anthurium*), Italian bellflower (*Campanula isophylla*), peace lily (*Spathiphyllum*), shrimp plant (*Beloperone guttata*), slipper flower (*Calceolaria*), urn plant (*Aechmea fasciata*), vriesia, wax or porcelain plant (*Hoya carnosa*), winter cherry (*Solanum capsicastrum*), zebra plant (*Aphelandra squarrosa*).

Houttuynia *H. cordata*, As., water pl. cult., ls. used in salads and soups.

Hovenia *H. dulcis*, As., Jap. raisin tree, fr. ed.

Howea *H. belmoreana, H. forsteriana*, Lord Howe Island, orn. palms much used as pot or room pls. (called 'Kentia').

Hoya *H. carnosa*, wax pl., Aus., a favourite hothouse pl.: other spp. cult.

Huckleberry *Gaylussacia baccata* and other spp., N.Am., ed. fr., cult.: **box —**, *G. brachycera:* **bear —**, *G. ursina:* **black —**, *G. resinosa, G. baccata, Vaccinium atrococcum:* **blue —**, *V. vacillans, V. pennsylvanicum:* **dwarf —**, *Gaylussacia dunosa:* **hairy —**, *Vaccinium hirsutum:* **squaw —**, *V. stamineum:* **sugar —**, *V. vacillans.*

Hold-me-tight (W.I.) *Achyranthes indica*, a weed, fr. w. spine-tipped bracts.

Hulver *Ilex montana*, E. N.Am.

HUMATA

Humata Trop. As. ferns w. small leathery ls.

Humea *See Calomeria.*

Humming bird flowers *Abutilon, Erythrina, Marcgravia* etc. — — **trumpet,** *Russelia,* scarlet fls. very attractive to humming birds. [Grant, K. A. & V. (1969) *Humming birds and their flowers,* 101 pp.]

Humulus Cult. orn. twiners. *See* hop.

Hungry rice *Digitaria exilis,* W.Af., cult., seeds eaten.

Hunter's nut (W.I.) *Omphalea megacarpa,* trop. Am., large seeds ed., cult.

Huntsman's cup or **horn** *Sarracenia purpurea,* N.Am.

Huon pine *Dacrydium franklinii,* Aus., yields a useful timber.

Huru Name for sandbox tree, *Hura crepitans.*

Hyacinth *Hyacinthus* spp. Gdn. or cult. hyacinths in their many forms are derived largely from *Hyacinthus orientalis,* E.Medit., As. Minor. — **bean,** *Dolichos lablab* (= *Lablab niger):* **grape** —, *Muscari* spp.: **orchid** —, *Dipodium ensiformis,* Aus.: **water** —, *Eichhornia crassipes:* — **vine,** *Clematis crispa,* N.Am.: **wild** — (N.Am.), *Camassia scilloides;* (S.Af.), *Lachenalia contaminata.*

Hydnocarpus oil *Hydnocarpus wightiana,* Ind., *H. anthelmintica,* Indo-China. Oil from seeds has been used in the treatment of leprosy but other drugs (sulphones) are now preferred.

Hydrangea *Hydrangea* spp. Shrubs sometimes tree-like or climbing: the common gdn. hydrangeas are derived from *H. macrophylla,* Jap.: fl. colour may vary w. soil, acid or alkaline. **Climbing** —, *H. petiolaris,* Jap., a strong deciduous climber reaching 25 m, climbs like ivy *(Hedera),* with adventitious rs.

Hydrastis *See* golden seal.

Hydrocleys Orn. water pls. w. large fls., trop. Am., cult.

Hydrocotyle *H. asiatica,* As., ls. eaten raw or cooked.

Hymenaea *H. courbaril,* trop. Am., W.I. locust, large tree yielding valuable timber, also copal found in the ground near the trees.

Hymenocallis Bulbous pls., mainly trop. Am., cult., large lily-like handsome or bizarre fls. *See* Willis.

Hymenophyllum Ferns, generally creeping, w. delicately membranous fronds, well suited for indoor cult. or fern cases.

Hypericum The large genus includes many good gdn. pls., such as *H. calycinum.* Several spp. have local med. uses.

Hyphaene *H. crinita,* Medit., dwarf fan palm, cult. for orn., yields veg. hair, used in upholstery.

Hypolepis Mainly creeping ferns w. soft papery fronds: *H. repens* readily becomes a weed in a fernery.

Hyssop *Hyssopus officinalis* and other spp.; *Agastache* spp. Some are tea substitutes: fls. of some are good nectar sources for hive bees. **Anise** —, *A. anisata,* N.Am., anise-scented: **giant** —, *A. urticifolia,* N.Am.: **hedge** —, *Gratiola:* **water** —, *Bacopa.*

I

Iberis Ann. or biennial herbs and subshrubs, cult. orn. pls., e.g. candytuft.
Ibogo *Tabernanthe iboga*, Congo, med., alkaloid from r.
Iburu *Digitaria iburua*, W.Af., seeds eaten like millet.
Icaco Or coco-plum, *Chrysobalanus icaco*, ed. fr., cult., trop.
Ice plant *Cryophytum crystallinum* (= *Mesembryanthemum*), popular bedding pl., brilliant fls., also a pot-herb, has a crystalline appearance: an allied sp., *C. roseum*, is used as a cover pl. in Calif. citrus groves.
Iceland poppy *Papaver nudicaule*, popular gdn. pl., many vars.
Ichu *Stipa ichu*, Chile, a good fodder grass, esp. in dry areas.
Idigbo *Terminalia ivorensis*, W.Af., a commercial timber, used for furniture.
Ignatius bean *Strychnos ignatii*, Philippines, seeds med.
Igusa *Juncus effusus* var. *decipiens*, Jap., cult. sedge, used for mats.
Ilama *Annona diversifolia*, C.Am., ed. fr., like custard apple.
Ilang-ilang *See* ylang-ylang.
Ilex *See* holly.
Iliau *Wilkesia gymnoxiphium*, Hawaii, orn, daisy-like fls., small tree.
Illawarra pine (Aus.) *Callitris cupressiformis:* — **palm,** *Archontophoenix cunninghamiana.*
Illipe nut *Shorea macrophylla*, *S. stenoptera*, *Isoptera borneensis*, sources of illipe fat.
Ilomba *Pycnanthus angolensis*, W.Af., a commercial timber.
Imbu *Spondias* sp., Braz., ed. fr., resembles a greengage.
Imbuya *See* embuia.
Immortelle (W.I.) *Erythrina* spp., esp. *E. umbrosa*, a shade tree for trop. crops: **mountain —,** *E. poeppigiana*, W.I.: **swamp —,** *E. glauca*, W.I.: **immortelles,** *see* everlastings.
Imphee (Zulu) Sweet-stemmed *Sorghum* used for chewing.
Inca wheat *Amaranthus caudatus*, seed used as a cereal by Amerindians.
Incaparina (C.Am.) Ed. high protein flour made from cotton seed.
Incarvillea Cult. orn. perens., esp. *I. delavayi* and *I. macrantha.*
Incense Used in some churches but use in med. now obsolete. Various fragrant resins may be used in its preparation, such as those from spp. of *Amyris*, *Boswellia*, *Bursera*, *Commiphora* and *Styrax.* **Cayenne —,** *Protium guianense:* — **cedar,** *Libocedrus decurrens:* — **tree,** *Protium heptaphyllum*, Braz.; *Canarium schweinfurthii*, trop. Af.; *Bursera simaruba*, W.I.: — **plant,** *Calomeria elegans:* — **wood,** *Amoora nitida.*
India rubber tree *Ficus elastica*, Ind. cult. orn. or shade tree, once a source of rubber.
Indian aconite *Aconitum ferox:* — **almond,** *Terminalia catappa:* — **arrow wood,** *Euonymus atropurpureus*, N.Am.: — **barberry,** *Berberis aristata:* — **bean tree,** *Catalpa bignonioides*, N.Am.: — **beech,** *Pongamia glabra:* — **bdellium,** *Commiphora mukul:* — **berry,** *Anamirta paniculata:* — **corn,**

INDIGO

Zea mays: — **blanket**, *Gaillardia:* — **cross**, *Tropaeolum majus:* — **fig,**
Opuntia ficus-indica: — **fire,** *Salvia coccinea,* N.Am.: — **garland flower,**
Hedychium: — **gum,** *Acacia, Feronia, Anogeissus, Combretum:* — **hawthorn,**
Rhaphiolepis: — **hemp,** *Apocynum cannabinum:* — **horse chestnut,** *Aesculus
indica:* — **laburnum,** *Cassia fistula:* — **liquorice,** *Abrus precatorius:*
— **madder,** *Rubia cordifolia, Oldenlandia umbellata:* — **mallow,** *Abutilon
avicennae:* — **mulberry,** *Morinda citrifolia:* — **olive,** *Olea cuspidata:* —
paint brush, *Castilleja* spp., N.Am.: — **pipe,** *Monotropa uniflora:* — **poke-
berry,** *Phytolacca acinosa:* — **potato,** *Ipomoea pandurata:* — **root,** *Asclepias
curassavica:* — **shot,** *Canna indica:* — **strawberry,** *Duchesnea indica:*
— **tobacco,** *Lobelia inflata,* N.Am.: — **turnip,** *Eriogonum longifolium.*

Indigo *Indigofera* spp. When natural indigo was imp. as a dye the crop was
grown mainly in Ind. and Sumatra, the imp. spp. being *Indigofera anil, I.
arrecta, I. sumatrana* and *I. tinctoria.* W. the introduction of aniline dyes cult.
for dye virtually ceased. Indigo may still be grown as a green manure crop
and soil improver to a small extent. **Bush** — (S.Af.), *Indigofera;* (Am.),
Dalea spp.: **Chinese green** —, *Rhamnus* spp., dye derived from b.: **false** —
(N.Am.), *Baptisia* spp., *Amorpha* spp.: **Java** —, *Indigofera arrecta, I. sumatrana:* **native** — (Aus.), *I. australis, Swainsona* spp.: — **weed** (W.I.),
Indigofera suffruticosa: **wild** — (W.I.), *I. anil.* var. *polyphylla:* **Yoroba** or
W.Af. —, *Lonchocarpus cyanescens.*

Inga The sweet pulp of the pods of several C.Am. spp. is eaten and relished:
some spp. grown as shade for coffee and other crops.

Ink berry *Phytolacca americana,* N.Am. cult. and naturalized elsewhere:
fr. w. inky juice: spread by birds; *Ilex glabra,* N.Am.: *Randia aculeata,* W.I.,
— **vine,** *Passiflora suberosa,* W.I.

Innocence *Houstonia caerulea,* N.Am., blue or white fls., cult.

Inoy (W.Af.) *Poga oleosa,* yields seed kernel oil and timber.

Insect powder plant *See* pyrethrum.

Insecticides, plant W. the appearance of the chlorinated and other synthetic
insecticides the veg. insecticides are of lesser importance now than formerly.
Pyrethrum, derris and tobacco or nicotine are still used. Others may be used
for special purposes. A large number of pls. are known to possess insecticidal
properties to some degree, e.g. spp. of – *Anabasis, Celastrus, Dolichos,
Haplophyton, Heliopsis, Lonchocarpus, Millettia, Pachyrhizus, Ryania, Schoenocaulon, Tephrosia* etc.

Insectivorous plants *See* carnivorous plants.

Interrupted fern *Osmunda claytoniana,* N.Hemisph., cult.

Inula Cult. orn. herbaceous pls., for fl. border or wild gdn.

Iochroma Attractive trop. Am. shrubs and small trees, tubular fls., cult.

Iodine bush *Mallotonia gnaphalodes,* W.I., C.Am.

Ionopsidium *I. acaule,* Portugal, small tufted ann., cult., rock gdn.

Ionopsis Trop. Am. orchids, cult., small fls. in racemes.

Ipecacuanha Drug consisting of the dried rhizome of *Cephaelis ipecacuanha*
(Braz. or Rio ipecacuanha) and *C. acuminata* (Cartagena, Nicaragua or Panama
ipecacuanha). Emetine and cephaeline are the main alkaloids present. The
drug may be used in bronchitis, croup and whooping cough. The name
ipecacuanha has been used for other pls. **Am.** —, *Gillenia stipulata:* **bastard**

—, *Asclepias curassavica:* **black** —, *Cephaelis emetica:* **Carolina** —, *Euphorbia ipecacuanha:* **false** —, *Richardia scabra:* **Goa** —, *Naregamia alata:* **greater** —, *Psychotria emetica:* **Ind.** —, *Cryptocoryne spiralis:* — **spurge**, *Euphorbia ipecacuanha:* **striated** —, *Cephaelis emetica:* **undulated** —, *Richardia* spp.: **white** —, *Ionidium ipecacuanha:* **wild** —, *Euphorbia ipecacuanha*, *Asclepias curassavica.*

Ipoh *See* arrow poisons.

Ipomoea Cult. orn. climbers, trop. and subtrop., including 'morning glory' (*I. tricolor*) and 'star ipomoea' (*I. coccinea*): the drug ipomoea, also called scammony or jalap, is the dried root of *Ipomoea orizabensis*, Mex. and is used for the production of ipomoea resin, a powerful purgative. Several other spp. have local med. uses. Seeds of some spp. ('morning glory') have been used for hallucinatory purposes.

Iré rubber *Funtumia elastica.*

Iresine Trop. Am. pls. w. orn. fol., favoured for 'bedding out'.

Iris This large genus (300 spp.) includes many popular gdn. pls. with showy fls.: numerous hybrids and cultivars: divided into several groups hort. Some spp. have local med. uses. **Algerian** —, *I. unguicularis:* **beachhead** —, *I. hookeri*, N.Am.: **bearded** —, *I. germanica:* **Cape** —, *Moraea* spp.: **German** —, *I. germanica:* **gladdon** —, *I. foetidissima:* **Jap.** —, *I. kaempferi:* **morning** — (Aus.), *Orthrosanthus multiflorus:* **native** — (Aus.), *Patersonia glabrata:* **N.Z.** —, *Libertia ixioides:* **red** —, *I. fulva:* **Siberian** —, *I. sibirica:* **snake's head** —, *Hermodactylus tuberosus:* **stinking** —, *Iris foetidissima:* **white** — (Aus.), *Diplarrhena moraea:* **wild** — (S.Af.), *Dietes* spp.

Iroko (Nigeria) *Chlorophora excelsa*, a much used valuable hardwood of W.Af., also called Af. teak, odum (Ghana) and mvule (E.Af.).

Ironbark Name used in Aus. for various spp. of *Eucalyptus*, many being commercial timbers, **grey** —, *E. paniculata:* **red** —, *E. crebra:* — **orchid**, *Dendrobium aemulum:* — **shrub** (W.I.), *Sauvagesia erecta*, Aus., *Bridelia exaltata.*

Iron cross plant *Begonia masoniana.*

Iron vine *Desmodium* spp., W.I.: — **weed**, *Vernonia* spp., N.Am.: — **wort**, *Sideritis lanata.*

Ironwood Name used in various countries for certain very hard woods, e.g. Borneo, *Eusideroxylon zwageri:* India, *Mesua ferrea:* N.Am., *Ostrya virginica*, used for mallets, tool handles etc.; *Olneya tesota*, wood too hard for ordinary tools, used for arrow heads by Amerindians: Af., *Casuarina, Copaifera, Lophira, Olea, Toddalia:* Persia, *Parrotia persica:* Aus., *Choricarpia subargentea.*

Isano oil (W.Af.) *Ongokea klaineana*, seeds yield drying oil.

Isoloma Trop. Am. orn. pls., cult., esp. *I. erianthum*, red tubular fls.

Isopogon Evergreen proteaceous shrubs, Aus., fls. in showy heads.

Ispaghul seeds *Plantago ovata*, As. Minor, Ind., seeds mucilaginous, med., used like psyllium seeds to counteract constipation, also called spogel.

Istli fibre *See* ixtle.

Ita palm *Mauritia flexuosa*, Guyana, ls. much used for thatching.

Italian alder *Alnus cordata:* — **bellflower**, *Campanula isophylla*, cult. orn. pl., a room pl.: — **corn salad**, *Valerianella eriocarpa:* — **cypress**, *Cupressus sempervirens:* — **millet**, *Setaria italica:* — **poplar**, *Populus nigra:* — **rye**,

Lolium italicum: — **senna,** *Cassia obovata:* — **whisk,** *Sorghum dochna* var. *technicum,* inflor. used for brushes and brooms.

Iva Wine or liqueur, C.Eur., flavoured with *Achillea moschata.*

Ivory nut *Phytelephas macrocarpa,* C.Am., palm, veg. ivory, used for orns.: — **wood,** *Siphonodon australis,* Aus. rain forest tree, used for carving.

Ivy *Hedera* spp., cult. evergreen climbers: name also used for other genera. **Boston** — (Am.), *Parthenocissus tricuspidata* (= *Ampelopsis veitchii*): **Canary Island** —, *Hedera canariensis,* orn. ls., popular room pl.: **Coliseum** —, *Cymbalaria muralis,* Medit.: **common** —, *Hedera helix,* Eur., cult., numerous vars., some variegated (*see Hedera* in Willis): **flowering** — (E.Af.), *Senecio macroglossus:* **Irish** —, *Hedera hibernica,* cult., useful ground cover under trees: **Jap.** —, *H. japonica:* **poison** —, *Rhus toxicodendron:* **tree** —, × *Fatshedera lizei,* cult. orn. (a bigeneric hybrid between *Fatsia* and *Hedera*).

Ixia *Ixia* spp., S.Af. bulbous pls. noted for their bright colours and fragrance, cult., many hybrids and cultivars. **Blue** —, *I. columnaris:* **green** —, *I. viridiflora.*

Ixora Evergreen shrubs or small trees, trop. and subtrop., many very handsome, widely cult., esp. *I. coccinea,* several vars. and hybrids. **Bearded** —, *I. barbata:* **Chinese** —, *I. chinensis:* **pink** —, *I. rosea:* **red** —, *I. coccinea:* **yellow** —, *I. lutea.*

Ixtle or **ixtli fibre** *Agave* spp., Mex., a coarse l. fibre much used in brush manufacture, the main spp. being *A. funkiana* (Jamauve ixtle) and *A. lecheguilla* (= *A. heteracantha*) (tula ixtle). Fibre from *Yucca* spp. may also be called ixtle (palma ixtle).

J

Jaborandi Name used in S.Am. for various aromatic or pungent pls. (*Piperaceae, Rutaceae*) that increase the flow of saliva when chewed. The drug jaborandi consists of the dried ls. of *Pilocarpus* spp., notably *P. microphyllus* (**Maranhan** —), *P. jaborandi* (**Pernambuco** —), *P. pennatifolius* (**Paraguay** —), *P. trachylopus* (**Ceara** —), *P. spicatus* (**Aracati** —), *P. racemosus* (**Guadeloupe** —).

Jaboticaba (Braz.) *Myrciaria,* esp. *M. cauliflora,* (Braz. grape), fr. ed., used for wine and jelly.

Jaca *See* jak.

Jacaranda *Jacaranda mimosifolia* (= *J. ovalifolia*), Braz., a widely grown orn. and street tree in warm cli., bears violet-blue fls. in great profusion. Name used for various S.Am. commercial timbers, *Dalbergia nigra, Jacaranda* spp., *Machaerium* spp.

Jack-ass clover *Wislizenia refracta* SW. U.S. a common desert ann.: — **in-a-box,** *Hernandia peltata,* W.I., l. juice a depilatory: — **in-a-box orchid,** *Caladenia roei,* Aus.: — **in-the-pulpit,** *Arum maculatum, Arisaema* spp.: — **fruit,** *see* jak.

Jacobaea *Senecio elegans,* S.Af., cult. orn. fl., several vars.

Jacobean lily *Sprekelia formosissima*, Mex., cult., large striking fls.

Jacobinia Trop. Am., cult. orn. fls., some are room pls.

Jacob's coat *Alternanthera bettzickiana*, Braz., cult., orn. colourful foliage: — **ladder**, *Polemonium caeruleum*, cult. orn. pl.: *Smilax herbacea*, N.Am.: — **staff**, *Verbascum thapsus*.

Jacquemontia *J. pentantha* (=*J. violacea*), trop. Am., cult. orn. climber.

Jaggery Crude or unrefined sugar prepared from juice of sugar cane or various palms, *Arenga, Borassus, Cocos* etc.

Jak Or jack, *Artocarpus heterophyllus*, trop. As., v. large ed. fr., one of the world's largest, may weigh over 40 kg, produced on trunk and larger branches: pulp and seeds eaten, usu. cooked, several vars. Tree yields a valuable timber, the standard wood of Sri Lanka.

Jalap *Ipomoea* (=*Exogonium*) *purga*, Mex., climber, wild and cult.: the drug consists of the tubercles which contain 9–18 % of resin, a powerful purgative. **Braz.** —, *I. tuberosa* (=*Piptostegia pisonis*): **false** —, *Mirabilis jalapa* (rs. used as adulterant): **Ind.** —, *Ipomoea turpethum*.

Jamaica bark *Exostema:* — **cherry**, *Muntingia calabura:* — **dogwood**, *Piscidia erythrina:* — **ebony**, *Brya ebenus:* — **fever plant**, *Tribulus cistoides:* — **honeysuckle**, *Passiflora laurifolia:* — **horse bean**, *Canavalia ensiformis:* — **mountain pride**, *Spathelia simplex:* — **pepper**, *Pimenta officinalis*, allspice: **⌐ plum**, *Spondias mombin* (= *S. lutea*), *S. purpurea:* — **quassia**, *Picraena excelsa:* — **sarsaparilla**, *Smilax regelii:* — **sorrel**, *Hibiscus sabdariffa*.

Jamba (Ind.) *Eruca sativa*, seeds yield ed. oil., substitute for rape.

Jamberberry *Physalis ixocarpa*, cult., frs. for jam or stewed.

Jambolan *Syzygium cumini* (= *Eugenia jambolana*), Java plum, cult., trop., fr. ed.

Jambos, wild (S.Af.) *Eugenia zeyheri*, frs. eaten.

Jamestown weed (Am.) *Datura stramonium.*

Janatsi (Jap.) *Debregeasia edulis*, fr. ed., useful fibre.

Japan or **Japanese alder** *Alnus japonica:* — **anemone**, *Anemone hybrida:* — **apricot**, *Prunus mume:* — **aralia**, *Fatsia japonica:* — **artichoke**, *Stachys sieboldii:* — **barberry**, *Berberis thunbergii:* — **buckthorn**, *Rhamnus japonica:* — **cedar**, *Cryptomeria japonica:* — **chestnut**, *Castanea crenata:* — **climbing fern**, *Lygodium:* — **clover**, *Lespedeza striata:* — **fleeceflower**, *Polygonum cuspidatum:* — **galls**, *Rhus semialata:* — **ginger**, *Zingiber mioga:* — **honeysuckle**, *Lonicera japonica:* — **hop**, *Humulus japonica:* — **hyacinth**, *Ophiopogon* spp.: — **knotweed**, *Polygonum cuspidatum:* — **lacquer**, *Rhus verniciflua:* — **laurel**, *Aucuba japonica:* — **lilac**, *Syringa amurensis:* — **maple**, *Acer palmatum:* — **medlar**, *Eriobotrya japonica:* — **pear**, *Pyrus pyrifolia:* — **pepper**, *Zanthoxylum piperitum:* — **persimmon**, *Diospyros kaki:* — **privet**, *Ligustrum japonicum:* — **raisin tree**, *Hovenia dulcis:* — **snowflower**, *Deutzia* spp.: — **toad lily**, *Tricyrtis hirta:* — **wax tree**, *Rhus succedanea:* — **wisteria**, *Wisteria floribunda:* — **yew**, *Taxus cuspidata.*

Japonica *Chaenomeles speciosa* early fl. orn. shrub, widely cult.

Jardinière or trough garden. Popular form of indoor gdn. in which house or room pls., maintained in individual pots or containers, are placed closely together in a trough, usually on a special stand or table. Contrasts in height,

JARRAH

l. colour, growth habit or texture of fol. are usually sought by careful arrangement. Fl. pls. may of course be introduced or changes made from time to time.

Jarrah *Eucalyptus marginata,* W.Aus., a hard, heavy commercial timber.

Jasmine *Jasminum* spp. climbing shrubs in warm and cold cli., handsome fls., often scented, some cult. for perfume. **Arabian** —, *J. sambac,* cult. trop.: **bastard** — (W.I.), *Cestrum:* — **box,** *Phillyrea:* **Cape** —, *Gardenia* spp.: **Carolina** —, *Gelsemium sempervirens,* cult.: **Catalonian** —, *J. grandiflorum:* **Cayenne** —, *Catharanthus roseus* (= *Vinca rosea):* **common** —, *J. officinale:* **false** — **root,** *Gelsemium sempervirens:* **grape** —, *Ervatamia coronaria:* **hill** —, *Trachelospermum fragrans,* Ind., cult.: **Italian** —, *J. grandiflorum:* **Maori** —, *Parsonsia,* N.Z.: **rock** —, *Androsace chamaejasme:* **Spanish** —, *J. grandiflorum:* **tree** —, *Holarrhena febrifuga:* **W.I.** —, *Faramea occidentalis:* **white** —, *J. officinale:* **wild** — (W.I.), *Tabernaemontana psychotriifolia, Cestrum latifolium:* **willow leaved** —, *Cestrum parqui:* **winter** —, *J. nudiflorum,* fls. yellow, widely cult.: **yellow** —, *J. humile,* Ind., cult.

Jatropha Several spp. cult., trop., fl. shrubs, e.g. *J. integerrima* (= *J. pandurifolia)* (fiddle-leaved *Jatropha,* w. red fls.), *J. podagrica, J. gossypiifolia, J. multifida:* some spp. w. local med. uses, e.g. *J. curcas* (physic nut).

Java almond *Canarium commune,* common shade or street tree, trop., kernel ed.: — **cardamom,** *Amomum maximum:* — **fig,** *Ficus benjamina:* — **grass,** *Polytrias praemorsa,* trop., lawn grass: — **indigo,** *Marsdenia tinctoria:* — **long pepper,** *Piper retrofractum:* — **plum,** *Syzygium cumini.*

Jelutong *Dyera costulata,* Malaysia, yields commercial timber, a much used, light weight hardwood: also, from tapping, a rubbery substance used in making chewing gum.

Jequerity seeds *See* crab's eyes.

Jequitiba *Cariniana* spp., Braz., a commercial timber.

Jereton *Didymopanax morototoni,* W.I., large tree, used for matches.

Jericho, rose of *Anastatica hierochuntia* (*see* Anastatica in Willis).

Jerusalem artichoke *Helianthus tuberosus,* a common veg. and stock feed: name inappropriate for an Am. pl., 'sunroot' a suggested alternative. — **cherry,** *Solanum pseudocapsicum:* — **cross,** *Lychnis chalcedonica:* — **rye,** *Triticum polonicum:* — **sage,** *Phlomis fruticosa.*

Jessamine *Jasminum* spp., *Plumeria alba:* **star** —, *Trachelospermum jasminoides:* **yellow** —, *Gelsemium sempervirens,* N.Am. (*see* jasmine).

Jesuits' bark *Cinchona officinalis* (quinine): — **nut,** *Trapa natans:* — **tea,** *Psoralea glandulosa.*

Jew's apple *Solanum melongena:* — **citron,** *Citrus medica* var.: — **mallow,** *Corchorus olitorius:* — **plum,** *Spondias cytherea* (= *S. dulcis).*

Jewel weed (Am.) *Phlomis, Impatiens.*

Jiggerwood (W.I.) *Bravaisia integerrima,* fine-grained, whitish wood.

Jimmy weed *Haplopappus laricifolius,* SW. U.S.

Jimson weed (Am.) *Datura stramonium, D. metel,* cosmop. weeds.

Jippi-jappa (Jam.) *See* Panama hat plant (*Carludovica insignis).*

Jiquitiba *Cariniana* spp., Braz. a valuable commercial timber.

Joannesia *Joannesia* spp., trop. Am. seed oil, med., good timber.

Job's tears *Coix lacryma-jobi,* trop. and subtrop., cult., the hard-shelled covering of the frs. (usu. grey) causes them to be valued for orn. purposes, beads, rosaries etc. Other (softer fruited) vars. grown for food in the East (Khasia hills and Burma).

Jobo *Spondias mombin* (= *S. lutea*), cult., trop., ed. fr. ('hog plum').

Joe pye weed *Eupatorium purpureum*, N.Am., and other spp.

John Crow's nose (W.I.) *Scybalium* spp. (= *Phyllocoryne*): **John-go-to-bed,** *Tragopogon pratensis:* **Johnny-jump-up,** *Viola tricolor:* **John's cabbage,** *Hydrophyllum virginianum.*

Johnson grass *Sorghum halepense*, has been used for forage, warm cli.: a troublesome weed, esp. w. sugar cane.

Joint weed or **wood** *Polygonella*, N.Am.

Jo-jo weed *Soliva pterosperma*, S.Am., troublesome lawn weed in Aus. and Calif., fr. has needle-sharp spines, penetrating the flesh.

Jojoba *Simmondsia chinensis* (= *S. californica*) SW. U.S., Mex., desert shrub, imp. browse pl., seeds ed., yield oil.

Jolly Brown (W.I.) *Solanum melongena*, brinjal, cult.

Jongkong *Dactylocladus stenostachys*, Borneo, a commercial timber.

Jonquil *Narcissus jonquilla*, cult. orn. fl., fragrant, used in perfumery.

Joseph's coat *Amaranthus gangeticus* (= *A. splendens*), trop., cult., colourful fol. variegated green, brown, red, yellow etc.

Joshua tree *Yucca brevifolia* and other spp., SW. U.S., Mex.

Jowar (Ind.) Guinea corn or sorghum.

Joyweed *Alternanthera* spp., trop. and subtrop.

Jubaea *J. chilensis* (=*J. spectabilis*), cult. orn. palm (Chilean wine or honey palm), kernels ed.

Judas' bag - *Adansonia digitata* (baobab): — **tree,** *Cercis siliquastrum*, Medit., cult. orn. tree, showy purple fls. appear before ls.: said to have been the tree on which Judas hanged himself.

Jug orchid (Am.) *Pterostylis recurva.*

Juglans *See* walnut.

Jujube *Zizyphus jujuba*, *Z. mauritiana*, Old world trop. and subtrop. ed. fr. used in many ways, often dried like dates: tree cult. in Orient for centuries, many vars.

Jumbie bead (W.I.) Name used for various orn. seeds, i.e. *Abrus precatorius*, *Adenanthera pavonina*, *Erythrina* spp., *Ormosia dasycarpa* etc.

Jumping bean The seeds of some trop. Am. *Euphorbiaceae* (Mex.), become infected with certain insect larvae. These make jerking movements, esp. when stimulated, as with heat, causing the seeds to move or 'jump'. 'Jumping' ceases when the larvae die or emerge. Such seeds are commonly sold as curiosities. They are mainly those of spp. of *Sebastiania*, esp. *S. pavoniana*, and the larvae those of a small moth, *Cydia saltitans*. Seeds of *Sapium* (*S. biloculare*) are also recorded as jumping beans.

Jump-up-and-kiss-me (W.I.) *Portulaca pilosa*, a useful edging pl.

June bean *Canavalia lineata*, N.Am.: — **berry,** *Amelanchier canadensis:* — **grass** (Am.), *Danthonia spicata*, *Phleum pratense:* — **rose** (Jam.), *Lagerstroemia speciosa.*

Jungle rice *Echinochloa colona*, a weed, bad in sugar cane in Jam.: *E. crusgalli*, seeds eaten, trop. and subtrop.

JUNIPER

Juniper *Juniperus* spp. Trees and shrubs, widely distributed in N. Hemisph.: fragrant, usu. highly coloured wood w. many uses: some spp. favoured for lead pencils: oil may be distilled from wood. The frs. of the common juniper (*J. communis*) are used for flavouring gin, liqueurs and cordials. **Calif.** —, *J. californica:* **Canary Isles.** —, *J. cedrus:* **Chinese** —, *J. chinensis;* **E.Af.** —, *J. procera* (pencils): **Grecian** —, *J. excelsa:* **Ind.** —, *J. macropoda:* **Mex.** —, *J. mexicana:* **pencil** —, *J. virginiana,* N.Am. oil, many hort. vars.: **Rocky Mt.** —, *J. scopulorum:* **Sierra** —, *J. occidentalis:* **Phoenician** —, *J. phoenicea,* Medit., wood used for house building in classical times.

Juno's tears *Verbena officinalis* (Vervain).

Jupiter's beard *Anthyllis barba-jovis, Sempervivum tectorum:* — **distaff,** *Salvia glutinosa.*

Jura turpentine *Picea abies* (=*P. excelsa*), common Eur. spruce.

Jute Or gunny, *Corchorus* spp., two spp. yield true jute or the jute of commerce – *C. capsularis* (white jute) and *C. olitorius* (tossa or upland jute). The former is more widely grown in Bengal (about 75 % of the acreage). It withstands water-logging in the later stages of growth better than *C. olitorius.* Jute substitute fibres are – **Bimlipatam** —, *Hibiscus cannabinus* and **Am., Chinese** or **Manchurian** —, *Abutilon avicennae.* Another is **Congo** —, *Urena lobata.*

K

Kaempferia *K. aethiopica,* trop. Af., rhizome aromatic, used as spice: some spp. med. or cult. orn. pls.: *K. galanga,* an oriental spice. *See* galangal.

Kaffir boom (S.Af.) *Erythrina caffra* and other spp. orn. red fls., red seeds used as beads: — **bread,** *Encephalartos* spp.: — **corn,** *Sorghum* spp.: — **lily,** *Clivia* spp., *Schizostylis coccinea:* — **orange,** *Strychnos spinosa:* — **plum,** *Harpephyllum caffrum, Odina* (=*Lannea*) *caffra:* — **potato,** *Plectranthus* (=*Coleus*) *esculentus:* — **thorn,** *Lycium afrum,* a hedge pl.; *Acacia* spp.

Kagne butter *Allanblackia oleifera,* trop. Af. seeds yield an ed. fat.

Kahakaha (N.Z.) *Collospermum hastatum, Astelia solandri.*

Kahikatea (N.Z.) *Podocarpus dacrydioides,* an imp. timber tree.

Kahua bark (Ind.) *Terminalia arjuna,* used in tanning.

Kaivum fibre (Ind.) *Helicteres isora,* a useful bast fibre.

Kaki Or Jap. persimmon, *Diospyros kaki,* choice ed. fr., widely cult., subtrop., fr. eaten fresh or dried, numerous vars.

Kaku oil (W.Af.) *Lophira alata* (=*L. procera*), oil from seeds.

Kalamet (Burma) *Mansonia gagei,* fragrant wood, a Burmese cosmetic.

Kalaw (Burma) *Hydnocarpus castaneus,* b. w. local med. uses.

Kalanchoe Cult. orn. fls., some spp. (e.g. *K. blossfeldiana*) good room pls.

Kale Or borecole, *Brassica oleracea* var. *acephala,* widely cult. hardy veg. and stock feed, many vars. 'curled', 'hardy sprouting', 'peren.', 'thousand head', etc.: variegated forms may be used for winter decoration. **Abyssinian** —, *Crambe abyssinica,* a potential oil seed crop: **Ind.** —, *Xanthosoma atrovirens:* — **rape,** *see* rape: **sea** —, *Crambe maritima.*

Kalkoentje (S.Af., 'little turkey') *Gladiolus* spp., *Tritonia* spp.

Kamala *Mallotus philippensis*, trop. As., red dye from frs.

Kamassi (S.Af.) *Gonioma kamassi*, Cape or Knysna boxwood.

Kanga butter *Pentadesma butyracea*, W.Af., ed. fat from seeds.

Kangaroo apple *Solanum aviculare:* — **fern**, *Microsorium diversifolium:* — **grass**, *Themeda australis* and other spp., *Anthistiria:* **black** — **paw**, *Macropidia fuliginosa:* **red and green** — **paw**, *Anigozanthos manglesii* (the floral emblem of W.Aus.): **red** — **paw**, *Anigozanthos rufus:* **green** — **paw**, *Anigozanthos viridis:* **yellow** — **paw**, *Anigozanthos flavidus:* — **thorn**, *Acacia armata* Aus., a spreading spiny shrub, useful barrier pl.: — **vine**, *Cissus antarctica*, Aus.

Kauluang *Sarcocephalus cordatus*, Thailand, Burma, a trade timber.

Kanniedood (S.Af.) *Aloe variegata*, partridge-breasted aloe, variegated ls.

Kansas gayfeather *Liatris callilepis*, cult. gdn. fl.: — **thistle**, *Solanum rostratum*, N.Am.

Kaolang (China) *Sorghum nervosum* or other spp., a much used cereal.

Kapa cloth *Broussonetia papyrifera*, *see* tapa cloth.

Kapok Or silk cotton, *Ceiba pentandra*, cult. trop. seed floss used in upholstery, cushions, life saving equipment etc.: inferior kapok from *Bombax malabaricum*, trop. As.

Kapong *Tetrameles nudiflora*, Thailand, a commercial timber.

Kapur *Dryobalanops aromatica*, *D. lanceolata* or other spp., Malaysia, imp. commercial, medium-weight hardwoods.

Karaka (N.Z.) *Corynocarpus laevigata*, cult. evergreen tree, fr. ed.

Karamu (N.Z.) *Coprosma lucida* and other spp. showy frs., cult.

Karanda (Ind.) *Carissa carandas*, shrub, fr. used for pickles or preserves.

Karanja (Ind.) *Pongamia glabra*, trop. Asia, seeds med., yield oil.

Kariba weed *Salvinia auriculata*. Soon after construction of the Kariba Dam in Rhodesia, conditions were ideal for the explosive growth of this floating water fern. With changing conditions its activity diminished and it became restricted to the more sheltered, less windy areas.

Kariyat (Ind.) *Andrographis paniculata*, med., bitter tonic and stomachic.

Karree (S.Af.) *Rhus lancea.*

Karri *Eucalyptus diversifolia*, W.Aus., a valued, hard, heavy, trade timber: — **tree** (N.Am.), *Paulownia tomentosa*.

Karroo buchu (S.Af.) *Diosma succulenta:* — **violet**, *Aptosimum indivisum*, *A. depressum.*

Katon *Sandoricum koetjape*, trop. As., a commercial timber.

Katsura *Cercidiphyllum japonicum*, Jap., a commercial timber.

Kaunghmu *Anisoptera scapula*, Burma, a commercial timber.

Kauri *Agathis australis*, a well known N.Z. tree, useful timber, source of kauri resin. **E.I.** —, *Agathis alba:* **Fijian** —, *A. vitiensis:* **Queensland** —, *A. robusta*, *A. palmerstonii*, *A. microstachya.*

Kava *Piper methysticum*, cult. South Sea Islands, r. used for national beverage.

Kawaka (N.Z.) *Libocedrus plumosa* (= *L. doniana*), a handsome dark wood.

Kawa-kawa *Piper methysticum*, *see* kava.

Keaki (Jap.) *Zelkova acuminata*, tough, elastic and durable timber.

Kecks or **kex** The dry hollow stems of hogweed or other umbellifers.

KEDAM

Kedam (Ind.) *Anthocephalus cadamba*, wood used for building, for tea boxes etc.

Kei apple (S.Af.) *Doryalis* (=*Aberia*) *caffra*, vigorous shrub w. large stiff thorns, a defensive hedge pl., yellow frs. used for preserves.

Kelly Or corn grass, *Rottboellia exaltata*, trop. fodder grass, also a weed.

Kempas *Koompassia malaccensis*, Malaya, a commercial timber.

Kenaf *Hibiscus cannabinus*, trop., esp. Ind., a commercial textile fibre or jute substitute: also called Deccan or ambari hemp, and Bimlipatan jute.

Kendal green Dye made by mixing woad and the yellow fls. of *Genista tinctoria*.

Kendyr fibre *Apocynum venetum*, Eur. As., stem fibre used locally for ropes fishing nets etc.: pl. also yields a seed floss.

Kenguel seed *Silybum marianum*, roasted seeds a coffee substitute.

Kennedia Aus. climbers, cult., fls. usu. red, some almost black.

Kentia palm *See Howea.*

Kentucky coffee tree *Gymnocladus dioica* (=*G. canadensis*), cult., seeds a coffee substitute; — **blue grass**, *Poa pratensis* (meadow grass), imp. agric. grass.

Keratto *Agave* spp., a Mex. fibre.

Kerguelen cabbage *Pringlea antiscorbutica:* occurs only on windswept Kerguelen and Crozet Isles.

Kermek *Limonium gmelinii* and *L. latifolium:* rs. for tanning in Russia, also gdn. pls.

Kerosene bush (Aus.) *Richea scoparia:* — **wood** (Aus.), *Halfordia scleroxyla*.

Kerria *Kerria japonica*, double fl. form much cult.: **white** —, *Rhodotypos scandens* (=*R. kerrioides*).

Keruing *Dipterocarpus* spp., well known trade timber, from some 20 spp., Malaysia.

Ketembilla *Doryalis* (= *Aberia*) *gardneri*, ed. fr., cult. (Ceylon gooseberry).

Keurboom (S.Af.) *Virgilia divaricata*, fragrant pink-mauve fls.

Kew pineapple The smooth cayenne pineapple widely distributed by Kew in early days: — **plant**, *Asystasia gangetica* cult. trop. climber, tubular yellow fls.

Keyaki (Jap.) *Zelkova serrata*, wood used for building and furniture.

Khahi weed (S.Af.) *Tagetes minuta: Inula graveolens*.

Khat Or qat, *Catha edulis*, dried ls. a stimulant and masticatory in NE.Af. and Arabia.

Khesari (Ind.) *Lathyrus sativus*, fodder crop and low grade pulse.

Khus-khus *Vetiveria zizanioides*, cult. As. trop., fragrant rs. distilled for oil (vetiver oil): used for fans, punkas, orn. articles etc.

Kiaat (S.Af.) *Pterocarpus angolensis*, a popular and handsome furniture wood.

Kibaazi (E.Af.) *Tephrosia noctiflora, T. ehrenbergiana*, low shrubs.

Kidney bean *Phaseolus vulgaris* (French or dwarf bean): — **fern**, *Trichomanes:* — **vetch**, *Anthyllis vulneraria:* — **wood** (N.Am.), *Eysenhardtia polystachya*.

Kif Moroccan word for cannabis or marihuana.

Kihopwe (E.Af.) *Jacquemontia tamnifolia*, a weed, ls. used as spinach.

Kikuyu grass *Pennisetum clandestinum*, used for pasture and lawns in warm cli.

Killarney filmy fern *Trichomanes radicans*.

Kilmarnock weeping willow *Salix caprea* var. *pendula*.

King cup *Caltha palustris* (marsh marigold) cult., many vars.: — **devil**, *Hieracium pratense*, an aggressive weed: — **fern**, *Todea barbara:* — **fisher daisy**, *Felicia bergeriana*, S.Af., bright blue fls., cult.: — **in-his-carriage orchid** (Aus.), *Drakaea glyptodon:* —'s **mantle**, *Thunbergia erecta*, cult.: — **of the meadow**, *Thalictrum polygamum*, N.Am.: — **plant**, *Anoectochilus regalis:* — **wood**, *Dalbergia cearensis*, Braz.; *Astronium fraxinifolium*, Braz., valued for inlaying and marquetry: — **of the woods orchid**, *Anoectochilus regalis*.

Kinnikinnik (N.Am.) *Arctostaphylos uva-ursi*, *Cornus*.

Kino An astringent resin-like substance obtained by tapping various trees, used med. and to some extent in local tanning. **Af.** —, *Pterocarpus erinaceus:* **Aus.** —, *Eucalyptus* spp., esp. *E. resinifera*, *E. rostrata*, *E. amygdalina:* **Bengal** —, *Butea monosperma* (=*B. frondosa*)*:* **E.Ind.** or **Malabar** —, *Pterocarpus marsupium:* **Jam.** —, *Coccolobus uvifera*.

Kirengeshoma *K. palmata*, Jap., tall peren. w. yellow fls., cult.

Kiri wood (Jap.) *Paulownia tomentosa*, orn. tree, wood used in Jap.

Kirschwasser Alcoholic beverage made from cherries.

Kisambale (E.Af.) *Lobelia fervens*, a spinach pl.

Kisidwe (W.Af.) *Allanblackia floribunda*, large forest tree, seeds yield ed. fat.

Kitaibelia *K. vitifolia*, E.Eur., mallow-like peren., showy fls., cult.

Kite tree (S.Af.) *Nuxia* (= *Lachnopylis*) *floribunda*.

Kittool or **kitul** (Ind.) *Caryota urens*, strong brush fibre from l. bases.

Kiwi fruit *Actinidia chinensis*, cult. commercially in N.Z.

Klapperbos (S.Af.) *Nymania capensis*.

Kleinhovia *K. hospita*, Indo-Malaysia, orn. or street tree w. showy pink fls.

Kleinia Af., Madag., Ind., Canary Isles, succulents w. red, orange or whitish fls., cult.

Knapweed *Centaurea nigra:* **brown** —, *C. nemoralis:* **great** —, *C. scabiosa*.

Knawel *Scleranthus annuus:* **perennial** —, *S. perennis*.

Kniphofia *See* red hot poker.

Knobthorn or **knobwood** (S.Af.) *Fagara* spp. esp. *F. capensis*, *F. davyi:* knobs on trunk often tipped w. thorns: also *Acacia* spp., e.g. *A. nigrescens*, *A. pallens*.

Knol-kohl Synonym of kohlrabi.

Knotweed or **knotgrass** *Polygonum aviculare; Illecebrum verticillatum:* — (**Am.**), *Paspalum distichum:* **giant** —, *Polygonum sachalinense:* **Jap.** —, *P. cuspidatum:* **sea** —, *P. maritimum*.

Kochia *K. scoparia*, cult. fol. pl., ls. turn red in autumn.

Koda millet *Paspalum scorbiculatum:* — **wood**, *Ehretia acuminata*.

Koelreuteria *K. paniculata*, China, cult. orn. tree.

Kohekohe (N.Z.) *Dysoxylum spectabile*, reddish wood used for cabinet work.

Kohlrabi *Brassica oleracea* var. *gongyloides*, a turnip-like veg. w. ls. on the swollen portion, widely grown.

Kohuhu (N.Z.) *Pittosporum tenuifolium*, cult. orn. tree, purple fls.

Kokam or **kokum butter** (Ind.) *Garcinia indica*, ed. fat from seeds, dried fr. rind used in curries.

Kokerite *Maximiliana regia*, large handsome S.Am. palm, cult., oil from frs.

Kokko *Albizia lebbek*, Ind., Burma, a commercial wood (E.Ind. walnut).

Kokoon (Sri Lanka) *Kokoona zeylanica*, inner b. used by jewellers for polishing.

Koksaghyz *Taraxacum bicorne* (= *T. koksaghyz*), 'Russian dandelion', considered for rubber during Second World War.

Kola *See* cola.

Kolkol (S.Af.) *Berzelia lanuginosa.*

Kolkwitzia *K. amabilis*, W.China, fl. shrub, pink and yellow fls.

Kongwa (E.Af.) *Commelina zambesica*, ls. used as spinach.

Konjaku (Jap.) *Amorphophallus rivieri*, ed. tubers, used for flour.

Korakaha *Memecylon umbellatum*, Ind., Sri Lanka, shrub w. showy, deep blue fls.

Koso or **kousso** *Hagenia abyssinica*, NE.Af., dried fls. med. (for tapeworm).

Kosumba (Ind.) *Schleichera oleosa* (= *S. trijuga*), hard durable wood used for agric. implements, carts, rice pounders etc.; fr. ed., seeds yield oil (Macassar oil).

Kowhai *Sophora microphylla*, tree or shrub w. golden fls., the national fl. or emblem of N.Z.: **red** —, *Clianthus puniceus.*

Krabak *Anisoptera curtisii* and allied spp., Thailand, a trade timber.

Kretek Indonesian cigarette flavoured w. powdered cloves.

Krobonko (W.Af.) *Telfairia occidentalis*, cult., large oily seeds used as food.

Kudzu vine (Jap.) *Pueraria lobata* (= *P. hirsuta*, *P. thunbergiana*) starchy rs. ed., cult. in many countries as fodder or cover pl., or for soil erosion control.

Kullam nut *Balanites orbicularis*, Somaliland, kernels yield oil.

Kumara (N.Z.) Sweet potato, *Ipomoea batatas.*

Kumbanzo *Holarrhena febrifuga*, b. med., febrifuge.

Kumbuk *Terminalia arjuna*, Ind., Sri Lanka, a commercial timber.

Kümmel *Carum carvi*, caraway.

Kumquat *Fortunella* spp. a small Oriental citrus fr., widely cult., many vars., the aromatic rind is eaten along w. the flesh: often preserved or candied: the round fruited forms are *F. japonica* and the oval *F. margarita*: **desert** — (Aus.), *Eremocitrus glauca*, used for drinks and preserves.

Kunzea Evergreen shrubs or small trees, Aus., N.Z., cult.: **silky** —, *K. pulchella.*

Kuraka nut (N.Z.) *Corynocarpus laevigatus*, seeds eaten by Maoris after roasting.

Kurakkan (Sri Lanka) *Eleusine coracana*, see ragi.

Kurchi bark (Ind.) *Holarrhena antidysenterica*, b. med., used locally for dysentry and as a substitute for ipecacuanha.

Kurdee *Carthamus tinctorius*, see safflower, imp. oil seed.

Kurrajong (Aus.) *Sterculia* or *Brachychiton* spp.

Kussum oil (Ind.) *Schleichera oleosa* (= *S. trijuga*), seeds yield oil (Macassar oil).

Kutira gum (Ind.) *Cochlospermum gossypium*, a gum of the tragacanth class w. similar commercial uses.

Kutki (Ind.) *Picrorhiza kurroa*, Himal., pl. med. (bitter tonic).

Kwatje (Indonesia) Delicacy made from water melon seeds.

Kweek grass (S.Af.) *Cynodon dactylon*, a lawn grass: also a weed.

Kwei (China) *Osmanthus fragrans*, cult., fls. for flavouring tea and confectionery.

Kweme nut (E.Af.) *Telfairia pedata*, see oyster nut.

Kydia *K. calycina*, Ind., small orn. tree w. pink or white fls., cult.

Kyetpaung (Burma) *Urceola esculenta*, a climbing shrub common in teak forests: contains rubber: fr. ed.

Kyor (Kashmir) *Sagittaria sagittifolia*, aquatic, ed. rhizomes.

L

Labdanum *See* ladanum.

Lablab bean *Dolichos lablab* (= *Lablab niger*), bonavist or hyacinth bean, widely cult. trop., a veg. (green bean) or pulse.

Labrador tea *Ledum palustre*, ls. used as tea, also *L. groenlandicum* N.Am.

Laburnum *L. anagyroides* (= *L. vulgare*) S. & C.Eur., cult. orn. shrub or small tree, showy fls., many vars.: poisonous, esp. seeds, dangerous to children: little sapwood, heartwood dark, close-grained, used for inlay work, turnery, musical instruments etc. **Cape** —, *Crotalaria capensis:* **Ind.** —, *Cassia fistula*, cult. orn. tree, trop., yellow fls. in hanging clusters: **Natal** —, *Calpurnia aurea:* **Nepal** —, *Piptanthus nepalensis:* **purple** —, + *Laburnocytisus:* **Scotch** —, *Laburnum alpinum*.

Labyrinth *See* maze.

Lac Although lac (or shellac) is an insect resin, secreted by the lac insect (*Laccifer lacca*), it is dependent upon and closely associated w. pl. life. The insect occurs naturally on many Ind. trees or pls. esp. kussum (*Schleichera oleosa*), ber (*Zizyphus mauritianus*), peepul (*Ficus religiosa*), palas (*Butea monosperma* (= *B. frondosa*)) and pigeon pea (*Cajanus cajan*).

Lacaena Fleshy-flowered C.Am. orchids, cult.

Lace-bark *Lagetta lintearia* (= *L. lagetto*), W.I., inner b. of lace-like character, removed and made into fancy articles: — **pine**, *Pinus bungeana*, China, grey bark peels off: — **tree** (N.Z.), *Plagianthus regius* (= *P. betulinus*), *Hoheria sexstylosa;* (Aus.), *Brachychiton discolor*.

Lace-leaf *Aponogeton fenestralis*, Madag., an aquatic favoured for trop. aquaria: perforated lace-like character of ls. due to patches of tissue between the veins dying away in the immature l.; also called Madagascar lace pl. — **grass**, *Eragrostis capillaris*, N.Am.: — **wood**, a trade term for 'quartered' wood of Eur. plane *Platanus hispanica* (= *P. acerifolia*).

Lachenalia S.Af. bulbous pls., cult., esp. *L. aloides* w. many vars.

Lachnaea S.Af. evergreen shrubs, often heath-like, cult.

Lacoocha *Artocarpus lacoocha*, Ind., ed. fr. 10–13 cm diam., cult.

Lacquer tree, Chinese or **Jap.** *Rhus verniciflua*, a natural varnish obtained by tapping: **Burmese** — —, *Melanorrhoea usitata:* **Pasto** —, *Elaeagia utilis*, Colombia.

Lactiferous, laticiferous or **latex-bearing plants** Widespread in the veg. kingdom, esp. in certain families such as the *Euphorbiaceae*, *Apocynaceae*, *Asclepiadaceae*, *Sapotaceae*, *Papaveraceae*, *Caricaceae*, *Compositae*, *Moraceae* etc. The latex of many has economic uses and may be extracted and used commercially, some of the better known being – plantation or Hevea rubber gutta-percha, chicle, jelutong, papain, opium, guayule rubber, castilloa rubber, Landolphia rubber, Funtumia rubber, chilte (Mex.) sorva (Braz.).

Lactuca *L. indica*, a pot-herb in the East, other spp. used similarly: *see* lettuce.

Lactucarium *Lactuca virosa*, the dried milky juice, a drug, now little used.

Ladanum Or labdanum, a fragrant resinous substance used in perfumery and in scenting soaps and tobacco. It is obtained from various spp. of *Cistus* or rock-rose in the Medit. esp. from *C. ladaniferus* (in Spain) and *C. creticus*. The twigs are boiled and the resin skimmed off the surface of the water.

Ladder fern *Nephrolepis*, handsome trop. ferns w. pinnate ls., cult., good basket ferns.

Ladino clover *Trifolium repens* var. *latum*, a large form of common white clover from Italy, rapid-growing and long-lived: the basic legume of California's irrigated pastures.

Lad's love *Artemisia abrotanum*, S.Eur., fragrant shrub, cult.

Ladies-in-a-boat *Rhoeo spathacea* (=*R. discolor*), C.Am., W.I.

Ladybird beads Name for 'crab's eyes' or 'jequerity seeds', *Abrus precatorius*.

Lady fern *Athyrium filix-femina*, cosmop. inc. Br., handsome, many forms, cult.: **Alpine** — —, *Athyrium alpestre*. — **lupine**, *Lupinus villosus*, N.Am.: — **of-the-night**, *Cestrum nocturnum*, cult., trop., scented fls. open at night; *Brunfelsia americana*, W.I. cult.

Lady's bedstraw *Galium verum*, Br.: — **comb**, *Scandix pecten-veneris*, Br.: — **delight**, *Viola tricolor:* — **eardrops**, *Brunnichia cirrhosa*, N.Am.: — **earrings**, *Impatiens capensis:* — **fingers**, *Anthyllis vulneraria; Hibiscus esculentus* (okra); *Musa* (cult. banana): — **leek**, *Allium cernuum:* — **mantle**, *Alchemilla vulgaris:* — **rue**, *Thalictrum clavatum:* — **seal**, *Tamus communis:* — **slipper**, *Cypripedium* spp.: — **smock**, *Cardamine pratensis:* — **thumb**, *Polygonum natans*, N.Am.; *P. persicaria:* — **tobacco**, *Antennaria* spp.: — **tresses**, *Spiranthes autumnalis*, Eur.; *S. sinensis*, Aus.

Laelia Handsome trop. Am. orchids, cult.

× **Laeliocattleya** Intergeneric hybrid orchids (*Cattleya* × *Laelia*), widely cult.

Lagerstroemia Cult. orn. fl. trees, trop. As., esp. *L. speciosa* and *L. indica*, some yield good timber.

Lagos rubber *Funtumia elastica*.

Lagurus *L. ovatus*, Medit., hare's tail, cult. orn. grass.

Lake cress *Armoracia aquatica*, N.Am.

Lalang *Imperata cylindrica* (=*I. arundinacea*), trop. grass weed, v. troublesome, ls. make good thatch.

Lallemantia *L. iberica*, As. Minor, *L. royleana* Ind., seeds yield drying oil.

Lamb-kill *Kalmia angustifolia*, N.Am., shrub may poison sheep.

Lamb's ears *Stachys olympica* (=*S. lanata*), cult. orn. fl.: — **lettuce**, *Valerianella locusta* (corn salad): — **quarters**, *Chenopodium album:* — **succory**, *Arnoseris:* — **tails**, *Lachnostachys*, Aus.

Lambertia Aus. shrubs, red or yellow fls., cult.

Lamprococcus Trop. Am. bromeliads, cult., orn. inflor.

Lamp wick *Phlomis lychnitis*, S.Eur., a subshrubby hoary pl.

Lamy butter *Pentadesma butyracea*, Sierra Leone, ed. fat from seeds.

Lancewood Name used for several strong elastic woods used for fishing rods and archery bows at one time, esp. *Oxandra lanceolata* and *Duguetia quitarensis* of the Caribbean region and both commercial woods. **Degame** —, *Calyco-phyllum candidissimum*, W.I.: **Honduras** —, *Lonchocarpus hondurensis:* **Moulmein** —, *Homalium tomentosum:* **N.Z.** —, *Pseudopanax crassifolius.*

Land cress Or Am. cress, *Barbarea praecox*, a salad pl.

Landolphia Several spp. in trop. Af. and Madag. have been exploited for rubber in the past.

Langsat *Lansium domesticum*, Malaysia, ed. tree fr., cult.

Lantana *L. camara*, trop. Am., a prickly stemmed shrubby weed in many warm countries; some vars. cult.

Lantern tree, Chilean *Crinodendron hookerianum*, cult., crimson urn-like fls.

Lapacho *Tabebuia avellanedae*, Argen., valuable fine-grained timber.

Lapageria *L. rosea*, Chile, orn. climbing shrub, ed. fr., cult.

Lapland rosebay *Rhododendron lapponicum.*

Laportea Trop. or subtrop. pls., many w. stinging hairs (cause severe pain).

Laptandra root *Veronica virginica*, N.Am., rhizome med.

Larch *Larix* spp. Deciduous trees of the N.Hemisph., several imp. for timber, e.g. **Common** or **Eur.** —, *L. decidua* (= *L. europaea*), extensive forests in the Alps, trees reach 40 m and 1 m diam. much used for telegraph poles, pit props, boat building: **Dunkeld** —, *L.* × *eurolepis*, a hybrid (*L. decidua* × *L. kaempferi*) cult.: **Jap.** —, *L. kaempferi* (= *L. leptolepis*), much cult. in Br., resists larch canker: **Siberian** —, *L. sibirica:* **Tamarack** or **Am.** —, *L. laricina:* **western** —, *L. occidentalis*, N.Am., the largest larch.

Larch bark The b. of the common or Eur. larch, *Larix decidua*, contains tannin (8–9 %) and has been used for tanning leather, esp. in Russia. Solid and powdered tanning extracts have been prepared from it in Russia and Sweden.

Larkspur *Delphinium ajacis* (= *Consolida ambigua*), S.Eur., a common gdn. ann., many vars., fls. used to garland mummies in ancient Egypt, **dwarf** —, *D. tricorne*, N.Am.: **Carolina** —, *D. carolinianum*, N.Am.: **forking** —, *D. consolida*, N.Am., cult.: **tall** —, *D. exaltatum*, N.Am.

Lasiopetalum Aus. evergreen shrubs, cult., the calyx being the orn. part of the fl.

Latex The milky juice of some pls., contained in special ducts or tissues. (*Alismataceae, Apocynaceae, Araceae, Asclepiadaceae, Euphorbiaceae, Compositae, Moraceae, Sapotaceae*, etc.)

Lattice leaf *See* lace leaf.

Lauan Or 'Philippine mahogany', commercial timber, a group name for lightweight spp. of *Parashorea, Pentacme*, and *Shorea:* **red** —, *Shorea negrosensis* and other spp.: **white** —, *Pentacme contorta, Parashorea plicata.*

Laurel Alexandrian —, *Calophyllum inophyllum*, trop. orn. tree: **bay** —, *Laurus nobilis*, Medit., the laurel of the ancients (crown of laurel leaves) ls. much used in flavouring: **bog** —, *Kalmia polifolia*, N.Am.: **Calif.** —, *Umbellularia californica:* **cherry** —, *Prunus laurocerasus*, widely cult. evergreen, windbreaks and hedges: **Chilean** —, *Laurelia serrata* (= *L. aromatica*), a commercial timber: **Chinese** —, *Antidesma bunias:* **Ecuador** —, *Cordia alliodora*, a commercial timber: **great** —, *Rhododendron maximum*, N.Am.: **ground** —, *Epigaea repens:* **Ind.** —, *Terminalia alata*, a commercial timber: **Jap.** —, *Aucuba japonica:* **magnolia** —, *Magnolia virginiana:* **mt.** —, *Kalmia* *latifolia:* **native** —, *Anopterus glandulosa,* Aus., Tasm.: **pig** —, *Kalmia angustifolia,* N.Am.: **Portugal** —, *Prunus lusitanica:* **seaside** —, *Phyllanthus epiphyllianthus,* W.I.: **sheep** —, *Kalmia angustifolia:* **spurge** —, *Daphne laureola:* **variegated** —, *Aucuba japonica:* **Versailles** —, *Prunus laurocerasus,* broad-leaved form: **W.I.** —, *Prunus occidentalis.*

Laurustinus *Viburnum tinus,* Medit., cult. evergreen shrub, winter fl.

Lavatera Several cult. orn. pls., inc. *L. trimestris,* a colourful ann.

Lavender *Lavandula* spp. cult. orn. fls.: — **oil,** *L. officinalis,* cult., obtained by steam distillation of the fl. tops, much used in perfumery, esp. for perfuming soap: **French** —, *L. stoechas,* Medit. a med. pl. from early times: — **cotton,** *Santolina:* **sea** —, *Limonium vulgare,* cult., fl. heads used as everlastings: **spike** —, *Lavandula latifolia* (= *L. spica*), S.Eur., yields spike lavender oil: **toothed** —, *Lavandula dentata:* **wild** —, *Heliotropium curassavicum,* W.I.

Lavandin Cult. hybrid lavender, cross between ordinary lavender and spike lavender, gives better yields of oil than its parents.

Lawyer, creeping *Rubus parvus,* N.Z.: — **vine,** *Rubus australis* N.Z.

Lead grass *Salicornia virginica,* N.Am.: — **plant,** *Amorpha canescens,* N.Am., once thought to indicate the presence of lead ore: — **tree,** *Leucaena glauca,* W.I.: — **wood,** *Combretum imberbe,* S.Af.: — **wort,** *Plumbago.*

Leaf cup *Polymnia uvedalia,* N.Am.: — **of life,** *Bryophyllum pinnatum* (= *B. calycinum*) produces young pls. on the margins of leaves even under dry conditions e.g. pinned to a wall.

Leaping spider orchid *Caladenia macrostylis,* Aus.

Leather flower *Clematis versicolor* and other spp.: — **leaf,** *Cassandra calyculata* E. N.Am., cult. fl. shrub; — **plant,** *Celmisia coriacea,* N.Z., the tough ls. have been made into garments: — **wood,** *Cyrilla racemiflora,* E. U.S.; *Dirca palustris,* E. U.S.; *Eucryphia* spp., Aus., Tasm., source of commercial honey.

Lecheguilla Or Ixtli fibre *Agave lecheguilla* (= *A. heteracantha*), Mex., a brush fibre.

Lecythis Many spp. yield strong, close-grained timber in trop. S.Am., some yield ed. nuts. (Paradise nut).

Ledger bark *Cinchona ledgeriana. See* cinchona.

Ledum *L. latifolium,* N.Am. and other spp. cult. orn. shrubs.

Leek *Allium porrum,* Eur., a widely cult. veg.: lower portion of blanched ls. is the part used; said to be a native of Switzerland; the national emblem of Wales; many vars., inc. at least one yellowish in colour. **Lady's** —, *A. cernuum:* **round-headed** —, *A. sphaerocephalum,* Br.: **sand** —, *A. scorodoprasum:* **stone** —, *A. fistulosum:* **wild** —, *A. tricoccum:* — **orchid,** *Prasophyllum elatum,* Aus.

Leersia Trop. and subtrop. marsh grasses, some used as fodder in As.

Leichhardt tree (Aus.) *Nauclea orientalis.*

Lekkerbreek (S.Af.) *Ochna pulchra,* handsome small tree.

Lemon *Citrus limon* (= *C. medica* var. *limonia*) a widely cult. citrus fr. of the subtrop.: fr. much used for beverages and a source of citric acid: oil of lemon, obtained from the peel, used for flavouring: seed oil used in soap making. — **grass,** *Cymbopogon citratus,* cult. trop., oil used in perfumery, esp. soap: — **mint,** *Mentha citriodora:* — **orchid** (Aus.), *Thelymitra antennifera:* — **scented verbena,** *Aloysia citriodora,* old time gdn. favourite: — **sumac,**

LEVER WOOD

Rhus aromatica, N.Am.: — **wood**, *Aspidosperma tomentosum*, Braz., useful timber; *Calycophyllum candidissimum*, W.I., a commercial timber: *Xymalos monospora*, S.Af. wood lemon-yellow in colour: *Pittosporum eugenioides*, N.Z.

Lent lily *Narcissus pseudonarcissus:* — **tree**, *Erythrina corallodendrum*, W.I.: —**en rose**, *Helleborus orientalis*.

Lentil *Lens esculenta* (= *L. culinaris*), a widely cult. and nutritious pulse crop, esp. in Ind. (dhal): a food pl. of great antiquity in the Medit.

Leopard bane *Doronicum pardalianches*, reputed to be poisonous; *Arnica acaulis:* — **flower**, *Belamcanda chinensis*, China, mottled orange-yellow fls., cult.: — **orchid**, *Thelymitra fuscolutea*, Aus.: — **tree**, *Caesalpinia ferrea*, trop. street tree w. spotted bark; *Flindersia collina*, Aus.: — **wood**, *F. maculosa*, Aus.; *Brosimum aubletii*, Guyana.

Lepironia *L. mucronata*, SE.As., a sedge cult. in China and Indonesia, the culms much used in basket and mat making after being beaten flat.

Leprosy gourd *Momordica charantia*, trop. veg., largely used w. curries.

Leptospermum Evergreen shrubs, mainly Aus., some cult. for orn. esp. *L. scoparium*, several vars., ls. have been used as tea (tea tree).

Leptotes S.Am. orchids, mainly Braz., allied to *Cattleya*, terete or subterete fol., cult.

Lerp Or Larap, an ed. manna (insect secretion); produced by insects on *Eucalyptus dumosa*, Aus.

Leschenaultia Shrubs, usu. heath-like, showy fls., cult.: **blue** —, *L. biloba*, W.Aus.: **red** —, *L. formosa*.

Lespedeza *L. striata*, Jap. clover, a useful fodder pl., some spp. cult. for orn.

Letterwood *Brosimum aubletii*, Guyana, used for violin bows and in turnery; *Piratinera guianensis*, trop. Am., a similar wood, also called 'snakewood'.

Lettuce *Lactuca sativa*, the most imp. salad pl.: place of origin uncertain: known to have been grown in the Orient and Medit. from early times: cult. in Br. by the middle of the 17th Century if not before. Two main categories, 'cabbage' and 'cos', and numerous vars., spring, summer, winter etc. **Acrid** or **bitter** —, *L. virosa:* **blue** —, *L. pulchella*, N.Am.: **lamb's** —, *Valerianella locusta* (corn salad): **miner's** —, *Claytonia sibirica*, N.Am.: **mt.** —, *Saxifraga erosa*, N.Am.: **opium** —, *Lactuca virosa:* **perennial** —, *L. perennis*, N.Am.: **prickly** —, *L. serriola:* **squaw** —, *Hydrophyllum occidentale:* — **tree**, *Pisonia grandis* var.: **wall** —, *Mycelis* (= *Lactuca*) *muralis:* **water** —, *Pistia stratiotes:* **white** —, *Prenanthes:* **wild** —, *Lactuca canadensis*, N.Am.: **willow-leaved** —, *L. saligna*.

Leucaena *L. glauca*, cult. or naturalized in trop., a hedge pl. or windbreak: if eaten freely by horses causes mane and tail hair to fall out: seeds used as beads.

Leucocoryne Bulbous pls. from Chile, blue and white fls., cult.

Leucojum Snowflakes, bulbous pls., C.Eur., Medit., cult., some resemble snowdrops.

Leucothoe Orn. shrubs, mainly N.Am., cult.

Levant galls *See* oak galls: — **garlic**, *Allium ampeloprasum:* — **madder**, *Rubia peregrina:* — **scammony**, *Convolvulus scammonia:* — **storax**, *Liquidambar orientalis*.

Lever wood *Ostrya virginica*, N.Am., a strong tough wood ('ironwood') favoured for tool handles, mallets etc.

145

LEWISIA

Lewisia Peren. herbs, W. N.Am., fls. often showy, cult.

Leycesteria *L. formosa*, Himal., cult. orn. shrub.

Liana or **liane** Feature of trop. veg. or rain-forest is the prevalence of large woody climbers (lianes or lianas) sometimes reaching a great length and the tops of tall trees. They may also spread in contorted masses on the ground. Many belong to the following genera – *Anamirta, Ancistrocladus, Anodendron, Bauhinia, Caesalpinia, Celastrus, Calamus, Cocculus, Derris, Distictis, Entada, Landolphia, Salacia, Securidaca, Toddalia, Willughbeia*.

Liatris Perens., N.Am., genus inc. several showy gdn. pls.

Libertia Perens. w. creeping rhi., S.Hemisph., orn. fls., cult.

Libocedrus See cedar, Chilean (*Austrocedrus chilensis*), incense (*Calocedrus decurrens*): both handsome evergreen conifers (old name *Libocedrus*).

Lightning shrub (S.Af.) *Clutia pulchella*, Natal natives use the pl. as a charm against lightning.

Lign-aloes See eagle wood.

Lignum-nephriticum *Eysenhardtia amorphoides*, C.Am., tree or shrub: the wood (or chips) placed in water produces a blue fluorescence: imported to Eur. for med. use in the 16th–18th Centuries.

Lignum-vitae *Guaiacum* spp., mainly *G. officinale* and *G. sanctum*, trop. Am., W.I., the heaviest commercial wood, used for making pulleys, bowls (for the game of bowls) and in engineering where a wood is required to withstand friction without firing: the main commercial kinds are Cuban, Jamaican, San Domingo and Puerto Rican. The tree is grown as an orn. tree in trop., bears blue fls. in great profusion. **Maracaibo — —**, *Bulnesia arborea*: **Paraguay — —**, *Bulnesia sarmienti*: **Queensland — —**, *Premna lignum-vitae*.

Ligustrum Cult. orn. shrubs inc. the privets, much used as hedges.

Lilac Or Syringa, *Syringa vulgaris*, E.Eur., numerous gdn. vars. one of the most popular fl. shrubs. **Calif. —**, *Ceanothus thyrsiflorus*: **common —**, *Syringa vulgaris*: **Himal. —**, *S. emodi*: **Ind. —**, *Melia*: **Josika —**, *Syringa josikaea*: **N.Z. —**, *Hebe hulkeana*: **— of the South** (Am.), *Lagerstroemia indica*: **Persian —**, *Syringa persica*, *Melia azedarach*: **Rouen —**, *S. chinensis*: **W.I. —**, *Melia sempervirens*: **wild —**, *M. sempervirens*, W.I.

Lilium Bulbs of some eaten in China and elsewhere. See lily.

Lillypilly (Aus.) *Syzygium* spp.

Lily *Lilium* spp. The true lilies, some 80 spp., are widely distributed in temp, regions of the N.Hemisph., and are freely cult. on account of their large, showy, often fragrant fls. There are many cultivars and hybrids. Numerous, other pls. or fls. have acquired the name 'lily'. **Af. —**, *Agapanthus*: **arum —**, *Zantedeschia aethiopica*: **atamasco —**, *Zephyranthes atamasco*: **amaryllis —**, *Hippeastrum equestre*: **Amazon —**, *Eucharis amazonica*: **annunciation —**, *Lilium candidum*: **belladonna —**, *Amaryllis belladonna*: **Berg —**, *Galtonia candicans*, cult.: **Bermuda —**, *Lilium longiflorum* var. *eximium*, bulbs once exported from Bermuda: **blackberry —**, *Belamcanda chinensis*: **blood —**, *Haemanthus*: **blue beard —**, *Clintonia borealis*: **boat —**, *Rhoeo spathacea* (=*R. discolor*): **Bourbon —**, *Lilium candidum*: **Brisbane —**, *Eurycles sylvestris*: **Can. —**, *Lilium canadense*: **Carolina —**, *L. michauxii*: **celestial —**, *Nemastylis geminiflora*, N.Am.: **Chilean —**, *Alstroemeria chilensis*: **Chinese sacred —**, *Narcissus*: **church —**, *Lilium longiflorum*: **climbing —**, *Gloriosa*

146

superba: **cobra** —, *Arisaema:* **corn** —, *Clintonia borealis:* **day** —, *Hemero-callis:* **Easter** —, *Lilium longiflorum* var. *eximium:* **Eucharist** —, *Eucharis grandiflora:* **fawn** —, *Erythronium:* **fire** —, *Cyrtanthus,* S.Af.; *Hippeastrum equestre,* Mex.: **flame** —, *Gloriosa,* S.Af.: **George** —, *Vallota speciosa,* S.Af.: **glory** —, *Gloriosa rothschildiana:* **gold band** —, *Lilium auratum:* **Guernsey** —, *Nerine sarniensis:* **Ifafa** —, *Cyrtanthus* sp., S.Af.: **impala** —, *Adenium multiflorum,* S.Af.: **Inanda** —, *Cyrtanthus sanguineus:* **Jacobean** —, *Sprekelia formosissima,* Mex.: **Kaffir** —, *Schizostylis coccinea:* **Knysna** —, *Vallota speciosa,* S.Af.: **leopard** —, *Lilium catesbaei,* N.Am.; *Belamcanda chinensis:* **Madonna** —, *Lilium candidum:* **Malagash** —, *Ammocharis falcata:* **Malta** —, *Sprekelia formosissima:* **Mariposa** —, *Calochortus:* **May** —, *Maianthemum bifolium:* **Michigan** —, *Lilium michiganense:* **Mozambique** —, *Gloriosa simplex:* **New Year** —, *Gladiolus cardinalis,* S.Af.: **orange** —, *Lilium bulbi-ferum:* **orange-red** —, *L. philadelphicum:* **paint-brush** —, *Haemanthus:* **palm** —, *Cordyline australis:* **Pancratian** —, *Pancratium trianthum,* bulb ed.: **Peruvian** —, *Alstroemeria aurantiaca:* **pig** — =arum —: **pine** —, *Lilium catesbaei,* N.Am.: **plantain** —, *Hosta:* **Poor Knight's** —, *Xeronema callistemon,* N.Z., from Poor Knight's Is.: **Pyrenean** —, *Lilium pyrenaicum:* **queen** —, *Phaedranassa,* Andes: **rock** —, *Arthropodium cirratum,* N.Z.: **royal** —, *Lilium superbum:* **snake** —, *Haemanthus,* S.Af.: **spider** —, *Hymenocallis:* **golden spider** —, *Lycoris aurea:* **spire** —, *Galtonia:* **St Bernard's** —, *Anthericum liliago:* **St Bruno's** —, *Paradisea liliastrum:* **St John's** —, *Clivia miniata,* S.Af.: **star** —, *Eucharis amazonica:* **swamp** —, *Zephyranthes:* — **thorn,** *Catesbaea spinosa,* W.I.: **tiger** —, *Lilium tigrinum:* **torch** —, *Kniphofia:* **trout** —, *Erythronium americanum:* **Turk's cap** —, *Lilium martagon, L. superbum:* **veld** —, *Crinum,* S.Af.: **Zephyr** —, *Zephyr-anthes candida:* **wood** —, *Trillium grandiflorum, Clintonia.*

Lily of the valley *Convallaria majalis,* N.Hemisph., popular gdn. pl.: **bush** —, *Pieris:* **false** —, *Maianthemum canadense:* — **orchid,** *Dendrobium mono-phyllum,* Aus.: — **tree,** *Clethra arborea,* Madeira: **wild** —, *Smilacina stellata,* N.Am.

Lima bean *Phaseolus lunatus,* trop. and subtrop., widely grown veg. and pulse crop.

Limay Trade name in Philippines for various dipterocarp. woods.

Limba *Terminalia superba,* W.Af., a commercial timber.

Lime *Tilia* spp., N. Hemisph. Many spp. cult. as orn. trees: several furnish commercial timber: the fls. are an imp. source of nectar to the hive bee, and dried may be used as a tisane or tea: the inner fibrous b. is used for matting in Eur.: lime wood is soft, even-textured, light in colour and weight and is favoured for special purposes, e.g. kitchen and dairy utensils, model-making, carving, artificial limbs etc., cuts well with and across the grain. The imp. commercial lime woods are – **Eur. lime,** *Tilia vulgaris* (= *T. europaea*), **Am.** — or **basswood,** *T. americana* and **Jap.** —, *T. japonica* or allied spp. **Common** —, *T. vulgaris* (= *T. europaea*): **Crimea** —, *T. euchlora:* **Man-churian** —, *T. mandshurica:* **Mongolian** —, *T. mongolica:* **pendant silver** —, *T. petiolaris:* **small-leaved** —, *T. cordata:* **white** —, *T. tomentosa.*

Lime (Citrus) *Citrus aurantiifolia,* trop., source of lime juice, a much used cordial: seed oil used for soap: the most trop. of the citrus frs. in cli. requirements. — **berry**, *Triphasia trifolia:* **Chinese** —, *Triphasia aurantiaca:* **musk** —, *Citrus microcarpa:* **myrtle** —, *Triphasia:* **Queensland wild** —, *Microcitrus inodora.*

Limelo Hybrid citrus fr. (lime × lemon), not a commercial fr.

Limequat Hybrid citrus fr. (lime × kumquat) resembles a lime: tree is more cold-resistant than lime.

Limnanthemum Aquatics, mainly trop. and subtrop., some v. attractive, cult., ed. tubers.

Limnanthes *L. douglasii*, N.Am., ('poached egg' fl.) cult. orn. ann., a good bee pl.

Limnobium *L. stoloniferum*, trop. Am. aquatic, cult., r. hairs used to show circulation of protoplasm.

Limnocharis Am. aquatics, cult., fls. yellow, long-stalked.

Limonium Sea coast pls. (sea lavender), cult., some of the best from Canary Isles.

Linaloe oil or **linaloa** *Bursera* spp., Mex., esst. oil distilled from wood.

Linaria Several spp. are popular gdn. pls., fls. w. wide colour range (toadflax).

Linden Lime tree, *Tilia.*

Lindera Aromatic trees and shrubs, N.Am., E.As.: frs. of *L. benzoin* used as a spice.

Linen Cloth made from flax, *Linum usitatissimum:* **China** or **Canton** —, *Boehmeria nivea.*

Ling *Calluna vulgaris*, common or Scottish heather, a shrub covering large areas of heathland in many countries in the N.Hemisph., affords food and cover for many animals and birds, esp. game birds: fls. an imp. source of honey of special flavour and characteristics: stems used for brooms or thatching. — **berry**, *Vaccinium vitis-idaea* (cowberry).

Lingue *Persea lingue*, Chile, a commercial timber.

Linsaya Ferns, mainly trop. many of handsome appearance, cult.

Linseed *Linum usitatissimum*, oil from the seeds a valuable drying oil used for paints, varnish, linoleum manufacture etc.: — **press-cake** is a good stock feed.

Linum Several spp. grown as gdn. pls.

Lion's ear *Leonotis leonurus*, S.Af., and other spp., name acquired because of a supposed resemblance of the inner part of the corolla to a lion's ear: — **foot**, *Prenanthes serpentaria:* — **tail**, *Leonotis nepetifolia.*

Lip fern *Cheilanthes*, N.Am.

Liparis Small fl. terrestrial orchids, cosmop., very few of orn. value or cult.

Liquorice *Glycyrrhiza glabra*, S.Eur., cult., the dried r. has a sweet taste almost free of bitterness: much used in confectionery: extracts are used in cough mixtures and in cough lozenges and pastilles, also used to counteract the taste of unpleasant medicines. **Anatolian** —, *G. glabra:* **Persian** —, *G. glabra* var. *violacea:* **Russian** —, *G. glabra* var. *glandulifera:* — **substitute**, *Periandra dulcis:* **Syrian** —, *G. glabra:* **wild** —, *G. lepidota*, N.Am.; *Rhynchosia phaseoloides*, W.I.

Lissochilus Terrestrial orchids, mainly Af., often showy, cult.

Litchi Or leechee, *Nephelium litchi* (= *Litchi chinensis*), S.China, ed. fr., cult., warm cli., eaten fresh, canned or dried: has jelly-like flesh and a brittle shell. ('litchi nuts').

Lithospermum Mainly Medit., some spp. orn., cult.

Litsea Ls. and b. of some spp. med.

Live-for-ever *Sedum purpureum*, Eur., cult.: **live-long,** *Sedum* spp.

Liverberry *Streptopus amplexifolius*, N.Am.

Livistona As. or Aus. palms, some, e.g. *L. chinensis* and *L. australis*, cult. as orn. pot pls.

Lizard orchid *Himantoglossum hircinum, Burnettia cuneata:* —'**s tail,** *Saururus cernuus*, N.Am., aquatic peren., cult.

Llala palm (S.Af.) *Hyphaene crinita*, large frs. used for fuel.

Llama *Annona diversifolia*, C.Am., ed. fr.

Loasa C. & S.Am. pls., some cult., orn. fls., may have stinging hairs.

Lobelia Several spp. cult. in the fl. gdn., some as edgings: many cultivars. The drug lobelia, lobelia herb or Indian tobacco consists of the dried ls. and stems of *Lobelia inflata*, E. U.S., cult. there and in Holland. It may be used for asthma and bronchitis. **Blue** —, *L. urens:* **great** —, *L. siphilitica:* **swamp** —, *L. glandulosa:* **water** —, *L. dortmanna:* **yellow** —, *L. lutea* (= *Parastranthus luteus*).

Loblolly bay *Gordonia lasianthus*, SE. U.S., evergreen tree, leathery ls., white fls., b. has been used in local tanning. — **magnolia,** *Magnolia grandiflora:* — **pine,** *Pinus taeda:* — **sweetwood** (W.I.), *Sciadophyllum:* — **tree,** *Cupania, Pisonia*.

Lobster claw *Heliconia*, N.Am.: — **plant,** *Beloperone guttata*.

Lockhartia Trop. Am. orchids w. small fls., some cult.

Locoweed *Astragalus mollissimus* and other spp., N.Am., poisonous to livestock, esp. horses (*loco* in Spanish signifies crazy), in the grass country of the mid-west U.S. and Can. *Oxytropis lambertii* and other spp. also involved.

Locust or **locust bean** Name applied to spp. in various genera, often w. large or mealy pods, *Astronium, Byrsonima, Ceratonia, Gleditsia, Hymenaea, Parkia, Robinia*. The common locust of the Medit. and of Biblical times is the carob, *Ceratonia siliqua*, with its ed. pods, and that of N.Am. the black locust, *Robinia pseudoacacia*, widely cult. for orn. and timber. **Af.** —, *Parkia filicoidea* and other spp.: — **bean gum,** *Ceratonia siliqua:* **bristly** —, *Robinia hispida*, N.Am.: **clammy** —, *R. viscosa*, N.Am.: **honey** —, *Gleditsia* spp., N.Am.: **water** —, *G. aquatica:* **W.Af.** —, *Parkia* spp.: **W.I.** —, *Hymenaea courbaril:* — **wood,** *Astronium fraxinifolium*, Braz.

Lodh bark (Ind.) *Symplocos racemosa*, ls. and b. used in dyeing, ls. med.

Loganberry *Rubus × loganobaccus*. A hybrid berry fr. said to be a natural cross between an Am. blackberry and a raspberry and to have originated in the gdn. of Judge Logan in Calif. in 1881: much favoured for jam: a thornless variety is popular.

Log fern *Dryopteris celsa*, N.Am.

Loggerhead weed *Spigelia anthelmia*, W.I.

Logwood *Haematoxylum campechianum*, C.Am., W.I., a small tree: dark heavy heartwood the source of a dye, mainly used for blues or blacks, first used in Br. at the time of Elizabeth I: used for ink: tree cult. in trop. for orn. or as a hedge: fls. an imp. commercial source of honey in the Caribbean region.

Lolagbola *Pterygopodium oxyphyllum*, W.Af., a commercial timber.

Loliondo *Olea welwitschii*, E.Af., a commercial timber.

LOMATIA

Lomatia Evergreen shrubs, Aus. and S.Am., some cult.

Lombardy poplar *Populus nigra* cultivar *italica*. Orn. tree of erect fastigiate or pyramidal form, brought from Italy to Br. in 1758.

Lonchocarpus Some spp. yield useful timber: the rs. of others may be used as fish poisons and have insecticidal properties due to rotenone. *L. cyanescens* is a dye pl. in W.Af. (Yoruba indigo).

London pride *Saxifraga spathularis* × *S. umbrosa*, Eur., a popular gdn. pl., many vars.

Long Jack *Flindersia oxleyana*, Aus., a strong fine-grained cabinet wood: *Triplaris surinamensis*, Guyana, cult. orn. shrub or tree w. clusters of large white fls. turning to pink: *Triplaris cumingiana*, trop. Am., tree w. bright red fls.

Long moss *Tillandsia usneoides*, trop. Am., hangs in long grey festoons.

Longan *Nephelium longana*, trop. As., small tree w. ed. fr. like an acid 'litchi', cult., esp. in S.China.

Lonicera Shrubs and climbers, N.Hemisph., many cult. for orn.

Loofah Or sponge gourd, *Luffa cylindrica* (= *L. aegyptiaca*), *L. acutangula*. The fibrous mesh of the mature dry fr., with seeds removed, may be used as a substitute bath sponge and in other ways: young frs. used as a veg.

Looking-glass tree *Heritiera littoralis*, *H. macrophylla*. Old World trop.

Loosestrife Yellow —, *Lysimachia vulgaris* and other spp., cult. gdn. pls.: **purple** —, *Lythrum salicaria*, N.temp., wild (in Br.), many gdn. forms.

Lopez root *Toddalia asiatica* (= *T. aculeata*), Ind., trop. As., climber, r. med. and yields a yellow dye.

Lopezia Cult. orn. fls., C.Am., interesting adaptations to ensure pollination. *See Lopezia* in Willis.

Lophopetalum *L. toxicum*, b. used for arrow poison in the Philippines.

Loquat Or Jap. medlar, *Eriobotrya japonica*, E.As., a widely cult. fr. tree of the subtrop., also grown for orn. and shade: yellow woolly frs. w. large brown seeds, eaten fresh, also preserved. **Wild** —, *Uapaca kirkiana*, W.Af., fr. ed.

Loranthus Parasites, Old World trop. and subtrop., some spp. troublesome on cult. tree crops (such as cocoa) and timber trees.

Lords and Ladies Or cuckoo-pint, *Arum maculatum*, Eur. inc. Br.

Lotus, sacred *Nelumbo nucifera* (= *Nelumbium nelumbo; N. speciosum*) of Egypt and Ind., a water lily w. large handsome fls. The lotus fr. of the ancients is considered to have been *Zizyphus lotus*. The genus *Lotus* contains spp. that are good pasture or forage pls.

Louisiana grass *Paspalum platycaule*, N.Am.

Louro *Ocotea* spp. Braz. or trop. S.Am. commercial timbers. — **inamui**, *O. barcellensis*, Braz.: — **preto**, *Ocotea* sp.: **red** —, *Ocotea rubra*, Braz., Guyana.

Lousewort *Pedicularis sylvatica*, Eur. inc. Br. It was once thought the presence of the pl. caused sheep to have lice: **Can.** —, *P. canadensis:* **swamp** —, *P. palustris*.

Lovage or **lovage angelica** *Levisticum officinale*, S.Eur., cult. ls. may be eaten after blanching like celery: fr. and r. used for flavouring. **Scotch** —, *Ligusticum scoticum*, may be used as a pot-herb.

Love apple Old name for tomato. — **bush**, *Bryophyllum pinnatum* (= *B. calycinum*), Madag., cult., plantlets are produced along the margins of the ls.: — **charm**, *Clytostoma callistegioides*, Am.: — **entangle**, *Sedum acre:* — **grass**, *Eragrostis:* — **in-idleness**, *Viola tricolor:* — **in-a-mist**, *Nigella damascena*, Medit., popular gdn. pl., several vars.: — **leaf**, *Bryophyllum pinnatum:* — **lies-bleeding**, *Amaranthus caudatus:* — **vine**, *Cuscuta* spp.

Loxa bark *Cinchona officinalis*, quinine b.

Lozane *Tephrosia macropoda*, S.Af., rs. once used as a fish poison on the Natal coast and as an insecticide for washing dogs.

Lucerne Or alfalfa (Am.), *Medicago sativa*, a widely cult. fodder pl. often grown under irrigation in dry areas: freely fed to poultry and to ostriches in S.Af.: made into hay, silage and meal: fls. a good nectar or honey source: ls. a source of commercial clorophyll. **Paddy's** — (Aus.), *Sida rhombifolia:* **tree** —, *Cytisus pullulans*, E.Eur., has good forage value, needs little water: **wild** —, *Stylosanthes mucronata.*

Lucky bean *Abrus precatorius*, trop.: *Afzelia quanzensis*, southern Af.: *Erythrina* spp.: — **nut**, *Thevetia peruviana* (= *T. neriifolia*), used as charm in W.I.

Lucmo or **lucumo** *Lucuma obovata*, Peru, wild and cult. tree, ed. fr. w. yellow mealy pulp.

Lucraban *Hydnocarpus venenatus*, Ind., seed oil used for skin complaints.

Luculia Deciduous fl. shrubs, cult., esp. *L. gratissima*, Himal., bears bunches of rose-pink fragrant fls.

Lucuma Several trop. S.Am. spp. have ed frs.

Luffa *See* loofah.

Luisia As. orchids, some cult., not v. showy.

Lumbang oil Or candle nut oil, *Aleurites moluccana*, trop. As., used for soap.

Lungwort *Pulmonaria officinalis*, Eur., cult. orn. fl. and old-time med. herb. **Golden** —, *Hieracium murorum:* **sea** —, *Mertensia maritima.*

Lunumidella *Melia composita*, Sri Lanka, a commercial timber, also called Ceylon cedar or Ceylon mahogany.

Lupin *Lupinus* spp. several spp. cult. as orn. gdn. pls., many hybrids, others grown as fodder or soil improvers (green manures). **Blue** —, *L. angustifolius:* **Egyptian** —, *L. termis*, seeds used as food by poorer classes (after treatment) in Egypt and S.Eur.: **tree** —, *L. arboreus*, Calif.: **white** —, *L. albus:* **wild** —, *L. perennis*, N.Am.: **yellow** —, *L. luteus*, Medit., seeds long used as a coffee substitute.

Lutqua (Ind.) *Baccaurea sapida*, Indomalaya, ed. acid frs.

Lychnis *L. chalcedonica*, E.Russia, common gdn. pl. (Jerusalem or Maltese cross).

Lygeum *L. sparteum*, N.Af., Medit., a grass used for mats, sails, ropes etc.

Lygodium Trop. climbing ferns, used locally for mats, basketry, fish traps etc.

Lyme grass *Elymus arenarius*, Br., maritime, a good sand binder, ls. waxy.

Lynx flower *Stanhopea hernandezii* (= *S. tigrina*), magnificent Mex. or C.Am. orchid, scented orange-yellow fls. blotched w. purple, up to 20 cm across.

Lyonia Cult. orn. shrubs resembling *Pieris*, N.Am., As.

Lysichitum Or Skunk cabbage, striking marsh pls., aroids w. large yellow or white spathes, cult. near water: notably *L. americanum* and *L. camtschatcense.*

Lysimachia Several spp. are popular gdn. pls., mainly yellow fls.

M

Mabee bark or **mabi** *Ceanothus* (=*Colubrina*) *reclinatus*, W.I. used for a beverage in Puerto Rico and Haiti.

Mabola *Diospyros discolor*, Malaysia, cult., tree w. ed. velvety fr., has dry greenish flesh.

Maca *Lepidium meyenii*, Peru, an Andean food pl. w. ed. fleshy r. which may be black, purple or yellow.

Macadamia nut Or Queensland nut, *Macadamia integrifolia* (=*M. ternifolia*), this popular Aus. dessert nut is grown in other countries, esp. Hawaii, where the kernels are processed (roasted or salted) and canned for export. **Rough-shelled — —**, *M. tetraphylla*.

Macaroni This typically Italian food product is made from hard or flint wheat (*Triticum durum*).

Macartney rose *Rosa bracteata*, China, vigorous, a defensive hedge pl.

Macary butter *Picramnia antidesma*, W.I.

Macassar oil *Schleichera oleosa* (= *S. trijuga*), Ind. oil from seeds an illuminant and hair dressing.

Macaw bush *Solanum mammosum*, prickly w. orn. yellow frs. — **palm,** *Acrocomia sclerocarpa*, ed. oil from fr. pulp and kernel; *Aiphanes erosa*, W.I., palm w. spiny trunk.

Mace Well known spice consisting of the ground aril of the nutmeg (*Myristica fragrans*), orange in colour, favoured for flavouring doughnuts, cakes, fish sauces, meat stuffings etc.

Mackay bean *Entada gigas* (=*E. scandens*), a well-known drift seed.

Macleania C. & S.Am. shrubs w. colourful fls., cult.

Macleaya Chinese herbaceous perens., large inflor. of small fls., cult.

Macqui *Aristotelia maqui*, Chile, a shrub w. small ed. frs. preserved and used for colouring wine.

Macrozanonia See *Alsomitra*.

Madagascar cardamom *Aframomum angustifolium:* — **clove,** *Ravensara aromatica:* — **copal,** *Trachylobium:* — **ebony,** *Diospyros haplostylis:* — **jasmine,** *Stephanotis floribunda*, cult. trop., waxen white fls., popular for bridal wreaths: — **mahogany,** *Khaya madagascariensis:* — **nutmeg,** *Ravensara aromatica:* — **plum,** *Flacourtia ramontchi*, small thorny tree w. ed. dark purple frs.: — **raffia,** *Raphia pedunculata:* — **rubber,** *Landolphia*.

Madar fibre Or mudar, *Calotropis gigantea*, Ind. and trop. As. a bast fibre, also yields a seed floss.

Madder *Rubia tinctorum*, Eur., formerly cult. for its much used red dye, later supplanted by synthetic dyes: still used for artists' colours – the 'rose madder' and 'madder brown' of the paint box. Madder was known to the Persians and Ancient Egyptians. **Field** —, *Sherardia arvensis*, Br.: **Ind.** —, *Rubia cordifolia:* **Khasia** —, *R. khasiana:* **Levant** —, *R. peregrina:* **Sikkim** —, *R. sikkimensis:* **wild** —, *R. peregrina*, Br.

Madeira broom *Genista virgata:* — **marrow,** *Sechium edule,* sou-sou: — **vine,** *Anredera* (=*Boussingaultia*) *baselloides,* trop. Am., long cult. in Madeira, orn. vine, ed. rs.: has become a weed in sugarcane in Natal.

Madia oil *Madia sativa,* S.Am. (tarweed) cult., seeds yield ed. oil.

Madras hemp *Crotalaria juncea,* sunn hemp: — **thorn,** *Pithecellobium dulce,* cult. as shade tree and hedge.

Madre de Cacao (W.I.) *Erythrina* spp. and *Gliricidia* spp., grown as shade for cocoa.

Madroña or **madroña laurel** *Arbutus menziesii,* Calif., evergreen tree, cult., allied to Eur. strawberry tree, *A. unedo.*

Madwort *Asperugo procumbens,* Br., name derived from madder for which its r. was used.

Mafootoo withe (W.I.) *Entada gigas,* sea bean.

Mafurra or **mafoureira** *Trichilia emetica,* E.Af., oil from seeds.

Magilp or **megilp** Oil used by artists, a mixture of linseed oil and mastic.

Magnolia Trees and shrubs, N.Am., Himal. and E.As., many w. large and magnificent fls., widely cult.: b. often aromatic: some afford useful timber. **Great-leaved** —, *M. macrophylla,* N.Am.: **laurel** —, *M. virginiana, M. grandiflora:* **mountain** —, *M. acuminata* N.Am., useful timber.

Maguey Or cantala fibre, *Agave cantala,* a commercial fibre, finer and softer but weaker than sisal, widely cult., trop.

Mahaleb or **mahaleb cherry** *Prunus mahaleb,* Eur., hard heavy wood used in turnery and for tobacco pipes.

Mahoe (W.I.) *Thespesia populnea; Hibiscus tiliaceus,* fibrous b. used for tying purposes: (N.Z.), *Melicytus ramiflorus.*

Mahogany The name mahogany, often w. a geographical prefix, has been used for a wide range of timbers that show some resemblance to true mahogany. The true or standard mahoganies fall into two groups – Am. (*Swietenia*) and Af. (*Khaya*). Among the former the following are the main commercial kinds – *Swietenia mahagoni,* W.I., the original or Cuban mahogany: *S. macrophylla,* C. & S.Am.: *S. candollei,* Venezuela. Among the true Af. mahoganies are – *Khaya ivorensis.* W.Af., *K. grandifoliola,* W.Af., *K. anthotheca,* W. & E.Af., *K. senegalensis,* W.Af., and *K. nyasica,* E. & C.Af.

The following are examples of other so-called mahoganies, and illustrate the many genera that may be involved. **Aus.** —, *Dysoxylum fraserianum:* **Bataan** or **Philippine** —, *Shorea polysperma:* **Braz.** —, *Carapa guianensis:* **Burma** —, *Pentace burmanica:* **ceda** —, *Guarea cedrata:* **Ceylon** —, *Melia composita:* **cherry** —, *Mimusops heckelii:* **E.I.** —, *Pterocarpus dalbergioides:* **Gaboon** —, *Aucoumea klaineana:* **Ind. white** —, *Canarium euphyllum:* **Rhodesian** —, *Guibourtia coleosperma:* **Sapele** —, *Entandrophragma cylindricum:* **white** —, *Eucalyptus acmenioides,* several other spp. of *Eucalyptus* are known as mahogany in Aus. [In *Kew Bull.,* 1936, pp. 193–210, some 200 'mahoganies' are listed.]

Maholtine *Wissadula rostrata,* trop. b. fibre, resembles jute.

Mahonia Evergreen orn. shrubs, As. and N.Am., cult., some w. local med. uses.

Mahwa Mowa or mahua, *Madhuca indica* (=*Bassia latifolia*), Ind., kernels yield oil: the ground oil cake 'mahwa meal' has been used for killing worms on lawns.

Maiden or **maiden tree** Horticultural term for a young tree, usu. a budded or grafted fr. tree, after one season's growth. —'s **blush,** *Momordica charantia,* W.I., young fr. a veg.; *Sloanea australis,* Aus.: — **pink,** *Dianthus deltoides,* Eur. inc. Br., cult.: — **plum,** *Comocladia dentata,* W.I.: —'s **tears,** *Silene vulgaris* (= *S. cucubalus*).

Maidenhair fern *Adiantum* spp. **Common** — —, *A. capillus-veneris* cosmop. inc. Br., numerous forms or cultivars, widely cult. — **tree,** *Ginkgo biloba,* an interesting Chinese tree freely cult., *see Ginkgo* in Willis, fr. ed.

Maidu *Pterocarpus pedatus.* Indo-China, a commercial timber.

Maigyee *Strobilanthes flaccidifolius,* Ind. to China, ls. and shoots yield a blue dye, cult.

Maire *Gymnelaea* sp. (= *Olea*), N.Z., hard heavy wood once used in wheelwrights' work, cabinet making etc.

Maize *Zea mays,* cult. throughout the trop. and subtrop. and a v. imp. food for man and his animals. Numerous industrial products are made from maize, esp. alcohol. Food products include maize or corn meal, cornflour, corn flakes, oil etc. There are countless local dishes made from fresh or dried maize, esp. in the U.S. Corn or maize oil is obtained from the embryos of maize separated during the preparation of maize starch and other products. It is similar to olive oil and is a constituent of diets intended to reduce high blood cholesterol levels (considered conducive to thrombosis). The ls. and stalks of maize are a good stock food and much used for silage. The following are the main subdivisions (or subspp.) into which maize is divided – *erythrolepis,* flour corn: *indentata,* dent corn: *indurata,* flint corn: *praecox,* pop-corn: *rugosa,* sweet corn. *See Zea* in Willis.

Makamaka *Ackama rosifolia,* N.Z., tree 6–12 m high, small fls.

Makoré *Mimusops heckelii,* W.Af., cherry mahogany, used for furniture.

Makua or **ma-plua** *Diospyros mollis,* Thailand, black dye from frs. used for silk.

Malabar gourd *Cucurbita ficifolia:* — **kino,** *Pterocarpus marsupium:* — **nightshade,** *Basella rubra:* — **nut,** *Adhatoda vasica:* — **rosewood,** *Dalbergia latifolia:* — **tallow,** *Vateria indica.*

Malacca cane *Calamus scipionum* and other spp., Malaysia, long favoured for walking sticks and valued on account of long internodes.

Malagueta pepper *See* Guinea grains.

Malambo *Croton malambo,* S.Am., b. med.

Malatti *Jasminum* spp., Ind., oil from fragrant fls. used in perfumery.

Malay apple *Syzygium malaccense,* Malaysia, ed. fr., several vars.

Male bamboo *Dendrocalamus strictus,* Ind., a much used bamboo, cult. — **fern,** *Dryopteris filix-mas,* N.Hemisph., a well known drug and anthelminthic, an extract being used for the expulsion of tapeworms.

Mallee Name in Aus. for dwarf or shrubby spp. of *Eucalyptus,* usu. w. many thin stems arising from a woody or enlarged base (lignotuber): often dominates the veg. in low rainfall areas. **Bull** —, a larger mallee which often becomes single-stemmed: **four-angled** —, *Eucalyptus tetraptera:* **fuchsia** —, *E. forrestiana:* **giant** —, *E. oleosa:* **red** —, *E. oleosa:* **whipstick** —, a smallgrowing mallee producing a number of thin stems.

Mallet bark *Eucalyptus occidentalis* var. *astringens*, Aus., used in tanning at one time.

Mallow Name used for various *Malvaceae*, esp. spp. of *Malva, Abutilon, Lavatera, Hibiscus* etc., includes many popular gdn. pls.: **common —**, *Malva sylvestris:* **curled —**, *M. crispa:* **Egyptian —**, *M. parviflora*, a pot-herb: **false —**, *Sphaeralcea:* **Ind. —**, *Abutilon indicum*, a source of fibre: **Jew's —**, *Kerria japonica:* **marsh —**, *Althaea officinalis:* **musk —**, *Malva moschata:* **poppy —**, *Callirhoe:* **prickly —**, *Sida spinosa:* **rose —**, *Hibiscus:* **sea —**, *Lavatera maritima:* **seashore —**, *Kosteletzkya:* **tree —**, *Lavatera:* **Venice —**, *Hibiscus trionum:* **wild —**, *Malvastrum*, S.Af.

Malope Cult. gdn. pls. w. large showy trumpet-shaped fls., Medit.

Malt Barley steeped in water to start germination and then kiln dried: used in beer and whisky making. **— extract**, has med. uses, mainly as a vehicle for cod-liver oil, also used for masking bitter tastes.

Maltese cross *Lychnis chalcedonica*, a popular gdn. pl.

Malukang fat *Polygala butyracea*, W.Af., cult., fat from seeds.

Malva Several spp. cult. as gdn. pls.

Mammee Mammee apple or St Domingo apricot, *Mammea americana*, W.I., cult., ed. fr., fls. used in preparing liqueur (eau de Creole): **Af. — apple**, *Mammea africana* (= *Ochrocarpus africanus*): **— zapote**, or marmalade plum, *Lucuma mammosa* (= *Calocarpum mammosum*), ed. fr. used for preserves, W.I.

Mammillaria Mainly small, rounded or cylindrical cacti, cult.

Mammoth tree *Sequoiadendron giganteum* (= *Sequoia gigantea; Wellingtonia gigantea*), W. U.S., the world's largest tree, or living thing, reaches 97 m in height and 27 m in girth. The thick b. prevents fire injury: wood soft, not strong.

Mamoncillo (Cuba) *Melicoccus bijugatus*, trop. Am., Spanish lime or genep, ed. fr.

Man-of-the-earth *Ipomoea pandurata*, N.Am.: **— orchis**, *Aceras anthropophorum*, Eur. inc. Br.

Mana grass (Sri Lanka) *Cymbopogon nardus*, oil-yielding.

Manac *Euterpe broadwayana*, Trinidad, a palm that grows in clumps.

Manchineel *Hippomane mancinella*, C.Am., W.I., coastal tree, cult. as windbreak: poisonous latex harmful to the eyes.

Manchurian jute *Abutilon avicennae:* **— walnut**, *Juglans mandschurica*.

Mandarin or **mandarin orange** *Citrus reticulata*, cult., trop. and subtrop., many vars.: **—'s hat**, *Holmskioldia sanguinea*, showy fls., cult.: **nodding —**, *Disporum maculatum*, N.Am.: **white —**, *Streptopus amplexifolius:* **yellow —**, *Disporum lanuginosum*.

Mandevilla *M. suaveolens*, Argen., cult. orn. trop. climber, fragrant white fls. 5 cm across.

Mandioca Manihot or cassava.

Mandrake *Mandragora officinalis*. S.Eur., known to the ancients and considered to have many virtues because of the frequent resemblance of the r. to the human figure: actually of no med. value. **Am. —**, *Podophyllum peltatum*.

Manetti Stock used for dwarf roses, suited to hot gravelly soils.

Manettia Trop. Am. climbing pls. w. handsome tubular fls., cult., esp. *M. inflata* and *M. bicolor*, fls. scarlet and yellow.

Mangabeira *Hancornia speciosa*, Braz., has yielded rubber, fr. ed.

Mangel-wurzel or **mangold** *Beta vulgaris* var. *macrorhiza*, a heavy yielding r. crop grown for livestock, esp. cattle.

Mango *Mangifera indica*, a common cult. fr. of the trop. and subtrop., esp. Ind. where there are countless vars., some v. choice and stringless: the fr. is canned and much used in chutney. **Wild** —, *Irvingia gabunensis*, *Cordyla pinnata*, trop. Af.

Mangosteen *Garcinia mangostana*, Malaysia, a choice trop. fr. very slow to grow, regarded by some as the best fr. of the trop. with its delicious flavour. **Af.** —, *Garcinia livingstonei*, fr. quite unlike true mangosteen.

Mangroves Woody pls. that occur in muddy swamps at the mouths of rivers, in lagoons and elsewhere in the trop. under tidal conditions. They do not occur on rocky, wind-swept or steeply sloping sandy beaches, where conditions are unsuitable for them. For mangroves to thrive a combination of mud flats and brackish water is needed. A large number of mangroves belong to the *Rhizophoraceae*, the most imp. genus being *Rhizophora*. Other mangrove genera are – *Aegiceras*, *Avicennia*, *Bruguiera*, *Carapa*, *Ceriops*, *Conocarpus*, *Laguncularia*, *Kandelia*, *Lumnitzera*, *Scyphiphora*, *Sonneratia*, etc. Usu. much branched w. aerial rs., both flying buttress and pillar rs.: aerating rs. rise from the mud in *Avicennia*, *Bruguiera*, *Sonneratia*, etc. Many show viviparous germination, the germinating seeds dropping into the mud. Growth then may be very fast, for the seedling to get its crown above high water level. One observer records a growth of twenty inches in as many hours with *Rhizophora mucronata* in E.Af.

Mangrove bark The b. of several different mangroves has been used in commercial leather tanning to some extent but the intense reddish colour it imparts to leather is a serious drawback. The tannin content of dry b. varies from 10 to 40 % or more. The extract is sometimes called cutch or mangrove cutch. Mangroves exploited for tan bark include spp. of *Rhizophora* (esp. *R. mucronata*), *Bruguiera*, *Ceriops*, *Avicennia*, *Kandelia*, *Carapa*, and *Heritiera*.

Manila hemp Or abaca, *Musa textilis*, a high quality fibre much in demand for maritime cordage and rope, produced mainly in the Philippines and to some extent in C.Am.

Manio *Podocarpus nubigenus*, *P. salignus*, Chile, commercial timber.

Manioc *See* cassava.

Manjack *Cordia collocca*, W.I. tree w. showy red cherry-like frs., ed. and fed to poultry.

Manna The word manna is used for various ed. materials directly or indirectly of veg. origin and usu., but not always, of a sugary nature. Some are insect or aphis exudations from insects feeding on certain pls. and are thus akin to honeydew, but are in a solid form when collected. Manna has been well known in Asia Minor from very early times. It is mentioned in the Bible and in the Koran. Some kinds of manna are still collected in certain areas and eaten, usu. in the form of sweetmeats. Other kinds are regarded as med. Biblical manna is believed to have been a lichen blown off or dislodged from rocks. Some fungi are also thought to have constituted manna.

The best known manna and the only one of commercial imp. is manna from the manna ash (*Fraxinus ornus*) of the Medit., obtained by tapping, mainly in

Sicily. It is a sugary substance and is used med. as a mild laxative. Tamarisk manna from at least four spp. of *Tamarix*, is well known in the near East and is an insect secretion. Manna may also be obtained from the Eur. larch (*Larix decidua*) and from certain other conifers inc. several spp. of *Pinus*. Other pls. known to be responsible for manna include spp. of *Alhagi* (*A. camelorum*), *Astragalus, Atraphaxis, Cotoneaster, Eucalyptus, Leptospermum, Myoporum, Quercus* and *Salix*. [Harrison, S. G. (1950) Manna and its sources. *Kew Bull.*, **5**, 407–17.]

Manna croup Preparation once made in parts of Eur. from the seeds of floating sweet grass or manna grass, *Glyceria fluitans*, and used as food.

Manoao *Dacrydium biforme*, N.Z., wood used locally for building, fences, railway sleepers.

Mansonia *M. altissima*, W.Af., a commercial timber, used for furniture.

Mantle or **mantle leaf** The special humus collecting type of l. in staghorn ferns. *See Platycerium* in Willis.

Manuka Or kanuka, *Leptospermum scoparium*, N.Z., a well known N.Z. pl., fls. white or coloured.

Manutu grass *Thalassia* sp., W.I., a submerged marine aquatic.

Manzanilla *Crataegus stipulosa, C. mexicana*, C.Am.

Manzanita *Arctostaphylos manzanita* and other spp.

Maori calceolaria *Jovellana sinclairii*, N.Z.: — **honeysuckle**, *Knightia excelsa, Maoutia puya:* — **onion**, *Bulbinella:* — **potato**, *Thelymitra longifolia:* — **privet**, *Geniostoma ligustrifolium*.

Maoutia *M. puya*, As., Polynesia, yields good fibre, used for nets, bags etc.

Maple *Acer* spp. Some maples are imp. for timber, others as orn. trees and shrubs, valued for their foliage and autumn colouring and often grown as street trees. Maple wood is in general hard, heavy and close-grained. It is much used for flooring, esp. for dance floors and squash courts, and for interior decoration and furniture. The maples yielding commercial timber include the following – **Norway** or **Eur.** —, *Acer platanoides:* **Pacific** or **Oregon** —, *A. macrophyllum:* **soft** or **red** —, *A. rubrum, A. saccharinum:* **rock** —, *A. saccharum, A. nigrum:* **Jap.** —, *Acer* spp.: **Queensland** —, *Flindersia brayleyana*. **Bird's eye** or **'fiddle-back'** — is a figured form of the wood used in veneers.

Some other maples are – **great** or **Scottish** —, *Acer pseudoplatanus* (*see* sycamore): **hedge** —, *A. campestre*, Eur. inc. Br., the tough elastic wood used in turnery, for cutlery handles and tobacco pipes (Ulmer pipes): **Montpelier** —, *A. saccharum*, E. N.Am. trees tapped in early spring, the juice yielding maple syrup and maple sugar.

Maqui *See* macqui.

Marana *Ackama paniculata, Pseudoweinmannia lachnocarpa*, Aus. rain forest trees.

Marang *Artocarpus odoratissimus*, Philippines, a large ed. fr., resembling jak fr.

Maranta Trop. Am. orn. fol. pls., cult., some are house pls.

Maraschino cherry or **marasco** *Prunus acida* var. *marasca*, cult., fr. used in the manufacture of maraschino (Italian liqueur).

Marattia Trop. orn ferns, mainly large, cult.

Marble wood Or zebra wood, *Diospyros* spp., notably *D. marmorata, D. kurzii* and *D. oocarpa*, Andamans: streaked or marbled ebony.

Mare's tail *Hippurus vulgaris,* aquatic, almost cosmop.

Margosa tree *See* neem.

Marguerite Or moon-daisy, *Chrysanthemum leucanthemum* (= *Leucanthemum vulgare*), cult. orn. fl.

Marianthus Aus. twining shrubs, attractive fls., cult.

Marigold Common or **pot** —, *Calendula officinalis,* S.Eur., cult. many vars. **Af.** —, *Tagetes erecta,* Mex., cult.: **bur** —, *Bidens tripartita:* **corn** —, *Chrysanthemum segetum,* a weed: **French** —, *Tagetes patula,* Mex., cult.: **marsh** —, *Caltha palustris,* Eur.; *C. leptosepala,* N.Am.: **Scotch** —, *Calendula officinalis:* **sweet** —, *Tagetes lucida,* S.Am.

Marihuana (Mex.) The drug cannabis, *Cannabis sativa.*

Mariposa lily *Calochortus kennedyi,* and other spp., S.Calif., scarlet fls.

Marjoram Common or **sweet** —, *Majorana hortensis* (= *Origanum majorana*), S.Eur., used as a flavouring from classical times: a much used and reliable seasoning for all meat dishes, stews, soups, fish etc., used w. canned meats. The distilled oil is also used, esp. w. sausages. **Pot** —, *Origanum vulgare,* Eur., cult. as a flavouring herb and also used in home remedies.

Marking nut *Semecarpus anacardium,* trop. As., juice of the fr. mixed w. lime has been used for marking linen: green frs. used for bird-lime.

Marlock *Eucalyptus* spp., Aus.

Marmalade plum *Calocarpum sapota* (=*C. mammosum; Lucuma mammosa*), C.Am., cult. trop., brown ed. fr. w. reddish flesh and one large seed: used for preserves.

Maroola plum *Sclerocarya caffra,* S.Af., fr. makes good jelly, oil from kernels.

Maroon hood orchid *Pterostylis cucullata,* Aus.

Marram grass Or beach grass, *Ammophila arenaria* (=*A. arundinacea*), valuable for binding and consolidating drifting sand, cult. in Br. and many countries. Used for thatch, baskets, brooms, mats, etc.

Marri *Eucalyptus calophylla,* W.Aus.

Marrow, vegetable *Cucurbita pepo,* numerous vars. cult.

Marsdenia Trop. shrubs and climbers, some orn. and cult. *M. cundurango,* Colombia, b. med. and used for snakebite: *M. erecta* SE. Eur., juice blisters the skin: *M. tenacissima* Ind., yields fibre used for bow strings: *M. tinctoria,* a substitute for indigo in Java at one time.

Marsh betony *Stachys palustris,* Eur. inc. Br.: **— fern,** *Thelypteris, Acrostichum:* **— grass,** *Spartina:* **— mallow,** *Althaea officinalis:* **— marigold,** *Caltha palustris:* **— miller,** *Blechnum pyramidatum,* Am.: **— rose,** *Orothamnus zeyheri:* **— rosemary,** *Ledum,* Am., *Limonium:* **— wort,** *Apium nodiflorum, A. inundatum* (floating).

Marsilea Widely dist. aquatics, cult., some eaten by Aus. aborigines.

Martynia Some spp., e.g. *M. diandra* and *M. fragrans* are notable for their heavily armed frs. which become entangled in the coats of grazing animals.

Marups *See* simaruba.

Marvel of Peru False jalap or four o'clock, *Mirabilis jalapa,* trop. Am., cult., also a weed: rs. formerly used as jalap. Fls. open in evening.

Marvello hair Chinese silkworm gut from silkworms fed on *Liquidambar formosana.*

Maryland pink root *Spigelia marylandica,* med., anthelminthic.

Mascarenhasia *M. elastica,* Madag., E.Af., has yielded rubber.

Masdevallia S.Am. orchids, large handsome fls., cult.

Masindi (E.Af.) *Cynodon transvaalensis,* a lawn grass.

Massachusetts fern *Dryopteris simulata,* N.Am.

Massoy bark *Cryptocarya* (= *Massoia) aromatica,* New Guinea, the aromatic b. yielding an esst. oil is an article of commerce in the Far East.

Mast Fr. of beech, oak, chestnut etc. At one time land in Br. was valued according to the amount of mast it produced, i.e. food for swine.

Masterwort *Astrantia major,* Eur., *A. maxima,* E. Caucasus, cult. orn. pls.

Mastic *Pistacia lentiscus,* an ancient resin of the E.Medit. used as a masticatory and mentioned by Theophrastus: obtained by puncturing the b. of this small tree, mainly on the island of Chios (Scio): used for high grade varnish, esp. picture varnish: a solution w. chloroform, ether or alcohol on cotton wool used as a filling for carious teeth. **Am.** —, *Schinus molle:* **Barbados — tree,** *Mastichodendron sloaneanum:* **Bombay —,** *Pistacia mutica* and other spp.: **E.I. —,** *P. khinjuk:* — **tree** (W.I.), *Bursera.*

Masticatories A number of pls. are regularly used as masticatories in different parts of the world, apart from the traditional masticatory of the contemplative Br. farmer or farm worker, viz. a piece of straw! The most extensively used masticatory in modern times must be chewing gum, esp. on the N.Am. Continent or wherever there is American influence. The basis of chewing gum is chicle or the rubbery latex of certain trop. Am. trees, notably *Achras zapota* of C.Am. In S.Am. a well known masticatory and stimulant is coca l. (*Erythroxylum coca)* long used by the Amerindians and the source of cocaine. Tobacco might also be regarded as a masticatory as it is often chewed. In the Medit.-the use of mastic as a masticatory, esp. for sweetening the breath, goes back to very early times. In Ind. and other trop. As. countries betel or areca nut is the traditional and much used masticatory, always so obvious from the way in which it stains the mouth and saliva red. The leaf of betel pepper (*Piper betle)* is chewed along with betel nut. Certain kinds of cutch are also used as masticatories in the East. On the Af. continent the best known masticatory is the kola nut of west trop. Af., held in high esteem as a stimulant, especially by those undergoing physical strain. In central Aus. aborigines chew ls. and twigs of *Duboisia hopwoodii.*

Masur (Ind.) Lentil, *Lens esculenta:* — **wood,** a trade name for prettily marked wood of Eur. birch.

Mat daisy *Raoulia parkii,* N.Z. — **grass,** *Nardus stricta,* Eur., in Br., a tough wiry grass, common, not a good pasture grass: — **grass** (Aus.), *Hemarthria incinata;* (China), *Cyperus tegetiformis.*

Matac *Ascelepias curassavica,* W.I., an orn. pl., also a weed.

Matai *Podocarpus spicatus,* N.Z., a commercial timber w. many uses: — or **Chinese water chestnut,** *Eleocharis dulcis,* crisp ed. corms.

Match heads *Comesperma ericinum,* Aus.: — **me,** *Acalypha macrophylla,* cult. orn. shrub, warm cli., large ls. blotched w. red and brown.

Matches The wood of poplar (*Populus* spp.) is extensively used in match making. Many other woods are used, such as ash, *Bombax,* cypress, *Didymopanax morototoni* (Trinidad), pine, spruce and W.I. birch (*Bursera).*

Maté Or Paraguay tea, *Ilex paraguensis,* a national beverage, enjoyed by over 20 million S. Americans, is made by brewing the ls. of this small tree or shrub, now extensively cult. Apart from its stimulating effect, due to caffeine, it is believed to have other physiological effects.

Mathers *Vaccinium ovalifolium*, N.Am.

Matico *Piper angustifolium*, Peru, and probably other spp., ls. used med. and exported.

Mats or **mat-making materials** These cover a wide range, esp. those used by the inhabitants of the less developed countries in the trop. and subtrop. Grasses and sedges are extensively used, as is bamboo or split bamboo in As. countries. Palm ls. of many different kinds are favoured and coir is much used. Among the sedges used and sometimes specially cult. for mat-making are the following – *Juncus communis*, Jap.: *Scirpus triqueter*, China: *Cyperus tegetiformis*, Taiwan: *Cyperus malaccensis*, China: *Cyperus tegetum*, Ind., Burma. Rush and reed mats are made in some countries. Mats made from the inner b. or bast of the lime (*Tilia*), or 'Archangel' mats were at one time imported to Br. from Russia for hort. purposes. The best straw mats are those made from rye straw, preferably hand-threshed straw. It is superior to wheat straw.

Matrimony The name is used in Jam. for a dish made from star apple and orange juice: — **vine**, *Lycium halimifolium*, SE.Eur., spiny shrub w. red frs., cult.

Mattipaul *Ailanthus malabarica*, a scented resin from incisions in the b., burned in Hindu temples.

Maurandia Mainly trop. Am. climbers w. large handsome fls., cult.

Mauritia Tall trop. Am. palms w. fan-shaped ls., cult for orn.: some yield fibre or oil.

Mauritius hemp *Furcraea gigantea*, fibre resembles sisal: — **raspberry**, *Rubus rosifolius*.

Maw seed Or poppy seed, *Papaver somniferum*, cult., an oil seed, the small grey seed used with confectionery (sprinkled on bread, buns, or cakes), fed to cage birds: oil used for artists' colours.

Mawah oil Geranium oil.

Maxillaria Trop. Am. orchids, cult.

May apple *Podophyllum peltatum*, N.Am., fr. ed.; *Passiflora incarnata:* — **berry**, *Rubus strigosus*, N.Am. ,Am. wild raspberry: **Burbank's — berry**, *Rubus microphyllus:* — **blob**, *Caltha palustris:* — **bush, Calif.**, *Heteromeles arbutifolia:* —**flower**, *Epigaea repens*, N.Am.: **flowering** —, *Spiraea prunifolia* and other spp.: —**fly orchid**, *Acianthus caudatus*, Aus.: **native** —, *Phebalium squamulosum*, Aus.: — **pops**, *Passiflora incarnata:* —**weed**, *Anthemis cotula*.

Maze Or labyrinth, a series of interconnecting, more or less concentric paths surrounded by evergreen hedges (box, yew, privet, etc.) difficult to get out of, once entered; started in Br. in the days of Queen Elizabeth I: the maze at Hampton Court, near London, is well known.

Mazzard *Prunus avium*, Euras. gean or wild cherry.

Mbocaya Or Paraguay cocopalm, *Acrocomia totai*, fr. pulp and kernel oil used locally for soap.

Mboga *Sesuvium portulacastrum*, E.Af., a spinach pl.

Mbura *Parinari curatellifolium*, trop. Af., ed. fr., provides good building poles.

Meadow beauty *Rhexia* spp., N.Am.: — **crane's bill**, *Geranium pratense:* — **crocus**, *Colchicum autumnale:* — **fern**, *Dryopteris thelypteris*, N. Hemisph.: — **fescue**, *Festuca pratensis:* — **foxtail**, *Alopecurus pratensis:* — **grass**, annual, *Poa annua:* — —, **common**, *P. pratensis:* — —, **rough**, *P. trivialis:* — —, **wood**, *P. nemoralis:* — **parsnip**, *Thaspium*, N.Am.: — **pink**, *Dianthus deltoides:* — **rue**, *Thalictrum* spp.: — **saffron**, *Colchicum autumnale:* — **sage**, *Salvia pratensis:* — **saxifrage**, *Saxifraga granulata:* — **sweet**, *Filipendula* · *ulmaria*.

Mealberry *Arctostaphylos uva-ursi*, N.Am.

Mealies (S.Af.) Or mielie, *Zea mays*, maize.

Mecca galls *See* oak galls.

Meconopsis Mainly As., w. showy fls. (blue poppies) cult.

Médaké *Arundinaria japonica*, Jap., widely cult. orn. bamboo.

Medemia *M. nobilis*, a Madag. palm, trunk yields a sago, ls. for baskets.

Medick Or burweed, *Medicago* spp., inc. many useful fodder or pasture pls.

Medinilla Old World shrubs or climbers, attractive fls., cult., esp. *M. magnifica*.

Medlar *Mespilus germanica*, Eur., improved vars. cult for fr. which needs to be stored or 'bletted' before eating. **Bronvaux** —, a graft hybrid between medlar and hawthorn: **Jap.** — =loquat: **Medit.** —, *Crataegus azarolus*, fr. ed. but acid: **white** —, *Vangueria esculenta*, trop. Af. fr. ed.: **wild** —, *Vangueria* spp., trop. and S.Af.

Melaleuca Aus. trees and shrubs, cult., beauty of fls. due to conspicuous stamens, united in bundles.

Melastoma Trop. Old World hairy shrubs, showy fls., cult.

Melegueta pepper *See* Guinea grains.

Mel grass *Ammophila arenaria*, marram grass.

Melianthus *M. major*, S.Af., handsome fol., cult., fls. rich in nectar.

Melick grass *Melica altissima*, cult., orn., grass: **wood** —, *M. uniflora*.

Melilot *Melilotus* spp., sweet clover: **white** —, *M. alba:* **yellow** —, *M. officinalis:* good forage and honey pls.

Meliosma Cult. trees and shrubs, mainly E.As., fls. usu. small, fragrant, in panicles.

Melon *Cucumis melo*, the melon has been cult. from v. early times, esp. in the Medit. Its origin is lost in antiquity. The many vars. grown commercially in warm temp. regions are divided hort. into two main groups, the 'netted' and the 'cantaloupe' or scaly-skinned melons, also called rock melons. It is not always easy to draw a hard and fast line between the two groups. Melons may have green, yellow or reddish flesh. They are usu. eaten raw but in some countries may be used as a veg. or pickled when in the immature state. — **cactus**, *Melocactus:* **Chinese preserving** —, *Benincasa hispida* (=*B. cerifera*): —**ette**, *Melothria pendula*, N.Am.: — **gene**, eggplant or brinjal: **tree** —, *Carica papaya:* **water** —, *Citrullus lanatus* (=*C. vulgaris*).

Memorial rose (Am.) *Rosa wichuraiana*. Orient, trailing, fls. white, cult.

Menduro (E.Af.) *Balanites maughamii*, ed. fr.

Meni oil *Lophira* spp., Nigeria, oil from kernels.

Menkulang *Tarrietia* (=*Heritiera*) *simplicifolia* and allied spp., Malaya, a trade timber.

Menthol Substance extracted from the oil of various mints, esp. Jap. peppermint, *Mentha arvensis* var. *piperascens*. Synthetic menthol is produced from thymol. Menthol is used in cases of bronchitis and sinusitis and in perfuming cigarettes.

Meranti Group name (trade name) for the timber of spp. of *Shorea* from Malaya, Sarawak, Brunei, Sabah and Indonesia. They are classified as *light red meranti, dark red meranti, yellow meranti* and *white meranti*. All are lightweight hardwoods and much used. No fewer than two dozen botanical spp. may be involved.

Merawan *Hopea mengarawan* and allied spp., Malaya, a lightweight commercial hardwood.

Merbau *Intsia bijuga* and *I. palembanica*, Malaya, E.I., a hard, heavy commercial timber (Borneo teak).

Mercury *Mercurialis* spp.: **annual** —, *M. annua*, a weed: **dog's** —, *M. perennis*, common in woods in Br.

Mermaid weed *Proserpinaca palustris*, N.Am., an aquatic w. minute white fls., cult.

Merry bells *Uvularia* spp., N.Am., showy bell-shaped fls. in spring.

Mersawa *Anisoptera laevis* and allied spp., Malaya, a lightweight commercial timber.

Mertensia Several spp. cult., mainly as rock gdn. pls.

Mescal *See* peyote.

Mesembryanthemum Succulents, mainly S.Af. w. colourful fls., many cult., popular for bedding, rockeries and sunny situations, e.g. *M. tricolor* and *M. crystallinum* (ice plant).

Mespilus, snowy *Amelanchier ovalis* (= *A. integrifolia*), E.Eur., fl. shrub w. clusters of white fls., cult.

Mesquite *Prosopis chilensis* (= *P. juliflora*), S. U.S., C.Am., a shrub or small tree w. pods that are a valuable feed for livestock, introduced to other countries: yields a hard, durable timber and fls. are a good honey source. — **grass,** *Bouteloua* spp.

Messmate (Aus.) Certain spp. of *Eucalyptus* with fibrous or stringy b.

Metasequoia *M. glyptostroboides*, a Chinese tree related to *Taxodium*, cult., a relic of the veg. of the past, previously known only from fossil remains: has been called 'dawn redwood'.

Metrosideros Mainly N.Z. shrubs or climbers, cult., some w. orn. fls. w. long stamens.

Meu *Meum athamanticum*, C. & S.Eur., meu, spignel or bald money, once cult. in Br. as an aromatic herb or veg.

Mexican apple *Casimiroa edulis:* — **avocado,** *Persea drymifolia:* — **elder,** *Sambucus mexicana:* — **flame vine,** *Senecio confusus:* — **hat,** *Rudbeckia:* — **lily,** *Hippeastrum reginae:* — **linaloe,** *Bursera delpechianum:* — **mahogany,** *Swietenia humilis:* — **mock orange,** *Philadelphus mexicanus:* — **orange blossom** or **flower,** *Choisya ternata:* — **orange daisy,** *Tithonia rotundiflora:* — **poppy,** *Argemone mexicana*, a weed: — **sarsaparilla,** *Smilax* sp.: — **scammony,** *Ipomoea orizabensis:* — **star,** *Milla biflora*, white star-shaped fls.: — **sunflower,** *Tithonia diversifolia*, heads 15 cm across: — **thistle,** *Cirsium conspicuum:* — **tiger flower,** *Tigridia*.

Mezcal (Mex.) Distilled liquor from *Agave*.

Mezereon *Daphne mezereum*, Eur. inc. Br., cult. orn. shrub, b. med.

Mfukufuku (E.Af.) *Brexia madagascariensis*, fr. ed.

Mfungu (E.Af.) *Celosia argentea*, ls. used as spinach.

Mho (E.Af.) *Markhamia hildebrandtii*, orn. fl. tree.

Mi *Madhuca longifolia*, Ind., an oil seed.

Miami mist *Phacelia purshii*, N.Am.

Michaelmas daisy *Aster* spp. many hybrids and cultivars.

Michauxia *M. campanuloides*, E.Eur., cult., orn. fls.

Michelia *M. champaca*, trop. As., cult. for fragrant fls., some spp. yield timber.

Michigan lily *Lilium michiganense*, N.Am.

Miconia Inc. trop. Am. orn. fol. pls.

Microlepis Trop. Am. ferns, cult.

Midge orchid *Prasophyllum* spp., Aus.

Midsummer daisy *Erigeron* spp.

Mignonette *Reseda odorata*, Medit., a well known fragrant gdn. pl.: **tree** —, *Lawsonia inermis*, henna: — **vine**, *Anredera* (= *Boussingaultia*) *baselloides*, S.Am., fast-growing creeper w. white fragrant fls., Madeira vine: **wild** —, *Reseda lutea*: **W.I.** —, *Lawsonia inermis*.

Mikania *M. cordata*, trop. Am. A related sp., *M. micrantha*, is a trop. weed.

Mile tree (W.I.) *Casuarina equisetifolia*.

Milfoil *Achillea millefolium*, a common grassland pl.: **water** —, *Myriophyllum* spp.

Milk pea *Galactia* spp., N.Am., has latex, rare in Leguminosae: — **maids**, *Burchardia umbellata*, Aus.: — **purslane**, *Euphorbia supina*, N.Am.: — **thistle**, *Silybum marianum*, *Sonchus* spp.: — **tree**, various trees w. latex, e.g. *Brosimum*, *Mimusops*: — **vetch**, *Astragalus*: — **weed**, *Euphorbia* spp., *Asclepias* spp. some, e.g. *A. incarnata*, yield a useful fibre: — **wood**, *Mimusops* S.Af.: —**wort**, *Polygala* spp. once thought to increase the flow of milk in cows: **sea** —**wort**, *Glaux maritima* (*see Glaux* in Willis).

Millet There are several different kinds of millet, imp. food pls. (human and animal) in many parts of the world, esp. the dry trop. and subtrop. Many will yield good crops where the rainy season is relatively short. **Af.** —, *Pennisetum spicatum*: **Babala** — (S.Af.), *Pennisetum typhoides*: **barnyard** —, *Echinochloa frumentacea*: **bulrush** —, *Pennisetum americanum* var.: **common** —, *Panicum miliaceum*: **foxtail** —, *Setaria italica*: **gero** —, *Pennisetum spicatum*: — **grass**, *Milium effusum*: **German** —, *Setaria italica*: **great** —, *Sorghum bicolor*: **Hungarian** —, *Setaria italica*: **Indian** — (N.Am.), *Oryzopsis hymenoides*: **Italian** —, *Setaria italica*: **Jap.** —, *Echinochloa frumentacea*: **Kodo** —, *Paspalum scrobiculatum*: **little** —, *Panicum miliare*: **Milo** —, *Sorghum bicolor*: **pearl** —, *Pennisetum typhoides*: **proso** —, *Panicum miliaceum*: **Sanwa** —, *Echinochloa frumentacea*: **Shama** —, *Echinochloa colona*: **spiked** —, *Pennisetum typhoides*.

Miltonia Am. orchids w. showy large fls., cult.

Mimosa Name used for the genus *Mimosa* and for some spp. of *Acacia*. The terms *mimosa bark* or *mimosa extract* are used by the leather industry or tanners for black wattle b. or extract. *Mimosa pudica* (sensitive plant) is a common trop. weed as is *Mimosa invisa* (giant sensitive pl.) very troublesome with sugar cane in Queensland: **creeping** —, *M. strigillosa:* **prairie** —, *Desmanthus illinoensis. See Mimosa* in Willis.

Mimulus Many spp. or their hybrids are gdn. pls., fls. often gaily marked: *M. luteus,* monkey musk, exists in several forms.

Minaret flower *Leonotis leonurus,* S.Af., cult.

Mind-your-own-business *Soleirolia* (=*Helxine*) *soleirolii,* Corsica, cult., v. small fls. and ls. spreads rapidly.

Mingimingi (N.Z.) *Cyathodes juniperina.*

Miniature gardens These constitute a popular form of gardening for many people with restricted space. Small pls. are used and grown in single containers and an attempt made to reproduce gdn. features on a small scale, e.g. paths, pools, bridges, mossy turf etc. Small cacti and succulents are often used for miniature or dish gdns. They do not need constant watering and have the advantage that they may be safely left during the owner's absence.

Mini-bush *Menziesia pilosa,* N.Am.: — **root,** *Ruellia tuberosa,* W.I., r.med.

Mint There are two main forms of mint commonly grown for culinary purposes in Br., the young leafy tops being used for flavouring vegs. such as boiled potatoes and peas and for cream cheese, chutney, vinegar etc. Mint is the most popular flavouring agent in domestic cookery in Br. but not in most countries. The two mints used are *Mentha spicata* (=*M. alopecuroides*) and *Mentha rotundifolia,* the latter w. hairy ls. and commonly called woolly mint, lamb's mint, round-leaved or apple-scented mint. It is considered by connoisseurs to have the better flavour. Mint tea is a popular Arab beverage in N.Af. **Am. wild** —, *Mentha canadensis:* **bergamot** —, *Mentha aquatica,* N.Am.: **cat**—, *Nepeta mussinii:* **corn** —, *Mentha arvensis:* **country** — (Sri Lanka), *Mentha javanica:* **fern** —, *Monarda menthifolia:* **field** —, *Mentha arvensis:* **horse** —, *M. longifolia; Monarda fistulosa,* Am.: **lamb's** —, *Mentha rotundifolia:* **lemon** —, *M. citriodora; Melissa officinalis:* **mountain** —, *Pycnanthemum* spp., N.Am.: **pepper** —, *Mentha piperita:* — **bush** (Aus.), *Prostanthera* spp.: **round-leaved** —, *Mentha rotundifolia:* **spear** —, *M. spicata:* **water** —, *M. aquatica:* **wood** —, *Blephilia hirsuta,* N.Am.: **woolly** —, *Mentha rotundifolia.*

Mint bush *Prostanthera* spp., Aus.: — **flower,** *Eupatorium riparium, E. glandulosum,* N.Am., fragrant small white fls., weeds.

Mirabelle plum *See* myrobalan plum.

Miraculous berry *Synsepalum* (=*Sideroxylum*) *dulcificum* W.Af., a shrub or small tree, the fleshy fr. not itself sweet, has an extraordinary action on the palate causing everything, even a lemon, to taste sweet for some time afterwards. The active principle, which has been isolated, modifies the action of the 'sweet' and 'sour' taste receptors and is regarded as having considerable potential as a sugar substitute without side effects. It has been prepared experimentally in pill form. — **fruit,** *Thaumatococcus daniellii,* W.Af., the jelly-like aril is intensely sweet causing other foods to taste sweet. — **vine,** *Momordica charantia,* W.I.

Miro *Podocarpus ferrugineous,* N.Z., a commercial timber.

Miscanthus *M. sinensis,* orn. grass w. feathery panicle or variegated ls.

Missanda *Erythrophleum guineense, E. ivorense,* trop. Af., a commercial timber.

Missey-moosey *Pyrus americana,* N.Am.

Missouri currant *Ribes odoratum*, N.Am.: — **gooseberry**, *R. missouriense:* — **gourd**, *Cucurbita foetidissima*, N.Am.

Mistletoe Name applied to a number of parasitic pls. belonging to different genera. **Am.** —, *Phoradendron* spp.: — **cactus**, *Rhipsalis:* **common** or **Eur.** —, *Viscum album*, grows on various trees inc. apple, hawthorn, lime, poplar and willow; Christmas supplies in Br. are commonly imported from France (*see Viscum* in Willis): **dwarf** —, *Arceuthobium pusillum*, N.Am.: **false** —, *Phoradendron flavescens*, N.Am.; *Loranthus* spp., Af.: **N.Z.** —, *Loranthus micranthus:* **red** —, *Elytranthe tetrapetala:* **W.I.** —, *Phoradendron trinervium; Phthirusa caribaea*, a troublesome pest that will quickly smother small trees.

Mistol *Zizyphus mistol*, Argen., ed. fr.

Mitchell grass *Astrebla pectinata*, a good pasture grass in Aus., drought-resistant.

Mitraria *M. coccinea*, Chile, cult, orn. climber w. bright red frs.

Mitrewort *Mitella* spp., N.Am.: **false** —, *Tiarella* spp., N.Am.

Mkani fat (E.Af.) *Allanblackia stuhlmannii*, ed. fat from kernels.

Mkwamba (E.Af.) *Flueggea virosa*, frs. used by native women, alleged to promote fertility.

Moa (Ind.) *Madhuca* (= *Bassia*) *latifolia*, kernels yield ed. oil, fleshy fls. ed.

Moabi *Mimusops djave*, W.Af., a commercial timber.

Mobola plum *Parinari curatellifolium* (= *Parinarium mobola*), S.Af.

Moccasin flower *Cypripedium* spp., N.Am.

Mock orange Or syringa, *Philadelphus coronarius*, SE. Eur., E.As., long cult., scented fls., several vars., other spp. cult. **Am.** — —, *Prunus caroliniana*, *Styrax americana:* **Aus.** — —, *Pittosporum undulatum:* — **privet**, *Phillyrea* spp., evergreen shrubs. —**er nut**, *Carya tomentosa*, N.Am.

Moghat root *Glossostemon bruguieri*, Egypt, r. med., sold in bazaars.

Mohle flowers *Jasminum sambac*, trop. As., oil from fragrant fls. used in perfumery.

Mohur tree Name for flamboyant, *Delonix regia*.

Molasses Or treacle, consists of the uncrystallizable sugars from cane juice in sugar manufacture: — **grass**, *Melinis minutiflora*, trop. Af., cult., trop. fodder grass.

Mole plant *Euphorbia lathyrus*, the presence of this pl. in a gdn. is alleged to keep moles away.

Moluccella *M. laevis*, Medit., gdn. pl., large calyx, dries well, used as everlasting: also called Molucca balm, shell fl. and Bells of Ireland.

Moly *Allium moly*, Medit., fls. yellow, cult.

Mombin Red —, *Spondias purpurea:* **yellow** —, *S. mombin*.

Momordica Trop. orn. climbers w. handsome frs., cult.

Monanthes Dwarf succulent pls., Canary Isles., cult.

Monarch-of-the-East *Sauromatum guttatum; S. venosum*, cult. — — — **veld**, *Arctotis* (= *Venidium*) spp., S.Af.

Monarda N.Am. aromatic pls. w. bright fls., cult., esp. *M. didyma*.

Mondurup bell *Darwinia meeboldii*, Aus.

Moneywort *Lysimachia nummularia:* **Cornish** —, *Sibthorpia europaea*.

Monkey apple *Annona glabra* (= *A. palustris*), W.I.; *Anisophyllea laurina*,

MONKSHOOD

trop. Af.; *Strychnos* spp. large frs. w. ed. pulp: — **bread,** *Adansonia digitata,* fr. pulp ed.: — **cocoa,** *Theobroma angustifolia:* — **comb,** *Pithecoctenium,* trop. Am.: — **face tree,** *Mallotus philippensis:* — **flower,** *Mimulus luteus, M. guttatus* and other spp.: — **guava,** *Diospyros mespiliformis,* W.Af.: — **gun,** *Ruellia tuberosa,* W.I., a weed, capsule explodes: — **nut,** *Arachis hypogaea; Hicksbeachia pinnatifolia,* Aus.: — **orange,** *Strychnos* spp., S.Af.: — **pot,** *Lecythis* spp., large woody fr.; *Cariniana* spp.: — **plum,** *Diospyros* spp., esp. *D. dichrophylla,* S.Af.; — **puzzle tree,** *Araucaria araucana:* — **rope,** *Rhoicissus capensis,* S.Af. and other spp.; *Dalbergia obovata,* S.Af.: —'s **hand tree,** *Cheirostemon platanoides,* Mex., long red stamens united at base: — **tail,** *Xerophyta retinervis,* S.Af.

Monkshood *Aconitum* spp. *A. napellus,* Eur. *A. wilsonii,* China, cult. gdn. pls. — **pepper tree,** *Vitex agnus-castus,* Medit.

Monochaetum Trop. Am. shrubs. cult., fls. w. unusual stamens.

Monolena Succulent stemless pls., Peru, Andes, cult.

Monos plum *Anamomis umbellulifera,* W.I., pleasant ed. juicy fr.

Monstera *M. deliciosa* (=*M. pertusa*), trop. Am. cult. orn. pl., a house pl., fr. ed., unusual perforated ls. *See* Willis.

Montanoa *M. bipinnatifida,* Mex. handsome fol. pl. subtrop., other spp. cult.

Montbretia *Crocosmia* spp. and hybrids, S.Af., popular gdn. pls.

Moon-beam *Tabernaemontana coronaria,* Ind., cult. orn. shrub: — **creeper,** *Ipomoea* (= *Calonyction*) *aculeata:* — **daisy,** *Chrysanthemum leucanthemum:* — **flower,** *Ipomoea bona-nox,* cult. trop. climber w. large white fls., 15 cm across, opening at night: — **seed,** *Menispermum canadense,* E. N.Am. cult. climber: — **trefoil,** *Medicago arborea:* — **vine,** *Ipomoea* spp.: — **wort,** *Botrychium lunaria.*

Moor grass Blue — —, *Sesleria caerulea:* **purple** — —, *Molinia caerulea,* a striped form grown for orn.

Mooseberry *Viburnum edule,* N.Am., fr. esteemed for jam: — **wood,** *Acer pennsylvanicum, Dirca palustris, Viburnum alnifolium.*

Mopane *Copaifera mopane,* C. & S.Af., tree covers vast arid areas: *Cochlospermum mopane,* S.Af.

Mora *Mora excelsa* (=*Dimorphandra mora*), trop. Am., a hard, durable commercial timber.

Moraea S.Af. bulbous pls., some orn., cult., some poisonous to livestock.

Morass weed (W.I.) *Ceratophyllum.*

Moreton Bay chestnut *Castanospermum australe,* widely grown street and orn. tree: — — **fig,** *Ficus macrophylla,* Aus., cult., for orn. and shade: — — **pine,** *Araucaria cunninghamii,* Aus.

Moriche *Mauritia setigera,* W.I., a tall palm.

Morina Thistle-like pls., some orn. and cult., e.g. *M. longiflora, M. persica.*

Morinda Trop. trees, several yield local dyestuffs.

Mormodes C. & S.Am. orchids, some w. very large fls., cult.

Mormon tea *Ephedra trifurca,* SW. U.S., a desert pl.

Morning brides *Chaenactis douglasii,* N.Am.

Morning glory *Ipomoea* (= *Pharbitis*) *purpurea,* a widely cult. climber in warm cli., fls. trumpet-shaped, usu. blue, some vars. of other colours. **Arrow-**

leaved — —, *Ipomoea sagittata*, N.Am.: **beach** — —, *I. pes-caprae*, trop.:
Cairo — —, *I. cairica*: **red** — —, *I. angulata*, trop. weed: **seven-leaved**
— —, *I. heptaphylla*, N.Am.: **wild** — —, *Evolvulus arizonicus*, N.Am.:
yellow — —, *Ipomoea tuberosa*.

Morrel Black —, *Eucalyptus melanoxylon*, W.Aus.: **petty** —, *Aralia racemosa*,
N.Am., large aromatic rs.

Mortel (W.I.) *Erythrina umbrosa*, a shade tree for trop. crops.

Mortiña *Vaccinium mortinia*, Ecuador, fr. ed., sold on markets.

Moschatel *Adoxa moschatellina*, Eur., As.

Mosquito bush *Ocimum americanum* and other spp., the aromatic dried pl.
burned as a mosquito repellent; *O. micranthum*, W.I.; *Hyptis suaveolens*,
W.I.: — **orchid**, *Acianthus exsertus*, Aus.: — **wood**, *Mosquitoxylum
jamaicense*, Jam.

Moss campion *Silene acaulis*, cult. orn. fl.: **club** —, *Lycopodium:* — **fern**,
Selaginella: — **locust**, *Robinia hispida*, N.Am.: — **pink**, *Polemonium*, N.Am.:
Spanish long —, *Tillandsia*, N. & C.Am.: **staghorn** —, *Lycopodium:*
— **verbena**, *Verbena tenuisecta*, N.Am.: —**y cyphel**, *Cherleria sedoides:*
— **of Tillaea**, *Crassula tillaea*.

Mother-in-law's-tongue *Sansevieria trifasciata*, orn. ls., a room pl.:
— **of-the-evening**, *Hesperis matronalis:* — **of thousands**, *Saxifraga
stolonifera* (= *S. sarmentosa*), cult., has creeping red stolons bearing young
pls.; name also used for *Cymbalaria muralis* and *Soleirolia soleirolii*, S.Eur.:
— **of thyme**, *Satureja acinos:* — **wort**, *Leonurus cardiaca*.

Mount Cook lily *Ranunculus insignis*, N.Z., large white fls.

Mountain ash *Sorbus* (= *Pyrus*) *aucuparia*, As., Eur. inc. Br., wild and cult.
orn. tree, many vars.: — —, **S.Af.**, *Ekebergia meyeri;* — —, **N.Am.**, *Sorbus
scopulina:* — **avens**, *Dryas octopetala:* — **berry**, Aus., *Cyathodes parvifolia:*
— **blaeberry**, *Vaccinium uliginosum:* — **bluebell**, *Campanula divaricata*,
N.Am.: — **buttercup**, *Ranunculus graniticola*, Aus.: — **camellia**, *Stewartia
ovata*, N.Am.: — **cypress**, *Widdringtonia cupressoides*, S.Af.: — **damson**,
Quassia (= *Simarouba*) *amara:* — **devil**, *Lambertia formosa*, Aus.: — **ebony**,
Bauhinia variegata: — **elder**, *Sambucus pubens*, N.Am.: — **everlasting**,
Antennaria dioica: — **flax**, *Phormium colensoi*, N.Z.: — **grape**, *Guettarda:*
— **holly**, *Nemopanthus mucronata:* — **Juneberry**, *Amelanchier bartramiana*,
N.Am.: — **lettuce**, *Saxifraga micranthidifolia*, N.Am.: — **laurel**, *Kalmia
latifolia*, N.Am.: — **lilac**, *Prostanthera lasianthos*, Aus.: — **lover**, *Paxistima
myrsinites*, N.Am.: — **mahogany**, *Cercocarpus*, N.Am.: — **maple**, *Acer
spicatum*, N.Am.: — **mint**, *Pycnanthemum;* — **papaw**, *Carica cundinamarcensis:* — **plum**, *Ximenia*, W.I.: — **pride**, *Brownea latifolia*, *Spathelia
simplex*, W.I.: — **rice**, *Oryzopsis* spp., N.Am.: — **rose**, *Brownea latifolia*,
W.I.: — **rose-bay**, *Rhododendron catawbiense*, N.Am.: — **spray**, *Holodiscus
discolor*, N.Am.: — **tea**, *Gaultheria procumbens*, N.Am.: — **tobacco**, *Arnica
montana*, N.Am.

Mournful widow *Scabiosa atropurpurea*, SW. Eur., cult. ann., fls. crimson,
many vars. used in funeral wreaths in some countries.

Mouse ear *Myosotis*, Am., *Cerastium*, *Holosteum:* — **tail**, *Myosurus minimus*,
Arisarum proboscideum, spathe has a long tail.

Moutan *Paeonia suffruticosa*, orn. gdn. pl.

MOXA

Moxa *Crossostephium artemisioides*, China, cult., the soft woolly mass collected from young ls. used med. like cottonwool.

Moxie plum *Gaultheria hispidula*, N.Am.

Moya grass (Ind.) *Pennisetum alopecuros*, considered for paper making.

Mpepa (E.Af.) *Flagellaria guineensis*, scrambles over trees, fr. ed., stems used for fish traps.

Mpesi (E.Af.) *Trema guineensis* ls. contain tannin, used for preserving fishing nets.

Mrs Robb's bonnet *Euphorbia robbiae*, Eur.

Msasa (E.Af.) *Ficus exasperata*, ls. resemble sandpaper on both sides.

Msoko (E.Af.) *Cassia abbreviata*, orn. fl. shrub.

Mucilage Viscid gum-like substance often produced in ducts or canals: common in water pls. and certain seeds, e.g. quince, flax, psyllium, *Brassica*, *Anthemis*.

Mud grass *Pseudoraphis spinescens*, Aus.: — **rush**, floating, *Scirpus fluitans*: — **wort**, *Limosella aquatica*.

Mudar fibre (Ind.) *Calotropis procera*, *C. gigantea*, a stem fibre, also yields seed floss.

Muduga oil (Ind.) *Butea frondosa*, oil from seeds med. (vermifuge).

Mugga (Aus.) *Eucalyptus sideroxylon*, cult.

Mugongo *Ricinodendron rautanenii*, S. trop. Af., a commercial timber.

Mugwort *Artemisia*, *Galium*.

Muhimbi *Cynometra alexandri*, Uganda, a v. hard commercial wood.

Muhlenbergia Some spp. are useful fodder grasses.

Muhugwe or **muhuhu** (E.Af.) *Brachylaena hutchinsii*, used for flooring blocks, a commercial timber.

Muhuti (E.Af.) *Erythrina abyssinica*, orn. fl. tree,

Mukawa (E.Af.) *Carissa edulis*, orn. fl. shrub.

Mukombokombo (E.Af.) *Gardenia urceoliformis*, orn. fl. shrub.

Mukuba (E.Af.) *Craibia elliotti*, orn. fl. tree.

Mukumari (E.Af.) *Cordia abyssinica*, a commercial timber.

Mukuyu (E.Af.) *Ficus* spp.

Mula or **muli** *Raphanus sativus*, Chinese or Oriental radish.

Mulatto plait (W.I.) *Sporobolus indicus*, a wiry weed-grass used for stuffing mattresses: —**'s ear**, *Enterolobium cyclocarpum*, W.I.

Mulberry *Morus nigra*, Orient, the **common** or **black** —, cult. for thousands of years: introduced to Br. in the 16th Century, cult. for fr. Wood hard and heavy, darkens w. age, has been used for furniture, snuff boxes, inlaying. The best mulberry for feeding silkworms and the one used in China is *Morus alba*: — **fig**, *Ficus sycomorus*, Egypt, ed. fr.: **French** —, *Callicarpa americana*, N.Am.: **Ind.** —, *Morinda citrifolia*, red dye from rs.: **paper** —, *Broussonetia papyrifera*, E.As.: **red** —, *Morus rubra*, N.Am.: **white** —, *Morus alba*, China, widely cult.

Mulberries on sticks orchid (Aus.) *Prasophyllum* spp.

Mulga *Acacia aneura*, Aus., the v. hard dark heartwood was used by aborigines for boomerangs and spear shafts, now used for fancy articles: **curly-barked** —, *Acacia cyperophylla*.

Mulla-mulla (Aus.) *Ptilotus exaltatus* and other spp.

Mullein *Verbascum thapsus* and other spp.: **common** —, *V. thapsus*, cult., also a weed: has med. uses: **dark** or **black** —, *V. nigrum*, once used in witchcraft and incantations, also for dyeing hair: — **foxglove**, *Seymeria macrophylla*, N.Am.: — **pink**, *Lychnis coronaria:* **purple** —, *Verbascum phoeniceum*, cult.: **white** —, *V. lychnitis*.

Mume Jap. apricot, *Prunus mume*, ed. fr. and orn. fl. tree, several vars.

Mummy wheat Wheat was well known to the Ancient Egyptians and specimens have been obtained from the tombs of the Pharaohs. So-called mummy wheat, ordinary wheat darkened or stained black to make it look old, has been sold by unscrupulous vendors to tourists and visitors to the Pyramids. Needless to say such mummy wheat may germinate when sown.

Mung Or green gram, *Phaseolus aureus* (= *P. radiatus*), an imp. Ind. pulse crop, beans green, yellow or black: name also applied to black gram, *Phaseolus mungo*.

Mungarima (E.Af.) *Ochna nandiensis*, orn. fl. shrub.

Muninga *Pterocarpus angolensis*, a commercial timber from Tanzania.

Munjeet Or Indian madder, *Rubia cordifolia*, rs. yield a red dye.

Mununga (E.Af.) *Ekebergia capensis*, a local shade tree.

Muringa (E.Af.) *Cordia holstii*, orn. fl. tree.

Murraim berries *Tamus communis*, black bryony.

Murraya *M. koenigii*, Ind., curry leaf tree, ls. much used in curries.

Murumuru *Astrocaryum murumuru*, Braz., tall palm, nuts yield oil. w. uses similar to coconut oil.

Musakwa (E.Af.) *Vernonia holstii*, orn. fl. shrub.

Muscadine *Vitis rotundifolia*, N.Am.

Muscatel The fr. of a special var. of red grape dried on the vine in Spain and Turkey.

Muscovado Raw or unrefined cane sugar.

Musine *Croton megalocarpus*, E.Af., a commercial timber and local shade tree.

Musizi *Maesopsis eminii*, trop. Af., a commercial timber.

Musk or **musk plant** *Mimulus moschatus*, Eur. inc. Br., widely grown in Br. in Victorian times as a fragrant cottage or house pl., but since about 1914 the scented form or variety appears to have been lost to cult. — **mallow**, *Abelmoschus moschatus* (= *Malva moschata*), cult. trop. aromatic seeds distilled for oil, used in perfumery: **Maori** —, *Mimulus repens*, N.Z.: — **root**, *Ferula sumbul:* **swamp** —, *Mazus radicans*, N.Z.: — **tree**, *Olearia argophylla*, Aus., wood w. musk-like odour.

Muskit *See* mesquite.

Musquash root *Cicuta maculata*, N.Am., poisonous.

Mussaenda *See* ashanti blood.

Musschia Madeira pls. w. large unusual attractive fls., cult.

Mustard *Brassica nigra*, black mustard, the seeds are the main source of the condiment but the seeds of white mustard, *B. alba*, may be used to some extent (mixed w. black). In terms of quantity there is more mustard used throughout the world than any other flavouring or spice, except pepper (*Piper*). For mustard powder the seeds are finely ground and sifted (except for French mustard). For pickles and chutney, the seeds may be used whole. For mustard pastes turmeric may be added to intensify the yellow colour.

Mustard seedlings are used as salad, as 'mustard and cress'. Mustard is grown as a green manure or soiling crop. The fls. are a good nectar source for the hive bee but the honey crystallizes rapidly. **Abyssinian** —, *Brassica carinata*, a pot-herb: **ball** —, *Neslia paniculata*: **brown** —, *Brassica nigra*: **buckler** —, *Biscutella laevigata*, Medit.: — **bush**, *Hermannia vesicaria*, S.Af.: **Chinese** —, *Brassica chinensis; B. japonica*: **clown's** —, *Iberis amara*: **garlic** —, *Alliaria petiolata*: **hare's ear** —, *Conringia orientalis*: **hedge** —, *Sisymbrium officinale*: **hoary** —, *Hirschfeldia incana*: **Ind.** —, *Brassica juncea*: **leaf** —, *B. juncea*: **Mithridate** —, *Thlaspi arvense*: **Sarepta** —, *B. besseriana*: **tower** —, *Arabis glabra*: **treacle** —, *Erysimum cheiranthoides*: — **tree**, *Salvadora persica*, thought to be the mustard tree of Scripture: **tumble** —, *Sisymbrium altissimum*: **white** —, *Brassica alba*: **yellow** —, *B. alba*.

Muster John Henry *Tagetes minuta*, N.Am., a weed, called khaki weed in S.Af.

Mutisia S.Am. erect or climbing shrubs, large fl. heads, cult.

Mutoo (E.Af.) *Dombeya quinqueseta*, an orn. fl. tree.

Mutton bird sedge *Carex trifida*, N.Z.

Mvule (E.Af.) *Chlorophora* (= *Maclura*) *excelsa*, a valuable commercial timber called *iroko* and *odum* in W.Af.

Myall *Acacia pendula*, Aus., hard, scented, close-grained wood, used in turnery and for fancy articles, tobacco pipes etc.

Myoporum Trop. trees and shrubs, some cult. for orn.: *M. laetum* yields useful timber in N.Z.

Myrabolans The dried frs. of certain Ind. spp. of *Terminalia*, long esteemed in leather tanning and exported to other countries. Br. tanners sometimes refer to them as 'myrabs'. They yield a soft mellow leather, but are generally blended w. other tans. *Terminalia chebula* is the most imp. sp. There are several commercial grades of myrabolans. The average tannin content of a good commercial sample is about 32 %. Myrobalan plum, also called myrobalan, mirabelle, or cherry plum, *Prunus cerasifera*, frs. candied in France.

Myrica *See* myrtle.

Myriophyllum Submerged fresh water aquatics, some favoured for aquaria and fish ponds.

Myrmecodia Old World epiphytes with swollen bases occupied by ants. *See* Willis.

Myrmecophila Trop. Am. orchids w. hollow pseudobulbs occupied by ants.

Myrmecophilous plants Those pls. which exhibit symbiosis with ants, i.e. afford shelter and sometimes food for ants. Usually the ants bestow some obvious benefit to the pl., such as keeping away leaf-eating insects or animals. Sometimes extrafloral nectaries provide the ants with food (cherry laurel). In *Acacia sphaerocephala*, C.Am., ants hollow out the large thorns and live inside; the sausage-shaped yellow food bodies at the tips of the leaflets are used by the ants. In *Cercospora peltata*, trop. Am. (trumpet tree) the ants live in the hollow stems and feed on food bodies produced on the lower side of the base of the l. stalk. Other pls. harbouring ants belong to the following genera or families – *Bombacaceae, Clerodendron, Cuviera, Duroia, Humboldtia, Hydnophytum, Korthalsia, Macaranga, Maieta, Myrmecodia, Nauclea, Rubiaceae, Triplaris*. Myrmecophilous pls. may sometimes be associated with habitats of low nutrient status.

Myrobalans *See* myrabolans.

Myrrh The aromatic resinous exudation (oleo-gum-resin) from the trunks of several species of *Commiphora*, notably *C. molmol*, in Somalia and Arabia: well known from Biblical times and often referred to w. frankincense. It is now of little importance compared with former times. It has been used med. as an ingredient of purgative pills: **garden —**, *Myrrhis odorata* Eur., old-time herb used in home remedies, also for flavouring brandy.

Myrtle *Myrtus communis*, S.Eur., the common myrtle, cult. orn. shrub w. fragrant fls.: esteemed by the Ancients, an emblem of love and peace favoured for wedding bouquets, an oil may be distilled from b., ls. and frs. **Aus. —**, *Leptospermum, Rhodamnia, Syzygium:* **bog —**, *Myrica gale:* **Cape —**, *Myrsine africana:* **crepe —**, *Lagerstroemia indica*, cult. orn. shrub in warm cli.: **downy —**, *Rhodomyrtus tomentosa:* **honey —**, *Melaleuca* spp., Aus.: **lemon-scented —**, *Backhousia citriodora*, Aus., yields esst. oil: — **lime**, *Triphasia trifolia*, a trop. hedge pl., frs. for pickles: **sand —**, *Leiophyllum buxifolium*, E. N.Am., cult. orn. shrub: **scrub —**, *Banksia* spp., Aus.: **Tasm. —**, *Nothofagus cunninghamii*, good timber, often beautifully marked: **wax —**, *Myrica cerifera*, N.Am.; *M. cordifolia*, S.Af., wax from berries used by early settlers (candles); **willow —**, *Agonis* spp. Aus.

Mysore thorn *Caesalpinia decapetala* (=*C. sepiaria*), Ind. a prickly shrub, used as a hedge.

Mzimbeet *Androstachys johnsonii*, E. & S.Af., Madag., a hard, heavy timber alleged to resist termites.

N

Na (Sri Lanka) *Mesua ferrea*, trop. As., v. hard wood, ironwood, once used for lances.

Naartje (S.Af.) Name used for local tangerine, *Citrus nobilis* var. *deliciosa.*

Naboom (S.Af.) *Euphorbia ingens* and other spp.

Naegelia *See* Smithiantha.

Nail galls Nail-like galls on ls. of lime (*Tilia*) and other trees: **silver — root**, *Paronychia argentea*, Medit., med. herb.

Naked boys (Br.) *Colchicum autumnale:* — **Indian** (W.I.) *Albizia caribaea*, an orn. shade tree bare of ls. for part of year: — **ladies** (Br.) *C. autumnale.*

Nal (Ind.) *Phragmites australis* (=*P. communis, P. karka*) considered for paper making.

Nam-nam (Ind.) *Cynometra cauliflora*, fleshy subacid young pods used in preserves or pickled.

Namaqualand beauty *Pelargonium incrassatum*, S.Af.

Nandina *N. domestica.* Orn. evergreen shrub, much cult. in Jap.

Nankeen Durable kind of cotton cloth originally made in China (Nankin) from buff-coloured cotton: later made in Eur., colour copied with dyes.

Nanking cherry *Prunus tomentosa* or Manchu cherry, frs. sweet and juicy.

Nanmu wood (China) *Persea nanmu*, a much used wood, esp. for coffins.

Nannorrhops *N. ritchieana*, Ind., a low palm, ls. used for baskets, brooms, mats, etc., seeds as beads.

Nannyberry *Viburnum lentago*, N.Am.

Nap-at-noon *Ornithogalum umbellatum*, Br., Star of Bethlehem.

Napier fodder *Pennisetum purpureum*, trop. Af., elephant grass, cult. for fodder trop. and subtrop.

Napoleon's buttons *Napoleona heudelotii*, trop. Af., strange red fls.: — **hat**, *Bauhinia variegata*, cult. orn. shrub, trop.; *Napoleona imperialis*, trop. Af. tree, cult.

Napuka (N.Z.) *Veronica speciosa*, evergreen shrub, purple fls., cult., several vars.

Nara or **narras** *Acanthosicyos horrida*, SW. Af., a remarkable sand dune pl. w. ed. pumpkin-like seeds, rs. may reach 12 m in length.

Naranjilla *Solanum quitoense*, Ecuador, Peru, orange fr. (5 cm diam.) has juicy ed. pulp, used for drinks and sherbets.

Narcissus Bulbous pls. of Eur., N.Af. and W.As., distinguished by a characteristic corona: many spp. and hybrids cult., e.g. *N. pseudonarcissus* (daffodil), *N. poeticus* (poet's narcissus), *N. jonquilla* (jonquil) fls. used in perfumery.

Narcotics Sleep-inducing drugs (*Cannabis, Hyoscyamus, Papaver*, etc.).

Nard grass *Nardus stricta*, Eur. inc. Br. a tough wiry grass of heath and moorland.

Nardoo (Aus.) *Marsilea drummondii*, the 'frs.' (sporocarps) of this fern ally used as food by aborigines and early settlers.

Narinjin Bitter glucoside largely responsible for the bitter flavour of grapefruit and shaddock or pomelo. It has been extracted and used for flavouring beverages and confections.

Narrawa burr (Aus.) *Solanum cinereum*.

Naseberry Sapodilla.

Nasturtium *Tropaeolum majus*, S.Am., showy fls., widely cult. unripe seeds pickled and used as a substitute for capers.

Natal grass *Rhynchelytrum repens* (= *Tricholaena rosea*), a good pasture and hay grass, cult. in other countries: — **laburnum**, *Calpurnia aurea:* — **lily**, *Crinum:* — **mahogany**, *Trichilia, Kiggelaria:* — **plum**, *Carissa macrocarpa* (= *C. grandiflora*), thorny evergreen shrub much used for hedges, large red. ed. fr. favoured for preserves: — **primrose**, *Thunbergia atriplicifolia*.

National flowers or **floral emblems** Many countries have adopted a locally well known or outstanding fl. as the country's national fl. or floral emblem. Some of these have been in existence for a very long time as in the case of Br. Others have been selected in comparatively recent times, such as those for Australia, New Zealand and Rhodesia. Many of the less advanced countries have not yet selected a floral emblem but may do so in due course. Some are represented on postage stamps.

The following are among the established national fls. or floral emblems. **Aus** – New South Wales, Waratah (*Telopea speciosissima*): Northern Territory, Sturt's desert rose (*Gossypium sturtianum*): S.Aus., Sturt's desert pea (*Clianthus formosus*): Tasmania, blue gum (*Eucalyptus globulus*): Victoria, heath (*Epacris impressa*): W.Aus., kangaroo paw (*Anigozanthos*). **Br.** – England, rose: Scotland, thistle: Ireland, shamrock: Wales, leek. **Can.** – maple leaf: Alberta, wild rose: Br. Columbia, dogwood: Manitoba, crocus: New

Brunswick, blue violet: Newfoundland, pitcher plant: Northwest Territory, mountain avens: Nova Scotia, mayflower: Ontario, Trillium: Prince Edward Island, lady's slipper: Quebec, white lily: Saskatchewan, prairie lily: Yukon, purple fireweed. **Jap.** – chrysanthemum: **Rhodesia** – gloriosa lily (*Gloriosa superba*). **S.Af.** – protea. **Switzerland** – edelweiss. **U.S.** – Alabama, golden rod: Alaska, forget-me-not: Arizona, giant cactus: Arkansas, apple blossom: Calif., Californian poppy: Colorado, columbine: Connecticut, mountain laurel (*Kalmia*): Delaware, peach blossom: Florida, orange blossom: Illinois, wood violet: Indiana, zinnia: Iowa, wild rose: Kansas, sunflower: Kentucky, golden rod: Louisiana, magnolia: Maine, pine: Maryland, black-eyed Susan (*Rudbeckia*): Massachusetts, trailing arbutus: Michigan, apple blossom: Minnesota, lady's slipper: Mississippi, magnolia: Missouri, downy hawthorn: Montana, butterwort (*Lewisia*): Nebraska, golden rod: Nevada, sage bush: New Hampshire, purple lilac: New Jersey, violet: New Mexico, yucca: New York, rose: North Carolina, ox-eye daisy: North Dakota, prairie rose: Ohio, scarlet carnation: Oklahoma, mistletoe (*Phoradendron*): Oregon, holly grape (*Mahonia*): Pennsylvania, mountain laurel: Rhode Island, violet: South Carolina, jasmine (*Gelsemium*): Texas, bluebonnet (*Lupinus*): Utah, Sego-lily (*Calochortus*): Vermont, red clover: Virginia, flowering dogwood: Washington, rhododendron (*R. macrophyllum*): W. Virginia, rhododendron (*R. maximum*): Wisconsin, violet: Wyoming, Indian paint brush (*Castilleja*): Hawaii, Hibiscus (*H. rosa-sinensis*).

Natural lacquer or **varnish** Terms given to the sap of certain trees applied direct to the article or surface to be varnished without the use of any solvent, the lacquerware of China and Japan being the best known, the tree in question being *Rhus verniciflua*. For Burmese lacquerware the sap of *Melanorrhoea usitata* is used.

Naval stores American term for rosin and turpentine derived from pine trees, originating from the early days of the 17th Century when the wooden vessels then in use needed large quantities of tar and pitch.

Navelwort *Hydrocotyle* spp.; *Umbilicus rupestris*.

Nazingu *Mitragyna* spp., Uganda, a commercial timber.

Necklace fern *Asplenium flavellifolium*, Aus.: — **poplar**, *Populus deltoides:* — **seeds**, *see* beads: — **tree**, *Ormosia:* — **weed**, *Veronica peregrina*, *Actaea* spp., N.Am.

Nectar The fresh nectar of fls. is little more than a weak solution of sucrose (cane sugar) generally containing 60–90% water with small quantities of other substances such as esst. or volatile oils, flavourings, gums and traces of mineral matter. In the case of nectar collected by the honey bee these lesser ingredients become of importance as the water is driven off the nectar in the hive and the nectar converted into honey, for they determine the aroma and flavour of the honey and account for the distinctive characteristics of honeys from different floral sources or regions.

Nectar may be secreted copiously in some fls. but sparsely in others. In the former category is the honeysuckle, certain orchids, the tulip tree (*Liriodendron*) in N.Am. and some of the Cape proteas or sugar bushes. In early days at the Cape, before the advent of cheap sugar, it was a common practice to boil the fls. in water and concentrate the liquid to the consistency of thick syrup for table use. On the other hand, many fls. often of great importance to the beekeeper, secrete only very small quantities of nectar. But the honey bee, with her specialized tongue or sucking mechanism, is able to absorb minute quantities, even when the surface appears only moist.

The secretion of nectar by the nectary is very dependent upon environmental conditions such as temperature, humidity and, in particular, soil moisture. W. many pls. secretion ceases under drought conditions. Some pls. such as willows and white clover (*Trifolium repens*) will secrete nectar at quite low temperatures. The concentration of the nectar in the nectary may vary considerably, being usually low in the early morning and increasing as the day advances. The age of the pl. may also affect secretion in some spp. W. heather or ling (*Calluna vulgaris*) young pls. are considered to secrete nectar more freely than old ones. W. some pls. cool nights followed by fine hot days provide the best conditions for nectar secretion. The word nectar is sometimes used commercially for certain fr. juices, e.g. mango, passion fr. and guava.

Nectarine Although the nectarine is distinctive from the peach in appearance and flavour it is closely allied to it and belongs to the same sp., *Prunus* (=*Amygdalus*) *persica*. The main difference is that the nectarine has a smooth instead of a downy skin. Its climatic and cult. requirements are similar.

Needle, Adam's *Yucca filamentosa* and other spp.: **bush —**, *Hakea leucoptera*, Aus.: **— furze**, *Genista anglica*: **— grass**, *Aristida*, *Stipa* and some other grasses: **— whin**, *Genista anglica*.

Neem, nim Or margosa, *Azadirachta indica*, a large tree of great importance in Ind., cult. for shade and timber, many med. and insecticidal uses.

Negro's head *Phytelephas macrocarpa*, fr., (vegetable ivory palm): **— pepper**, *Xylopia aethiopica*, W.Af.

Nelumbo Cult. as orn. aquatics (water lilies); *N. pentapetala* (=*N. lutea*), yellow lotus of N.Am., seed ed.; *N. nucifera*, As., sacred lotus, many vars. *See* Willis.

Nemesia *N. strumosa*, S.Af., cult. orn. fl., many strains, brilliant colours, much used for edgings and window boxes: other spp. cult.

Nemophila *N. menziesii*, N.Am., popular gdn. ann.: other spp. cult.

Neoregelia S.Am. bromeliads cult. as orn., fol. pls., some as house pls.

Nepal paper Or bhutia paper, *Daphne cannabina*, Himal., made from fibrous b.: may also be made from *Edgeworthia gardneri*.

Nepenthes *See* pitcher plant.

Nephrolepis Handsome trop. ferns (ladder ferns), popular greenhouse subjects.

Nerine Bulbous S.Af. pls. resembling *Amaryllis*, cult.

Neroli Aromatic or esst. oil distilled from the fls. of the sweet and bitter orange.

Nerve root *Cypripedium acaule*, N.Am.

Net bush *Grevillea intricata*, W.Aus., orn. cult.

Nettle Name used for various pls., usu. w. stinging hairs. **Common stinging —**, *Urtica dioica*, Eur., has stinging hairs, ls. may be used as spinach and are a commercial source of chlorophyll, stems yield a strong fibre, a common weed: **dead —**, *Lamium* spp.: **devil's —**, *Laportea* spp.: **dog —**, *Urtica urens*: **dwarf bush —**, *Urtica incisa*, N.Z.: **false —**, *Boehmeria* spp.: **giant —**,

Laportea gigas, tree, Ind. to Aus., and other spp., esp. *L. moroides,* Aus., the stinging power of the ls. is very great and even dangerous: **hedge —,** *Stachys* spp.: **hemp —,** *Galeopsis tetrahit:* **Nilgiri —,** *Girardinia* spp., stems yield a silky fibre used locally: **Roman —,** *Urtica pilulifera,* Medit.: **Spanish —,** *Bidens* spp., W.I.: **small —,** *Urtica urens:* **spurge —,** *Cnidoscolus* spp., N.Am.: **— tree,** *Celtis* spp. esp. *C. australis,* S.Eur. and As.: **— vine,** *Tragia volubilis,* W.I.: **W.I. —,** *Urera caracasana, U. baccifera,* W.I., powerful stinging hairs: **wood —,** *Laportea canadensis,* N.Am.

Newbouldia *N. laevis,* W.Af., used as a hedge or boundary pl.

New Jersey blueberry *Vaccinium caesariense:* **— — tea,** *Ceanothus americanus:* **— Orleans moss,** *Tillandsia usneoides:* **— York fern,** *Dryopteris noveboracensis:* **— York iron weed,** *Vernonia noveboracensis.*

New Zealand bluebell *Wahlenbergia albomarginata:* **— — bur,** *Acaena:* **— — cabbage,** *Cordyline australis:* **— — daisy bush,** *Olearia:* **— — flax** or **hemp,** *Phormium tenax,* ls. yield a commercial fibre, widely cult.: **— — laburnum,** *Sophora tetraptera:* **— — pincushion,** *Raoulia.*

Ngai camphor *Blumea balsamifera,* Ind., Malaysia, obtained by distillation, med.

Niangon *Tarrietia utilis,* W.Af., a commercial timber.

Niaouli oil *Melaleuca viridiflora,* New Caledonia, steam distillation of ls., a substitute for *Eucalyptus* oil.

Nicandra *N. physaloides,* Peru, a vigorous ann. w. blue fls., cult., shoo-fly.

Nicaragua wood *Dalbergia, Caesalpinia, Haemotoxylon.*

Nicker bean *Caesalpinia bonducella, Entada gigas,* both are drift seeds, may be brought to Eur. from trop. by Gulf Stream.

Nicotiana Or flowering tobacco, *Nicotiana alata, N. sylvestis* and other spp., C. & S.Am., cult. orn. fls., very fragrant. *See* tobacco.

Nicotine Commercial nicotine, used as an insecticide, is prepared largely from the refuse of tobacco factories and from *Nicotiana rustica.*

Nicuri palm Or Ouricuri, *Syagrus* (=*Cocos*) *coronata,* Braz., ed. pulp of fr. yields oil, as do kernels.

Nidularium Trop. Am. bromeliads w. orn. fol., esp. *N. innocentii,* ls. purple below, cult.

Nierembergia Am. perens., many procumbent, showy fls., cult.

Nigella Anns., Medit., W.As., orn. fls., cult.; *N. damascena,* love-in-a-mist; *N. hispanica,* fennel fl.

Niger seed *Guizotia abyssinica,* an oil seed and much favoured for cage birds.

Nigger heads *Enneapogon nigricans,* Aus., a grass w. fluffy dark seed heads: *Rudbeckia serotina,* N.Am. an aggressive weed: **— teats,** *Rudbeckia bicolor,* N.Am.: **— mouth,** *Pedilanthus tithymaloides,* W.I. sometimes used as a hedge.

Night-flowering cactus *Selenicereus grandiflorus, Epiphyllum hookeri,* sweet-scented fls. open in the evening and fade by morning. **— — cereus,** *Hylocereus lemairei,* W.I.: *H. undatus,* N.Am.: **— — cestrum** or **jasmine,** *Cestrum nocturnum,* N.Am.

Nightshade Black —, *Solanum nigrum,* in some countries ls. used as spinach (Greece), ripe black frs. eaten raw and made into jam (Natal), regarded as poisonous in unripe state: **deadly —**, *Atropa belladonna*, frs. poisonous, pl. med., belladonna: **enchanter's —**, *Circaea lutetiana*, a weed of shady places.

Nikau palm *Rhopalostylis sapida*, N.Z.

Nile grass *Acroceras macrum*, a creeping peren. pasture grass, cult.

Nim *See* neem.

Nimble Will *Muhlenbergia diffusa*, *M. schreberi*, N.Am., often occur as weeds.

Ninde (Malawi) *Aeolanthus gamwelliae*, trop. Af., ls. and shoots distilled yield an oil rich in geraniol resembling palma rosa oil.

Niopo snuff *Piptadenia peregrina*, prepared from pods by Amerindians on the Rio Negro, Brazil.

Nipa palm *Nypa fruticans*, trop. As., young fl. stalks tapped for sugary juice, used for jaggery or alcohol, an imp. industry in the Philippines.

Nipplewort *Lapsana communis*.

Nispero Loquat, *Eriobotrya japonica*.

Nit grass *Gastridium ventricosum*, Eur., an ann. grass.

Nits-and-lice *Hypericum drummondii*, N.Am.

Nivenia Dwarf subshrubs w. pretty fls., S.Af., cult.

Nodding alder *Alnus pendula*, Jap., small tree, wood used in turnery and for combs: **— anemone**, *Anemone cernua*, China and Jap., med.: **— bells**, *Streptocarpus*, S.Af.: **— brome**, *Bromus anomalus*, W. N.Am., a good fodder grass.

Nolana Showy anns. from Chile and Peru, large fls., cult.

Nolina N. & C.Am. pls. related to *Dasylirion*, mainly 6–9 m high, cult.

Noltea *N. africana*, S.Af., orn. evergreen shrub w. white fls., cult.

Nomenclature That branch of systematic botany which deals with the giving of names to pls. is known as pl. nomenclature. It is distinct from pl. identification and classification. The common names of pls. vary w. different languages whereas a properly constituted botanical or scientific name is international. In order to stabilize the use of botanical names an international set of rules of nomenclature has been drawn up. This is known as the 'International Code of Botanical Nomenclature' and governs the formation and use of all scientific pl. names with the exception of the names of cultivars. These are dealt with by a separate code, more recently devised, and known as the 'International Code of Nomenclature for Cultivated Plants'. Interested readers should consult these publications.

An important ruling of the International Code of Botanical Nomenclature is that scientific pl. names must be in Latin or latinized. This removes the difficulty of the multiplicity of different languages which would make the use of common names between different nationalities unsatisfactory. It is considered that as Latin is a dead language its use removes the possibility of national bias.

The present system of naming pls., the binomial or binary system, was standardized by the Swedish botanist Linnaeus (1707–78). Each name consists of two parts, a generic name and a specific name or epithet, analogous to the names of people where these consist of a surname (corresponding to the genus or generic name) and the first name or Christian name, corresponding to the specific name. There are many advantages in the modern binomial system of nomenclature. Prior to the important work of Linnaeus, pls. received names, especially from herbalists, which sometimes amounted to short descriptions or diagnostic phrases. Such 'phrase names' were subject to continual change as other related spp. were discovered. As more and more spp. became known, the more cumbersome did such phrase names become. Their replacement by names of two words came as a great relief. As an example of the clumsiness of these early names one may quote the common elder which had the following names – *Sambucus*, Fuchs 1542: *Sambucus fructu in umbella nigro*, C. Bauhin, 1623: *Sambucus caule perenni ramosa*, Linne 1737: *Sambucus caule arboreo ramoso floribus umbellatis*, Royen 1740. These were mercifully replaced by *Sambucus nigra* L. in 1753.

Some pls. did appear to have binomial names prior to 1753 or Linnaeus' time, but this was accidental as they were in reality very short descriptive phrases of two words only. The original authorship of many of the old two word names of pls. accepted by Linnaeus in 1753 is disguised, for Linnaeus adopted them and he is treated as author. A good example is the common foxglove, *Digitalis purpurea*, a name attributed to Linnaeus although actually given by Fuchs in 1542.

The specific name normally agrees in gender with the generic name, e.g. *Rosa alba*, *Linum usitatissimum*. However, as most trees are regarded as feminine, the feminine form of the specific epithet is commonly given, e.g. *Fagus sylvatica*, *Populus alba* etc.

With each binomial pl. name the name of the author or authors of that name is added, the author being the person who gave the first valid description of the pl. Authors names are commonly abbreviated. They may be dropped in popular works, such as this, for reasons of space. As new specific (or generic) names are published they are collected and listed in the *Index Kewensis* or 'Kew index' for which a supplement is issued every five years. This enables botanists in any country to be kept up-to-date w. new pl. names or the names of newly discovered pls. Generic names are always written with a capital letter and it is now universal practice to use a small letter for all specific names.

For the name of a new pl. to be accepted according to the International Rules the description of it must be validly and effectively published. It must be published in a printed form and available on sale (or offered in exchange), printed in a book or recognized scientific periodical. Names that have not been effectively published have no standing botanically. Valid publication requires that a name must be accompanied by an adequate description of the pl. to which it refers. Such descriptions must be in Latin (since 1935) so that botanists of all nationalities may understand them. There are those who feel that, with Latin now taught less and less in schools and that with English (or American) rapidly becoming a world language, simple English might be a better medium for publication and be more readily understood by present-day and future botanists.

When a name given to a genus by a pre-Linnaean author is taken over by Linnaeus or a subsequent author it is indicated thus – *Mercurialis* (Tourn.) L. (named by T. accepted by L.). When a name was given, but not published, by a botanist, and subsequently published by another, it is shown thus – *Leersia* Soland. ex Sw. (given by Solander, in ms., and published by Swartz). When a name is published by a person writing in someone else's publication, it is indicated by 'in', e.g. L.C. Rich. in Michx. means given by Richard in Michaux's *Flora*. When a sp. is transferred from one genus to another it retains its specific name if possible, i.e. if the new genus does not already have a sp. of that name, and the author of the first is indicated in brackets, e.g. *Cheiranthus tristis* L. may become *Matthiola tristis* (L.) R. Br.

With regard to the various groups into which pls. may be classified families, if large, may be divided into subfamilies, tribes and subtribes, genera into subgenera, sections, subsections and series, and spp. into subspp., var., subvar., forms and cultivars.

It is not possible here to discuss the important role of the type specimen, i.e. the herbarium specimen on which the original description was based, and the complex questions of typification, priority, synonymy, double citation, name changes, conservation, cultivars etc. Such matters are discussed lucidly, in simple terms, in Jeffrey, C. (1968) *An introduction to plant taxonomy*, 128 pp.

Nomocharis Bulbous pls. from Asia w. showy fls., cult.

Nondo *Ligusticum canadense*, N.Am.

None-so-pretty *Saxifraga spathularis* × *umbrosa*, cult.; *Silene armeria*.

Nongo *Albizia grandibracteata*, *A. zygia*, Uganda, a commercial timber.

Nonsuch *Medicago lupulina* (black medic), Eur. inc. Br., common in pastures.

Noogoora bur (Aus.) *Xanthium spinosum*, troublesome in sheeps' wool.

Noon flower *Disphyma australe; Tragopogon pratensis* (goat's beard).

Nopalea Cacti. w. flattened joints like *Opuntia*, *N. coccinellifera* a host pl. for the cochineal insect.

Northern fern *Blechnum boreale*.

Norfolk Island Hibiscus *Lagunaria patersonii*, evergreen tree w. large pink or mauve fls., cult.: — — **pine**, *Araucaria heterophylla* (=*A. excelsa*), popular orn. tree, esp. in Medit. area, also a pot pl.

Normandy cress Name for Am. or land cress.

Nose-burn *Tragia* spp., Texas.

Nothofagus Or southern beech, beech-like trees of Aus., N.Z. and S.Am., some imp. for timber.

Notholaena Widely distributed ferns, some cult.

Notylia C. & S.Am. orchids, fls. usu. small, white, scented and present in profusion.

Nsambya (E.Af.) *Markhamia platycalyx*, orn. fl. tree, cult., coppices well for poles.

Ntonga nut *Cryptocarya latifolia*, S.Af., kernel oil used locally by Africans.

Num-num *Carissa bispinosa*, S.Af., a heavily armed shrub w. ed. red. frs.

Nun's hood orchid *Phaius grandiflorus* (=*P. tankervilliae*), cult.

Nut Botanically a hard one-seeded indehiscent fr.: popularly an edible or oil-yielding seed (or fr.), usu. w. a hard or brittle shell, exceptions being the groundnut (*Arachis*) or paper-shell almond. — **grass**, *Cyperus rotundus*, a bad trop. and subtrop. weed.

Nuts, edible or **dessert** Nuts gathered from wild trees or pls. have been used as food by man in many parts of the world from the earliest times, their richness in protein and oil or fat giving them high food value as human nutriment. Their popularity as food has continued throughout the ages and today many different kinds of nuts are cult. on a large scale and are of considerable commercial importance.

Among the ed. nuts that occur or are cult. in the trop. or warm cli. are the **Brazil** — (*Bertholletia excelsa*), the **paradise** or **sapucaia** — (*Lecythis usitata*), the **butter** or **swarri** — (*Caryocar nuciferum*) and the **cashew** — (*Anacardium occidentale*). These are all native to trop. S.Am. although the cashew is freely grown (and naturalized) in other parts of the trop., esp. the Ind. sub-continent which produces the bulk of the world's supply of cashew nuts. The **Queensland** or **macadamia** — (*Macadamia*) is of Aus. origin, as the name indicates, but is extensively grown commercially in Hawaii. Other trop. nuts include the **coconut**, **peanut**, **Java almond** (*Canarium commune*), **pili** — (*Canarium* spp.) of the Philippines and **E.Af. oyster** — (*Telfairia pedata*). Many other nuts of warm cli. are used locally for food such as the **water chestnut** (*Trapa*) in China and the **yeheb** — (*Cordeauxia edulis*) of Somalia.

The ed. nuts of temp. or subtemp. countries include such well known nuts as **almonds, walnuts, chestnuts, hazelnuts** (including **Barcelona** — and **cob** —) and the **pistachio** —, cult. in both the Old and the New Worlds. Pine kernels or pignolias (*Pinus pinea*) are produced in quantity in Medit., esp. Italy. They were esteemed as far back as Roman times and were in fact part of the ration of Roman soldiers in Br. Several different ed. pine kernels occur in N.Am., being a favourite food of the Amerindian. The **pecan** — and various **hickories** (*Carya* spp.) are other well known ed. nuts of N.Am. Ed. **acorns** (*Quercus*) and **beech seeds** or **mast** (*Fagus*) may be less used now than formerly as human food. [Howes, F. N. (1948) *Nuts, their production and everyday uses*, 264 pp.]

The following are other ed. nuts or seeds and frs. referred to as 'nuts'. **Areca** —, *Areca catechu*: **Aus. chest**—, *Castanospermum australe*: **Baroba** —, *Diplodiscus paniculatus*, Philippines: **Ben** —, *Moringa oleifera*: **betel** —, *Areca catechu*: **bladder** —, *Staphylea*: **bonduc** —, *Caesalpinia crista*: **bread** —, *Brosimum alicastrum*: **candle** —, *Aleurites triloba*: **Chile** —, *Araucaria araucana*: **chufa** —, *Cyperus esculentus*: **clearing** —, *Strychnos potatorum*: **cohume** —, *Attalea cohune*: **coquilla** —, *Attalea funifera*: **dika** —, *Irvingia gabonensis*, W.Af.: **double coconut**, *Lodoicea maldivica*: **doum palm** —, *Hyphaene thebaica*: **fox** —, *Euryale ferox*: **Gaboon** —, *Coula edulis*: **gasso** —, *Manniophyton africanum*, Congo: **gevuina** —, *Gevuina avellana*, Chile: **ginkgo** —, *Ginkgo biloba*: **gnetum** —, *Gnetum gnemon*, trop. As.: **gru-gru** —, *Acrocomia sclerocarpa*, W.I.: **helicia** —, *Helicia diversifolia*, Aus.: **illipe** —, *Shorea* spp.: **inoi** —, *Poga oleosa*, W.Af.: **ivory** —, *Phytelephas macrocarpa*: **jack** —, *Artocarpus heterophyllus*: **jajoba** —, *Simmondsia chinensis* (= *S. californica*): **Jamaica cob** —, *Omphalea triandra*: **Jesuit's** —, *Trapa natans*: **Karaka** —, *Corynocarpus*: **kaya** —, *Torreya nucifera*: **kola** —, *Cola* spp.: **kubili** —, *Cubilia blancoi*, Philippines: **ling** —, *Trapa bicornis*: **Lunan** —, *Otophora fruticosa*: **manketti** —, *Ricinodendron rautanenii*, trop. Af.: **marking** —, *Semecarpus anacardium*: **Ngapi** —, *Pithecellobium lobatum*: **nicuri palm** —, *Syagrus* (= *Cocos*) *coronata*, Braz.: **olive** —, *Elaeocarpus ganitrus*: **owusa** —, *Plukenetia conophora*: **physic** —, *Jatropha curcas*: **pig** —, *Conopodium majus*, Eur.: **Quan-**

dong —, *Eucarya* (= *Fusanus*) *acuminata*, Aus.: **Queensland** —, *Macadamia, Macrozamia:* **rose** —, *Hicksbeachia pinnatifolia:* **rush** —, *Cyperus esculentus:* **shea** —, *Butyrospermum parkii:* **snake** —, *Ophiocaryon paradoxum:* **soap** —, *Sapindus* spp.: **taccy** —, *Caryodendron orinocense:* **tagua** —, *Phytelephas macrocarpa:* **tallow** —, *Ximenia americana:* **tiger** —, *Cyperus esculentus,* ed. tubers: **yeheb** —, *Cordeauxia edulis:* **Zulu** — *Cyperus esculentus.*

Nutmeg *Myristica fragrans,* Moluccas or Spice Isles, tree cult. in trop. The ripe nutmeg fr. is yellow and resembles a large apricot as its hangs on the tree. The spice consists of dried kernels finely ground while mace is the dried arillus of the seed. Nutmeg is much used in domestic cookery and by the food manufacturers. It blends well with sweet foods such as cakes and puddings but is also used to flavour meat products. Nutmeg oil may be used med. as a carminative. **Ackawai** —, *Acrodiclidium camara,* Guyana, Braz., frs. med.: **Bombay** —, *Myristica malabarica,* seeds only slightly aromatic and of no real value: **Braz.** —, *Cryptocarya moschata:* **calabash** —, *Monodora myristica,* W.Af.: **false** —, *Pycnanthus angolensis* (= *P. kombo*), trop. Af.: **Ind.** —, *Myristica malabarica:* **Madag.** —, *Ravensara aromatica:* **mountain** —, *Myristica fatua,* Moluccas: **Otoba** —, *M. otoba,* Colombia, fat from seeds: **Peruvian** —, *Laurelia aromatica:* **W.Af.** —, *Monodora myristica:* **wild** —, *Virola surinamensis,* W.I.

Nux vomica Seeds of *Strychnos nux-vomica,* a deciduous tree of Ind., Burma and Sri Lanka. They yield strychnine.

Nuxia *N. verticillata,* ls. used in Madag. for flavouring *betsabetsa* the national alcoholic beverage made from sugar cane juice.

Nyala tree *Xanthocercis* (= *Pseudocadia*) *zambesiaca,* a large handsome tree of the low veld.

Nyatoh *Palaquium* spp. Malaysia, Indon., a commercial timber.

O

Oak *Quercus* spp. The two common Eur. oaks or 'English oak' are *Q. robur* (pedunculate oak) and *Q. petraea* (sessile oak), wild and cult., in Br. The wood is famous for its strength and durability and was all-important in the days of wooden sailing ships and 'men of war'. The wood is still much used in boat building and for railway wagons, fencing, house building, furniture, shop fittings etc. Other oaks important in the timber trade (in Br.) and rated as commercial timbers are – **Am. red** —, *Q. rubra* and *Q. falcata* mainly: **Am. white** —, *Q. alba, Q. prinus* and other spp.: **Eur.** —, as above, may be known by various regional names, French, Polish, Slavonian etc.: **Jap.** —, *Q. mongolica* and other spp., widely used, more easily worked than English oak: **Persian** —, *Q. castaneifolia,* Iran: **Turkey** —, *Q. cerris,* Eur.

Other oaks, i.e. true oaks or spp. of *Quercus,* sometimes imp. locally for timber include – **bear** —, *Q. ilicifolia,* N.Am.: **Calif.** —, *Q. lobata:* **chestnut** —, *Q. prinus:* **cork** —, *Q. suber:* **holly** —, *Q. ilex:* **iron** —, *Q. obtusifolia,* N.Am.: **jack** —, *Q. marylandica:* **Kermes** —, *Q. coccifera:* **live** —, *Q.*

virginiana and other evergreen oaks: **manna—**, *Q. cerris:* **Oregon white —**,
Q. garryana: **possum —**, *Q. nigra:* **post —**, *Q. obtusiloba:* **Quebec —**, *Q.
alba:* **scarlet —**, *Q. coccinea,* N.Am.: **shin —**, *Q. gambelii:* **shingle —**,
Q. imbricaria, N.Am.: **sweet acorn —**, *Q. ballota,* Spain: **valonea —**, *Q.
macrolepis:* **willow —**, *Q. phellos,* N.Am.: **yellow —**, *Q. velutina,* N.Am.,
yields a dye, quercitron.

Trees or woods that have acquired the name 'oak' because of some resem-
blance to true oak (*Quercus*) include the following - **Af. —**, *Oldfieldia africana;
Lophira alata* (=*L. procera, L. lanceolata*): **Aus. —**, *Eucalyptus regnans:*
Braz. —, *Posoqueria latifolia,* commonly used for walking sticks: **bull —**,
Casuarina lehmanii: **Ceylon —**, *Careya arborea, Schleichera trijuga:* **desert —**,
Casuarina decaisneana: **forest —**, *C. torulosa:* **Patana —**, *Careya arborea,*
Sri Lanka: **river —**, *C. cunninghamiana:* **she —**, *Casuarina* spp.: **silky —**,
Grevillea robusta: **Spanish —**, *Inga laurina,* W.I.: **swamp —**, *Casuarina
glauca.*

Oak bark Oak b. from the two common oaks (*Quercus robur* and *Q. petraea*)
has long been an imp. tanning material, esp. in Eur. countries. In Br. it has
been the traditional material of the leather tanner from very early times. The
famous English leathers of the past, known throughout the world for their
high quality, were produced on a basis of good quality hides and skins with
oak tannage. The merits of oak tannage apply particularly to heavy leather,
i.e. sole leather. A drawback of oak tanning is the slowness of the process.
Good quality commercial b. contains 12–14 % tannin (air-dried). The spent
b. or 'tan' was used for hort. purposes, covering roads that crossed horse
racing tracks, circus rings etc. A tanning extract is made from well-matured
oak wood (usu. waste wood or scraps) in some Eur. countries. **— fern,**
Gymnocarpium dryopteris: **— feeding silkworms**, in N. China the silkworm
Antheraea pernyi feeds on the ls. of *Quercus mongolica* and in Jap. *Antheraea
yama-mai* on ls. of *Quercus dentata:* **— galls** are the best known of the veg.
galls (due to insect attack) that have been used in leather tanning. They are
derived mainly from *Q. infectoria* or allied spp. in As. Minor and the E. Medit.
and are known as Levant, Mecca, Aleppo or Turkish galls. They may contain
36–58 % tannin. Oak galls, known as Morea, Greek Marmora or Italian galls,
may be derived from *Q. ilex;* **— leaf geranium,** *Pelargonium quercifolium.*

Oat *Avena sativa,* the common oat, a widely grown cereal in temp. countries:
origin uncertain: was known to the ancient lake dwellers of Eur.: much used
as human food, oatmeal porridge, breakfast foods (rolled oats), scones,
biscuits, oatcakes etc. A valuable food for animals, esp. horses. Numerous
vars. cult. Other spp. of *Avena* are or have been grown to some extent.
Bristle —, *Avena strigosa:* **Hungarian —**, *A. orientalis:* **naked —**, *A. nuda:*
red or **Algerian —**, *A. byzantina:* **small —**, *A. strigosa.*

Other *Gramineae,* some of them weeds, that show some resemblance to cult.
oats, have acquired the name of 'oats'. **Bearded — grass,** *Avena barbata:*
black — —, *Stipa avenacea,* N.Am.: **false —,** *Arrhenatherum elatius:*
— grass, *Avena, Arrhenatherum, Helictotrichon:* **swamp —,** *Trisetum pennsylvanicum,* N.Am.: **water —,** *Zizania* spp., *Danthonia spicata:* **wild —,** *Avena
fatua, Uniola latifolia, Danthonia* spp.: **yellow —,** *Trisetum flavescens,* N.Am.
Obada *Ficus vogelii,* W.Af., latex a source of rubber at one time.

Obeche Or Wawa, *Triplochiton scleroxylon*, W.Af., a light commercial wood, much used for plywood and veneers, stains well, often stained like mahogany, has similar grain.

Obedient plant *Physostegia virginiana*, orn. gdn. pl., if the fls. are moved laterally on their pedicels they retain the new position.

Oca *Oxalis tuberosa* (= *O. crenata*), a starchy r. crop, Andean highlands: — **quina**, *Ullucus tuberosus*, Andes, ed. tubers, cult.

Ochna Old World trees and shrubs, some cult., mainly for showy frs.

Ochra *See* okra.

Ocimum Several spp. yield thymol-containing esst. oils.

Oconee bells *Shortia galacifolia*, N.Am., cult. rock gdn. pl., white fls., ls. go crimson.

Ocotillo Or coach whip, *Fouquieria splendens*, Mex., wax from stems, cult. as a hedge (thorny).

October flower *Polygonella polygama*, N.Am.

Odoko *Scottellia coriacea*, W.Af., a commercial timber.

Odontadenia Climbing shrubs, trop. Am., showy fls., cult.

Odontoglossum S.Am. orchids, cult., numerous hybrids.

Odum (Ghana) *See* iroko.

Oenothera Cult. orn. fls. *O. biennis*, the common evening primrose.

Ofram *Terminalia superba*, W.Af., a commercial timber.

Ogea *Daniellia ogea* (=*D. thurifera*), W.Af. tree yields a commercial timber and copal.

Ogeeche lime *Nyssa ogeche*, N.Am. frs. preserved, fls. a good honey source.

Ohia *Syzygium malaccense*, E.I., ed. fr., tree cult.

Ohio buckeye *Aesculus glabra*, N.Am.

Oil cake The residue from the crushing and pressing of oil seeds for oil: the most imp. are cotton, groundnut, copra, linseed, palm kernel, rape, soya, sesame and sunflower: another use is as fertilizer or manure, esp. when unsuited for animal feeding, e.g. castor oil cake.

Oil grass A number of trop. grasses are aromatic and are distilled for esst. oil which is used in perfumery, esp. for soap and in other ways. The more imp. are – Citronella grass, *Cymbopogon nardus:* lemon grass, *C. citratus* and *C. flexuosus* (E.Ind. lemon grass): palma-rosa, *C. martinii:* vetivert, *Vetiveria zizanioides*.

Oil palm *Elaeis guineensis*, W. & C.Af., cult. there and in other trop. countries. Palm oil is the oil extracted from the fleshy pericarp of the fr., palm kernel oil that from the kernels, much used for margarine. Palm oil is extensively used for food in W.Af. and in the manufacture of soap, candles, tin-plate etc. There are many vars., 'Deli' being a large-fruited cultivar.

Oils, vegetable Veg. oils, which serve mainly as storage products in the pl., constitute one of the largest and most imp. groups of economic pl. products. They are imp. as food for man and for some animals, in addition to which many of them have imp. industrial uses. The veg. oils fall into two main groups – the fixed or fatty oils and the esst. or volatile oils, both of these groups being capable of further subdivision. The fixed oils are sometimes considered under the following heads – non-drying, semi-drying and drying, the first category containing the greatest number with such commercially imp. oils as coconut, palm oil and palm kernel oil, cotton seed oil, groundnut, olive and maize oil. These are used in vast quantities for ed. purposes, notably margarine and cooking oils and fats. Large quantities are used in soap making, in the treatment of veg. fibres and for other industrial purposes. Linseed oil is the most extensively used of the drying oils, with uses mainly in the paint, varnish and linoleum industries.

The distinction between veg. oils and fats is merely a physical one for they are similar chemically. In some instances temperature determines whether the oleaginous product is liquid or solid, for instance coconut oil is liquid in the trop. but solid in cold climates. Cocoa fat or butter becomes liquid from the heat of the human body. This accounts for its value pharmaceutically.

Fixed or **fatty oils** – babassu, bacury, beech nut, ben, Brazil nut, candlenut, carapa or andiroba, cashew kernel, castor, chaulmoogra, cherry kernel, cocoa or cacao butter, coconut (copra), cohune, colza, cotton seed, croton, curcas, groundnut, grape seed, grapefruit seed, hazelnut, hemp seed, illipe (*Shorea*), kapok seed, lemon seed, lime seed, linseed, macassar, maize, maw seed (poppy), melon seed, mustard, Niger or ramtil, oiticica, olive, orange seed, palm oil (Af.), palm kernel (Af.), peach, perilla, pistachio, plum kernel, pumpkin seed, rubber seed (*Hevea*), sesame or gingelly, soybean, sunflower, tea seed, tomato seed, tung, walnut.

The **essential oils** are largely used in perfumery and flavouring and some for medicinal purposes. Large quantities of some esst. oils, such as geranium and citronella are used for perfuming soap. **Essential oils** – almond (bitter), angelica, anise or aniseed, basil, bay, bergamot (orange), birch, bois de rose, boronia, buchu, cade, cajuput, camphor, cananga, caraway, cardamon, carrot, cassia, cassie, catnip, cedarwood, celery, chamomile, chenopodium, cinnamon, citron, citronella, clary, clove, costus, coriander, cubeb, cumin, dill, eucalyptus, fennel, gardenia, garlic, geranium, ginger, ginger grass, grapefruit, hyacinth, jasmine, jatamansi, jonquil, juniper, lavender, lemon, lemon grass, lime or limette, linaloe, lovage, mace (nutmeg), mandarin, marjoram, mignonette, myrtle, narcissus, neroli (citrus), niaouli, nutmeg, orange, origanum, orris, palmarosa, parsley, patchouli, pennyroyal, peppermint, petitgrain, pimento, pine needle, rose, rosemary, rue, sage, sandalwood, sassafras, savin, savory, spearmint, spike (lavender), star anise, storax, tangerine, tarragon, tea tree, thyme, tuberose, turmeric, turpentine, vetivert, wintergreen, wormseed, ylang-ylang. [Guenther, E. (1949–52) *The essential oils*, vols. 3–6.]

Oiticica *Licania rigida*, Braz., seeds yield a drying oil: name used for other trees, e.g. *Clarisia* (= *Soaresia*) *racemosa* and *Licania tomentosa*.

Okan *Cylicodiscus gabunensis*, W.Af., a commercial timber.

Okari *Terminalia okari*, Polynesia, tree w. ed. fr.

Ok-gue (China) *Ficus pumila*, fr. used for jelly.

Okoume Or Gaboon mahogany, *Aucoumea klaineana*, W.Af., also yields a resin.

Okra Or Ochra, ladies' fingers or gumbo, *Hibiscus esculentus* (= *Abelmoschus esculentus*) young frs. mucilaginous, a common veg. in warm cli., much used in soups and stews: **Chinese —**, *Luffa acutangula*, W.I.: **wild —**, *Hibiscus vitifolius*, W.I.

Okwen *Brachystegia leonensis* and other spp., trop. Af., commercial timber.

Old maid Or periwinkle, *Catharanthus roseus* (=*Lochnera rosea, Vinca rosea*) Madag., cult.: — **man**, southern wood or lad's love, *Artemisia abrotanum*, fragrant, cult.: — **man's beard**, *Clematis* spp.; *Chionanthus virginiana; Tillandsia usneoides*, W.I. a common epiphyte, hangs in festoons from trees, used for stuffing pillows and mattresses, also called New Orleans moss: — **witch grass**, *Panicum capillare*, N.Am.

Oldenlandia Some are trop. weeds, e.g. *O. bojeri.*

Oleander *Nerium oleander*, Medit., fl. shrub, widely cult., subtrop., many vars.; *N. odorum*, Persia to China, also cult.: **yellow** —, *Thevetia peruviana* (= *T. neriifolia*).

Oleandra Trop. ferns, w. creeping or climbing jointed stems, cult.

Olearia Evergreen Aus. and N.Z. trees and shrubs, many cult.

Oleaster *Elaeagnus* spp. N. Hemisph., cult. orn. shrubs.

Olibanum Or gum olibanum, a name for frankincense.

Olive *Olea europaea*, of ancient cult. in Medit. countries as a source of food or ed. oil: the best grades of oil (cold pressed) are used as salad oil and for cooking: the oil is employed in canning sardines and has med. uses on the preparation of liniments, ointments and plasters. Olives are preserved in brine and pickled, while stuffed olives are a delicacy. **Calif.** —, *Umbellularia californica:* **Chinese** —, *Canarium album* and other spp.: **Ceylon** —, *Elaeocarpus serratus:* **Hawaiian** —, *Osmanthus sandwicensis:* **Indian** —, *Olea cuspidata*, useful wood: **Java** —, *Sterculia foetida:* **Russian** —, *Elaeagnus* spp.: **sand** —, *Dodonaea viscosa*, S.Af.: **wild** —, *Olea africana* (= *O. verrucosa*), S.Af.

Olivillo *Aextoxicon punctatum*, Chile, a commercial timber.

Olona *Touchardia latifolia*, Hawaii, yields a strong fibre used for fishing nets and lines.

Omphalodes Mainly Medit. pls., cult., usu. in rock gdns.

Omu *Entandrophragma candollei*, W.Af., a commercial timber.

Oncidium A large genus of trop. Am. orchids (over 500 spp.), some cult.

Oncoba *O. echinata*, trop. Af., seeds yield med. oil. *O. spinosa*, trop. Af., ed. fr. pulp. *O. kraussiana*, S.Af. orn. fl. shrub, cult.

One-day-flower *Tradescantia virginiana*, popular gdn. pl., many vars.

Onion *Allium cepa*, a widely grown veg. in warm and temp. cli. for food or flavouring: was known to the Ancient Egyptians: numerous vars., special vars. are grown for the production of dehydrated onion or onion powder, now an imp. trade with the advantage over fresh onions in keeping qualities. **Egyptian** —, *A. cepa* var. *proliferum*, bulbils at top of scape often w. fls.: **everready** — = Welsh onion: **Jap. bunching** —, *A. fistulosum:* **multiplier** —, *A. cepa* var. *aggregatum:* **Negi** —, = Jap. bunching —: — **orchid**, *Orthoceras* spp., Aus.: **potato** —, *Allium cepa*, produces many lateral bulbs: **sea** —, *Urginea maritima*, Br.: **top** —, *A. cepa* var. *proliferum*, inflor. w. fls. and bulbils; **tree** — = Egyptian —: **Welsh** —, *A. fistulosum:* **wild** —, *A. cernuum*, *A. textile*, N.Am.

Ononis *O. fruticosa*, S.Eur., and other spp., cult. orn. shrubs.

Onosma Hairy pls., mainly Medit., rs. of *O. echioides* once used for dyeing woollens.

Opepe *Sarcocephalus* (= *Nauclea*) *diderrichii*, W.Af., a much used timber.

Opera girls Or dancing girls, *Mantisia saltatoria*, Ind., strange striking fls.

Ophiopogon *O. japonicus*, Jap., ed. tubers: a carpeting pl. in Italy: other spp. cult.

Ophrys Old World orchids, fls. often resemble insects, cult.

Opium poppy *Papaver somniferum*, cult., Persia to China: in most countries the production of opium is by licence or under official control. Opium consists of the dried latex from the unripe seed capsules, obtained by scarifying them. It contains about 25 different alkaloids, the most imp. being morphine (9–17 %), a powerful analgesic and narcotic, much used to relieve pain or anxiety.

Opopanax *Opopanax* spp., As. Minor, yield 'gum opopanax', a scented oleo-resin used in perfumery: name for cassie, *Acacia farnesiana*, used in perfumery.

Opossum wood *Halesia carolina*, N.Am.; *Quintinia sieberi*, Aus.

Opuntia Mainly large cacti, C. & S.Am., cult.: some have become troublesome weeds in other warm countries, Aus., S.Af., Medit., St Helena etc.: some have ed. frs. (prickly pears) or are used for defensive hedges. Spineless forms are a useful emergency food for livestock in drought prone areas. *See Opuntia* in Willis.

Orache *Atriplex hortensis*, an ann. veg. w. large ls. used like spinach: there are vars. w. red or copper-coloured ls., which turn green on cooking: **sea-beach —**, *A. arenaria*, N.Am.

Orange *Citrus sinensis* (=*C. aurantium* var. *sinensis*), the **common sweet —** is thought to have originated in S.China: now widely cult. in trop. and subtrop. There are numerous cvs. sometimes divided into groups such as Spanish, Medit., blood, navel etc. There is a large export trade in oranges from many countries and large quantities are converted to orange juice as a beverage or vitamin source. Orange peel oil is used as a flavouring. Oil obtained from the seeds is used in soap making. The bitter or Seville orange is the one favoured for marmalade. **Azapa —**, a well-known Chile orange: **Bigarade —** = Seville **—**: **bitter —** = Seville **—**: **Bergamot —**, esst. oil from peel: **cocoa —**, seedling oranges that come up in Trinidad cocoa plantations: **Jap. —**, *Citrus reticulata:* **kidglove —** = tangerine **—**: **king —**, *C. reticulata:* **mandarin —**, *C. reticulata:* **mock —**, *Philadelphus* spp.: **osage —**, *Maclura pomifera:* **Satsuma —**, a seedless Jap. orange: **Seville —**, *Citrus aurantium:* **Shamuti —** = Jaffa **—**: **tangerine —**, *C. reticulata:* **— thorn,** *Citriobatus pauciflorus:* **trifoliate —**, *Poncirus trifoliata*, used for hedges and as a citrus stock: **Valencia —**, a famous Spanish cv. of *C. sinensis*.

Orange creeper *Marianthus ringens*, W.Aus., orn., cult.

Orchard grass Am. name for *Dactylis glomerata*, cocksfoot, imp. agric.

Orchid There are a large number of orchids distributed all over the world, esp. in the high rainfall trop. The family Orchidaceae consists of over 12,000 spp. (*see* Orchidaceae in Willis). Many of the most handsome orchids are native to C. & S.Am. and numerous cultivars or hybrid forms have been produced from them by breeders and fanciers. Orchids fall broadly into two main groups, terrestrial and epiphytic. The epiphytic forms abound in warm cli. and terrestrial in colder cli. such as that of Br.

The following include some of the better known or more interesting of the native Br. orchids – **Bee orchid**, *Ophrys apifera:* **bird's nest —**, *Neottia nidus-avis:* **bog —**, *Hammarbya paludosa:* **butterfly —**, *Platanthera* spp.:

coral root —, *Corallorhiza trifida:* **dwarf** —, *Orchis ustulata:* **early purple** —, *Orchis mascula:* **fen** —, *Liparis loeselii:* **fly** —, *Ophrys insectifera:* **frog** —, *Coeloglossum viride:* **Jersey** —, *Orchis laxiflora:* **green-winged** —, *Orchis morio:* **helleborine** —, *Cephalanthera* spp.; *Epipactis* spp.: **lady's tresses** —, *Spiranthes* spp., *Goodyera* sp.: **lady's slipper** —, *Cypripedium calceolus:* **lizard** —, *Himantoglossum hircinum:* **man** —, *Aceras anthropophorum:* **monkey** —, *Orchis simia:* **musk** —, *Herminium monorchis:* **poor man's orchid** —, *Schizanthus* spp., *Iris* spp.: **pyramid** —, *Anacamptis pyramidalis:* **small white** —, *Pseudorchis albida:* **soldier** —, *Orchis militaris:* **spider** —, *Ophrys* spp.: **sweet scented** —, *Gymnadenia conopsea:* **twayblade** —, *Listera* spp.: **— tree**, *Bauhinia variegata; Amherstia nobilis.*

Orchis Name for any terrestrial orchid. The name of a genus of terrestrial orchids of temp. regions, some cult.

Ordeal poisons In former times or in uncivilized countries persons suspected of witchcraft or crime were made to swallow certain poisonous substances. If they died they were considered to have been guilty but if they vomited the poison they were considered to be innocent. The following are some of the pls. used as ordeal poisons. *Antiaris toxicaria*, Upas tree, celebrated ordeal poison tree of Java: the sap from the b. contains a virulent poison used for poisonous darts or arrows. *Cerbera tanghin*, a small tree w. milky latex, the frs. constituting the famous ordeal poison of Madag., *Erythrophleum*, 'sassy bark', an infusion of the b. was once a common ordeal poison in W.Af., also used for poisoning arrows and as a fish poison. *Physostigma venenosum*, Calabar bean or ordeal bean: the bean contains an alkaloid, physostigmine, used in ophthalmic medicine. *Strychnos* spp. the seeds, commonly poisonous, have been used in trop. Af.

Oregano A condiment or flavouring consisting of the dried ls. of *Origanum* (marjoram) or *Lippia* spp.: popular in Italy for flavouring pizza, and in C. & S.Am. for Mexican or Latin-American dishes.

Organ-pipe cactus *Lemaireocereus thurberi, L. schottii.* SW. U.S.

Orixa *O. japonica,* a shrub w. scented ls., a hedge pl. in Japan.

Ornithogalum Old World bulbous pls., cult. *See* chincherinchee.

Orobanche Rs. of several spp. used as food by Amerindians.

Orpine Or live-long, *Sedum telephium,* old-time (Elizabethan) gdn. pl., many vars.

Orris root *Iris florentina,* Eur., Medit., the dried r. smells like violets and is used in perfumery. This pl. is considered to be the *Fleur-de-lis* of heraldry.

Ortanique or **ortanique orange** Chance citrus hybrid which arose in Jam. in the 1920s and was soon cult. commercially: believed to be a cross between a sweet orange and a tangerine; name said to have arisen this way – *orange/ tang*erine/*unique*!

Osage orange *Maclura pomifera* (= *M. aurantiaca*), S. & C. U.S. small thorny tree, cult. as orn. and hedge pl., wood has been used for bows.

Osier *Salix* spp. Osiers are thin coppice shoots of certain willows used for wicker-work and in basket-making. The shoots are raised in special osier beds and are usually of one season's growth only and used w. or without the b. The spp. employed (in Br.) are the common osier (*Salix viminalis*), purple osier (*S. purpurea*) and forms of the almond-leaved willow (*S. amygdalina*) which are known by various names such as black mauls, black Hollander, green sucklings and glibskins.

Osmanthus Evergreen shrubs, As., fragrant fls., cult.: fls. of *O. fragrans* used to perfume tea in China.

Osmunda fibre *Osmunda* spp. used in hort., esp. for orchids.

Oso-berry *Osmaronia* (= *Nuttallia*) *cerasiformis*, Calif., cult., orn. shrub, v. early flowering.

Ostrich fern *Matteuccia struthiopteris*, N.Hemisph., bears fertile and sterile fronds.

Ostrowskia *O. magnifica*, C.As., showy peren., bell-shaped fls., 15 cm across.

Oswego tea *Monarda didyma*, N.Am., ls. used like tea, cult. gdn. pl.

Otaheite apple *Syzygium malaccense*, W.I.; *Spondias* spp.: — **chestnut**, *Inocarpus edulis*, small tree, Pacific region, cult., fr. ed. like a chestnut: — **gooseberry**, *Phyllanthus distichus*, fr. acid, mainly used for jams and pickles: — **myrtle**, *Securinega durissima*: — **potato**, *Dioscorea bulbifera* var. *sativa* cult.

Otto of Rose Or attar of rose, *Rosa* spp., much used in perfumery: the aromatic oil is obtained by distillation of rose petals, mainly those of the damask rose, *Rosa damascena* (chiefly in E.Eur.) and to a less extent the cabbage rose, *Rosa centifolia*, the Provence rose, *R. centifolia* var. *provincialis* and the musk rose, *R. moschata*.

Otu *Cleistopholis patens*, E. & W.Af., a commercial timber.

Ouricuri palm (Braz.) *Syagrus coronata*, *Attalea excelsa*, yield a kernel oil.

Ourisia *O. coccinea*, Chile, cult., a rock gdn. pl.

Overlook bean (W.I.) *Canavalia ensiformis*, a trop. pulse.

Owala *Pentaclethra macrophylla*, W.Af., an oil seed.

Owl clover *Orthocarpus purpurascens*, SW. U.S.

Oxalis *Oxalis* spp. Some spp. are cult. as gdn. pls.: the tubers of some are ed.: several have become bad weeds, notably the S.Am. spp., *O. articulata*, *O. latifolia* and *O. corymbosa*, multiplying rapidly by means of small bulbils, when hoeing merely spreads the pest. Cape sorrel, *Oxalis pes-caprae*, has become a troublesome weed in Medit. orange groves.

Ox-eye *Heliopsis helianthoides*, N.Am.; *Buphthalmum* spp., Eur. — — **daisy**, *Chrysanthemum leucanthemum*, common in Br.: —**lip**, *Primula elatior* and other spp.: **sea** — —, *Borrichia frutescens*, N.Am.: —**tongue**, *Picris echioides* and other spp.

Oxford ragwort *Senecio squalidus*, an introduced weed in Br., spreading.

Oxytropis Genus inc. some cult. gdn. perens., mainly rock gdn. pls.

Oyster leaf *Mertensia maritima*, Br. — **nut**, *Telfairia pedata*, E.Af., the large flat seed of a gourd w. ed. kernel: — **plant**, *Tragopogon porrifolius*, Medit., salsify, a veg., rs. eaten: — **plant, Spanish**, *Scolymus hispanicus*, rs. eaten: — **plant, W.I.**, *Rhoeo spathacea*, a small succulent pl. often grown for edging, the bracts subtending the fls. resemble a half-open oyster shell: **vegetable** —, *Tragopogon porrifolius*.

Ozone fibre (Am.) *Asclepias incarnata*, swamp milkweed, fibrous b.

P

Pachira Handsome trop. Am. trees, dense fol., large showy fls. and frs., cult.

Pachycereus Large Mex. cacti w. diurnal fls., ed. fr., cult.

Pachyphytum Mex. succulents w. pendent fls., related to *Echeveria*, cult.

Pachyrhizus *P. erosus*, yam bean, cult. trop., seed and r. ed.

Pachysandra *P. terminalis*, Jap., a good cover pl. for shade, cult.

Paco-paco *Wissadula spicata*, trop. Am., b. fibre used for ropes.

Padauk *Pterocarpus* spp. a hard, close-grained reddish wood mainly from the Andamans, esteemed for furniture and other purposes: **Andaman** —, *P. dalbergioides:* **Burma** —, *P. macrocarpus:* **W.Af.** —, *P. soyauxii.*

Paddle wood *Aspidosperma excelsum*, Guyana, trop. Am., a strong elastic wood used for paddles, tool handles etc.

Paddy Rice, *Oryza sativa*, the unhusked grain: the term may be used for rice generally.

Paeony or **peony** *Paeonia* spp., N.Hemisph., large showy fls., cult., two main groups – herbaceous and shrubby: many cultivars and hybrids: some have local med. uses. **Tree** —, or moutan, *P. suffruticosa* (=*P. moutan*), cult. for centuries in Orient, many vars.: **yellow** —, *P. lutea*, China, cult.

Pagoda flower *Clerodendrum paniculatum*, red fls., As., cult.: — **tree**, *Plumeria rubra*, *Sophora japonica*, *Ficus indica.*

Pahautea *Libocedrus bidwillii*, N.Z., a timber tree.

Paich-ha *Euonymus maackii*, Jap., a boxwood substitute.

Paigle Old name in Br. for cowslip, *Primula veris* or buttercup, *Ranunculus.*

Pain killer *Dissotis rotundifolia*, E.Af., ls. chewed to relieve coughs.

Paint brush Devil's — —, *Pilosella aurantiaca:* **Indian** — —, *Castilleja* spp., N.Am.: — — (S.Af.), *Haemanthus* spp.: —**ed cup**, *Castilleja*, N.Am.: **downy** — —, *C. sessiliflora*, **scarlet** — —, *C. coccinea:* —**ed feather**, *Vriesia carinata:* —**ed grass**, *Phalaris arundinacea:* —**ed lady**, *Castilleja* spp., N.Am.; *Gladiolus* spp., S.Af.: — **root**, *Lachnanthes tinctoria:* — **trillium**, *Trillium undulatum.*

Painter's palette *Anthurium.*

Pak-choi Or chinese cabbage, *Brassica chinensis.*

Palamut *See* valonea.

Palas Or bastard teak, *Butea monosperma* (=*B. frondosa*), Ind., handsome fl. tree w. many uses, red juice yields Bengal kino, fls. yield a dye, wood durable under water.

Paldao *Dracontomelon dao*, Philippines, a commercial timber.

Palisander Name for wood of Braz. spp. of *Dalbergia*, *Jacaranda* and *Machaerium.*

Palisota Trop.Af. pls. w. colourful frs. (blue or scarlet), cult.

Palms The palms (Palmae), like the grasses, are one of the most imp. groups or families of pls., esp. in the warmer parts of the world. There are some 2500 spp. in 217 genera. Economically they are of v. great importance. In some parts of the trop. they supply a large part of the human diet and other necessities of life. Two of the most imp. palms in this respect are the coconut palm (*Cocos nucifera*) and the Af. oil palm (*Elaeis guineensis*) both widely cult. in the trop. and the source of large quantities of ed. oil extensively used in margarine manufacture. Coir fibre is also a product of the coconut palm. The date palm is of course of great food value to many people. Carnauba wax, the most extensively used veg. wax, is the product of a palm (*Copernicia cerifera*, Braz.). Palm stems or trunks are much used in local building as are the ls. for thatching, mats, baskets etc. (*See* Palmae in Willis.)

Palms are often a conspicuous feature of trop. vegetation. Some reach a considerable size and are of great beauty. It is not surprising that many are cult. for orn. both in the trop. and subtrop. Others are favoured as pot pls. or house pls. in temp. countries. There are two main groups of palms, those w. feather or pinnate ls. and those w. fan-shaped ls. The former predominate by about 2 to 1. **Am. oil** —, *Corozo oleifera* (= *Elaeis melanocarpa*): **Alexandra** —, *Archontophoenix alexandrae*, Aus.: **bacaba** —, *Oenocarpus bacaba:* **bamboo** —, *Areca madagascariensis:* **Bangalow** —, *Archontophoenix cunninghamiana:* **betel** —, *Areca catechu:* **black roseau** —, *Bactris major*, W.I.: **cabbage** —, *Roystonea oleracea; Livistona australis:* **cabbage palmetto** —, *Sabal palmetto:* **cane** —, *Chrysalidocarpus lutescens*, many small stems: **caranday** —, *Copernicia australis:* **carnauba wax** —, *Copernicia cerifera*, Braz.: **coco-de-mer** —, *Lodoicea maldivica:* **coconut** —, *Cocos nucifera:* **cohune** —, *Attalea cohune*, Braz.: **coquito** —, *Jubaea chilensis:* **corozo** —, *Corozo oleifera:* **curly** —, *Howea belmoreana:* **date** —, *Phoenix dactylifera:* **double coconut** —, *Lodoicea maldivica:* **doum** —, *Hyphaene thebaica:* **dwarf fan** —, *Hyphaene humilis:* **ejow** —, *Arenga pinnata:* **fan** —, *Borassus flabellifer:* **fishtail** —, *Caryota mitis:* **gingerbread** —, *Hyphaene coriacea:* **gomuti** —, *Arenga pinnata:* **gru-gru** —, *Acrocomia sclerocarpa:* **honey** —, *Jubaea chilensis:* **ivory** —, *Phytelephas macrocarpa:* **Kentia** —, *Howea belmoreana:* **kittool** —, *Caryota urens:* **nikau** —, *Rhopalostylis sapida*, N.Z.: **nypa** —, *Nypa fruticans:* **oil** —, *Elaeis guineensis*, W.Af.: **palmetto** —, *Sabal* spp.: **palmyra** —, *Borassus flabellifer:* **Panama hat** —, *Carludovica insignis* (not a true palm): **pataua** —, *Oenocarpus bataua* Braz.: **paxiuba** —, *Socratea exorrhiza*, trop. Am.: **peach** —, *Guilielma gasipaes*, C.Am., fr. eaten cooked: **Pejobaye** — = peach —: **piassaba** —, *Leopoldinia piassaba*, *Attalea funifera:* **raffia** —, *Raphia farinifera:* **ratan** —, *Calamus* spp.: **royal** —, *Roystonea* (= *Oreodoxa*) *regia:* **sago** —, *Metroxylon sagus*, *M. rumphii:* **saw palmetto** —, *Serenoa serrulata:* **sealing-wax** —, *Cyrtostachys lakka*, Sumatra, bright red l. sheaths: **sugar** —, *Arenga pinnata:* **tagua** —, *Phytelephas macrocarpa:* **talipot** —, *Corypha umbraculifera:* **thatch** —, *Thrinax parviflora:* **toddy** —, *Caryota urens:* **uricury** —, *Cocos coronata:* **vegetable ivory** —, *Phytelephas macrocarpa:* **wax** —, **Columbian**, *Ceroxylon andicola:* **wax** —, **carnauba**, *Copernicia cerifera:* **wine** —, *Jubaea chilensis* and other palms.

Palma Christi Castor oil plant, *Ricinus communis:* — **fibre**, *Samuela carnerosana*, Mex., used for twine: — **rosa**, or ginger grass oil, *Cymbopogon martinii*, trop., cult., used in perfumery, esp. for soap.

Palmetto *Sabal palmetto*, S. U.S., and other spp. fibre used for brushes: **saw** —, *Serenoa repens*, petioles sharply serrate, S. U.S., stem fibre used for brushes and in upholstery.

Palmiet *Prionium serratum*, S.Af., a large sedge-like pl., grows near or in streams in some areas (Palmiet river).

Palmiste *Roystonea oleracea*, W.I., young growing point of stem used as veg.

Palmyra *Borassus flabellifer*, Ind. *See Borassus* in Willis.

Palosapis *Anisoptera thurifera* and other spp., Philippines, a trade timber.

Paloverde *Cercidium* spp. SW. U.S., desert shrubs or small trees, conspicuous in water courses.

Palsy-wort *Primula veris*, cowslip, once thought to cure the palsy.

Palta Avocado pear (Spanish).

Palu (Ind.) *Mimusops hexandra*, hard, durable wood.

Pameroon bark *Trichilia moschata*, W.I.

Pampas grass *Cortaderia selloana* (=*C. argentea*), large orn. grass, widely cult.

Pampelmousse *See* Pompelmous.

Panama hat plant *Carludovica insignis* (=*C. palmata*), C.Am., the palm-like young ls. used in making Panama hats, so called because Panama was once the point of distribution. The industry is now imp. in Ecuador where many thousand people are engaged in it and the pl. extensively cult.

Pancratium Bulbous pls. w. large white fls., cult.

Panda *P. oleosa*, W.Af., a tree, oil from seeds used in cooking.

Pandanus *See* screw-pine.

Pandorea Aus. evergreen climbers, showy pink or white fls., cult.

Panga-panga *Millettia stuhlmannii*, E.Af., a commercial timber.

Pangium *P. edule*, Malaya, seeds eaten after treatment to remove harmful properties.

Pangola grass Or pongola grass, *Digitaria decumbens*, S.Af., grown for forage and hay, esp. in W.I. and S. U.S.

Panic grass *Panicum* spp., N.Am.: **hairy** — —, *Panicum effusum*, Aus. — —, *Paspalidium gracile*, Aus.: **wiry** — —, *Entolasia stricta*, Aus.

Panirband (Ind.) *Withania coagulans*, fr. used to coagulate milk.

Pansy *Viola* spp., a favourite gdn. fl., long cult. many vars. and strains: **field** —, *V. arvensis*: **Trinidad** —, *Torenia asiatica*, ann. w. blue fls.

Papain *See* papaw.

Papapsco *Acer* spp., a name used for figured maple wood.

Papaw *Carica papaya*, a melon-like fr., widely cult. in the trop. and subtrop., tree of v. rapid growth: fr. much used in fr. salad; the green frs. are scarified for their milky juice which is dried and yields papain, a drug valued for its digestive action. It is also a meat tenderizer and an ingredient of chewing gum. **Am.** —, *Asimina triloba*, SE. U.S., ed. fr.: **mountain** —, *Carica cundinamarcensis*, S.Am., frs. used in preserves or candied.

Paper Most of the world's paper is made from pulpwood, mainly obtained from coniferous trees but many other pl. materials may be used in different parts of the world, the main consideration being that such materials as are suitable are available in bulk and at no very great cost. Some of the best grades of paper are made from cotton and linen rags and from esparto grass. Other grasses are used, also bamboo, cereal straw and sugar cane bagasse for certain

PARSLEY

kinds of paper. — **bark,** *Melaleuca* spp., Aus., thin papery b.; *Streblus asper,* trop. As., used in paper making: — **birch,** *Betula papyrifera,* N.Am.: —**chase tree** or dhoby's tree, *Mussaenda frondosa,* large white sepals, cult.: — **flower,** *Psilostrophe cooperi,* SW. U.S.: **large** — **flower,** *Thomasia macrocarpa:* — **mulberry,** *Broussonetia papyrifera,* China, cult., as coppice, b. used: — **tree,** *Edgeworthia papyrifera,* Orient., b. used for paper; *Streblus asper,* trop. Aus.

Paphinia Trop. Am. orchids w. large beautiful fls., cult.

Papoose root Or squaw root, *Caulophyllum thalictroides,* N.Am., rs. med.

Paprika Spice made from *Capsicum, C. frutescens* or *C. annuum,* esp. in the Balkan countries and in Spain, Morocco and Calif. It is much used for flavouring meat dishes, esp. Hungarian goulash and is claimed to have a high vitamin content.

Papyrus *Cyperus papyrus* (=*C. antiquorum*), Nile region and other Af. rivers: an ancient Egyptian writing material, thin sections of the stem being pressed together: v. buoyant, used for rafts and boats.

Para grass *Brachiaria mutica,* a trop. fodder grass, sometimes a weed: — **nut,** *Bertholletia excelsa:* — **rubber,** *see* rubber plants.

Paradise flower *Caesalpinia pulcherrima, Solanum wendlandii,* cult. trop.: **grains of —,** *see* Guinea grains, *Aframomum melegueta,* W.Af.: — **tree,** *Quassia* (=*Simarouba*) *glauca,* trop. Am.: — **nut,** *Lecythis* spp., trop. Am.

Paraguay tea *See* maté: — **palm,** *Acrocomia sclerocarpa,* pericarp and kernel oil used in cooking.

Parakeelia (Aus.) *Calandrinia* spp., fleshy pls. w. brilliant fls.

Parana pine *Araucaria angustifolia* (=*A. brasiliana*), a commercial timber.

Parasol flower *Holmskioldia sanguinea,* Ind., shrub. w. unusual scarlet fls., cult.

Parchment bark *Pittosporum crassifolium,* N.Z. tree w. felted ls. and purple fls.

Pareira root or **pareira brava** *Chondrodendron tomentosum,* S.Am., root med.: **false — —,** *Cissampelos pareira:* **white — —,** *Abuta rufescens:* **yellow — —,** *Aristolochia glaucescens.*

Parilla, yellow *Menispermum canadense,* N.Am., moonseed, rhi. med.

Parkia *P. biglobosa,* W.Af., Af. locust bean, mealy pod pulp ed.

Parkinsonia *P. aculeata,* Jerusalem thorn, good barrier hedge in dry trop.

Parnassus, Grass of *Parnassia palustris,* Eur. inc. Br., boggy places, cult.

Parodia Round or cylindrical cacti, S.Am., cult.

Parrot apple *Clusia* spp., W.I.: — **beak,** *Pedicularis racemosa,* N.Am.: — **beak orchid,** *Pterostylis nutans,* Aus.: — **bill,** *Clianthus puniceus,* Aus., masses of scarlet lobster-claw fls.: — **bush,** *Dryandra sessilis,* W.Aus., orn. cult.: — **feather,** *Myriophyllum brasiliense,* S.Am., much used in aquaria: — **food,** *Goodenia ovata,* Aus.: — **heads,** *Sarracenia psittacina,* N.Am.: — **plant,** *Crotalaria cunninghamii,* Aus.: — **weed,** *Bocconia frutescens,* W.I.

Parrotia *P. persica,* As. Minor, tree w. orn. autumn fol., cult.

Parsley *Petroselinum crispum* (=*P. hortense, P. sativum*); **common parsley,** thought to be native to Sardinia, is perhaps the most popular flavouring herb in English-speaking countries. There are several forms. In the popular 'curled' form, the ls. are deeply divided and the segments turned back. In 'fern-leaved' parsley the l. is very finely divided but the segments not turned back, giving a fern-leaf appearance. In turnip-rooted parsley the r. develops like a parsnip or carrot, and is the part used. Parsley may be dried (for winter use) or a flavouring oil distilled from it. **Bur** —, *Caucalis latifolia:* **corn** —, *Petroselinum segetum:* **cow** —, *Anthriscus sylvestris:* **dog** —, *Aethusa cynapium:* **desert** —, *Lomatium dissectum,* N.Am.: — **fern,** *Cryptogramma crispa:* **fool's** —, *Aethusa cynapium:* — **haw,** *Crataegus marshallii,* N.Am.: **hedge** —, *Torilis japonica* (=*Caucalis anthriscus*): — **piert,** *Aphanes* (=*Alchemilla*) *arvensis:* **stone** —, *Sisum amomum:* **water** —, *Oenanthe sarmentosa,* N.Am. rs. ed.: **wild** —, *Lomatia silaifolia,* Aus.

Parsnip *Pastinaca sativa,* Eur., a widely cult. veg. in temp. countries: the long fleshy ed. tap root varies in size and shape w. different vars. and is greatly affected by cultural conditions. **Cow** —, *Heracleum sphondylium,* Eur.; *H. lanatum,* N.Am., r. ed.: **Peruvian** —, *Arracacia xanthorhiza* (*A. esculenta*), ed. tubers: **water** —, *Sium* spp.

Parson-in-the-pulpit orchid *Glossodia major:* —'**s bands orchid,** *Eriochilus cucullatus,* Aus.

Partridge berry *Gaultheria procumbens,* N.Am., ls. distilled for oil, med.: *Mitchella repens,* N.Am., ls. med.: — **breasted aloe,** *Aloe variegata,* cult., a house pl.: — **pea,** *Cassia fasciculata,* N.Am.: — **wood,** *Andira inermis* and *Caesalpinia granadillo,* both trop. Am. and commercial timbers.

Paspalum Several spp. are imp. pasture grasses, esp. *P. dilatatum,* widely cult. for hay and pasture, a weed in irrigation channels in Aus. *P. notatum* is a useful lawn grass in E.Af.

Pasque flower *Pulsatilla vulgaris,* wild in Br. (chalk soils), cult.

Passion flower *Passiflora* spp. This large genus has over 500 spp. mainly trop. Am.: many are highly decorative climbers and widely cult. The name was given by early missionaries in S.Am. because of a fancied resemblance in the fls. to the implements of Crucifixion. — **fruit,** many spp. of *Passiflora* have ed. frs. and are cult. for fr.: the most imp. is *P. edulis,* purple granadilla, commercially cult. in subtrop. for pulp and juice which are canned. *P. quadrangularis* is much grown in trop. The yellow granadilla, *P. edulis* f. *flavicarpa* is much grown in Hawaii.

Pataua *Oenocarpus pataua, Jessenia pataua,* Braz., fr. pulp yields ed. oil.

Patchouly *Pogostemon patchouly* (=*P. cablin*), Ind., oil distilled from ls. used in perfumery, esp. popular in the Orient. **Malayan** —, *P. heyneanus:* **Chinese** —, *Microtaena cymosa.*

Patience *Rumex patientia,* Eur., formerly used as spinach.

Patrinia Perens., temp. As., orn. fls., cult.

Pattern wood *Alstonia congensis,* a soft white wood, also called alstonia.

Paulownia *P. tomentosa* (=*P. imperialis*), China, orn. fl. tree, cult.

Pau santo *Kielmeyera coriacea,* Braz. ground b. a substitute for ground cork.

Pawpaw *See* papaw.

Paxiuba palm *Iriartea ventricosa,* trop. Am., stem thickened half way up.

Pea *Pisum sativum,* **culinary** or **garden** —, cult. from ancient times, origin uncertain, seeds used as a veg. and in soups, sold fresh, canned, dried, split, as flour or from deep freeze. In 'sugar peas' where the pods lack fibre or 'parchment' the pod is eaten along with the peas. The field pea, *Pisum sativum* var. *arvense* is also widely cult. in temp. countries as animal food. **Asparagus** —, *Tetragonolobus purpureus:* **bitter** —, *Daviesia* spp.: **black eye** —, *Dolichos* spp., W.I.: **blue** —, *Hovea* spp., Aus.; *Clitoria ternata* trop.; *Psoralea pinnata*, S.Af.: **butterfly** —, *Clitoria mariana:* **chick** —, *Cicer arietinum:* **Congo** —, *Cajanus cajan:* **coral** —, *Abrus precatorius:* **cow** —, *Vigna unguiculata:* **Darling** —, *Swainsona greyana*, Aus.: **desert** —, *Clianthus formosus* Aus.: **dogtooth** —, *Lathyrus sativus:* **dun** — = *field* —: **everlasting** —, *Lathyrus latifolius:* **field** —, *Pisum sativum* var. *arvense:* **flame** —, *Chorizema cordatum*, Aus.: **flat** —, *Platylobium formosum*, Aus.: — **flower**, *Centrosema, Clitoria:* **glory** —, *Clianthus formosus:* **grey** — = field —: **Indian** —, *Lathyrus sativus:* **Lord Anson's** —, *Lathyrus magellanicus:* **maple** — = field —: **marsh** —, *Lathyrus palustris:* **mutter** — = field —: **no-eye** —, *Cajanus cajan*, W.I.: **parrot** —, *Dillwynia* spp.: **partridge** — = field —: — **bush**, *Brachysema lanceolatum:* **pigeon** —, *Cajanus cajan:* **purple** —, *Hovea* spp., Aus.: **Riga** —, *Lathyrus sativus:* **rosary** —, *Abrus precatorius:* **scarlet** —, *Kennedia* spp.: **sea** —, *Lathyrus japonicus:* **Spanish** —, *Cicer arietinum:* **sweet** —, *Lathyrus odoratus:* **Tangier** —, *Lathyrus tingitanus:* — **tree**, *Caragana arborescens, Sesbania* spp.: **wedge** —, *Gompholobium* spp.: **wild** —, *Lathyrus sylvestris:* **winged** —, *Tetragonolobus purpureus.*

Peace lily *Spathiphyllum* spp., *S. wallisii*, Colombia, has a bright green spathe, a room pl.

Peach *Prunus* (= *Amygdalus*) *persica*, the most widely grown of the stone frs.: canned more extensively than any other fr. with the possible exception of the pineapple: the tree is considered to be of hybrid origin and to have originated in China where it was known over 2000 years ago. It is thought to have reached Eur. at an early period via the overland trade routes. **Af., negro, Guinea** or **wild** —, *Nauclea latifolia* (= *Sarcocephalus esculentus*), W.Af.: **myrtle** —, *Hypocalymma robustum*, Aus., orn. evergreen shrub, cult.: — **palm** or **nut**, *Guilielma* (= *Bactris*) *gasipaes*, C. & S.Am., fleshy ed. fr. pulp used in various ways, kernels yield oil: **wild** —, *Kiggelaria africana*, S.Af.: — **wood**, *Haematoxylon brasiletto*, Braz. ('Braz. wood'); *Caesalpinia echinata.*

Peacock flower *Caesalpinia pulcherrima*, Barbados pride, cult. trop.; *Delonix* (= *Poinciana*) *regia*, flamboyant: *Moraea villosa*, S.Af.: — **tiger flower**, *Tigridia pavonia*, Mex., cult., many vars.

Peanut Or groundnut, *Arachis hypogaea*, believed to be of S.Am. origin: an imp. crop in trop. and subtrop.: kernels used as human and animal food and a source of oil employed as a salad or cooking oil and for margarine and soap. Nuts used in confectionery and for nut butter: shells also have uses (insulation). **Hog** —, *Amphicarpaea monoica*, N.Am., vine w. cleistogamous fls. and underground ed. frs.

Pear *Pyrus communis*, Eur., W.As., cult. from early times: grown commercially for fresh fr. and for canning, also dried: the close-grained wood is used for carving, inlay work and drawing instruments: stained black it may be used in place of ebony, esp. for piano keys. **Af.** —, *Manilkara obovata*, W.Af.: **aguacate** —, = avocado: **alligator** — = avocado: **avocado** —, *see* avocado: **Bolwyller** —, a hybrid between pear and whitebeam, fr. red w. sweet yellow flesh: **balsam** —, *Momordica charantia:* **Chinese** —, *Pyrus pyrifolia:*

garlic —, *Crateva gynandra:* **hard** —, *Olinia, Strychnos,* S.Af.: **Jap.** —, *Pyrus pyrifolia:* **Nigerian** —, *Guarea cedrata:* **Persian** —, *Pyrus glabra:* **red** —, *Scolopia mundii,* S.Af.: **thorn** —, *Scolopia zeyheri,* S.Af., hard heavy wood was used in wagon making: **white** —, *Apodytes dimidiata,* S.Af., strong elastic wood: **wild** —, *Dombeya rotundifolia,* S.Af.: **wooden** —, *Xylomelum pyriforme* Aus.: **woolly** —, *Xylomelum* spp., Aus.

Pearl barley The barley grain w. the outer tissue ground off, reducing the grain to small round pellets, used in domestic cookery for soups, stews, puddings etc. — **bush,** *Exochorda racemosa,* China, orn. shrub, cult.: — **cud-weed,** *Anaphalis margaritacea,* N.Am., involucral bracts pure white: — **fruit,** *Margyricarpus setosus,* Andes, low shrub w. white frs., cult.: — **millet,** *Pennisetum typhoides:* —**wort,** *Sagina procumbens* and other spp.

Peat In temp. zones in many parts of the world peat occurs consisting largely of partly decomposed mosses, notably *Sphagnum* (*S. cymbifolium*). It is much used for fuel and horticultural purposes.

Pecan nut *Carya pecan* (=*C. illinoensis*), N.Am., a popular dessert nut, cult. commercially in N.Am. and elsewhere: somewhat resembles a walnut and has similar uses.

Pectin Chemically pectins are polyuronides and are present in frs. of many kinds in both the ripe and unripe condition. Their economic significance is that they impart to jams or preserves the ability to 'set' or form a gel on cooling, esst. with jellies.

Peel, candied *Citrus medica,* the preserved rind of the citron, or other citrus frs., lemon or orange.

Peepul (Ind.) Pipul, pipal or bo tree, *Ficus religiosa,* a well-known and sacred tree: characteristic prop-like rs. support the tree: water drips from l. tips.

Pegwood *Cornus sanguinea, Euonymus europaeus.* Cobbler's or shoemaker's pegs also made from hornbeam and birch: clothes pegs from willow.

Pei-mu (China) *Fritillaria roylei,* corms med.

Peribaye palm *See* peach palm.

Pelican flower or **plant** *Aristolochia gigas, A. grandiflora,* fl. bud resembles a pelican.

Pellaea Mainly small ferns, cult., often in hanging baskets.

Pellitory *Anacyclus pyrethrum,* Medit., r. med. — **of-the-wall,** *Parietaria officinalis* (=*P. judaica*), a common weed.

Peluskins Field peas, *Pisum sativum* var. *arvense.*

Pe-mou oil *Fokienia hodginsii,* Indo-China, oil distilled from r. used in perfumery.

Penang lawyer *Licuala* spp. Malaya, small palms, stems used for walking sticks when heavy walking sticks were popular: the name is a corruption of the Malay name 'loyak' or 'loyar'.

Pencil cedar *Juniperus virginiana,* N.Am., easily worked reddish wood imp. for making pencil cases or slats: from the waste an oil is distilled. **E.Af.** — —, *J. procera,* has similar uses. — **flower,** *Stylosanthes,* N.Am.: — **orchid,** *Dendrobium beckleri,* Aus.

Penda *Xanthostemon oppositifolius,* Aus. timber tree.

Peniocereus *P. greggii,* Mex., a cactus w. a turnip-like r., cult.

Penny cress *Thlaspi arvense, T. perfoliatum:* — **flower,** *Lunaria annua:* — **piece,** *Pouteria multiflora,* W.I., tree, fr. has large seed: — **pies,** *Umbilicus rupestris.*

Pennyroyal *Mentha pulegium,* S.Eur., a prostrate aromatic mint-like herb, cult., oil has med. uses. **Am.** —, *Hedeoma pulegioides,* aromatic, med.; *Pycnothamnus rigidus,* N.Am., shrubs, ls. used as tea: **false** —, *Isanthus brachiatus,* N.Am.: **mock** —, *Hedeoma pulegioides,* N.Am.

Pennywort Several pls. w. round ls. have acquired the name in Br. e.g., *Hydrocotyle vulgaris, Cymbalaria muralis, Sibthorpia europaea, Umbilicus rupestris.* **Am.** —, *Obolaria virginica:* **Asiatic** —, *Hydrocotyle asiatica:* **marsh** —, *Hydrocotyle vulgaris:* **wall** —, *Umbilicus rupestris* (= *U. pendulinus):* **water** —, *Hydrocotyle,* N.Am.

Pentas Herbs or subshrubs, trop. & C.Af., Madag., some cult. esp. *P. lanceolata.*

Pentstemon or **penstemon** *Penstemon* spp., many orn. spp. cult.

Penwiper plant *Notothlaspi rosulatum,* N.Z., ls. at first covered w. white hairs.

Peperomia or **pepperelder** Large genus (about 1000 spp.) mainly trop. Am., many spp. are popular house or foliage pls. and tolerate a dry atmosphere, such as *P. caperata* and *P. argyreia.*

Pepino Or melon pear, *Solanum muricatum,* C.Am., subshrub w. ed. frs.: some cult. vars. are seedless.

Pepper *Piper nigrum,* ordinary pepper, i.e. black or white pepper, is regarded as the world's most imp. spice in that greater quantities are used than any other spice. Where seasoning is concerned it is considered that no other spice can do so much for so many different kinds of food. The spice consists of the ground dried berries of a trop. vine and has been used in Ind. from time immemorial. For black pepper the dried 'berry' is ground with the black skin attached and for white pepper without it. Red or cayenne pepper is an entirely different spice and consists of ground dried chillies (*Capsicum* spp.). **Af.** —, *Xylopia aethiopica,* W.Af.: **Ashanti** or **Benin** —, *Piper guineense:* **bell** — = chillie: **betle** —, *Piper betle:* **bird** —, *Capsicum frutescens* (= *C. minimum):* **bush** —, *Clethra alnifolia:* **cayenne** —, *Capsicum* spp.: **cherry** —, *Capsicum frutescens:* **Ethiopian** —, *Xylopia aethiopica:* **goat** —, *Capsicum frutescens:* — **grass,** *Lepidium* spp., N.Am.: **green** —, *Capsicum frutescens:* **Guinea** — = Guinea grains: **Ind. long** —, *Piper longum:* **Jam.** — = pimento: **Jap.** —, *Zanthoxylum piperitum:* **Java** —, *Piper cubeba* (cubebs): **Kawa** —, *Piper methysticum:* **long** —, *Piper longum:* **Madag.** —, *Piper nigrum:* **melagueta** — = Guinea grains: **native** —, *Drimys lanceolata:* **negro** —, *Xylopia aethiopica:* **pod** —, *Capsicum* spp.: **poor man's** —, *Lepidium virginianum,* N.Am.: — **root,** *Dentaria* spp., N.Am.: — **and salt,** *Erigenia bulbosa,* N.Am.: **sweet** —, *Capsicum annuum:* — **tree,** *Schinus molle,* orn. tree, subtrop., frs. may be used as a spice; *Kirkia wilmsii,* S.Af.; *Micropiper excelsum,* N.Z.; *Pseudowintera axillaris,* N.Z.: — **vine,** *Ampelopsis arborea:* **wall** —, *Sedum acre:* **W.Af. black** —, *Piper clusii:* **wild** —, *Piper capense,* S.Af.

Pepper berry tree *Cryptocarya obovata,* Aus. rain forest tree.

Pepperidge Or tupelo, *Nyssa sylvatica,* N.Am.

Peppermint *Mentha piperita,* Eur., cult., there are two main forms, the 'black' and the 'white', grown for distillation of oil, the former yielding more oil of less delicate aroma than the 'white'. Peppermint oil is much used for flavouring confectionery, esp. 'mint' sweets, also with chewing gum and liqueurs (crème de menthe) and to flavour pharmaceutical or dental preparations. Jap., Chinese, Taiwan or Braz. peppermint oil is derived from *Mentha arvensis* var. *piperascens* and is an entirely different oil, rich in menthol. The name peppermint is used in Aus. for certain spp. of *Eucalyptus:* **broad-leaved** —, *E. dives:* **small-leaved** —, *E. nicholi.*

Pepperwort *Lepidium* spp.

Pera do campo (Braz.) *Eugenia klotzschiana,* pear shaped aromatic frs.

Perching lily *Astelia* spp.: — kohueu, *Pittosporum cornifolium,* N.Z.

Pereskia Woody or vine-like cacti, cult., esp. *P. bleo* and *P. aculeata,* some w. ed. fr.

Perfume plants *See* oils, esst.

Peribaye palm *Guilielma utilis,* trop. S.Am., fr. pulp eaten when boiled, tastes like sweet potato (*Ipomoea*).

Perilla oil *Perilla ocymoides* (=*P. frutescens*), *P. nankinensis,* China, Jap., Ind., seeds yield a drying oil similar to linseed, used for waterproofing paper, umbrellas etc. and for lacquerware.

Peristeria Trop. Am. orchids, cult.

Periwinkle Greater —, *Vinca major,* Eur.: **lesser** —, *V. minor,* Eur.: both cult. and naturalized in many countries, inc. Br.: **Madag.** —, *Catharanthus roseus* (=*Vinca rosea*), cult. trop. and subtrop. and a weed.

Pernambuco wood *Caesalpinia echinata,* Braz., used in turnery and for violin bows.

Pernettya *P. mucronata,* S.Am. and other spp. cult. orn. shrubs, colourful frs.

Peroba wood *Paratecoma peroba,* Braz., used for building and furniture.

Perowskia *P. atriplicifolia,* Himal. cult. orn. shrub w. blue fls., freely visited by hive bees for nectar.

Persian berries Or yellow berries, *Rhamnus infectoria,* the unripe frs. were once an imp. source of yellow dye in Eur. — **clover** *Trifolium resupinatum:* — **insect powder,** *Chrysanthemum coccineum:* — **lilac,** *Melia azedarach.*

Persimmon Jap. — or kaki, *Diospyros kaki,* a much esteemed dessert fr. like a tomato in appearance, cult. commercially in subtrop.: fr. commonly dried in China. **Am.** —, *Diospyros virginiana,* fr. ed., wood strong and tough, much used.

Peru balsam *Myroxylon pereirae,* San Salvador, C.Am., the scorched and wounded trunk of the tree yields this vanilla-scented aromatic resin: brought to Eur. by the Spaniards in the 16th Century and credited w. great med. value, now little used, mainly for perfuming ointments. — **bark,** *Cinchona:* — **cotton,** *Gossypium peruvianum:* — **daffodil,** *Hymenocallis:* **marvel of** —, *Mirabilis jalapa,* cosmop. in trop.: — **nasturtium,** *Tropaeolum tuberosum:* — **nutmeg,** *Laurelia serrata:* — **parsnip,** *Arracacia xanthorhiza:* — **swamp,** lily, *Zephyranthes candida.*

Pescatoria Trop. S.Am. orchids w. handsome fls., cult.

Petit grain oil *Citrus aurantium* var. *bigaradia,* the oil distilled from the ls. and twigs of the bitter orange, naturalized in some warm countries.

Petrea Trop. Am. climbing shrubs, orn., cult., esp. *P. volubilis* 'purple wreath', a beautiful pl., star-like purple fls. w. large blue bracts.

Pe-tsai Or Chinese cabbage, *Brassica chinensis.*

Pettigree or **pettigrue** *Ruscus aculeatus,* butcher's broom.

Petty morrel *Aralia racemosa,* N.Am.: — **spurge,** *Euphorbia peplus:* — **whin,** *Genista anglica.*

Petunia Present day gdn. petunias, used so much for bedding, are hybrids derived mainly from *Petunia integrifolia* and *P. nyctaginiflora,* S.Am.

Peyote, Peyotl Or mescal buttons, *Lophophora* (= *Echinocactus) williamsii,* Mex., the dried crown of the cactus is chewed by Amerindians during religious ceremonies. A narcotic present (anhalonin) causes hallucinations and a sense of well-being.

Phacelia Anns., mainly N.Am., some cult., esp. *P. tanacetifolia,* a good nectar pl. for the hive bee.

Phaenocoma *P. prolifera,* a hoary S.Af. shrub w. handsome crimson fls., cult.

Phaius Old World orchids, many spp. in cult.

Phalaenopsis Handsome Asiatic orchids, cult.

Pharbitis Trop. climbing pls., related to *Ipomoea,* several cult., esp. *P. learii,* blue dawn fl.

Pheasant's eye *Adonis autumnalis,* Eur., fls. blood red w. black eye, cult. — — **narcissus,** *Narcissus poeticus, N. majalis.*

Phellodendron As. trees w. orn. fol., cult., some have local med. uses.

Phenomenal berry Cult. hybrid berry fr., cross between a raspberry and dewberry.

Philadelphus Shrubs, N. Hemisph., many cult., e.g. mock orange.

Phillyrea Evergreen shrubs, dense fol., *P. angustifolia,* Medit., long cult.

Philodendron Cult. orn. fol. pls., esp. *P. elegans,* a room pl.

Philotheca Aus. heath-like shrubs, **mauve —,** *P. salsifolia.*

Phlogacanthus Handsome Ind. shrubs or subshrubs, cult.

Phlomis Usu. intensely hairy or woolly pls., Medit., As., cult.

Phlox *Phlox* spp. mainly Am., the **common garden —** is *P. drummondii,* **blue —,** *P. divaricata,* N.Am.: **creeping —,** *P. stolonifera,* N.Am.: **mountain —,** *P. subulata,* N.Am.: **perennial —,** *P. paniculata,* N.Am.: **wild —,** *Sutera grandiflora,* S.Af.

Phoenix Af. and As. palms inc. the date palm: *P. canariensis* is a widely grown orn. palm in subtrop., esp. Medit.

Phormium *See* New Zealand flax.

Photinia Oriental trees and shrubs, cult., esp. *P. serrulata,* Chinese hawthorn.

Phygelius *P. capensis,* S.Af., cult. orn. shrub.

Phyllocactus Cult. orn. cacti, trop. and subtrop.

Phyllodoce Cult. orn. shrubs, Eur., N.Am.

Phyllostylon *P. brasiliensis,* trop. Am., timber like boxwood.

Physic nut *Jatropha curcas,* a shrub widely cult. in trop., often as a hedge or boundary pl.: seed yields a purgative oil w. various uses.

Phyteuma Cult. rock gdn. and border pls., Eur.

Piassava fibre Or Pissaba, this coarse broom or brush fibre is derived from the fibrous stems or l. bases of various trop. palms. With its toughness, elasticity and imperviousness to moisture it is well suited for bass brooms and mechanical road sweepers. The finer grades are used for scrubbing and other household brushes. Originally the fibre came to Br. as packing material on ships from Brazil arriving at London docks, where it was discarded. A

London brush maker, one Thomas Hanks, noticed it and tried it out in brooms, finding it very successful. His first order was from a London street sweeping contractor who ordered six dozen of the new bass brooms. From that time (1864) onwards, the use of piassava fibre for bass brooms was greatly extended. W.Af. piassava arrived on the scene in 1886. **Af. piassava**, *Raphia hookeri*, *R. palma-pinus*: **Bahia** —, *Attalea funifera*: **Ceylon** —, *Caryota urens*, same as Ind. katool fibre, used for brushes: **Madag.** —, *Dictyosperma fibrosum* (= *Vonitra fibrosa*): **Para** —, *Leopoldinia piassaba*.

Picea *See* spruce.

Pichi *Fabiana imbricata*, Chile, Peru, evergreen shrubs, ls. med.

Pichon (W.I.) *Dodonaea viscosa*, seaside shrub and hedge pl.

Pichurim Or puchury bean, *Nectandra puchury*, Braz., aromatic seeds med.

Pick-a-back plant *Tolmiea menziesii*, N.Am., young pls. on mature ls., a room pl.

Pickpocket *Capsella bursa-pastoris*, N.Am., shepherd's purse.

Pickerel weed *Pontederia cordata*, N.Am., a widespread aquatic, seeds ed.

Picmoc Or lattan, *Desmoncus major*, W.I., prickly climbing palm, flexible stems, used for baskets.

Picotee Form or variety of carnation, *Dianthus caryophyllus*.

Picralima *P. klaineana*, trop. Af., seeds and b. med. (febrifuge).

Picrotoxin *Anamirta cocculus*. Ind. drug extracted from frs. used in cases of barbiturate poisoning.

Pie marker *Abutilon theophrasti*, Ind., naturalized in N.Am. — **crust**, *Jacquinia arborea*, W.I.

Pieris *P. floribunda*, SE. U.S., and other spp., evergreen fl. shrubs, cult.

Pifine grass *Panicum hemitomon*, N.Am., good fodder grass, prairies.

Pig balsam *Tetragastris balsamifera*, W.I.; it was once thought that pigs rubbing against the trees would heal wounds or sores by the resin: —**'s ear**, *Cotyledon orbiculata*: — **face**, *Carpobrotus* spp., Aus.: — **laurel**, *Kalmia angustifolia*, N.Am.: — **lily** = arum lily: — **nut**, *Carya glabra*, N.Am. and other spp.; *Conopodium majus*, Eur. tuberous rs. ed. when roasted: *Simmondsia chinensis* (= *S. californica*), N.Am.: — **weed**, *Amaranthus albus*, *A. retroflexus* and other spp., cosmop. weeds, used as pot-herbs or spinach: **Russian** — **weed**, *Oxyris amaranthoides*: **winged** — **weed**, *Cycloloma atriplicifolium*, N.Am.

Pigeon bean *Vicia faba* var. *equina*, a small-seeded form it is a staple food for racing pigeons: — **berry**, *Phytolacca americana; Duranta repens*: — **foot**, *Salicornia europaea*: — **grape**, *Vitis cinerea*: — **grass**, *Setaria glauca* and other spp.: — **pea**, *Cajanus cajan* (=*C. indicus*), an imp. trop. pulse crop, esp. in Ind. (dahl), numerous vars.: — **plum**, *Coccolobus laurifolia*, W.I., trop. Am.: — **wood**, *Hedycarya arborea*, N.Z.

Pignolia or **pignon** The ed. kernel of the stone pine, *Pinus pinea*.

Pilea *P. cadieri*, a popular house pl. w. silvery markings on ls.: *P. muscosa* (= *P. microphylla*), artillery pl., has explosive anthers.

Pilewort *Ranunculus ficaria*, Eur. inc. Br.: *Erechtites hieraciifolia* N.Am.

Pili nut *Canarium ovatum*, *C. luzonicum*, Philippines, ed. nut w. a thick shell.

Pillarwood *Cassipourea elliottii*, E.Af., a commercial timber.

Pillwort *Pilularia globulifera* and other spp., aquatics, mainly of warm regions.

Pilocarpine Drug (alkaloid) obtained from the ls. of *Pilocarpus* spp. *See* jaborandi.

Pimbina *Viburnum edule*, N.Am.

Pimelea Aus. shrubs, heads of bright fls., cult.

Pimento Or allspice, *Pimenta dioica* (=*P. officinalis*), W.I., trop. Am., cult. trop., the dried unripe but fully grown frs. are a much used spice. The flavour is considered to resemble a mixture of spices, hence the name, allspice. It is used ground or whole for flavouring soups, sausages, pickles, confectionery and liqueurs (Benedictine and Chartreuse): also called Jamaican pepper. — **leaves**, *Pimenta acris*, W.I., ls. distilled for Bay oil or Bay rum.

Pimpernel Common or **scarlet**, *Anagallis arvensis:* **blue** —, *A. monelli:* **bog** —, *A. tenella:* **false** —, *Lindernia* spp. N.Am.: **Rhodesian** —, *Wormskioldia longipedunculata*, a spinach pl.: **yellow** —, *Lysimachia nemorum*.

Pinang Malay name for betel nut.

Pinanga Trop. As. slender palms, cult.

Pinaster *Pinus pinaster*, S.Eur., maritime or cluster pine, imp. for rosin and turpentine.

Pin cushion *Raoulia millaria*, N.Z.; *Brunonia australis*, Aus.: — **bush**, *Hakea laurina*, Aus.: — **cactus**, *Mammillaria* spp., SW. U.S.: — **flower**, *Scabiosa* spp.; *Leucospermum* spp., S.Af.: **native** —, *Cotula squalida*, N.Z.: — **plant**, *Cotula barbata:* — **tree**, *Dais cotinifolia*, S.Af. an orn. small tree.

Pinder (W.I.) *Arachis hypogaea*, groundnut.

Pine *Pinus* spp. The pines are the most imp. group of conifers in terms of numbers and economic importance, esp. for timber. They are evergreen resinous trees, widely distributed in the N. Hemisph. Several spp. yield on tapping an oleo-resin which on distillation yields turpentine and rosin, the most imp. being the **Maritime** — of S.Eur. (*P. pinaster*), **Am. long-leaf** — (*P. palustris*) and **chir** — of Ind. (*P. roxburghii*). See *Pinus* in Willis.

The following pines yield commercial timber – **Am. pitch** —, *Pinus palustris:* **Austrian** —, *P. nigra:* **Can. red** —, *P. resinosa:* **Caribbean pitch** —, *P. caribaea:* **Corsican** —, *P. nigra* var. *maritima:* **Jack** —, *P. banksiana*, N.Am.: **lodgepole** —, *P. contorta* var. *latifolia*, W. N.Am.: **maritime** —, *P. pinaster*, S.Eur.: **ponderosa** —, *P. ponderosa*, W. N.Am.: **radiata** —, *P. radiata*, N.Am., widely cult.: **Scots** —, *P. sylvestris*, Eur.: **Siberian yellow** —, *P. sibirica:* **sugar** —, *P. lambertiana*, W. N.Am.: **western white** —, *P. monticola*, W. N.Am.: **yellow** —, *P. strobus*, E. N.Am. The common pine throughout Br. is the Scots pine, *Pinus sylvestris*. Timber is imported from Eur. and is known in the trade as red or yellow deal, redwood, or 'red', the home-grown timber being known as Scots pine.

Other pines or co-called pines include – **Aleppo** —, *P. halepensis*, Medit. wood much used by ancient Greeks and Romans: **Bhutan** —, *P. wallichiana*, Ind.: **Bunya-bunya** —, *Araucaria bidwillii*, Aus.: **celery** —, *Phyllocladus* spp., N.Z.: **cypress** —, *Callitris* spp.: **digger** —, *Pinus sabiniana*, N.Am.: **golden** —, *Pseudolarix amabilis*, China: **Huon** —, *Dacrydium franklinii:* **loblolly** —, *Pinus taeda:* **mallee** —, *Callitris* spp., Aus.: **Norfolk Island** —, *Araucaria heterophylla:* **Oregon** —, *Pseudotsuga douglasii:* **Parana** —, *Araucaria angustifolia:* **parasol** —, *Sciadopitys verticillata:* **screw** —, *Pandanus* spp.:

silver —, *Dacrydium colensoi:* **stone** —, *Pinus pinea*, S.Eur., ed. kernels: **Weymouth** —, *Pinus strobus*.

Pine bark The b. of some pines is rich in tannin and has been used in some countries for tanning leather, e.g. the Aleppo pine (*Pinus halepensis*) in Algeria and Tunis, giving a reddish leather. The loblolly pine has been used in N.Am. and the chir pine in the Himal. In N.Z. the b. of *Pinus radiata*, which is much grown, has shown some promise, esp. when mixed with black wattle.

Pineapple *Ananas comosus* (=*A. sativus*), trop. S.Am., one of the best known cult. dessert frs. of the trop. and subtrop., extensively grown for canning, now naturalized in many parts of Af. and As. trop. Introduced to Br. at the end of the 17th Century and cult. a great deal in hot-houses over a long period. With the introduction of fresh pineapples from the Azores about 1860 hot-house cult. in Br. gradually ceased. There are many vars. 'Smooth Cayenne' (ls. not prickly, favoured for canning) and 'Queen' being much grown. Pineapple juice is also canned. — **fibre**, from the ls. is strong and soft but difficult to extract: prepared commercially in the Philippines and Taiwan for export to Spain for fine embroidery: — **flower**, *Eucomis comosa*, S.Af. a pot pl.: — **guava**, *Feijoa sellowiana:* **wild** —, *Tillandsia fasciculata*, trop. Am., a showy pl., cult. [Collins, J. L. (1960) *The pineapple*, 294 pp.]

Pineapple weed *Matricaria matricarioides*, N.Am., ls. w. strong pineapple odour: its spread in Br. coincided w. the development of the motor car and it is thought its seeds were spread through mud on motor car tyres.

Pinedrops *Pterospora andromeda*, N.Am.

Pinellia Oriental tuberous pls., green spathes, cult.

Pinesap *Monotropa hypopitys*, N.Am.

Piney varnish *Vateria indica*, Ind., an oleo-resin obtained by tapping the trunk.

Pinguicula Insectivorous or carnivorous pls. inhabiting boggy areas. *See* carnivorous plants and Willis.

Pinguin fibre *Bromelia pinguin*, W.I., ls. yield a useful fibre.

Pink *Dianthus* spp. Many spp. cult.: gdn. pinks are largely of hybrid origin, many descended from *D. monspessulanus*, W. Medit. and *D. sinensis*. — **ball**, *Dombeya wallichii:* — **buttons**, *Kunzea capitata:* **Cheddar** —, *Dianthus gratianopolitanus* (=*D. caesius*): **Chinese** —, *D. sinensis:* **clove** —, *D. caryophyllus:* **Deptford** —, *D. armeria:* **fire** —, *Silene virginica*, N.Am.: **Ind.** —, *Dianthus sinensis*, *Spigelia marylandica:* **Jap.** —, *Dianthus heddewiggii:* **maiden** —, *Dianthus deltoides:* **prairie** —, *Silene subciliata*, N.Am.: — **root**, *Spigelia marylandica*, N.Am.: **rose** —, *Sabatia* spp.: **sea** —, *Armeria maritima:* — **star**, *Sabatia campestris*, N.Am.: — **weed**, *Polygonum aviculare* and other spp.: — **wood**, *Beyeria viscosa*, W.Aus.

Pinnay oil *Calophyllum inophyllum*, seed oil a common illuminant.

Pinweed *Lechea* spp., N.Am.

Pinxter flower *Rhododendron nudiflorum*, N.Am., the parent of many early fl. forms.

Pipal tree *See* peepul.

Pipe tree Syringa or lilac, branches w. large pith easily removed to make pipe stems: tobacco pipe bowl commonly made of briar root or cherry wood:

— **vine,** *Aristolochia sipho* and other spp.: — **wort,** *Eriocaulon aquaticum* (= *E. septangulare*), Br.

Pipsissewa *Chimaphila cisatlantica,* N.Am.

Pipul *See* peepul.

Piquia *Caryocar villosum,* Braz., fr. yields a veg. fat, tree once cult. in Malaya.

Pisang Malay name for banana.

Pisonia Glandular frs. stick to feathers of birds and may cause death (*see* Willis).

Pistache *Pistacia* spp.: **coolie** —, *Sterculia apetala,* W.I., seeds eaten.

Pistachio nut *Pistacia vera,* a much esteemed ed. or dessert nut of Medit. and near eastern countries: used in ice cream and confectionery and for garnishing on account of the attractive green colour of the kernel. — **galls,** insect galls, rich in tannin, may form on several spp. of *Pistacia,* the best known being those collected from *P. terebinthus* in the Medit. They may attain a large size and have a tannin content of 50–60 %.

Pistol bush *Adhatoda duvernoia,* S.Af.: — **plant,** *Pilea muscosa, see* artillery pl.

Pita fibre Or silk grass, *Bromelia* (= *Aechmea*) *magdalenae,* C. & S. trop. Am., fibre at one time exploited ('arghan').

Pitanga Braz. or Surinam cherry, *Eugenia uniflora,* W.I., cult., trop. and subtrop., a shrub w. ribbed cherry-like frs. of good flavour: may be grown as a hedge.

Pitaya or **Pitahaya** Name used in C. & S.Am. for the ed. frs. of various large cacti such as spp. of *Hylocereus* and *Lemaireocereus.* They are usu. larger than tunas, frs. of *Opuntia* spp. and considered superior. They may be eaten out-of-hand, used for drinks, sherbets, or preserved.

Pitcairnia Andean bromeliads w. usu. red or yellow fls. and prickly ls., cult.

Pitch Burgundy —, *Picea abies* (= *P. excelsa*), common Eur. spruce, the resin obtained by tapping, used for special varnishes and med. (for plasters) at one time: — **pine,** *Pinus palustris,* N.Am., an imp. timber.

Pitcher plant Term used for insectivorous pls. catching insects in special receptacles or pitchers. They belong to such genera as *Cephalotus, Darlingtonia, Heliamphora, Nepenthes, Sarracenia. See* carnivorous plants, also *Dischidia* and *Nepenthaceae* in Willis.

Pith plant Or shola, *Aeschynomene indica, A. aspera,* trop. As. and Af. semi-aquatic shrubs, pith used for 'topis' or sun-helmets and fancy articles: — **tree,** *Herminiera elaphroxylon,* ambatch, widely distrib. in trop. Af., esp. along Nile, used for floats and sun-hats.

Pithecoctenium Trop. Am., climbing shrubs w. showy fls., cult.

Pitomba *Eugenia luschnathiana,* Braz., Paraguay, tree w. juicy yellow ed. fr.

Pittosporum Old World evergreen trees and shrubs, cult.: *P. tobira,* widely grown on Riviera as a street tree.

Pituri *Duboisia hopwoodii.* Aus., ls. med. and a masticatory w. aborigines.

Pityrogramma Ferns, many w. gold or silver 'powder' on undersurface of ls., cult.

Pixie cap orchid *Acianthus fornicatus,* Aus.

Plane Common or **London** —, *Platanus* × *hispanica* (= *P.* × *acerifolia*), Eur. believed to be of hybrid origin, much planted in towns, withstands sooty conditions, b. peels off annually: **Am.** —, *P. occidentalis:* **Oriental** —, *P.*

PLANER TREE

orientalis, S.Eur., As. Minor, wood strong and tough, much used, resembles beech. In Am. and Scotland plane may be called sycamore.

Planer tree *Planera aquatica*, SE. U.S., or water elm, a useful timber tree.

Plantain *Plantago* spp. Some spp. are cosmop. and troublesome weeds, esp. on lawns: seeds often mucilaginous (*see* psyllium) and seed heads fed to cage birds (*see Plantago* in Willis). The word plantain is commonly used in English-speaking countries for coarse cooking bananas, staple foods in many parts of trop. Af. In Ind. the name plantain is used for *all* bananas. **Bastard** —, *Heliconia bihai*, W.I.: **buck's horn** —, *Plantago coronopus:* **common** —, *P. major:* **great** —, *P. major:* **hoary** —, *P. media:* — **lily**, *Hosta* spp.: **rattlesnake** —, *Goodyera pubescens:* **ribwort** —, *P. lanceolata:* **sea** —, *P. maritima:* **water** —, *Alisma plantago-aquatica:* **wild** —, *Heliconia bihai*, W.I.

Platter leaf *Coccoloba unifera*, W.I., N.Am., seaside grape, large ls.

Platycerium *See* staghorn fern.

Pleione Trop. As. orchids, Ind. to Formosa, large showy fls., cult.

Pleurisy root *Asclepias tuberosa*, N.Am., med., used by early settlers for coughs.

Pleurothallis Trop. Am. orchids, fls. often small, cult.

Ploughman's spikenard *Inula conyza*, Eur. inc. Br., dry chalky soils.

Plum There are several groups or classes of this well known fr. Some are notably dessert plums, others are favoured more for cooking and preserves. The origin of the well known English dessert plum is often obscure but they are considered to have originated mainly from eastern Eur. or As. Minor, from *Prunus domestica*. The bullace, however, is regarded as of English origin and to be descended from the wild plum of the hedgerows, *Prunus insititia*. Closely allied to the damsons and the bullace are the mirabelles, a class of plum imp. in France. Another group is the St Julien plum or plums, which in England are mainly used as stocks for grafting. Some authorities think the gages may be hybrids between *Prunus domestica* and *P. insititia*. The transparent gages are considered by many to be the best flavoured of all plums. The myrobalans or cherry-plums, *P. cerasifera*, with their cherry-like frs. constitute a very distinctive group. Another group are the so-called Japanese or saticine plums, *Prunus triflora*, often grown in the subtrop., well known vars. being Satsuma, w. blood-red flesh, Kelsey, Burbank etc. They are well suited for long distance transport. [Taylor, H. V. (1949) *The plums of England*, 151 pp.] **Alleghany** —, *Prunus alleghaniensis:* **Am. wild** —, *P. americana:* **apricot** —, *P. simonii*, N. China: **beach** —, *P. maritima*, N.Am.: **bird** —, *Berchemia discolor*, S.Af.: **Burdekin** —, *Pleiogynium solandri:* **Calif.** —, *Prunus salicina:* **Can.** —, *P. nigra:* **Chickasaw** —, *P. angustifolia:* **Chinese** —, *P. salicina:* **coco** —, *Chrysobalanus icaco:* **date** —, *Diospyros kaki:* **gingerbread** —, *Parinari macrophylla:* **governor** —, *Flacourtia indica, F. ramontchi:* **Guinea** —, *Parinari excelsa:* **hog** —, *Spondias* spp.: **Jam.** —, *Spondias purpurea:* **Java** —, *Syzygium cumini:* **kaffir** —, *Harpephyllum caffrum*, S.Af.: **Madag.** —, *Flacourtia ramontchi:* **marmalade** —, *Lucuma mammosa*, W.I.: **mobola** —, *Parinari curatellifolia*, S.Af.: **monkey** —, *Ximenia caffra:* **Natal** —, *Carissa grandiflora:* **Niger** —, *Flacourtia flavescens:* **Oklahoma** —, *Prunus gracilis:* **peach** —, *P. salicina:* **pigeon** —, *Coccolobus laurifolia:* **river** —, *Prunus americana:* **sapodilla** —, *Achras zapota:* **Sebesten** —, *Cordia myxa:* **sour** —, *Owenia acidula*, Aus.: **Spanish** —, *Spondias purpurea*, W.I.:

spiny —, *Ximenia americana:* **St Julian —**, *Prunus domestica* var. *juliana,* the prune: **wild —**, *Harpephyllum caffrum; Pappea capensis,* S.Af.: **wild goose —**, *Prunus hortulana,* N.Am.: **— wood,** *Santalum lanceolatum,* Aus.: **— yew,** *Cephalotaxus* spp.

Plumbago Old World shrubs or perens.: *P. capensis* w. soft blue fls. is widely grown, often as a hedge: in the trop. it does not tend to climb: **Ceylon —,** *P. zeylanica,* white fls., cult.: **red —,** *P. rosea,* Ind., cult.: **wild —,** *P. scandens,* W.I.

Plumboy (Am.) *Rubus* spp.

Plume grass Or ravenna grass, *Erianthus ravennae,* Eur.: **short-haired — —,** *Dichelachne scuirea,* Aus.: **— locust,** *Amorpha fruticosa,* N.Am.: **— poppy,** *Macleaya cordata, M. microcarpa,* gdn. pls.: **— weed,** *Carminatia tenuiflora.*

Plush Condoo *Planchonella laurifolia,* Aus. rain forest tree, sheds b. in patches.

Poached egg daisy *Myriocephalus stuartii,* Aus.: **— — flower,** *Limnanthes douglasii,* Calif., cult. gdn. fl.

Pod Popular term for frs. of Leguminosae or other similar frs.: **— mahogany,** *Afzelia quanzensis* C. & S.Af., good timber.

Podachaenium Two trop. Am. shrubs, cult.

Podalyria *P. calyptrata,* S.Af., shrub, pale rose fls., velvety calyx, cult.

Podo Trade name for timber from E.Af. spp. of *Podocarpus* (*P. gracilior, P. milanjianus, P. usambarensis*).

Podocarpus Trees, over 100 spp., widely distributed, esp. in S.Hemisph.: many yield valuable timber, often termed yellow wood: *P. macrophylla* is used for hedges in Jap.

Podophyllum *P. peltatum,* E. N.Am., the drug podophyllum consists of the dried rhizome and rs. from which a slow-acting purgative resin is extracted. **Indian —,** *P. emodi* (= *P. hexandrum*), a cult. gdn. pl.

Pogo *Poga oleosa,* W.Af., a commercial timber.

Pogonia *P. ophioglossoides,* N.Am., terrestrial orchid, cult.

Pohuehue *Muehlenbeckia complexa,* N.Z., a common coastal shrub.

Pohutukawa *Metrosideros tomentosa,* N.Z., strong durable timber, many uses.

Poi W.I. *Basella alba,* a spinach pl.

Poinciana, dwarf *Caesalpinia* (= *Poinciana*) *pulcherrima,* orn. fl. shrub, trop., cult.

Poinsettia *Euphorbia* (= *Poinsettia*) *pulcherrima,* Mex., widely cult. shrub w. crimson floral bracts, a popular house or pot pl., esp. at Christmas: one var. has white bracts.

Poison ivy *Rhus radicans,* N.Am., ls. cause severe dermatitis w. many people: **— peach,** *Diospyros dichrophylla,* S.Af.

Poisonous honey There are many instances where the honey derived from certain pls. by the hive bee has proved poisonous or injurious to human beings. The earliest record is probably that given by Xenophon in 400 B.C., whose soldiers were affected, losing their senses and suffering from vomiting and purging, the pl. in question being thought to have been *Rhododendron.* There are other cases where *Rhododendron* and some other genera in the Ericaceae have produced similar harmful effects. The mountain laurel (*Kalmia latifolia*) of N.Am. is a notable example. Other genera suspected are *Pieris, Andromeda,* and *Leucothoe.* The yellow jasmine (*Gelsemium semper-*

virens) is also known to have been a source of poisonous or harmful honey in the U.S. In S.Af. unwholesome honey has been due to spp. of *Euphorbia* and in N.Z. to hive bees collecting honeydew from the ls. of the tutu (*Coriaria arborea*) known to be a poisonous pl.

Poisonous plants Defined as those pls. which may cause a severe departure from normal health when a small quantity is eaten by a creature which is susceptible to its effects. Pls. poisonous to human beings and pls. poisonous to livestock constitute the two main categories. Some pls. commonly cause poisoning in both man and animals. Other may be troublesome or dangerous with only certain kinds of livestock. Horses and pigs may be susceptible to some poisonous pls. which are not troublesome to cattle because of their different body structure and digestive system. The digestive juices of the rabbit are able to neutralize certain veg. poisons which are quite lethal to man. The common foxglove (*Digitalis purpurea*) is one of the most poisonous of pls. but rarely causes poisoning because it is quite unpalatable to animals. Some poisonous pls. lose their toxicity when dried (e.g. in hay) such as buttercups (*Ranunculus* spp.) while others are equally toxic in the dry as in the fresh state.

In some countries poisonous pls. can be responsible for enormous losses with livestock. This has applied particularly in the past in countries such as Am., southern Af. and Aus., esp. when stock are being driven long distances to new grazing areas. In Br. and northern Eur. the pls. that cause sporadic poisoning with grazing animals are relatively few. Notable among them are laburnum and yew, the two most poisonous trees, also various other conifers, if eaten in sufficient quantity. This is most liable to happen in winter when pasturage is short or the ground covered with snow. Other pls. that may cause poisoning are water dropwort (*Oenanthe crocata*), hemlock (*Conium maculatum*), cowbane (*Cicuta virosa*), rhododendrons, ragwort and bracken under some conditions. Acorns and oak ls. in excess can poison horses readily, also cattle, but not pigs.

As far as human beings are concerned, the main danger of poisoning, apart from poisonous fungi, is with small children picking up and eating poisonous frs. and seeds such as deadly nightshade, mezereon and bryony, all attractive-looking fruits. The seeds of laburnum and yew can again be dangerous, also seeds of the thorn apple (*Datura stramonium*). **Poison berry** (W.I.), *Cestrum pallida;* (W.Aus.), *Gastrolobium parvifolium:* — **bulb,** *Crinum asiaticum;* (S.Af.), *Boophone disticha:* — **bush** (S.Af.), *Acokanthera spectabilis:* — **elder,** *Rhus vernix:* — **ivy,** *Rhus radicans, R. toxicodendron:* — **oak,** *Rhus* spp.: — **pea,** *Swainsona:* — **wood,** *Metopium toxiferum,* N.Am., causes severe dermatitis.

Pokaka *Elaeocarpus hookerianus,* N.Z.

Pokeweed, poke Or inkberry, *Phytolacca americana* (=*P. decandra*), N.Am., a weed in many countries, red juice of fr. has been used for colouring wine. **Venezuelan** or **W.I.** —, *P. rivinoides.*

Pokosola *Ochrosia elliptica,* Solomon Isles, kernels ed.

Polar plant *See* compass plant.

Polecat bush *Rhus aromatica,* N.Am.: — **geranium,** *Lantana sellowiana:* — **tree,** *Illicium floridanum,* ls. aromatic when crushed.

Polemonium Inc. many rock gdn. or border pls., e.g. *P. coeruleum,* Jacob's ladder, several vars.

Polyanthus Garden race of primulas originating in a cross between the primrose (*Primula vulgaris*) and the cowslip (*P. veris*).

Polybotrya Cult. orn. ferns, trop., dimorphic ls.

Polycynis Trop. Am. orchids, cult.

Polygala Many cult. orn. spp. in this large genus.

Polygonum Many spp. are common or cosmop. weeds, others cult. as gdn. pls. In England, *P. aubertii* (= '*P. baldschuanicum*'), the vigorous 'Russian vine' is called 'polygonum'. **Jap. —,** *P. cuspidatum,* a weed.

Polypodium Many ferns in this large genus are cult.

Polypody *Polypodium.*

Polystichum Genus incl. many cult. ferns.

Pomaderris Aus. and N.Z. shrubs or small trees, cult., *P. elliptica* used med.

Pomegranate *Punica granatum,* tree fr. of age-old cult., in some countries, esp. Medit.: the juicy frs. are popular in hot climates and are the source of the beverage grenadine. The b. has med. uses and the fr. shells, rich in tannin (26 %) are used in local leather tanning. Many vars., some dwarf or w. double fls., used as orn. or pot pls. **Wild —,** *Burchellia bubalina,* S.Af.

Pomelo Pompelmous or shaddock.

Pomerac *Syzygium malaccense,* orn. tree w. ed. fr., cult. trop.

Pompelmous Shaddock or pomelo, *Citrus grandis* (=*C. maxima*), the largest of the citrus frs., has a very thick rind, considered to be a parent of the grapefruit.

Pompion Pumpkin, *Cucurbita pepo.*

Pompon Name used for vars. of dahlias and chrysanthemums w. small globular fl. heads: **— tree,** *Dais cotinifolia,* S.Af., heads of small pinkish fls.

Pond apple *Annona glabra,* Florida, W.I., fr. ed. but insipid, tree used as stock for other *Annona* frs.: **— grass,** *Commelina diffusa,* W.I., a pantrop. weed: **— nut,** *Nelumbo nucifera:* **— pine,** *Pinus serotina,* SE. U.S.: **— spice,** *Litsea aestivalis:* **— weed,** *Potamogeton* spp., numerous spp. (*see Potamogeton* in Willis): **— weed, Can.,** *Elodea canadensis,* naturalized in many countries, cult., ponds and aquaria: **— weed, Cape,** *Aponogeton distachyus.*

Pony (W.I.) Or yellow poui, *Tecoma serratifolia,* fl. tree.

Poonac (Ind.) Crushed seeds, or seed cake after oil extraction, used in animal feeding and as a fertilizer.

Poor man's orchid *Schizanthus; Iris:* **— — weather glass, pimpernel,** *Anagallis arvensis,* fls. close on approach of bad weather.

Pop-ash *Fraxinus caroliniana,* N.Am.: **pops,** *Physalis pubescens,* W.I., a weed.

Poplar *Populus* spp. Poplars (cottonwood in U.S. and some called aspen) are widely distributed in the N.Hemisph. and much grown for timber, being fast-growing and easy to grow. The wood is generally light in weight and soft. It resists splintering and is extensively used for matches, also for match boxes, chip baskets for fr. and for wood pulp. Plywood and wood-wool are other uses. The poplars yielding commercial timber include – *P. alba,* white poplar: *P. balsamifera,* Can. poplar: *P. canescens,* grey poplar: *P. deltoides,* eastern cottonwood: *P. grandidentata,* Can. poplar: *P. tremula,* Eur. aspen: *P. tremuloides,* Can. aspen: *P. trichocarpa,* black cottonwood: *P. nigra,* black poplar. In the U.S. the trade name poplar refers to the tulip tree, *Liriodendron tulipifera*. Some poplars are grown for orn. or for screening, notable among them being the distinctive Lombardy poplar, *P. nigra* cv. *Italica*.

Poppy Several of the true poppies, *Papaver* spp., are grown as gdn. fls., anns. such as the opium and Iceland poppy or perens. such as the Oriental poppy. Some other pls. with poppy-like fls. have acquired the name of poppy. **Blue** —, *Meconopsis* spp.: **Calif.** —, *Eschscholzia, Platystemon:* **common** or **corn** —, *Papaver rhoeas:* **Flanders** —, *P. rhoeas:* **horned** —, *Glaucium flavum:* **Iceland** —, *Papaver nudicaule:* **Matilija** —, *Romneya coulteri:* **Mex.** —, *Argemone mexicana*, a weed: **opium** —, *Papaver somniferum:* **Oriental** —, *P. orientale:* **plume** —, *Macleaya cordata*, E.As.: — **mallow**, *Callirhoe papaver:* **prickly** —, *Argemone mexicana:* **Santa Barbara** —, *Hunnemannia fumariifolia:* — **seed,** *see* maw seed: **Tibetan** —, *Meconopsis:* **tree** —, *Romneya*, N.Am.: **violet-horned** —, *Roemaria hybrida:* **water** —, *Limnocharis humboldtii*, Braz., cult.: **Welsh** —, *Meconopsis cambrica:* **wood** —, *Stylophorum diphyllum*, N.Am.

Porcelain plant *Hoya carnosa*, Queensland, wax-like fls., a room pl.

Porcupine grass *Stipa spartea*, N.Am.; *Triodia irritans*, Aus.: — **plant**, *Solanum aviculare*, N.Z.: — **tree**, *Centrolobium paraense*, W.I., curious spiny frs.: — **wood**, *Cocos nucifera*, coconut, trade name for wood.

Poroporo *Solanum aviculare*, N.Z.

Porphyrocoma Two trop. Am. spp.: *P. lanceolata*, cult. orn. pl., purple fls. and red bracts.

Port Jackson fig *Ficus rubiginosa*, Aus., a widely grown shade tree: — **wine magnolia**, *Michelia figo*, China, shrub, fls. fragrant.

Portea Epiphytic bromeliads, Braz., colourful inflor., cult.

Porter bush *Synedrella nodiflora*, W.I., a pantrop. weed.

Portia tree (Ind.) *Thespesia populnea*, orn. or avenue tree, timber durable.

Portugal laurel *Prunus lusitanica*, evergreen shrub or small tree, cult.: — **quince**, *Cydonia oblonga*, cult. for fr.

Portulaca *P. grandiflora*, sun pl., many vars. and other spp. cult.: *P. quadrifida,* a bad trop. weed.

Posh-te *Annona scleroderma*, C.Am., ed. fr.

Posidonia balls In some parts of the world the action of the sea rolls the fibrous residue of the ls. of *Posidonia* into balls, perhaps the size of a tennis ball, which may be washed ashore. This happens w. *Posidonia oceanica* in the Medit. and *P. australis* in Aus.

Posoqueria Trop. Am. shrubs, fragrant fls., cult.

Possum grape *Cissus* spp.; *Vitis baileyana*, N.Am.: — **haw**, *Ilex decidua;* *Viburnum nudum*, N.Am.: — **oak**, *Quercus nigra*, N.Am.: — **wood**, *Halesia carolina*.

Postman, running *Kennedia prostrata*, Aus.

Pot Name for cannabis.

Potamogeton A large genus (over 100 spp.) of aquatics, cosmop., some used in aquaria to oxygenate the water.

Potato *Solanum tuberosum*. One of the world's most imp. food crops for man and his domestic animals: brought to Spain from S.Am. in the latter part of the 17th Century and soon reached other Eur. countries, inc. Br. Different vars.

already existed in the Andes at the time of its discovery. New vars., resistant to disease, are constantly being produced. The potato is cooked in various ways and is imp. in the commercial production of various products, such as dehydrated potato, potato flour, starch, alcohol, gum, 'potato chips' etc. The starch is imp. in the textile industry. Various alcoholic beverages are made from potato spirit, flavoured in various ways, among them the Scandinavian 'aquavie' or 'akrawit'. **Air** —, *Dioscorea bulbifera:* — **bean,** *Apios tuberosa:* **Carib** —, *Dioscorea bulbifera:* **Chinese** —, *Dioscorea batatas:* — **creeper,** *Solanum seaforthianum* and other spp., trop. Am., cult.: **Fendler** —, *Solanum fendleri:* **Hausa** —, *Solenostemon rotundifolius:* **Kaffir** —, *Plectranthus esculentus* or other spp., S.Af.: — **onion,** *Allium cepa* var., grows like a large shallot: — **orchid,** *Gastrodia sesamoides:* **Otaheite** —, *Diospyros bulbifera:* **sweet** —, *Ipomoea batatas:* **tree** —, *Solanum macranthum,* prickly shrub, Braz., cult.: — **vine,** *Ipomoea pandurata,* N.Am.: **wild** —, *Solanum fendleri:* — **yam,** *Dioscorea esculenta.*

Pot-herb Any pl. whose ls. or young shoots are cooked (boiled) and eaten, like spinach. The total number of pls. so used throughout the world must run into many hundreds or thousands (*see* spinach).

Potpourri Mixture of dried scented fls. or ls., specially prepared and kept in a large, wide-mouthed jar, which allows of a hand being inserted on removing the lid, very popular in Victorian times. Rose petals form the basis of most potpourri mixtures. These are collected from scented roses, usu. red, just as they open, and carefully dried, out of the sun. Various fragrant herbs and fls. may be added, such as rosemary, lavender, lemon-scented verbena etc., according to taste. Other aromatic products such as scented oils, cloves or cinnamon may also be added. Salt is commonly sprinkled into the mixture.

Potentilla Good gdn. or border pls., many hybrids.

Pottery tree of Para *Moquilea* (= *Licania*) *utilis,* the powdered b. baked w. an equal quantity of clay is used for making cooking vessels, which can withstand great heat. The b. of *Hirtella americana* of Guyana has similar properties.

Poui (W.I.) **Pink** —, *Tabebuia pentaphylla:* **yellow** —, *T. serratifolia* trop. Am. orn. fl. trees, cult.

Pounce *Tetraclinis articulata,* Medit., N.Af., resin once used as pounce or for glazing paper.

Pourouma *P. edulis,* Colombia, ed. frs.

Poverty bush *Eremophila* spp., Aus.: — **grass,** *Aristida dichotoma, Danthonia spicata, Sporobolus vaginiflorus,* N.Am.: *Eremochloa bimaculata,* Aus.: — **weed,** *Monolepis nuttalliana.*

Poyok (Sierra Leone) *Afrolicania elaeosperma,* seeds yield a drying oil.

Pracaxi fat *Pentaclethra filamentosa,* Braz., seeds yield a semi-solid fat.

Pradosia Trop. Am. trees, some spp. yield good timber.

Prairie The grass country east of the Rocky Mts.: also a French name for permanent pasture seed mixture (grass and clovers mainly). — **apple,** *Psoralea esculenta:* — **beard grass,** *Andropogon scoparius:* — **bulrush,** *Scirpus paludosus:* — **buttercup,** *Ranunculus rhomboideus:* — **clover,** *Petalostemon* spp.: — **coneflower,** *Ratibida* spp.: — **cup,** *Ceratochloa* (= *Bromus*) *unioloides:* — **cup-grass,** *Eriochloa contracta:* — **dock,** *Silphium terebinthinaceum:* — **false indigo,** *Baptisia leucantha:* — **grass,** *Sporobolus* *asper, Bromus, Spartina:* — **iris,** *Nemastylis geminiflora:* — **mimosa,** *Desmanthus illinoensis:* — **orchid,** *Habenaria leucophaea:* — **parsley,** *Polytaenia nuttallii:* **Queen of the** —, *Filipendula rubra:* — **rocket,** *Erysimum asperum:* — **rose,** *Rosa setigera:* — **sagewort,** *Artemisia frigida:* — **smoke,** *Anemone patens* var. *wolfgangiana:* — **spear grass,** *Poa* spp.: — **sunflower,** *Helianthus laetiflorus:* — **trefoil,** *Lotus americanus:* — **turnip,** *Psoralea esculenta:* — **willow,** *Salix humilis.*

Prayer plant *Maranta* spp., orn. ls., room pls.

Praying virgin orchid *Drakaea elastica,* W.Aus.

Preservation, cut flower Various factors are involved in the length of life of cut flowers. A suitably low temperature is probably the most imp. single factor. The end of their life may be, and often is, due to the blockage of the conducting vessels by colonies of bacteria or other organisms, or by latex or decomposition products. Reducing the rate of water loss and metabolism increases the length of life. Environmental conditions before cutting may have an effect. It has been shown experimentally that high pre-cutting temperatures for carnations and high light intensity for chrysanthemums resulted in short-lived flowers. Burning the ends of poppy stems (which have latex), dipping rose stems in boiling water or scraping the base of the stems and putting hollyhock stems in nitric acid prolonged cut flower life. Copper salts in the water proved effective with some fls. but not with others. Calcium nitrate was effective with some bulb cut fls. Alcohol effectively prolonged the life of delphinium, gladiolus, sweet pea and zinnia. Hydrazine sulphate has proved useful with carnations.

With the autumn preservation of beech ls. (*Fagus*) when they commence to turn yellow, the following method is recommended to prevent them becoming brittle and shrivelled. Immediately after cutting the branches are placed in a bucket of water in which a handful of salt has been dissolved, and left for 24 hours. They are then transferred to a glycerine solution, 1 part of glycerine to 3 parts of water, and left for 3 weeks or so until the solution is no longer absorbed. With this treatment the ls. are permanently soft and pliable.

For the preservation of pls. as herbarium specimens, *see* collecting.

Pretty maids *Saxifraga granulata* var. *plena*, orn. gdn. pl.

Prickly ash *Zanthoxylum* spp.: — **comfrey,** *Symphytum asperrimum:* — **cucumber,** *Cucumis metuliferus,* weed in sugar cane, Aus.: — **currant** *Ribes lacustre:* — **chaff flower,** *Achyranthes aspera:* — **heath,** *Pernettya:* — **lettuce,** *Lactuca scariola:* — **Moses,** *Acacia ulicifolia,* Aus.: — **pear,** *Opuntia* spp., frs. ed.: — **pole,** *Bactris,* W.I.: — **poppy,** *Argemone mexicana:* — **shield fern,** *Polystichum aculeatum:* — **weed,** *Desmanthus illinoensis.*

Pride of Barbados *Caesalpinia pulcherrima:* — — **Burma,** *Amherstia nobilis:* — — **India,** *Lagerstroemia speciosa* (= *L. flos-reginae*), *Koelreuteria paniculata,* China; *Melia azedarach:* — — **Persia,** *Melia azedarach.*

Prim (Am.) Privet, *Ligustrum.*

Prima vera *Tabebuia donellsmithii,* Mex., C.Am., a commercial timber.

Primrose *Primula vulgaris,* Eur., inc. Br. many improved or gdn. forms: **bird's eye** —, *P. farinosa:* **Cape** —, *Streptocarpus* spp.: **Chinese** —, *Primula sinensis:* **evening** —, *Oenothera biennis:* **Greenland** —, *Primula egaliksensis:* — **willow,** *Ludwigia,* W.I.

Primula Large genus, N. Hemisph., mostly Alpine, many spp. cult.

Prince Albert's Yew *Saxegothaea conspicua*, S.Am., a yew-like orn. tree, cult.: —'s **feather**, *Amaranthus hypochondriacus*, trop. Am., coloured ls., a popular gdn. pl., several vars.: **princess' feather**, *Polygonum orientale:* — **tree**, *Tibouchina semidecándra*, S.Am.: —'s **wood**, *Cordia gerascanthus*, trop. Am.; *Exostema caribaeum*, W.I., the bitter b. once had a reputation as a febrifuge.

Prisoner orchid *Caladenia ericksonae*, Aus.

Privet Common or **Eur.** —, *Ligustrum vulgare*, Eur. inc. Br., naturalized in N.Am. The evergreen privet, usu. grown for hedges, is *L. ovalifolium*, Jap.: **Chinese** —, *L. sinense:* **glossy** —, *L. lucidum:* **golden** —, *L. ovalifolium* cv. *Aureum*, ls. w. irregular yellow border: **Ibota** —, *L. ibota:* **mock** —, *Phillyrea:* **swamp** —, *Forestiera acuminata*, N.Am.: **variegated** —, *L. ovalifolium* cv. *Argenteum*, ls. edged w. yellow.

Privy vine *Antigonon leptopus*, Mex., widely cult. subtrop.

Proboscis flower *Martynia louisiana*, S. U.S., or unicorn pl.

Procession flower *Polygala vulgaris*, Eur. inc. Br., common milkwort.

Promenaea Braz. orchids w. large pretty fls., cult.

Promotion nut *Anacardium occidentale*, cashew nut.

Prophet plant *Arnebia echioides*, Medit., favoured for the rock gdn. or dry wall, fls. yellow.

Propolis Or bee glue, as it is sometimes called, is of veg. origin and is the resinous or sticky substance which bees use for closing up cracks or crevices in their nests or hives, thereby making them weather and draught proof. They also use it for bracing the combs. Propolis may be collected by bees from a number of different pls. but little is known with certainty regarding its sources. Propolis varies much in colour and in physical properties. Pls. recorded as sources of propolis include birch, alder, beech, willow, horse chestnut, the sticky buds of some poplars, the heads of sunflowers and buds of hollyhocks. To the beekeeper propolis is a nuisance for it interferes with his or her manipulation of combs in the hive. It is said to have been used in the varnish employed by famous Italian violin makers (e.g. Stradivarius). A belief held by old-time British beekeepers was that bees knew their mastei and would even follow him to the grave, bees having been observed flying round coffins. The real reason for this was that the bees were interested in the fresh varnish on the coffin as a source of propolis!

Proso millet *Panicum miliaceum*, quick-growing, much grown in Ind.

Prostanthera Aus. mint bushes, usu. strongly scented.

Protea Shrubs or small trees, mainly S.Af. (sugar bushes) many spp. with attractive fl. heads: fls. rich in nectar, once used as a source of sugar syrup: the national fl. of S.Af.: **giant** —, *P. cynaroides*, fl. heads 13–21 cm long and wide.

Prune Or dried plum. *Prunus domestica* var. *juliana*, vars. with high sugar content which will dry without fermenting: valued for nutritional, demulcent and laxative properties.

Prunus Includes many orn. trees and shrubs (flowering cherries, almond, *Prunus triloba*, etc.) and imp. temp. tree frs. such as the apricot, cherry, nectarine, peach, plum and prune. *See* separate headings and *Prunus* in Willis.

PSEUDERANTHEMUM

Pseuderanthemum Trop. shrubs, cult for fls. or orn. foliage.

Pseudotsuga Trees of N.Am., Orient (hemlock spruces). *P. menziesii* Douglas fir, is a widely cult. timber and orn. tree.

Psyllium Or flea seed, *Plantago psyllium*, *P. arenaria*, S.Eur., NW.Af. Seeds v. mucilaginous, used med. in treatment of chronic constipation: similar to ispaghul seed.

Pteris Many orn. ferns of this large genus cult., some popular as room pls, (*P. cretica*, *P. multifida*, *P. tremula*).

Pterostylis Orchids, mainly Aus. and N.Z., fls. usu. green or greenish.

Puccoon Red —, *Sanguinaria canadensis*, E. N.Am., bloodroot: **yellow** —, *Lithospermum canescens*, N.Am., red r., cult. orn. fl.

Puchurim nut *Acrodiclidium puchury*, trop. Am., W.I., med.

Pudding pipe tree *Cassia fistula*, cult., orn. tree, trop., attractive pods, med.

Pueraria *See* kudzu.

Puke weed *Lobelia inflata*, Am., med., emetic.

Pulasan *Nephelium mutabile*, E.I., cult. tree fr., resembles rambutan.

Pulmonaria Several spp. are cult. gdn. pls.

Pulque (Mex.) The intoxicating drink prepared from *Agave* juice.

Pulsatilla Several gdn. pls. allied to *Anemone*, feathery styles lengthen after fl.

Pulses Like the cereals, the pulses are an imp. group of food pls. serving the needs of mankind and having been cult. from the earliest times. They are equally imp. in warm and in temp. cli. Pulses are of special significance as human food in some As. countries, such as China and Ind., also in C. & S.Am. where the dwarf, French or haricot bean (*Phaseolus vulgaris*) in its numerous vars. often constitutes a staple diet. Compared with cereals as food, the pulses have the advantage of higher protein content. Many are imp. food for livestock. The following are among the better known pulses (*see also* separate headings) – adsuki bean, Bambarra groundnut, beans, Bengal or Mauritius bean, chick pea, cluster bean or guar, cowpea, field bean, field pea, French, kidney, dwarf or haricot bean, gram (Ind.) – golden, green and black, ground bean (*Kerstingiella geocarpa*), ground nut or peanut, horse gram, horse bean, lablab bean, lentil, lima or pole bean, lupin, moth bean, mung bean (*Phaseolus radiatus*), pea, rice bean, scarlet runner bean, white seeded forms used as butter beans, soy bean, sword or jack bean, tepary bean, urd.

Pultenaea Aus. evergreen shrubs, fls. usu. yellow, cult.

Pulu fibre *Cibotium glaucum*, this fern fibre was imported to Br. at one time from the Sandwich Isles.

Pulza *Jatropha curcas:* trop., physic nut.

Pumelo *See* pompelmous.

Pumilio pine oil *Pinus mugo* var. *pumilio*, C.Eur., obtained by distillation of ls., esp. in Austrian Alps, med. inhalant.

Pumpkin *Cucurbita pepo*, the origin of this widely grown veg. is uncertain. Frs. of the large vars., grown as stock food, may reach an enormous size and weigh between 50 and 90 kg, probably the world's largest or heaviest fr. **Seminole** —, *Cucurbita moschata*, an ancient food pl. of the Amerindian in Florida: **wild** —, *Cucurbita foetidissima*, N.Am.

Punah *Tetramerista glabra*, Malaya, a medium-weight commercial timber.

Punch and Judy orchid *Gongora* spp., trop. Am., strange fls., cult.

Puncture weed *Tribulus* spp. esp. *T. terrestris,* the three-pointed fr. a bane to cyclists.

Pupunha palm *Bactris minor,* Braz., peach palm, fr. ed.

Purging buckthorn *Rhamnus catharticus,* Eur. inc. Br.: — **cassia,** *Cassia fistula:* — **flax,** *Linum catharticum:* — **nut,** *Jatropha curcas,* trop.

Purple heart *Peltogyne paniculata* trop. Am., or other spp. a commercial timber, hard heavy, purplish when fresh: — **mat,** *Nama demissum,* SW. U.S.: — **pitcher plant,** *Sarracenia purpurea,* N.Am.: — **prairie clover,** *Petalostemum purpureum:* — **top** (Aus.), *Verbena.*

Purslane Common —, *Portulaca oleracea,* Ind., widely cult. esp. in Eur.: young ls. used in salad, older ls. like spinach: large-leaved and golden forms exist. **Horse** —, *Trianthema portulacastrum,* trop. Af., a pot-herb: **sea** —, *Halimione* (=*Atriplex*) *portulacoides; Honkenya* (=*Arenaria*) *peploides:* **tree** —, *Portulacaria afra,* S.Af.: **Surinam** —, *Talinum quadrangulare:* **water** —, *Lythrum* (=*Peplus*) *portula:* **winter** —, *Montia* (=*Claytonia*) *perfoliata,* naturalized in Br., ls. eaten in salad or boiled.

Pussley (W.I.) *Portulaca oleracea,* purslane.

Pussytoes *Antennaria rosea,* N.Am.: — **willow,** *Salix caprea* in Br., *S. humilis, S. discolor* in N.Am.

Putat (Java) *Pandanus* spp., cult., ls. used for weaving.

Puya Andean pls. w. long spiny ls. and showy fls., cult.

Pygmy pipes *Monotropsis odorata,* N.Am., r. highly scented.

Pyinkado *Xylia dolabriformis,* Ind., Burma, a commercial timber second only to teak in Burma, noted for strength and durability.

Pyinma *Lagerstroemia speciosa* (=*L. flos-reginae*), Ind., a notable orn. tree, yields commercial timber.

Pyracantha *P. coccinea,* S.Eur., As., cult. orn. shrub w. red berries, several vars.: **Chinese** —, *P. angustifolia:* **Nepalese** —, *P. crenulata.*

Pyrene oil Oil extracted from olive stones, used for making green soap in Greece.

Pyrethrum *Chrysanthemum cinerariifolium.* This well known insecticide consists of the powdered dried fl. heads. Extracts are prepared from it. A cult. field crop in many countries, esp. E.Af. and S.Am.: known as **Dalmatian** —, **Persian** —, *C. coccineum* (*C. roseum*) has also been used as an insecticide. **Garden** —, *C. coccineum,* early daisy-like fls., a commercial cut fl. in Br.

Q

Quackgrass *See* couchgrass.

Quail grass *Celosia argentea,* trop., used as spinach: — **plant,** *Heliotropium curassavicum,* S. U.S., W.I.

Quake grass or **quaking grass** *Briza media,* Eur. inc. Br.: **small** — —, *B. minor:* **large** — —, *B. maxima:* used as orn. grasses.

Quaker ladies *Houstonia caerulea,* N.Am., cult. gdn. pl.

Qualup bell *Pimelea physodes,* Aus.

Quamash *Camassia* spp., N.Am., bulbs were a food of the Amerindians.

Quamoclit Handsome climbing pl. w. red fls., cult. trop., esp. *Ipomoea sanguinea* (= *Quamoclit coccinea*) and *I. versicolor* (= *Q. lobata*).

Quandong *Santalum acuminatum*, Aus.; *Elaeocarpus grandis*, Aus.

Quaruba *Vochysia hondurensis*, trop. Am. and other spp., a commercial timber.

Quassia wood Or bitter wood, *Picrasma* (= *Aeschrion*) *excelsa*, C.Am., W.I.: quassia chips infused in water were used as a bitter, once used as an insecticide and for fly papers: **Surinam —**, *Quassia amara*, C.Am., W.I. the original quassia wood.

Quebec oak *Quercus alba*.

Quebracho *Schinopsis balansae*, *S. lorentzii*, S.Am., esp. Argen., the tannin extract prepared from the heartwood was for a long time the world's most imp. leather tanning material in terms of production. Over-exploitation of the trees and lack of regeneration, through cattle eating the seedlings, caused the industry to decline. Quebracho tans quickly and is valued in the production of sole leather.

Queen's delight *Stillingia sylvatica*, Am.: **— flower**, *Lagerstroemia speciosa* (= *L. flos-reginae*) cult. trop. v. showy fls.: **— lily**, *Phaedranassa*, Andes, showy fls., cult.: **Queen-of-the-meadow**, *Filipendula ulmaria*, Eur. inc. Br.: **— orchid**, *Thelymitra variegata*, Aus.: **— root**, *Stillingia sylvatica*: **— wreath**, *Petrea volubilis*, trop. climber w. showy fls. cult.

Queensland arrowroot *Canna edulis*, starch or arrowroot prepared at one time: **— asthma plant**, *Euphorbia pilulifera*, a trop. weed: **— blue grass**, *Dichanthium sericeum*: **— fig**, *Ficus cunninghamii*: **— hemp**, *Sida rhombifolia*: **— itch tree**, *Davidsonia pruriens*: **— kauri**, *Agathis brownii*: **— nut**, see Macadamia nut: **— wild lime**, *Microcitrus inodora*.

Quercitron *Quercus* spp., a yellow dye obtained from the inner b. of certain oaks, *Q. tinctoria*, *Q. velutina*, N.Am.

Quercus *See* oak.

Quesnelia Trop. S.Am. bromeliads, similar to *Aechmea*, cult.

Quetsche or **quetschen plum** *Prunus domestica* var. *pruneauliana* Eur., yields an inferior kind of prune.

Quick, Quick-thorn or **quickset thorn** *See* hawthorn: **— set hedge**, a hawthorn hedge: **— or kwitch grass**, *see* couch grass: **— stick** (Jam.), *Gliricidia sepium*, a shade tree, branches r. quickly: **— weed**, *Galinsoga*, N.Am.: **—silver weed**, *Thalictrum dioicum*, N.Am.

Quillaja bark *Quillaja saponaria*, Chile, Peru, soap tree, b. lathers in water and is a commercial source of saponin: used in pharmacy as an emulsifying agent, esp. for volatile oils: bark of *Q. poeppigii* is similarly used.

Quillings Trade term for pieces of broken quills of cinnamon b.

Quillwort *Isoetes lacustris* and other spp.

Quimbombo (Cuba) *Hibiscus esculentus*, or okra.

Quince *Cydonia oblonga* (= *C. vulgaris*) As. Minor, a small tree, ed. fr. used mainly in preserves, seeds mucilaginous. **Bengal —**, *Aegle marmelos*, bael fr.: **Chinese —**, *Chaenomeles* (= *Cydonia*) *sinensis*: **dwarf —**, *Chaenomeles japonica* (= *Cydonia maulei*): **Jap. —**, *Chaenomeles speciosa* (= '*Cydonia japonica*'), a well-known orn. fl. shrub.

Quinine The chief alkaloid in the b. of various spp. of *Cinchona*. For many

years it was extensively used against malaria, but was largely superseded by
less toxic and more effective anti-malarial drugs (*see* cinchona): — **berry**,
Cephalanthus natalensis: — **bush**, *Cowania mexicana:* — **tree**, *Rauvolfia
natalensis*, S.Af.

Quinoa *Chenopodium quinoa, C. nuttalliae* and *C. pallidicaule*, high altitude food
pls. in the Andes of Chile and Peru, seeds used as a cereal.

Quinsy berry Blackcurrant, *Ribes nigrum.*

Quisqualis *Q. indica*, trop. As., 'Rangoon creeper', cult., orn. pl.

Quiteria root *Spigelia pedunculata*, Colombia, r. used as vermifuge.

Quiver leaf *Populus tremuloides*, N.Am.

R

Rabbit bells *Crotalaria angulata*, Am.: — **berry**, *Shepherdia argentea*, N.Am.,
fr. ed.: — **brush** or **bush**, *Chrysothamnus nauseosus*, W. N.Am.: — **ear's
orchid**, *Thelymitra antennifera*, Aus.: —'**s foot clover**, *Trifolium arvense*,
Eur.: — **meat** (W.I.), *Alternanthera ficoidea*, popular food for rabbits:
— **orchid**, *Caladenia menziesii*, Aus.: — **pea**, *Tephrosia virginiana*, N.Am.:
— **tobacco**, *Pterocaulon undulatum*, *Evax* sp., N.Am.: — **vine**, *Teramnus
labialis*, W.I., a weed.

Radish *Raphanus sativus.* The origin of the common radish is uncertain. It
is mainly used in salads but eaten cooked in some countries: the ls. may be
used like spinach, or the young seed pods pickled. Some of the Oriental
radishes (Jap. radish) are v. large, like turnips, and are used as livestock food.
Chinese —, *Raphanus sativus:* **horse —**, *Armoracia rusticana* (=*Cochlearia
armoracia; A. lapathifolia*): **Jap.** — or daikon, *Raphanus sativus:* **sea —**, *R.
raphanistrum* subsp. *maritimus:* **wild —**, *R. raphanistrum.*

Raffia *Raphia farinifera* (= *R. pedunculata*), Madag., the main source of the
raffia used for tying in horticulture and in fancy work: stripped off young ls.:
also obtained from some W.Af. spp. of *Raphia.*

Rafflesia *R. arnoldii*, Sumatra. A parasitic pl. of the jungle (on rs. of *Vitis* spp.)
of interest on account of its enormous fl., the largest fl. of the world or an
honour it may share with the giant aroid *Amorphophallus titanum*, although
that so-called fl. is actually an inflor. The fl. of *R. arnoldii* is of a purplish
colour with a smell of decaying meat and visited by carrion flies. It is fleshy
and may be from 45 to 60 cm or more across and weigh 7 kg.

Ragged cup *Silphium perfoliatum*, N.Am., med.: — **orchis**, *Habenaria lacera*,
N.Am.: — **robin**, *Lychnis flos-cuculi*, Eur. inc. Br.; *Cleome* (=*Gynandropsis*)
speciosa, W.I.

Ragi (Ind.) *Eleusine coracana*, a common cereal in many trop. countries, esp.
Ind., also called finger millet.

Ragweed *Ambrosia* spp., N.Am., some cause hay fever.

Ragwort *Senecio* spp. **Common —**, *S. jacobaea*, a troublesome weed in
pasture, can be injurious to livestock: **golden —**, *S. aureus*, N.Am.: **Oxford
—**, *Senecio squalidus*, alien in Br.: **sea —**, *Cineraria maritima.*

Rai Or Indian mustard, *Brassica juncea*, an oil seed crop.

Railway creeper (Ind.) *Ipomoea palmata*, trop.: — **daisy**, *Bidens pilosa:* — **fence**, *Bauhinia pauletia:* — **sleepers** or **railroad cross ties**, in Br. and many other countries, Eur. red deal *Pinus sylvestris*, is much used after pressure creosoting, as are other coniferous woods: *Eucalyptus* is much used in Aus., and in Ind. teak, deodar, pyinkado, chir and sal: — **vine**, *Ipomoea pes-caprae.*

Rainbow leaf *Smodingium argutum*, S.Af.: **rain lily**, *Cooperia drummondii*, N.Am.: **prairie** — —, *C. pedunculata:* **rain tree**, *Samanea* (= *Enterolobium; Pithecellobium*) *saman*, trop. S.Am. widely cult. in trop. as orn. and shade tree: the ejection of moisture by insects (cicadas) on the tree accounts for the name : other so-called rain trees are spp. of *Andira*, *Oreodaphne* and *Koelreuteria.*

Raisin *Vitis vinifera*, culinary raisins are the small dried berries of a certain var. of white grape: sultanas or sultana raisins are from a seedless var.: — **tree**, *Hovenia dulcis*, Jap., ed. fr.

Rajah cane *Eugeissona minor*, Malaya, the stilt rs. were much used at one time for walking sticks and umbrella handles.

Rakkyo *Allium chinense*, much grown and esteemed for pickling in the Orient.

Rambutan *Nephelium lappaceum*, Malaya, tree w. ed fr., subacid pulp, much esteemed.

Ramie *Boehmeria nivea*, Orient, b. yields an imp. commercial fibre, also called China grass or rhea, a fine strong durable fibre: ls, white below: China and Jap.: **green** —, *B. nivea* var. *tenacissima*, ls. green below, Malaya.

Ramin *Gonystylus bancanus* and allied spp., Malaya, Sarawak, a much used lightweight commercial timber.

Ramonda Stemless herbs, S.Eur., showy fls. cult.

Ramontchi (Ind.) *Flacourtia ramontchi*, small spiny tree w. ed. fr. widely cult. in trop., often as a hedge, also called governor's plum.

Ramp *Allium tricoccum*, N.Am.

Rampion *Campanula rapunculus*, Medit., W.As., r. ed., ls. used in salads, sometimes cult.; *Phyteuma orbiculare*, Eur., fleshy rs. used as veg. and in salads.

Ram's head *Cypripedium arietinum:* — **horn**, *Proboscidea*, *Martynia:* — **horns**, *Orchis mascula.*

Ramsons *Allium ursinum.*

Ramsted (Am.) *Linaria vulgaris.*

Ramtil (Ind.) *Guizotia abyssinica*, an oil seed (Niger seed).

Randia Trop. As. and Af. shrubs, some w. very long corolla tubes, cult., some spp. w. ed. frs. or used locally in dyeing.

Rangoon creeper *Quisqualis indica*, a vigorous creeper, cult., trop.

Ranunculus, garden *Ranunculus asiaticus*. Other spp. also cult.

Raoul grass *Rottboellia exaltata*, Old World trop., a tall coarse grass also called 'itch grass' or 'corn grass', a troublesome weed in Louisiana sugar cane and W.I.

Rapanea Adopted as a trade name for timber of *Rapanea* spp., E.Af.

Rape *Brassica* spp. **Eur.** or **German** —, *B. napus* subsp. *oleifera*, much grown as a green feed for livestock; seed a source of oil (colza oil, an illuminant) and used for feeding poultry and cage birds: **Ind** — or **tori**, *B. napus* var. *dichotoma*, imp. oil seed crop on the plains of Ind. in the cold season. — **kale**, similar to ordinary kale but more foliaceous and less winter hardy.

Raphia *See* raffia.

Raphides Needle-shaped crystals in the tissues of pls., as in the Araceae, may cause skin irritation.

Raphiolepis *R. umbellata* (=*R. japonica*), Jap., orn. shrub, cult.

Rasamala *Altingia excelsa*, trop. As., Asian trade name for the liquid storax obtained from the tree.

Raspberry *Rubus* spp. The **common Eur.** — in its many forms or vars., is derived from *Rubus idaeus*, Eur. inc. Br. Am. raspberries are derived from *R. occidentalis* and *R. strigosus*. There are many other raspberries or so-called raspberries in other parts of the world – **Am. red** —, *R. strigosus:* **black** —, *R. occidentalis*, N.Am.: **Ceylon** —, *R. albescens:* **Ind.** —, *R. lasiocarpus*, little flavour: **Mauritius** —, *R. rosifolius:* **Mysore** —, *R. albicans:* **purple flowering** —, *R. odoratus*, N.Am.: **purple cane** —, *R. parviflorus:* **Queensland** —, *R. probus:* — **rose**, *R. odoratus*, N.Am.: **strawberry** —, *R. illecebrosus*, Jap., cult.: **W.I.** —, *R. rosifolius:* **yellow Himal.** —, *R. ellipticus*.

Rasp fern (Aus.) *Doodia* spp., small decorative ferns: —**y root orchid**, *Rhinerrhiza divitiflora*, Aus.

Rata *Metrosideros robusta*, N.Z., hard, heavy, durable timber w. many uses.

Rati (Ind.) *Abrus precatorius*, 'crabs eyes', orn. seeds.

Rat's bane (W.Af.) *Dichapetalum toxicarium* (=*Chailletia toxicaria*) kernel used in Sierra Leone to poison rats and other animals, sold on markets as 'broke back': — **tail**, *Antholyza ringens:* — — **cactus**, *Aporocactus flagelliformis* (=*Cereus flagelliformis*), Mex., crimson fls., cult.: — — **fescue**, *Vulpia bromoides*, Aus.: — — **grass**, *Sporobolus elongatus*, Aus.: — — **orchid**, *Dendrobium teretifolium*, Aus.: **rat root**, *Chiococca alba*, N.Am., rs. have rat-like odour.

Rattan cane Or rotan cane, the stripped stems of various climbing palms in trop. As., notably spp. of *Calamus*, *Daemonorops*, *Korthalsia*, *Plectocomia* etc. These canes are very strong and elastic and have many uses, whole or split (for baskets and fancy articles). They may be seen as chimney sweeps' and drain sweepers' rods (*see* Burkill, I. H. (1935) *Dictionary of the economic products of the Malay Peninsula*, 2402 pp.)

Rattle bells or **box** *Crotalaria*, N.Am.: — **bush**, *Crotalaria incana*, W.I.: — **pod**, *Crotalaria retusa:* **red** —, *Pedicularis palustris* Br.: — **top**, *Cimicifuga* spp.: — **weed**, *Cassia bauhinioides*, SW. U.S.; *Baptisia tinctoria*, S. U.S.: **yellow** or **hay** —, *Rhinanthus minor*, Br.

Rattlesnake fern *Botrychium virginianum*, N.Am.: — **grass**, *Glyceria canadensis*, N.Am.: — **master**, *Liatris scariosa*, *L. squarrosa*, *Eryngium yuccifolium*, reputed cures for snakebite: — **plantain**, *Goodyera* spp., N.Am.: — **root**, *Prenanthes* spp.: — **weed**, *Daucus pusillus*.

Rauli *Nothofagus procera*, Chile, a commercial timber.

Raupo *Typha angustifolia*, N.Z.

Rauvolfia or **Rauwolfia** *Rauvolfia serpentina*, a shrub of Ind., Burma, Thailand and Indonesia, the r. containing several alkaloids, notably reserpine, and long used in local medicine: *R. vomitoria*, trop. Af. (Af. rauvolfia) also yields reserpine and it may be made synthetically. The drug has been used for hypertension (high blood pressure) and in chronic mental illness. It can produce undesirable side effects, nasal congestion, lethargy etc.

Ravenna grass Or plume grass, *Erianthus ravennae*, N.Am.

Ravine orchid *Sarcochilus fitzgeraldii*, Aus.

Ravison French name for charlock or similar seeds (oil seeds).

Raw beef tree *Terminalia arjuna*, Ind. or E.I., b. red and juicy when cut.

Rayo (W.I.) *Cordyline terminalis*, trop. As., orn. ls., cult., many vars.: **white —**, *Dracaena fragrans*.

Razor grass (W.I.) *Scleria pterota*, a sedge.

Red acaroid *Xanthorrhoea australis:* — **alder**, *Alnus rubra:* — **ash**, *Fraxinus pubescens:* — **bay**, *Persea carolinensis:* — **beaks orchid**, *Lyperanthus nigricans*, Aus.: — **bean**, *Kennedia rubicunda, Dysoxylum muelleri*, Aus.: — **beech**, *Nothofagus fusca:* — **birch**, *Betula nigra:* — **box**, *Tristania conferta:* — **buckeye**, *Aesculus pavia:* — **bud**, *Cercis canadensis:* — **cardinal**, *Erythrina arborea:* — **cedar**, *Toona ciliata:* — **clover**, *Trifolium pratense:* — **currant**, *Ribes rubrum; Coprosma quadrifida*, Aus.; *Rhus legatii*, S.Af.: — **elm**, *Ulmus alata:* — **fir**, *Abies magnifica:* — **gum**, *Eucalyptus rostrata; Liquidambar styraciflua:* — **head**, *Asclepias curassavica*, W.I.: — **hot cat's tail**, *Acalypha hispida:* — **hot poker**, *Kniphofia* spp., S.Af.; *Norantea guianensis*, W.I.: — **ironwood**, *Lophira alata*, W.Af.: — **ivory**, *Phyllogeiton* (=*Rhamnus*) *zeyheri*, S.Af.: — **leg grass**, *Bothriochloa ambigua*, Aus., a weed: — **locust**, *Daubentonia punicea:* — **mangrove**, *Rhizophora mangle:* — **maple**, *Acer rubrum:* — **moonseed**, *Cocculus carolinus:* — **oak**, *Quercus rubra:* — **pear**, *Scolopia mundtii:* — **pine**, *Dacrydium cupressinum:* — **poppy**, *Papaver rhoeas:* — **posy**, *Boophone disticha:* — **puccoon**, *Sanguinaria canadensis:* — **sandalwood**, *Adenanthera pavonina, Pterocarpus santalinus:* — **rattle**, *Pedicularis palustris:* — **root**, *Ceanothus americanus, Lachnanthes caroliana* (= *L. tinctoria*): — **spider lily**, *Lycoris radiata:* — **star clusters**, *Pentas coccinea:* — **thistle**, *Emilia coccinea:* — **top grass**, *Agrostis alba, Triodia flava*, N.Am.; *Rhynchelytrum repens* (=*Tricholaena rosea*), S.Af.: — **treasure**, *Crassula falcata*, S.Af.

Redwood Name given to several different woods that are red or reddish in colour, Calif. redwood being perhaps the best known. **Af. —**, *Adina microcephala:* **Andaman —**, *Pterocarpus indicus:* **Braz. —**, *Brosimum paraense, Caesalpinia brasiliensis:* **Calif. —**, *Sequoia sempervirens*, an enormous tree reaching over 100 m in height and 3–8 m in diam. at the base w. v. thick b.: wood soft: **Ind. —**, *Soymida febrifuga; Chukrasia tabularis:* **W.I. —**, *Guarea guidonia:* **Zambesi —**, *Baikiaea plurijuga*.

Reed The common reed, *Phragmites australis* (=*P. communis*), common in wet places, more or less cosmop.: much used for thatching. The Spanish reed, *Arundo donax*, S.Eur., is noted for its use in musical instruments which goes back for at least 5000 years. It is also called bamboo reed, Danubian reed, giant reed, Italian reed and Provence reed. **—mace**, *see* bulrush.

Rehmannia Large fl. herbs, Orient., cult., esp. *R. angulata*.

Remijia S.Am. trees and shrubs, allied to *Cinchona*, b. of some contains quinine.

Renanthera Trop. As. orchids, showy fls., cult., esp. *R. coccinea*.

Rengas *Gluta renghas*, Malaya, tree, latex causes blisters or sores: sawdust or axe chips may also be harmful.

Rennet, vegetable *Withania coagulens*, berries used as rennet for coagulating milk in Afghanistan.

Resin bush (S.Af.) *Euryops multifidus.*

Resins, vegetable The term resin is not easy to define as the natural resins differ so much among themselves; nevertheless they constitute a distinct group of pl. products easily recognizable in practice. Unlike the gums they are insoluble in water (an age-old use being the caulking of boats) and are inflammable, hence their early use as torches in some countries. Their use in varnishes or protective coatings also goes back to early times. The Egyptians smeared their mummy cases with resin and the Incas of S.Am. used embalming resins. Fragrant resins such as frankincense and myrrh were used in Biblical times and were imp. articles of trade in Medit. countries and As. Minor before the Christian era. They also had med. uses and were employed in making perfumes and unguents.

Pl. resins occur very widely in the veg. kingdom, in any part of the pl., but are seldom present in sufficient quantity to make collection worthwhile. Resin is generally secreted in pl. tissue in special cavities or canals lined with a layer of secretory cells (resinogenous layer). These canals are often of a much branched or anastomosing nature w. the result that when one is severed resin flows to the wound from a considerable distance. It is this that makes tapping for resin w. pines and other resin-producing trees possible. The three most important families for the bulk or commercial production of resin are the Pinaceae (pine resin, kauri resin), Leguminosae (copals) and Dipterocarpaceae (dammars). The Burseraceae are also notable as a resin- or balsam-yielding family and the Umbelliferae for scented and medicinal resins such as asafoetida, galbanum, ammoniacum etc.

W. the advent of synthetic resins great changes took place in the demand for and utilization of natural resins, particularly in paint and varnish manufacture where the copals and dammars have been much used. For some purposes a combination of natural and synthetic resins has proved v. successful. Pine resin or rosin, after distillation of the turpentine, is produced in greater quantity and is more extensively used (in a variety of industries) than any other natural resin. Large quantities are used in the paper industry for paper sizing, to impart lustre and weight and to hinder the absorption of ink or moisture. In combination with alkalis rosin is also used in soap manufacture. The main producing countries are the SE. U.S. and France (the Landes region). Some Medit. countries and Ind. produce lesser amounts.

The following may be included among the more notable veg. resins – acaroid or accroides, amber, ammoniacum, asafoetida, benzoin, Canada balsam, colophony or pine resin, copal, copaiba, dammar, dragon's blood, elemi, euphorbium, frankincense, guaiacum, kauri, lac (insect resin, *see* lac), ladanum, mastic, myrrh, natural lacquers, Peru balsam, rosin (*see* colophony), sandarac, storax, tolu balsam. (Howes, F. N. (1949) *Vegetable gums and resins,* 188 pp.)

Restharrow *Ononis repens:* **prickly** —, *O. spinosa:* **small** —, *O. reclinata.*

Restrepia Trop. Am. orchids, free fl., cult.

Resurrection plant Several pls. that appear dead after drought or a dry season but revive again w. rain or moisture have acquired this name. The best known is probably the Rose of Jericho, *Anastatica hierochuntica*. Another well-known one is *Selaginella lepidophylla* of S. Texas and Mex. It curls up into a tight ball with drought, but with rain expands again and continues to grow. Both these pls. are sold as novelties. *Myrothamnus flabelliformis* is a resurrection pl. of southern Af. Resurrection ferns include *Polypodium polypodioides*, N.Am. and *Cheilanthes* (=*Nothoclaena*) *eckloniana*, S.Af. The Br. moss *Leucobryum glaucum* sometimes behaves like a resurrection pl.

Retamo wax *Bulnesia retamo*, S.Am., a shrub abundant in parts of Argen.: the wax has properties similar to carnauba wax.

Retem *Retama* (=*Lygos* or *Genista*) *raetam*, As. Minor, a desert shrub, wood makes good charcoal.

Retting The rotting away of soft tissues, usu. by long immersion in water: jute and other fibres are prepared in this way.

Rhapis Oriental palms mainly dwarf w. reed-like stems, orn., cult., esp. *R. excelsa*, ground rattan cane.

Rhatany root Or Peruvian rhatany, *Krameria triandra*, r. med., astringent and tonic: **Braz.** or **Para —**, *K. argentea*.

Rhazya *R. orientalis*, Orient, pretty blue fls. cult.

Rhea *See* ramie.

Rhine berry *Rhamnus catharticus*.

Rhinoceros bush Or rhenoster bos, *Elytropappus rhinocerotis*, S.Af., a characteristic pl. of the Karoo and a weed in some places, ousts desirable fodder pls.

Rhipsalis Epiphytic cacti w. small fls., trop. Am., As. and Af.

Rhode Island bent grass *Agrostis vulgaris*.

Rhodes grass *Chloris gayana* S.Af., cult. for pasture and fodder in warm cli.

Rhodesian ash Or wild ash, *Burkea africana:* — **copalwood**, *Guibourtia* (=*Copaifera*) *coleosperma*, a commercial timber: — **wisteria**, *Bolusanthus speciosus*.

Rhodochiton *R. atrosanguineus*, Mex. orn. climber w. blood red fls.

Rhododendron Along with the rose the rhododendron has done more to beautify gdns. of temp. regions than any other shrub. The genus, which now includes the azaleas (deciduous and with 5 instead of 10 stamens) occurs mainly in the N.Hemisph. The Chinese–Tibetan–Himal. region is the main centre and the source of the most beautiful spp. The many gdn. hybrids are derived largely from the following spp. – *arboreum, catawbiense, caucasicum* and *ponticum*, and to a less extent from *fortunei, griffithianum, thomsonii* and *maximum*. **Dahurian —**, *R. dauricum*, Siberia, long cult., the earliest to fl.: **Pontic —**, *R. ponticum*, Spain and Portugal, As. Minor etc. naturalized else where, inc. Br. The wood of the larger rhododendrons is hard, strong and close-grained, used for tool handles, that of *R. maximum*, N.Am., for tobacco pipes. *See Rhododendron* in Willis.

Rhodostachys S.Am. bromeliads w. spiny ls. and blue fls., cult.

Rhodotypos *R. scandens* (=*R. kerrioides*), Jap., cult. orn. shrub, white fls.

Rhoeo *R. spathacca*, C.Am., boat lily, cult. orn. foliage pl., ls. purple on under surface.

Rhubarb *Rheum rhaponticum*, ordinary or culinary rhubarb, cult. in most temp. countries for the fleshy l. stalks eaten stewed w. sugar as dessert, also canned, made into jam and used for home-made wine. Med. rhubarb is derived mainly from *Rheum palmatum* and *R. officinale.* Its action is that of a mild purgative, similar to senna. **Chinese** —, *R. palmatum* var. *tanguticum:* **E.I.** —, old name for Chinese —: **English** —, *R. rhaponticum:* **Himal.** or **Ind.** —, *R. emodi:* **monk's** —, *Rumex alpinus:* **prickly** —, *Gunnera manicata:* **Siberian** —, *Rheum rhaponticum:* **Tibetan** —, *R. officinale:* **Turkey** —, old name for Chinese rhubarb: **wild** —, *Rumex hymenosepalus,* Am.

Rhynchostylis Indo-Malayan orchids w. large fls., cult.

Ribgrass *Plantago lanceolata,* a troublesome weed esp. on lawns: — **wort,** *Plantago* spp.

Ribbon fern *Pteris cretica, Vittaria* spp.: — **grass,** *Phalaris arundinacea* var. *variegata,* a variegated form of reed canary grass, an old gdn. favourite, also called gardener's garters: — **wood** or **tree,** *Plagianthus betulinus,* N.Z., b. fibre used by Maoris; *Euroschinus falcata,* Aus., a local furniture wood.

Rice *Oryza sativa.* Perhaps the world's most imp. food pl. for it feeds more human beings than any other cereal or food pl. It is cult. in the trop. and subtrop. where conditions are suitable. There are basically two main types of rice, *lowland* rice grown under water and *upland* rice grown like any other cereal. There are numerous vars., mainly with starchy grain, but some are glutinous and used in the East for sweetmeats. In the Orient rice is the basis of alcoholic beverages. In Western countries rice or rice flour is used in cooking and for breakfast foods. It is also a commercial source of starch. **Can. wild** —, *Zizania aquatica:* — **cutgrass,** *Leersia oryzoides,* N.Am.: — **flower,** *Pimelea linifolia,* Aus.: — **grass,** *Spartina* × *townsendii:* **Indian** — (N.Am.), *Zizania aquatica:* **jungle** —, *Echinochloa colonum* (sharma millet): **Manchurian water** —, *Zizania aquatica:* **mountain** —, *Oryzopsis* spp., N.Am.: — **paper,** *Tetrapanax papyriferum,* paper-like material made from pith, used by Chinese artists (has nothing to do with the cereal rice). **Malayan** — **paper,** *Scaevola koenigii:* **Tuscarora** —, *Zizania aquatica:* **W.Af.** —, *Oryza glaberrima,* an indigenous cult. rice: **wild** —, *Zizania aquatica,* Can.

Richweed (Am.) *Pilea* spp.

Rimu *Dacrydium cupressinum,* N.Z., an imp. commercial timber: **mountain** —, *D. laxiflorum,* N.Z., perhaps the smallest conifer, useful for checking soil erosion.

Ringworm bush *Cassia alata,* shrub, cosmop., trop., ls. used for ringworm in W.I., in W.Af., l. juice used for skin complaints: — **root,** *Rhinacanthus communis.*

Ripple grass (Am.) *Plantago lanceolata, P. major.*

Rivea *R. corymbosa,* Mex., seeds med., used in rituals.

River bells (S.Af.) *Phygelius* spp.: — **bank grape,** *Vitis vulpina* N.Am.: — **birch,** *Betula nigra:* — **cane,** *Phragmites australis* (= *P. communis):* — **marble,** *Calophyllum antillarum,* W.I., hard durable wood: — **oak,** *Casuarina stricta,* N.Am.; *Heterodendrum deifolium,* Aus.: — **plum,** *Prunus americana:* — **weed,** *Podostemum* spp., N.Am.

Roanoke bells *Mertensia virginica,* N.Am.

Roast beef plant *Iris foetidissima* Br., large orange seeds in capsule.

Robin-run-in-the-hedge *Glechoma hederacea:* —**'s plantain,** *Erigeron pulchellus,* N.Am.: — **red breast bush,** *Melaleuca lateritia,* Aus.

Robinia False acacia or black locust (Am.), *Robinia pseudoacacia* N.Am., widely cult., as a timber or orn. tree in other countries inc. Br.: wood hard and tough, exceptionally durable in the ground, tree often grown for coppice: fls. fragrant, a good nectar or honey source in some countries.

Roble High class furniture wood from various botanical sources, *Tabebuia pentaphylla*, C.Am., and other spp., *Torresea caerensis*, S.Am., *Platymiscium pinnatum*, S.Am., W.I.

Rocambole Or giant garlic, *Allium scorodoprasum*, bulbs used for flavouring.

Rochea Cult. orn. fls. S.Af. *R. coccinea*, w. waxy scarlet fls. the best known.

Rock balsam *Peperomia*, W.I.: — **brake**, *Polypodium vulgare:* **curled — brake**, *Cryptogramma crispa:* — **broom**, *Genista:* — **carrot**, *Thapsia edulis:* — **cress**, *Arabis:* — **dammar**, *Hopea odorata:* — **daisy**, *Perityle*, SW. U.S.: — **elm**, *Ulmus thomasii:* — **foil**, *Saxifraga:* — **fern**, *Cheilanthes*, Aus.: — **jasmine**, *Androsace:* — **orchid**, *Dendrobium speciosum*, Aus.: — **pink**, *Dianthus kitaibelii:* — **rose**, *Helianthemum, Cistus*, cult. orn. shrubs favoured for dry sunny places: — **rose, annual**, *Tuberaria guttata:* — **sage**, *Lantana involucrata*, W.I.: — **spray**, *Cotoneaster horizontalis:* — **tobacco**, *Primulina tabacum*, China, tobacco scented: — **trumpet**, *Macrosiphonia*, N.Am.

Rocket Or dame's violet, *Hesperis matronalis*, common gdn. pl.: **dyer's —**, *Reseda luteola:* **garden —**, *Eruca sativa:* **London —**, *Sisymbrium irio:* **prairie —**, *Erysimum asperum:* **purple —**, *Iodanthus pinnatifidus:* **salad —**, *Eruca sativa*, oil and fodder crop in N.Ind.: **sand —**, *Diplotaxis muralis:* **sea —**, *Cakile maritima:* **wall —**, *Diplotaxis tenuifolia:* **wall —, annual**, *D. muralis:* **white —**, *D. erucoides:* **yellow —**, *Barbarea stricta*.

Rocky Mt. cherry *Prunus melanocarpa:* — — **fir**, *Abies lasiocarpa:* — — **flax**, *Linum lewisii:* — — **grape**, *Mahonia aquifolium:* — — **white pine**, *Pinus flexilis:* — — **Woodsia**, *Woodsia scopulina*.

Rodgersia Cult. gdn. pl., E.As., fls. in large panicles.

Rodriguezia Trop. Am. orchids w. showy fls., cult.

Rodwood *Eugenia monticola*, W.I.

Rogation flower *Polygala vulgaris*, milkwort, once used for orn. in Rogation week.

Roko *See* iroko.

Romneya *R. coulteri*, Calif., large white fls., cult.

Romulea Old World genus of crocus-like pls., cult.

Rondeletia *R. odorata*, C.Am., W.I., cult. orn. shrub w. scented vermilion fls.

Rooi els (S.Af.) Red alder, *Cunonia capensis*, reddish wood has been used for furniture: — **kappie** (S.Af.), *Satyrium carneum:* — **stompie**, *Mimetes lyrigera*, S.Af. shrub, b. once used for tanning.

Room plants *See* house pls.

Root crops or **root vegetables** *Temperate* or *subtemperate:* artichoke (Chinese and Jerusalem), beetroot, carrot, celeriac, chervil, chicory, Chinese yam, garlic, gobo or ed. burdock, horse radish, leek, oca-quina (Peru), okra, onion, parsley, turnip-rooted parsnip, Peruvian parsnip, potato, radish, ratala (*Plectranthus*, Sri Lanka), rocambole, salsify, scorzonera, shallot, sugar beet, swede, tiger nut, turnip. *Tropical:* arrowroot, cassava, cocoyam (dasheen, eddoes or tannia), edible canna, Fiji arrowroot, ginger, sweet potato, turmeric, yam-bean, yam.

Many of those listed as temp. may be, and often are, grown at the higher altitudes in the trop.

Rope Among the better-known veg. fibres used for making rope are hemp, Manila hemp or abaca, sisal, phormium or N.Z. flax, sunn hemp, Mauritius hemp, cotton and coir (in Ind.): — **bark**, *Dirca palustris*, N.Am.: — **mangrove**, *Hibiscus tiliaceus* (= *Pariti tiliaceum*), pantrop.: — **vine**, *Anodendron paniculatum*, *Tanaecium jaroba*, W.I.

Roridula Two spp. of S.Af. insectivorous pls. *See Roridulaceae* in Willis.

Rosary, seeds used for *See* beads. — **pea**, *Abrus precatorius*.

Rose *Rosa* spp. Roses and rhododendrons have probably done more to beautify temp. gdns. than any other two groups of pls. The genus *Rosa* consists of some 250 spp. Many of these are in cult. and there are endless hybrids, vars. or cultivars. New vars. or forms are being provided constantly by breeders and enthusiasts. Below are some of the better-known spp. Among the economic uses of the rose is the production of otto or attar of rose, esteemed in perfumery, and rose-hip syrup, used as a pleasant tasting vitamin source, esp. w. children. Another use, a minor one, is the dried fragrant petals as a basis of potpourri. **Abyssinian** —, *Rosa sancta*: **apple** —, *R. pomifera*: **Arkansas** —, *R. arkansana*: **Austrian** —, *R. lutea*: **Ayrshire** —, *R. arvensis*: **Banksian** —, *R. banksiae*: **burr** —, *R. microphylla*: **cabbage** —, *R. centifolia*: **Calif.** —, *R. californica*: **Cherokee** —, *R. laevigata*: **Chinese** —, *R. indica*: **damask** —, *R. damascena*: **dog** —, *R. canina*: **Eglantine** —, *R. rubiginosa*: **Fortune's** —, *R. fortuneana*: **French** —, *R. gallica*: **Fujiyama** —, *R. nipponensis*: **Irish** —, *R. hibernica*: **Jap.** —, *R. rugosa*: **Macartney** —, *R. bracteata*: **musk** —, *R. moschata*: **Otto** or **Attar of** —, *R. damascena* mainly: **pasture** —, *R. humilis*: **prairie** —, *R. setigera*: **Ramanas** —, *R. rugosa*: **Scotch** —, *R. spinosissima*.

Apart from the true roses or spp. of *Rosa* there are numerous pls. that have acquired the name of 'rose' because of the resemblance of the fl. to a rose or for some other reason. — **acacia**, *Robinia hispida*, Am., orn. shrub w. rose-coloured fls.: **Af. dog** —, *Oncoba kraussiana*, S.Af.: **Alpine** —, *Rhododendron ferrugineum*: **Andes** —, *Bejaria racemosa*: — **apple**, *Syzygium* (= *Eugenia*) *jambos*, *S. malaccense*, E.I., ed. fr.: — **bay**, *Nerium oleander*, *Epilobium* (= *Chamaenerion*) *angustifolium*: — **campion**, *Lychnis*: **changeable** —, *Hibiscus mutabilis-*cult. orn. shrub: **Chinese** —, *Hibiscus rosa-sinensis*: **Christmas** —, *Helleborus*: — **crown**, *Sedum rhodanthum*, N.Am.: **desert** —, *Adenium*: **elf** —, *Claytonia virginiana*: — **gentian**, *Sabatia angularis*, N.Am.: **guelder** —, *Viburnum opulus*: **Jam.** —, *Blakea trinervia*: — **of Jericho**, *Anastatica hierochuntica*, see resurrection pl.: **Juno's** —, *Lilium candidum*: — **leek**, *Allium canadense*: **Lenten** —, *Helleborus orientalis*: **Malay** —, *Syzygium malaccense*: — **mallow**, *Hibiscus rosa-sinensis*: **native** —, *Boronia serrulata*, Aus.: — **pink**, *Sabbatia angularis*, N.Am.: **river** —, *Bauera rubioides*, Aus.: **rock** —, *Cistus*, *Helianthemum*: — **root**, *Sedum rosea*: — **of Sharon**, *Hibiscus syriacus*: **stock** —, *Sparmannia africana*, S.Af.

Roseau Black —, *Bactris major*, W.I., a many-stemmed palm w. long black spines: **white** —, *Gynerium sagittatum*, W.I., a stout reed, 3–6 m, used for thatching.

Roselle *Hibiscus sabdariffa*, cult. trop., red fleshy calyx used for jellies, sauces and a beverage: ls. used as spinach: stem yields a fibre, like kenaf.

Rosemary *Rosmarinus officinalis*, S.Eur., evergreen shrub, old gdn. favourite: ls. dried and used for flavouring soups, stews, etc. and in potpourri: oil used in cosmetics and soaps: **bog** —, *Andromeda polifolia:* **marsh** —, *Limonium vulgare* (= *Statice limonium*): **wild** —, *Eriocephalus africanus*.

Rosewings *Calopogon pulchellus*, N.Am., tuberous rooted terrestrial orchid.

Rosewood Several timbers that show some resemblance to the original Braz. rosewood (*Dalbergia*), often seen in antique furniture, bear the name. **Af.** —, *Guibourtia demeusei:* **Aus.** —, *Dysoxylum fraserianum, Heterodendron oleiferum:* **Bahia** —, *Dalbergia nigra:* **Braz.** —, *D. nigra:* **Honduras** —, *D. stevensonii:* **Ind.** —, *D. latifolia:* **Madag.** —, *D. greveana:* **New Guinea** —, *Pterocarpus indicus;* **Rio** —, *Dalbergia nigra:* **W.Af.** —, *Pterocarpus erinaceus*.

Rosin Or colophony, is pine resin obtained by tapping the trunks of various pines, notably the longleaf pine *Pinus palustris*, SE. U.S. and maritime or cluster pine, *P. pinaster*, SW. France: also the Aleppo pine, *P. halepensis*, Medit. and chir pine, *P. roxburghii*, Ind. Rosin has many uses, esp in paper sizing.

Rotang or **rotan** Malay name for a climbing palm: *see* rattan.

Rotenone The active principle in certain insecticidal pls. such as *Derris* and *Lonchocarpus*.

Rouge plant Or safflower, *Carthamus tinctorius*, cult.: *Rivina humilis*, W.I., cult., frs. bright scarlet.

Roupala Handsome trop. Am. trees and shrubs, cult.

Roundwood *Pyrus americana*, N.Am.

Rowan, roan Or mountain ash, *Sorbus aucuparia*, Eur. inc. Br., wild and cult. orn. tree, withstands harsh conditions, characteristic of Scottish highlands: wood resembles pear or apple.

Royal fern *Osmunda regalis*, Euras.: — **palm**, *Roystonea* (= *Oreodoxa*) *regia*, cult. trop., much used for avenues.

Rozelle *See* roselle.

Rubber plants Rubber, or caoutchouc, is the coagulated latex of various trees, shrubs, climbers or other pls., mainly of warm cli. Many of these pls. have been commercial sources of rubber in the past but at present practically the only commercial source of natural rubber is that obtained from *Hevea brasiliensis*, commonly called plantation, Hevea or Para rubber. As the name indicates the tree is native to Braz., but it has been widely cult. in other trop. countries, esp. the East, since the latter part of last century, its introduction to the East having been largely due to the Royal Botanic Gdns., Kew. The tree is quick-growing and erect, requiring trop. conditions with a rainfall of not less than about 180 cm. It prefers a soil of pH 5.5–6.5. There are now many improved high-yielding forms or cultivars of the tree. One of the reasons why *Hevea* has gained ascendancy over other rubber-yielding trees is because it shows 'wound response' while other rubber trees do not, i.e. tapping stimulates the further production of latex. The collected latex from tapping is coagulated by means of acid and made into sheet rubber on the plantation, or the latex may be exported in its liquid form to manufacturing countries.

The number of industrial and domestic uses of rubber is legion but the largest use in terms of tonnage is in the motor tyre industry. Each year sees more motor vehicles on the roads of the world and more tractors and trailers in use, all requiring tyres and tyre renewal. The Second World War acted as a great stimulus to the successful production of synthetic rubber, particularly in the U.S. It is favoured for special purposes, e.g. where rubber comes in contact with oil or petrol. Properties which account for the many and special uses of natural rubber are first its elasticity, combined with pliability, toughness, resilience to abrasion, impermeability and resistance to chemicals. An ·increasing amount of crumb rubber, or special formulation rubber, is prepared on estates. Crumb rubber has been used in bitumen mixes for road surfacing.

The exploitation of the so-called wild or jungle rubbers, especially in Af. and trop. Am., was at its height in the latter part of the 19th Century but ceased in the early part of the 20th Century, due to the rapid ascendancy of cult. plantation of Hevea rubber. Disruption of supplies of plantation rubber during war-time led to some further exploitation but this was only temporary. The following are some of the rubbers, other than Hevea, that have been exploited or have been of special interest in the past – **Abbo rubber** (W.Af.), *Ficus vogelii:* **Af.** —, *Funtumia elastica* and other spp.: **Assam** —, *Ficus elastica:* **Bitinga** —, *Raphionacme utilis:* **Bolivian** —, *Sapium aucuparium:* **Borneo** —, *Willughbeia firma:* **Ceara** —, *Manihot glaziovii* and other spp.: **C.Am.** —, *Castilloa elastica:* **chilte** —, *Cnidoscolus:* **Chinese** — **tree,** *Eucommia ulmoides:* **Colombia scrap** —, *Sapium* spp.: **Colorado** —, *Actinea richardsonii:* **Congo** —, *Ficus vogelii:* **couma** — (Braz.), *Couma* spp.: **Dahomey** —, *Ficus vogelii:* **E.Af.** —, *Landolphia* spp.: **Esmeralda** —, *Sapium jenmanii:* **Etanga** —, *Bitinga:* **Guayule** — (C.Am.), *Parthenium argentatum:* **golden rod** —, *Solidago* spp.: **India** or **Indian** —, *Ficus elastica:* **Intisy** —, *Euphorbia intisy:* **Iré** — (W.Af.), *Funtumia* spp.: **koksaghyz** —, *Taraxacum bicorne,* Russia: **krim-saghyz** —, *Taraxacum megalorhizon:* **Landolphia** — (trop. Af.), *Landolphia* spp.: **Lagos** —, *Funtumia elastica:* **Madag.** —, *Mascarenhasia, Cryptostegia, Marsdenia* and *Landolphia* spp.: **Mangabeira** —, *Hancornia speciosa:* **Manicoba** — =Ceara —: **Mex.** — = Guayule —: **Micrandra** —, *Micrandra* spp.: **milkweed** —, *Asclepias* spp.: **Orinoco scrap** —, *Sapium jenmani:* **Panama** —, *Castilloa elastica:* **Pernambuco** —, =Mangabeira —: **Rangoon** —, *Urceola maingayi:* **silk** —, *Funtumia elastica:* **tau-saghyz** —, *Scorzonera tausaghyz:* **tekesaghyz** —, *Scorzonera acanthoclada:* **tirucalli** —, *Euphorbia tirucalli:* **Tongking** —, *Bleekrodea tonkinensis:* **Ule** —, = Panama —: **vine** —, *Carpodinus* spp., *Cryptostegia* spp.: **W.Af. tree** —, *Funtumia elastica.*

Rubus Shrubby or herbaceous pls., often prickly, mainly N.Hemisph. some are imp. berry frs. *See Rubus* in Willis.

Rudbeckia N.Am. perens., cult., large fl. heads, conical disc often of contrasting colour.

Ruddy hood orchid (Aus.) *Pterostylis rufa.*

Ruddles Old name for marigold.

Rue *Ruta graveolens,* S.Eur. Culinary herb (perennial) w. strong flavour used for seasoning (sausages) but less so now than formerly, also for treating croup in poultry.

Ruellia Handsome free-fl. pls., many are good pot pls., e.g. *R. macrantha*.

Rukam *Flacourtia rukam*, Malaysia, spiny tree w. ed. fr.

Rum The potable spirit made from sugar cane: — **cherry**, *Prunus serotina*.

Running postman *Kennedia prostrata*, W.Aus.

Rupture wort *Herniaria glabra*, Br. name from fancied remedial powers.

Rush *Juncus effusus* and other spp., usu. in damp places. In Br. many of the former uses for rushes have died out, e.g. bags, saddles, beehives etc.: still used for seating chairs. Pith used for rush lights before the introduction of candles: **beak —**, *Rhynchospora:* **bul—**, *Typha:* **bulbous —**, *Juncus bulbosus:* **chair maker's —**, *Scirpus americanus:* **flowering —**, *Butomus umbellatus:* **— grass**, *Sporobolus*, N.Am.: **scouring —**, *Equisetum*, N.Am.: **sea —**, *Juncus maritimus:* **spike —**, *Eleocharis:* **wood —**, *Luzula*.

Rusha grass Or geranium grass. *Cymbopogon martinii*, Ind., yields esst. oil.

Russelia C. & S.Am., orn. fl. shrubs, cult., esp. *R. juncea* and *R. sarmentosa*.

Russian olive *Elaeagnus angustifolia*, Euras.: — **pigweed**, *Axyris amaranthoides:* — **thistle** or **tumbleweed**, *Salsola pestifera:* — **vine**, *Polygonum aubertii* (= '*baldshuanicum*'), fast-growing climber, quickly covers unsightly objects.

Rusty-back fern *Ceterach officinarum*, Br.: — **shield bearer**, *Peltophorum inerme* (=*P. ferrugineum*), trop. As., cult. orn. tree w. yellow fls. and masses of rust-coloured pods.

Rutabaga Swede or Swedish turnip, *Brassica napo-brassica*.

Rutin Med. product used for treating certain haemorrhagic disorders and obtained from various pl. sources, e.g. fls. of *Sophora japonica* from China, buckwheat, ls. of *Eucalyptus macrorrhyncha*.

Ryania *R. speciosa*, W.I., shrub, the stem wood yields an insecticide.

Rye *Secale cereale*. Imp. cereal in many temp. countries, esp. N.Eur., where poor sandy or acid soils prevail, unsuited for other cereals. Rye bread of many kinds is made from it, also potable spirit. It makes a good stock food. **Giant** or **Jerusalem —**, *Triticum polonicum*, C.Eur., flour unsuited for bread or macaroni: **wild —**, *Elymus* spp.

Ryegrass *Lolium* spp. Imp. agric. grasses, also used for playing fields and lawns. **Italian —**, *Lolium multiflorum* a valuable fodder grass and widely cult.: **perennial —**, *Lolium perenne*, a pasture grass of rapid growth, nutritious and high yielding, good for hay, grass drying or silage, several strains: **Russian wild —**, *Elymus junceus*, drought- and cold-resistant: **Wimmera —**, *L. rigidum*.

S

Sabadilla Or sevadilla, *Schoenocaulon* (=*Sabadilla*) *officinale*, C. & S.Am., seeds insecticidal, extract used in parasiticide ointments.

Sabicu *Lysiloma sabicu* and other spp., trop. Am., a mahogany-like wood.

Sabre bean *Canavalia ensiformis*, trop., widely cult.

Saccharum Tall trop. and subtrop. grasses, imp. in sugar cane breeding.

Saccoladium Trop. As. orchids, cult.

Sachaline knotweed *Polygonum sachalinense*, Sachalin Isle, a large orn. and forage pl., 2.5–3.5 m high.

Sack tree (Sri Lanka) *Antiaris toxicaria*.

Sacky-sac *Inga laurina*, W.I., ed. pod sold on markets.

Sacred plants In many parts of the world certain pls. or trees are held sacred or supposed to possess supernatural properties, but with the advance of education the tendency is for such beliefs to be relinquished. A good example of a sacred pl. is the sacred lotus or water lily of the East, *Nelumbo nucifera* (= *Nelumbium nelumbo*). The so-called 'sacred bamboo' of China is *Nandina domestica*, a shrub, with much the aspect of a bamboo with cream flowers and red berries. It is used in temples. Sacred basil, *Ocimum sanctum*, is sacred to the Hindu religion.

A number of trees are sacred throughout Ind. including the peepul, *Ficus religiosa*, of great significance to devout Buddhists, palas, *Butea monosperma* (= *B. frondosa*), esp. sacred to Brahmins, the night-fl. jasmine *Nyctanthes arbor-tristis*, and the temple tree, *Plumeria acutifolia*. In trop. Af. some tribes venerate the sausage tree, *Kigelia pinnata*.

Saxaoul *Haloxylon ammodendron*, W. & C.As., the hard heavy wood (sinks in water) is used for fuel and yields a green dye.

Safflower *Carthamus tinctorius*, cult. as an oil seed crop in Ind. and elsewhere: the yellow fls. at one time used as a yellow dye and saffron substitute.

Saffran hout (S.Af.) *Elaeodendron croceum*.

Saffron *Crocus sativus*, Eur., the world's most expensive spice, being very laborious to collect. It consists of the dried stigmas of the saffron crocus, prepared largely in Spain, some 2000 fls. being needed to produce 28.3 g (or 1 oz) of saffron: also used as a dye in ancient times: a popular colour for clothes with the women of Ancient Greece. It is used for flavouring and colouring certain foods, saffron cakes, buns or loaves being traditional in Cornwall.

Sagapenum *Ferula* spp., a fragrant oleo-gum-resin, once had med. uses.

Sage *Salvia officinalis*, S.Eur., a shrub: ordinary or culinary sage, a much used flavouring in Eur. and U.S., esp. for fatty meats, sausages, duck, goose and in stuffings: an imp. ingredient in spice formulations, esp. in Germany. Many sages or spp. of *Salvia* are grown in the fl. gdn. and for bedding. **Blue —**, *Salvia azurea*, N.Am.: **— brush**, a type of vegetation in SW. U.S. characterized by spp. of *Artemisia* noted for their fragrance: **— bush**, *Helichrysum diosmifolium*: **clary —**, *Salvia sclarea*, oil used in toilet waters: **Greek —**, *Salvia triloba*, used like ordinary sage or as an adulterant: **grey —**, *Atriplex canescens*, W. N.Am.: **Jerusalem —**, *Phlomis fruticosa*: **meadow —**, *Salvia pratensis*: **pineapple-scented —**, *Salvia rutilans*: **red-tipped —**, *Salvia horminum*, Medit. has coloured bracts: **red** or **scarlet —**, *S. coccinea*, *S. splendens*: **— rose** (W.I.), *Turnera*: **purple-leaved —**, an orn. form of culinary sage: **three-lobed —**, *S. triloba*, Medit., ls. used as tea substitute: **W.I. —**, *Lantana* spp.: **wood —**, *Teucrium scorodonia*: **— wood** (S.Af.), *Buddleja salvifolia*, *Tarchonanthus camphoratus*.

Sago palm Prickly — —, *Metroxylon rumphii*: **spineless — —**, *M. sagus*: In the E.I. and Malaya both are utilized for sago production, *M. sagus* being more extensively used and cult. The starch is removed from the stem pith with water and then dried and processed. Some other palms (*Arenga, Caryota*) have been used in the same way. Sago-like products have also been obtained from some cycads (*Cycas* and *Encephalartos*).

Saguaragy bark *Colubrina rufa*, Braz., med., used for fevers.

Saguaro Local name for the giant cactus, *Carnegiea gigantea*, S. U.S.

Sahagunia *S. strepitans*, Braz., yields good timber: ed. fr.

Sainfoin *Onobrychis viciifolia* (= *O. sativa*), Eur. inc. Br., wild and cult. mainly on chalk soils for sheep: pretty pink fls. a good nectar or honey source.

Saint Andrew's Cross *Hypericum:* — **Barbara's herb**, *Barbarea vulgaris:* — **Bernard's lily**, *Anthericum liliago:* — **Bruno's lily**, *Paradisea liliastrum:* — **Dabeoc's heath**, *Daboecia cantabrica:* — **George's herb**, *Valeriana officinalis:* — **Helena gum wood**, *Commidendrum gummiferum:* — **Ignatius' bean**, *Strychnos ignatii*, Philippines, seeds yield strychnine: — **John's bread**, *Ceratonia siliqua* (carob): — **John's bush**, *Psychotria nervosa*, W.I.: — **John's wort**, *Hypericum perforatum:* — **Joseph's lily**, *Lilium candidum:* — **Lucie cherry**, *Prunus mahaleb*, wood used for cherry tobacco pipes: — **Martin's fl.**, *Alstroemeria:* — **Patrick's cabbage**, *Saxifraga spathularis*, on mt. rocks in Ireland: — **Peter's cabbage**, *Saxifraga virginiensis*, N.Am.: — **Peter's wort**, *Ascyrum stans*, N.Am.

Saintpaulia *S. ionantha*, E. trop. Af., Af. or Usambara violet: a v. popular pot and house pl. derived from other species: fls. normally violet blue: many vars. and colour forms.

Saké or **saki** National alcoholic beverage of Jap., made from rice.

Sal *Shorea robusta*, Ind. a valuable timber, resembles teak.

Salab-misri *See* salep.

Salad oil Nominally olive oil, but other oils such as groundnut and maize used.

Salal *See* shallon.

Salep Or salab-misri. The dried starchy tubers of certain orchids, used for culinary or med. purposes, mainly in the East. **Ind.** —, *Eulophia campestris* and *E. nuda* mainly: **Persian** —, *Dactylorhiza latifolia* and *D. laxiflora:* **Eur.** —, *Orchis* spp.

Saligna gum *Eucalyptus saligna*, Aus., cult. in S.Af. and elsewhere for timber.

Saligot Or water chestnut, *Trapa natans*.

Salix *See* willow.

Sallow A name for certain willows (*Salix caprea*, *S. atrocinerea*, etc.): — **thorn**, *Hippophae rhamnoides*.

Sally, white (Aus.) *Eucalyptus pauciflora*.

Salmon berry *Rubus nutkanus*, N.Am., fr. ed.: *R. spectabilis*, N.Am. fr. ed.

Salmwood *Cordia alliodora*, trop. Am., W.I., a commercial timber, also called Ecuador laurel.

Salpiglossis S.Am., herbaceous pls. w. large fls., cult., modern vars. largely derived from *S. sinuata*, w. silky funnel-shaped fls.

Salpinga Trop. S.Am. pls. w. attractive colourful fol., cult.

Salsify *Tragopogon porrifolius*, Eur. a veg. w. a long fleshy tap root, the part eaten: young ls. used in salads: **Spanish** —, *Scorzonera hispanica*.

Salsilla (W.I.) *Bomarea edulis*, a handsome climbing pl. w. ed. tubers.

Saltbush *Atriplex* spp., *Kochia* spp., Aus., often grow in saline arid areas and are a good emergency stock food in drought.

Salt cedar *Tamarix pentandra*, SW. U.S.: — **fish wood**, *Sterculia caribaea;* *Machaerium robiniifolium*, W.I.: — **marsh grass**, *Puccinellia maritima* and other spp. a stoloniferous coastal grass useful for colonizing coastal mud in Br.: — **marsh mallow**, *Kosteletzkya virginica*, N.Am.: — **tree**, *Halimo-dendron argenteum*, Siberia: — **water grass**, *Spartina*: —**wort**, *Glaux maritima*, *Salsola kali*, *Salicornia*, N.Am.

Salvia Many spp. of this large genus are gdn. pls. others have med. uses.

Saman Or rain tree, *Samanea saman*, cult., trop., of rapid growth.

Sambo (W.I.) *Cleome*.

Samphire *Crithmum maritimum*, Eur. inc. Br., a seaside pl., fleshy ls. used in salads and pickles: **golden** —, *Inula crithmoides*, a pot-herb: **marsh** —, *Salicornia* spp.

Sampson (W.I.) or **Sampson palm** *Bactris sworderiana*.

Sanchezia Handsome trop. Am. perens. w. showy fls., cult.

Sand binding plants The only satisfactory means of fixing shifting sand dunes is by establishing veg. because mechanical measures, although useful for a time, are never lasting. The total area of the world occupied by sand dunes, mainly near the sea coasts or in deserts or semi-deserts, is enormous. It has been estimated by one writer at about 13,000 million hectares or nearly twice the total area of the U.S. (*U.S. Dept. Agric. Yearbook*, 1948, p. 70.)

The most satisfactory pls. to use in sand dune fixation vary according to local conditions and climate. Deep rooting, drought-resistant grasses or herbaceous pls. that will increase rapidly by means of creeping rhizomes or stolons, which r. and give rise to leafy growth, are among the best early colonizers. Different grasses are used under temp., subtrop. and trop. con-ditions. Marram grass (*Ammophila arenaria*) has been used with great success in many areas. Once an initial grassy or herbaceous cover has been obtained shrubs or woody vegetation, of a more permanent nature with deeper or more extensive rooting systems, are established. Among woody spp. that have proved useful in many countries are *Acacia cyanophylla*, *Acacia cyclops*, *Casuarina*, *Eucalyptus camaldulensis* and *Tamarix*. Other useful sand binders are *Cassia mimosoides* and the sensitive plant (*Mimosa pudica*).

Sand box tree *Hura crepitans*, trop. Am., orn. tree w. explosive fr. widely cult., trop. (*see Hura* in Willis): — **burr**, *Cenchrus tribuloides:* — **cherry**, *Prunus pumila*, N.Am.: — **dropseed**, *Sporobolus cryptandrus*, N.Am.: — **grass**, *Triplasis, Catapodium*, N.Am.: — **grape**, *Vitis rupestris*, N.Am.: — **jack**, *Quercus incana:* — **leek**, *Allium scorodoprasum:* — **myrtle**, *Leiophyl-lum buxifolium*, N.Am.: — **pear**, *Pyrus pyrifolia:* — **spurrey**, *Spergularia* spp., N.Am.: — **vine**, *Gonolobus* (=*Ampelasmus*) spp.: — **verbena**, *Abronia latifolia*, N.Am.: — **wort**, *Arenaria, Minuartia, Moehringia*, N.Am.

Sandalwood True or **Ind.** —, *Santalum album*, the fragrant wood distilled for oil, used med. and in perfumery: wood made into cabinets: **Af.** —, *Excoecaria africana:* **Aus.** —, *Eucarya acuminata* and *E. spicata*, distilled for oil: **red** —, *Adenanthera pavonina*, orn. tree, trop.; *Pterocarpus santalinus:* **W.I.** —, *Amyris balsamifera*.

Sandarac This resin is obtained from a small coniferous tree, *Tetraclinis articulata*, of N.Af. (mts. of Algeria and Morocco). Aus. sandarac is similar but is obtained from spp. of *Callitris*. The main use of sandarac is in the preparation of special varnishes, e.g. picture varnish. Formerly sandarac was used as pounce, before the arrival of blotting paper.

Sanders wood, red *Pterocarpus santalinus*, Ind. and SE. As., a good furniture wood.

Sandfly bush *Zieria smithii*, Aus.

Sandpaper tree *Cordia ovalis*, Angola: the stems of *Equisetum* have been used like sandpaper, also ls. of some spp. of *Ficus, Dillenia, Curatella, Aphananthe* etc.

Sang *Panax quinquefolius*, ginseng, N.Am.

Sanguinaria *See* bloodroot.

Sanguisorba *S. obtusa*, Jap., cult. orn. fl.

Sanicle, wood *Sanicula europaea*, Eur., once used med.

San Pedro (Peru) *Trichocereus pachanoi*, used med.

Sansevieria *See* bowstring hemp.

Sant pods (Sudan) *Acacia arabica*, used in leather tanning in N.Af.: were used in this way by Ancient Egyptians: dried pods contain about 30 % tannin.

Santa Maria *Parthenium hysterophorus*, SW. U.S., a weed: name also used for timber of *Calophyllum brasiliense*, C.Am. a commercial wood.

Santol *Sandoricum koetjape* (= *S. indicum*), S.E. Asia, ed. tree fr., not highly esteemed.

Santolina *S. chamaecyparissus*, S.Eur., cult., once used med. for ringworm.

Santonica *See* santonin.

Santonin Drug obtained from the unopened fl. heads of *Artemisia cina* (santonica) from Russian Turkestan and other spp. of *Artemisia*, notably Ind.: much used at one time for the expulsion of roundworms and threadworms but now largely superseded by other drugs.

Sanvitalia Cult. orn. fls., Mex., notably *S. procumbens*, grown as ann.

Sapele *Entandrophragma utile*, W.Af., imp. mahogany-like commercial timber: **heavy —**, *E. candollei*, W.Af., also called 'omu'.

Sap-green *Rhamnus catharticus:* **— wood**, younger outer wood of a trur ҁ or branch.

Sapodillo *Achras zapota* (= *Manilkara achras*), trop. Am., ed. fr. (chiko) widely cult.: coagulated resinous latex used for chewing gum: the tree yields a commercial timber.

Saponaria Many spp. cult. in the fl. gdn. (soapworts).

Saponin *See* soap pls.

Sapote Or marmalade plum, *Calocarpum sapota* (= *C. mammosum*), C.Am. an esteemed ed. tree fr., widely cult., trop.: **black —**, *Diospyros ebenaster*, Mex.: **green —**, *Calocarpum viride*, C.Am.: **white —**, *Casimiroa edulis*, C.Am. fr. the size of an orange, aromatic yellow pulp: **yellow —**, *Lucuma salicifolia*, Mex., C.Am., wild and cult. tree fr.

Sappan wood *Caesalpinia sappan*, trop. As., heartwood formerly a source of yellow dye.

Sapu (Sri Lanka) *Michelia. M. champaca*, fragrant fls. distilled for perfume: wood much used.

Sapucaia nut (Braz.) *Lecythis* spp., a choice dessert nut resembling the Brazil nut (*Bertholletia*) from large forest trees in the Amazon region: frs. large and woody (monkey pots).

Saraca *S. indica,* small tree, fls. in large clusters, scented at night, used in temples.

Sarcanthus Trop. As. orchids, cult.

Sarcochilus Trop. As. and Aus. orchids, cult.

Sarcocolla Or gum sarcocolla, *Astragalus* spp., Persia, used med. in the past.

Sarracenia N.Am. insectivorous pls., cult. *See Sarracenia* in Willis.

Sarsaparilla *Smilax* spp., C. & S.Am., the dried r. considered to have tonic properties: used less now than formerly: **true** or **Jam.** — *Smilax ornata* and *S. utilis.* Other kinds, mainly locally used, include – **Costa Rican** —, *Smilax ornata:* **false** —, *Hardenbergia violacea,* Aus.: **Honduras** —, *Smilax regelii:* **Mex.** —, *S. media, S. aristolochiifolia:* **Para** —, *S. spruceana:* **Tampico** —, *S. aristolochiifolia:* **wild** —, *S. glauca,* N.Am., *S. cumanensis,* W.I.: — **substitute,** *Menispermum canadense.*

Sarson *Brassica campestris* var. *sarson,* an imp. Ind. oil seed crop grown in the cold season.

Sasa Jap. name for dwarf bamboos: a genus of dwarf bamboos, some cult.

Sasimua (E.Af.) *Hypericum revolutum,* orn. fl. shrub.

Sassafras *Sassafras albidum* (= *S. officinale*), SE. U.S., wood distilled for oil, used med. and for flavouring. **Braz.** —, *Aniba panurensis* and *A. amazonica,* wood distilled for Braz. sassafras oil: **Chinese** —, *Sassafras tsumu:* **grey** —, *Doryphora aromatica,* Aus.: **Nepal** —, *Cinnamomum cecidodaphne:* **Tasm.** —, *Atherosperma moschatum:* **yellow** —, *Doryphora sassafras* Aus. rain forest tree.

Sassy bark *Erythrophleum suaveolens* (= *E. guineense*), W.Af. *See* ordeal poisons.

Satin flower *Sisyrinchium* spp., Am., some cult.: — **bush,** *Podalyria sericea,* S.Af.: — **walnut,** *Liquidambar styraciflua,* E. N.Am., a soft, easily worked furniture wood.

Satin wood Sri Lanka or E.I., *Chloroxylon swietenia,* much favoured for furniture and cabinet work. **Af.** —, *Fagara macrophylla,* trop. Af., a commercial timber: **Braz.** —, *Euxylophora paraensis:* **Jam.** or **W.I.** —, *Fagara flava:* **Nigerian** —, *Distemonanthus benthamianus:* **scented** —, *Ceratopetalum apetalum,* Aus.

Satiné *Brosimum paraense,* trop. Am., a commercial timber.

Satyrium Af. and As. tuberous rooted orchids, cult.

Sau (Sri Lanka) *Albizia moluccana* or other spp., quick-growing orn. and timber tree.

Sauerkraut German food prepared from shredded cabbage with the addition of salt and of spices or flavouring. Fermentation takes place.

Sauromatum *S. guttatum,* E.I., the tuber may be grown indoors, without soil or water as a curiosity, the large inflor. appearing before the leaves.

Sausage tree *Kigelia* spp., trop. Af. bears pendulous frs. on long stalks resembling a large sausage in shape, cult. for orn. in trop.

Savanna flower (W.I.) *Mandevilla hirsuta:* — **grass** (W.I.), *Axonopus compressus.*

Savin *Juniperus sabina,* C. & S.Eur. oil distilled from ls. used in pharmaceutical preparations.

Savonette *Lonchocarpus sericeus,* W.I., a timber tree w. showy pink fls.

Savory Summer —, *Satureja hortensis,* Medit., an ann. flavouring herb w.

aromatic ls., used fresh or dried, esp. with beans or pulses: traditional with broad beans (*Vicia*): a common ingredient in mixed herbs: has a sharp taste, like sage. **Winter** —, *Satureja montana*, S.Eur., a peren. flavouring herb used like summer savory: a useful herb but perhaps less popular than summer savory and not so flavourful.

Savoy cabbage Cabbage notable for its winter hardiness: has wrinkled ls.

Sawdust Sawdust is produced in enormous quantities from wood working industries. Its economic disposal is sometimes difficult. The main use is as fuel, in special grates. Other uses include – as litter for bedding down livestock, protecting dance floors, use on butchers' floors, packing material, insulation, chipboard, stuffing (cushions, dolls, toys etc.), wood flour for use in explosives, linoleum and plastics, an ingredient in many compositions and flooring compounds, an absorbent, mulching with other materials for hort. or agric. use.

Saw grass *Cladium* spp., Am.: — **greenbriar**, *Smilax bona-nox*, N.Am.: — **palmetto**, *Serenoa serrulata* and other spp., N.Am.: —**wort**, *Serratula tinctoria*, Eur. inc. Br.: —**wort, alpine**, *Saussurea alpina*, Eur. inc. Br.

Saxifrage *Saxifraga* spp. a large genus (370 spp.), mainly alpine, many hybrids and cultivars, much grown in rock gdns. (*See Saxifraga* in Willis): **burnet** —, *Pimpinella saxifraga*: **golden** —, *Chrysosplenium oppositifolium*: **Irish** —, *Saxifraga rosacea*: **meadow** —, *S. granulata*: **pepper** —, *Silaum silaus*.

Scabious *Scabiosa* spp., mainly Medit., over 100 spp. several cult. as orn. fls. **devil's bit** —, *Succisa pratensis* (= *Scabiosa succisa*): **field** —, *Knautia arvensis* (= *Scabiosa arvensis*): **red Indian** —, *Scabiosa atropurpurea*, N.Am.: **wild** —, (S.Af.), *Scabiosa africana*.

Scabrin *Heliopsis scabra*, Mex., an insecticidal oil w. properties resembling pyrethrum has been obtained from the r.

Scabweed Term used in N.Z. for mat-forming pls., such as occur in dry stream beds, on stony flats and on screes and rocky places in mt. regions, e.g. *Raoulia australis*, *R. haastii*, *Scleranthus uniflorus* etc.

Scaevola Trop. and subtrop. coastal pls., esp. Aus., cult., *S. frutescens* widely distrib. in trop. As.

Scald Common name for dodder (*Cuscuta*).

Scaly bark (Aus.) Certain spp. of *Eucalyptus*.

Scammony root *Ipomoea orizabensis*, Mex., the dried r. is the source of the resin or drug known as ipomoea, a drastic purgative. **Levant** —, *Convolvulus scammonia*, formerly used, now of little commercial importance.

Scaphosepalum Trop. Am. orchids, cult.

Scarborough lily *Vallota speciosa*, a S.Af. bulbous pl. w. scarlet fls. The name arose from the fact that some 200 years ago a ship carrying some bulbs from S.Af. was wrecked off the Yorkshire coast and the bulbs were washed ashore. They were collected by residents of Scarborough and grown, hence the name. There are now several cultivars.

Scariole Old name for endive.

Scarlet bell Or flame tree, *Spathodea campanulata*, cult. trop.: — **cordia**, *Cordia sebestena*, widely cult. trop.: — **ipomoea**, *Ipomoea hederifolia*, cult. trop.

Scarlet runner *Kennedia prostrata*, Aus.: — — **bean**, *Phaseolus coccineus* (=*P. multiflorus*), grown in temp. countries as green beans, esp. Br., where it thrives and tolerates cool summers, being native to the higher altitudes in C.Am.

Scented cups *Gardenia capensis*, S.Af.: — **fern**, *Mohria caffrorum*, S.Af.: — **orchis**, *Gymnadenia conopsea:* — **orchid**, *Habenaria dilatata*, N.Am. ('scent bottle'): — **polypody**, *Polypodium pustulatum:* — **verbena**, *Lippia citriodora*.

Schisandra Climbing shrubs, mainly As., orn. frs., cult.

Schizanthus Showy anns., Chile, cult., many hybrids and cultivars showing great diversity of colour (butterfly flower or poor man's orchid).

Schizopetalon *S. walkeri*, Chile, cult. orn. fl.

Schizophragma Climbing deciduous shrubs w. aerial supporting rs., live ivy, cult.

Schizostylis *S. coccinea*, S.Af., red fls., cult., esp. as a pot pl. (kaffir lily).

Schomburgkia Trop. Am. and W.I. orchids, cult., many hybrids.

Schotia Handsome shrubs or small trees, S.Af., red or pink fls., cult., esp. on the Riviera.

Scilla Old World bulbous pls., among the most showy spring fls. in temp. countries.

Scindapsus Cult. trop. orn. fol. pls.: *S. aureus* a hothouse climber w. variegated ls.

Scoke (Am.) *Phytolacca americana*, pokeweed.

Scootberry *Streptopus americanus*, N.Am.

Scopolia *S. carniolica*, C.Eur., a source of hyoscyamine and atropine: **Jap.** —, *S. japonica*.

Scorpion everlasting *Helichrysum scorpioides*, Aus.: — **grass**, *Myosotis* spp. (forget-me-not): — **senna**, *Coronilla emerus*, C. & S.Eur. shrub: — **weed**, *Phacelia crenulata*, SW. U.S.

Scorzonera *S. hispanica*, Spain, a culinary r. veg.: long tapering tap root: cooked like salsify.

Scotch or **Scottish asphodel** *Tofieldia pusilla:* — **attorney** (W.I.), *Clusia* spp., trees that eventually destroy their host pls.: — **bonnets**, *Marasmius oreades:* — **broom**, *Sarothamnus scoparius:* — **fir**, *Pinus sylvestris:* — **lovage**, *Ligusticum scoticum:* — **maple**, *Acer pseudoplatanus*, sycamore: — **mist**, *Galium sylvaticum:* — **thistle**, *Onopordum acanthium*.

Scrambled eggs *Corydalis* spp., N.Am.

Scratch coco (W.I.) *Colocasia esculentum*, dasheen: — **bush**, *Urera* spp., small trees w. stinging hairs.

Screw auger (Am.) *Spiranthes cernua:* — **bean**, *Prosopis* spp., N.Am.: — **palm** (Aus.), *Pandanus* spp.: — **pine**, *Pandanus* spp. trop. sea coast or marsh pls. resembling palms; used for thatch, live fences, frs. ed. (*See Pandanaceae* in Willis): — **stem**, *Bartonia paniculata*, N.Am.: — **tree**, *Helicteres isora*, E.I., twisted fr. used med.

Scrub apple *Angophora*, Aus.: — **myrtle**, *Backhousia*.

Scuppernong *Vitis rotundifolia*, N.Am.

Scurf pea *Psoralea* spp., N.Am.

Scurvy grass *Cochlearia officinalis*, *Barbarea verna*, ls. eaten.

Scutch grass (Am.) *Cynodon dactylon*.

Scutellaria This large genus includes many gdn. fls.

Scuticaria Handsome trop. S.Am. orchids, cult.

Sea aster *Aster tripolium:* — **bean,** *Entada gigas* (= *E. scandens*) the large flat brown seeds, 4–5 cm across, are well known ocean drift seeds and have been picked up on the coasts of Br., brought by the Gulf Stream; *Mucuna gigantea,* also called sea bean: — **beet,** *Beta maritima:* — **blite,** *Suaeda maritima,* S. *fruticosa:* — **buckthorn,** *Hipphophae rhamnoides:* — **burdock,** *Xanthium echinatum:* — **coconut,** *Manicaria saccharifera,* common in S.Am. ocean drift: — **daffodil,** *Pancratium maritimum,* Medit.: — **grape,** *Coccoloba uvifera,* W.I.: — **heath,** *Frankenia laevis* and other spp.: — **holly,** *Eryngium maritimum,* rs. may be preserved in sugar: — **hollyhock,** *Hibiscus palustris,* N.Am.: — **island cotton,** *Gossypium barbadense:* — **kale,** *Crambe maritima:* — **lavender,** *Limonium* spp.: — **lyme grass,** *Elymus arenarius:* — **myrtle,** *Baccharis halimifolia,* trop. Am.: — **oats** (Am.), *Uniola paniculata* used for fixing sand dunes: — **onion,** *Urginea maritima:* — **ox eye,** *Borrichia frutescens,* N.Am.: — **pink,** *Armeria maritima:* — **poppy,** *Glaucium flavum:* — **purslane,** *Honkenya* (= *Arenaria) peploides:* — **ragweed,** *Senecio cineraria:* — **rocket,** *Cakile maritima:* — **samphire,** *Crithmum maritimum.*

Seal flower *Dicentra spectabilis,* 'bleeding heart'.

Seaside alder *Alnus maritima,* SE. U.S.: — **bean,** *Canavalia maritima,* W.I.: — **croton,** *Croton punctatus,* N.Am.: — **gentian,** *Eustoma exaltatum,* S. U.S.: — **golden rod,** *Solidago sempervirens,* N.Am.: — **grape,** *Coccoloba uvifera,* W.I. frs. used for jelly: — **ground cherry,** *Physalis maritima,* N.Am.: — **heliotrope,** *Heliotropium curassavicum,* N.Am.: — **lavender** (W.I.), *Mallotonia gnaphaloides:* — **mahoe,** *Hibiscus tiliaceus,* W.I., yields fibre for local use: — **petunia,** *Petunia parviflora,* Am: — **pink** *Sabatia stellaris,* N.Am.: — **plantain,** *Plantago oliganthos:* — **verbena,** *Verbena maritima,* N.Am.: — **yam,** *Ipomoea pes-caprae.*

Sebestena *Cordia sebestena,* W.I. orn. fl. tree, cult.

Sebestens *Cordia myxa,* Medit., Ind., ed. fr.

Sedge Name used mainly for spp. of the large genus *Carex* (over 1500 spp.) and to a less extent other genera in the *Cyperaceae,* such as *Cyperus, Cladium, Rhynchospora* etc. Many are used in mat and basket making and some for thatching, such as the fen sedge, *Cladium mariscus,* in Br. The starchy rhizome of some sedges is ed. Other sedges may be troublesome weeds. They vary in size from a few cm upwards, one of the largest being the giant sedge of New Zealand, *Gahnia xanthocarpa.*

Sedum Many spp. in this large N. temp. genus cult. as gdn. pls.

Seepweed (Am.) *Suaeda suffrutescens,* N.Am., an indicator of alkaline soils: thrives on salt flats.

Segra seed (W.I.) *Fevillea cordifolia.*

Selaginella Vast genus of pls., resembling mosses, mainly trop. Am., many cult.

Selangan *Shorea guiso* and other spp., Borneo, a commercial timber.

Self or **selfs** Hort. term used for fls. entirely of one colour, e.g. carnations: — **heal,** *Prunella vulgaris.*

Semaphore plant *Desmodium gyrans,* see telegraph plant.

Seminole tea *Asimina reticulata,* N.Am., a med. tea made from the fls. by Seminole Amerindians.

Semolina Starchy food product made from hard or flint wheat.

Sempervivum Fleshy stemless pls., C. & S.Eur., cult., includes the houseleek.

Sempilor *Dacrydium elatum*, Sarawak, a commercial timber.

Sen *Acanthopanax ricinifolius*, Jap., a commercial timber.

Senat seed *Cucumis melo* var. *agrestis*, once exported from the Sudan as an oil seed.

Senecio One of the largest genera of fl. pls. w. 2000–3000 spp., inc. many gdn. pls. and some weeds. *See Senecio* in Willis.

Senega root Or snake root, *Polygala senega*, N.Am., med. (bronchitis), was used by Amerindians as a snakebite remedy.

Senna *Cassia senna*, NW. Af., Ind. Dried ls. and pods a much used aperient, employed in proprietary medicines: two main trade forms, **Alexandrian** (=*C. acutifolia*) and **Tinnevelly** (=*C. angustifolia*). **Aleppo** —, *Cassia italica:* **Am.** —, *C. marylandica:* **bastard** —, *Coronilla valentina*, ls. much used to adulterate senna: **bladder** —, *Colutea arborescens:* **coffee** —, *Cassia occidentalis:* **dog** —, *C. obovata:* **Italian** —, *C. obtusifolia:* **prairie** —, *C. fasciculata:* **ringworm** —, *C. alata:* **scorpion** —, *Coronilla emerus*, cult. orn. shrub: **Spanish** —, *Cassia obovata:* **wild** —, *C. marylandica.*

Sensitive plants Several pls. are sensitive to touch, i.e. exhibit some form of movement when touched, the common 'sensitive plant', *Mimosa pudica* (trop. and subtrop.) being perhaps the best known. The ls. immediately droop when touched. A close ally is the 'giant sensitive plant', *Mimosa invisa*, one of the worst weeds of sugar cane in Queensland. It grows through and over the cane. Other pls. with sensitive ls. that may behave similarly are certain spp. of *Aeschynomene*, *Biophytum* (esp. *B. sessile*), *Cassia*, *Neptunia* (notably *N. oleracea*) and *Schrankia* (*S. uncinata* and *S. aculeata;* sensitive briar of N.Am.).

Many climbing pls. have sensitive tendrils. In the rapid-growing climber *Cobaea scandens* the tendril nutates with great rapidity and is highly sensitive to contact, as may be seen by rubbing one side and watching it for 5 minutes. Sensitive stamens occur in *Berberis*, *Centaurea*, *Portulaca*, *Sparmannia* and many *Compositae:* sensitive stigmas in *Martynia*, *Mimulus* and *Strobilanthes*. The sensitive fern is *Onoclea sensibilis*, N.Am., N.As.

Sepetir *Pseudosindora palustris* and other spp., Sarawak, a lightweight commercial timber.

Sequins (S.Af.) *Geissorhiza* spp., bright ixia-like fls.

Sequoia *See* redwood and mammoth tree.

Serapias Terrestrial orchids, Medit., cult.

Seraya Dark and **light red** —, *Shorea* spp. N. Borneo, commercial timber: **yellow** —, *Shorea gibbosa* and other spp.: **white** —, *Parashorea plicata.*

Serpentaria or **serpentary** *See* snakeroot.

Serpolet oil *Thymus praecox*, wild thyme.

Serradella or **serradilla** *Ornithopus sativus*, Medit., an ann. fodder pl., also a green manure and cover crop on sandy soils.

Service berry Or Juneberry, *Amelanchier canadensis*, N.Am., cult. orn. tree, ed. fr.

Service tree, wild *Sorbus torminalis*, Eur. inc. Br., orn. tree, cult., wood used in turnery and carving: **bastard** — —, *Sorbus hybrida*, N. & C.Eur., cult. orn. tree.

Sesame or **sesamum** *Sesamum indicum* (= *S. orientale*), an imp. trop. oil seed crop, esp. in Ind. Oil has properties and uses similar to olive oil. Seed may be sprinkled on bread, rolls or confectionery like poppy seed: the interest in low cholesterol diets, in connection with thrombosis, led to greater use of seed and oil. **Black** —, *Hyptis spicigera*, trop. Af.: **fake** —, *Ceratotheca sesamoides*, W.Af.

Sesbania Some spp. cult. as orn. or shade trees or as green manure pls.: some e.g. *S. aculeata* Ind. yield a b. fibre.

Setcreasia Orn. fol. pls., esp. *S. purpurea* and *S. striata*, room pls.

Sevadilla *See* sabadilla.

Seven year vine (W.I.) *Ipomoea tuberosa*.

Shack-shack (W.I.) Or shake-shake, a name given to various pods with seeds that rattle, *Crotalaria, Albizia, Leucaena* etc. **blue** — —, *Crotalaria verrucosa:* **common** — —, *Albizia lebbek:* **yellow** — —, *Crotalaria retusa*.

Shad or **shadbush** *Amelanchier alnifolia*, N.Am.

Shaddock *See* pompelmous.

Shake-shake *See* shack-shack.

Shakespeare, plants referred to by William Shakespeare (1564–1616) was by birth a countryman and grew up in the West of England amidst its wealth of wild fls. Like many country folk of his time he was familiar with, or seemingly knowledgeable in regard to, the everyday pls. and fls. of his day, both wild and cult. The following are among the pls. or fls. referred to by him – aconite, almond, anemone, apple, apricot, ash, aspen, bachelor's buttons, balm, barley, bay tree, beans, beech, bilberry, birch, bitter-sweet, blackberry, box, brier, broom, bulrush, burdock, burnet, cabbage, carnation, caraway, carob or locust, carrot, cedar, chamomile, cherry, chestnut, clover, colocynth, columbine, cork, corn, corn cockle, cowslip, crocus, crown imperial, currant, cypress, daffodil, daisy, darnel, date, dewberry, dock, dogberry, ebony, eglantine, elder, elm, fennel, fern, fig, filbert, flag, flax, fumitory, furze, garlic, ginger, gooseberry, gorse, gourd, grape, grass, harebell, hawthorn, hazel, heath, hemlock, hemp, henbane, holly, honeysuckle, hyssop, iris, ivy, knotgrass, ladies' smock, larkspur, laurel, lavender, leek, lemon, lettuce, lily, lime, ling, mace, mallow, mandrake, marigold, marjoram, medlar, mint, mistletoe, moss, mulberry, mushroom, mustard, myrtle, narcissus, nettle, nutmeg, oak, oats, olive, onion, orange, oxlip, palm, pansy, parsley, peach, pea, pear, peony, pepper, pig nut, pine, pink, plane, plantain, plum, pomegranate, poppy, potato, primrose, prune, pumpkin, quince, radish, raisin, reed, rhubarb, rice, rose, rosemary, rue, rush, rye saffron, samphire, savory, sea holly, sedge, senna, speargrass, strawberry, sycamore, thistle, thyme, turnip, vetch, vine, violet, walnut, wheat, willow, wormwood, yew. [Ellacombe, H. N. (1896) *The plant-lore and garden-craft of Shakespeare*, 383 pp.]

Shaking grass Or quaking grass, *Briza* spp., cult. orn.

Shallon bush *Gaultheria shallon*, W. N.Am., fr. ed., cult. as cover for game.

Shallot *Allium ascalonicum*, long cult. for flavouring and culinary purposes or pickling in temp. and subtemp. countries.

Shambalit *Merendera persica,* bulbs med. in Iran and Afghanistan.

Shame plant *Mimosa pudica,* sensitive pl.: *Neptunia plena,* W.I.

Shamrock The true shamrock is generally considered to be black medic or non-such, *Medicago lupulina,* but white clover, *Trifolium repens,* is commonly used as shamrock. The wood sorrel, *Oxalis acetosella,* which also has trifoliolate ls., has also been used.

Shantung cabbage Or pe-tsai, *Brassica chinensis.*

Shave grass (N.Am.) *Equisetum hyemale,* stems used for polishing.

Shaving brush tree *Pachira insignis.*

She balsam *Abies fraseri,* N.Am.: — **oak,** *Casuarina:* — **silk grass,** *Agave barbadensis.*

Shea butter *Butyrospermum paradoxum* (= *B. parkii*), trop. Af., esp. W.Af., a savannah tree, ed. fat from kernels much used.

Sheep bane *Hydrocotyle,* W.I.: — **beard,** *Urospermum picroides,* Medit., a bristly pl.: —'s **berry,** *Viburnum lentago,* N.Am.: —'s **bit scabious,** *Jasione montana:* —'s **ears,** *Helichrysum appendiculatum,* S.Af.: —'s **fescue,** *Festuca ovina:* —'s **laurel,** *Kalmia angustifolia* or other spp., N.Am.: —'s **sorrel,** *Rumex acetosella,* a troublesome weed of acid soils: **vegetable —,** *Raoulia* spp., N.Z.

Shell bush *Nautochilus labiatus,* S.Af.: — **flower,** *Moluccella laevis,* Medit. cult.; *Tigridia pavonina; Chelone:* — **ginger,** *Alpinia speciosa.*

Shellac Or lac, is a resinous substance prepared from the secretion of the lac insect (*Laccifer lacca*) which lives on various trees in the Asiatic trop., esp. Ind. Imp. among these are *Schleichera oleosa, Zizyphus jujuba* and *Butea monosperma* (= *B. frondosa*). Shellac declined in importance with the advent of synthetic resins.

Shepherd's clock *Tragopogon pratensis:* — **club,** *Verbascum thapsus:* — **cress,** *Teesdalia nudicaulis:* — **needle,** *Scandix pecten-veneris; Bidens pilosa,* N.Am.: — **purse,** *Capsella bursa-pastoris,* a cosmop. weed: — **rod,** *Dipsacus pilosus:* — **weather glass,** *Anagallis arvensis.*

Shield fern *Dryopteris filix-mas:* **hard — —,** *Polystichum aculeatum:* **prickly — —,** *Aspidium aculeatum:* — **flower,** *Aspidistra.*

Shin leaf *Pyrola* spp., N.Am.

Shingle Many different woods are used for shingles, those that may be easily split and are durable being favoured. Western red cedar (*Thuja plicata*) yields a very long lasting shingle. — **oak,** *Quercus imbricaria, Casuarina stricta:* — **plant,** *Monstera acuminata,* C.Am., large cordate ls.: — **tree,** *Acrocarpus fraxinifolius,* trop. As. esp. Ind.: — **wood** (W.I.), *Nectandra.*

Shisham wood Or sisso, *Dalbergia sissoo,* Ind.

Shittim wood Of Scripture is believed to have been *Acacia tortilis* var.

Shivery grass *Briza minor,* orn., occurs in most trop. countries.

Shoe buttons *Syngonanthus flavidus,* N.Am.: — **flower,** *Hibiscus rosa-sinensis:* **—maker's bark,** *Brysonima* spp. W.I., b. once used in leather tanning: — **strings,** *Tephrosia virginiana,* N.Am.

Shola pith *Aeschynomene aspera* and *A. indica,* Ind., the soft pithy stems of these semi-aquatic shrubs are used in making trop. sun helmets and fancy articles.

Shooting star *Dodecatheon* spp., N.Am., cult. orn. fls.

Shore weed *Littorella lacustris*, Eur.: — **grindelia**, *Grindelia robusta:* — **lupin**, *Lupinus littoralis*, W. U.S., r. ed.: — **podgrass**, *Triglochin maritima*, N.temp., young ls. used as veg. —**wort**, *Mertensia maritima*.

Shortia Beautiful dwarf perens., N.Hemisph., cult.

Shovel weed *Capsella bursa-pastoris*, shepherd's purse.

Shower of gold *Cassia fistula*, cult. orn. tree, trop., yellow fls.; *Galphimia* (= *Thryallis*) *glauca*, Mex., cult. orn. shrub w. masses of small yellow fls.

Shrimp plant *Beloperone guttata*, Mex., inflor. w. attractive shrimp-coloured bracts: a v. popular house pl.

Shude Or shudes, rice husks and polishings or rice bran, used for cattle feeding or mixing w. oil cake.

Siam weed *Eupatorium odoratum*, a trouble some trop. weed.

Siberian saxifrage *Bergenia* spp. cult. orn. fls.: some spp. have attracted attention as tanning materials, the r. and l. being rich in tannin, esp. *B. crassifolia*, badan or Russian saxifrage: — **yellow pine**, *Pinus sibirica*.

Sickle acacia *Acacia harpophylla*, Aus.: — **pod**, *Arabis canadensis:* — **senna** or **weed**, *Cassia tora*, pantrop.

Sida Several spp. yield stem fibres used locally in trop. and subtrop., some spp. are weeds.

Sidalcea Popular gdn. pls., W. N.Am.

Side oats Or mesquite grasses, *Bouteloua*, N. & S.Am., valuable fodder grasses: form a large part of the herbage of the prairie.

Side saddle plant *Sarracenia purpurea*, N.Am.

Sierra Leone butter Or Lamy butter, *Pentadesma butyracea*, W.Af., fat from kernels: — — **copal**, *Copaifera copallifera* (=*C. guibourtiana*).

Silene Some spp. are useful rock gdn. and border pls.

Silk cotton tree *Ceiba pentandra* the seed floss is commercial kapok: **Ind.** — — —, *Bombax malabaricum*, the light wood is used for matches: **yellow** — — —, *Cochlospermum gossypium*, Ind. cult. orn. tree: — **grass**, *Bromelia magdalenae*, C.Am. l. fibre resembles pineapple fibre; *Oryzopsis hymenoides*, N.Am.: — **tassel**, *Garrya elliptica*, N.Am.: — **tree**, *Albizia julibrissin*, trop. As.: — **vine**, *Periploca graeca*, SE.Eur. cult.: — **weed**, *Asclepias* spp., name derived from silky seed floss.

Silkworm food plants As is well known, the common silkworm (*Bombyx mori*) feeds on the ls. of the mulberry, *Morus alba*. There are other silkworms. The tussore or tussur silkworm (*Antheraea paphia*) feeds on the ls. of pollarded trees of *Terminalia tomentosa* and *T. catappa* in Ind. The Manchurian tussore silkworm (*Antheraea pernyi*) feeds on ls. of certain oaks (*Quercus*). The Eri silkworm (*Attacus ricini*) is reared on ls. of the castor oil pl., *Ricinus communis*, in parts of trop. As. Af. wild silk or Anape silk is obtained from *Anaphe infracta* which feeds on the ls. of the tree *Bridelia macrantha*. Other silkworms may feed on *Ailanthus*, *Cajanus*, *Careya*, *Dodonaea*, *Ficus*, *Liquidambar*, *Melastoma*, *Michelia* and *Symphonia*.

Silky bent grass *Apera* spp.: — **brown top**, *Eulalia fulva:* — **camellia**, *Stewartia malachodendron*, N.Am.: — **oak**, *Grevillea robusta; Cardwellia sublimis*, Aus., a commercial timber, spp. of *Musgravea oreocallis:* — **willow**, *Salix sericea*, N.Am.

Silphium Coarse perens. of N.Am., some cult. as gdn. pls.: *S. perforatum*, a potential peren. fodder pl. in Soviet Union. — **of the Ancients,** *Ferula narthex.*

Silver bell *Halesia carolina,* N.Am., cult. orn. tree: — **berry,** *Elaeagnus argentea,* N.Am.: — **bush,** *Anthyllis barba-jovis,* SW.Eur., Medit.: — **fern,** *Ceropteris* (=*Pityrogramma*) *calomelanos:* — **fir,** *Abies, Picea:* — **leaf,** *Leucadendron argenteum,* S.Af.; *Panax fruticosus* var., trop.: — **maple,** *Acer saccharinum,* N.Am.: — **oak,** *Grevillea robusta,* a widely cult. Aus. tree: — **leaf poplar,** *Populus alba:* — **rod,** *Solidago, Asphodelus:* — **tails,** *Ptilotus obovatus,* Aus.: — **thistle,** *Onopordon acanthium:* — **tree,** *Leucadendron argenteum,* S.Af., a well known tree of the Cape, ls. used for painting upon: — **wattle,** *Acacia dealbata,* Aus., cult.: — **weed,** *Potentilla anserina, Argyreia.*

Silverballi *Nectandra* spp., Guyana, a hard, heavy timber.

Silver-grain The grain or figure seen in wood, cut radially, when wide or fairly wide medullary rays are present, conspicuous in oak (*Quercus*), *Grevillea* and other Proteaceae.

Silver-grey wood *Terminalia bialata,* Ind., a commercial timber.

Silversword (Hawaii) *Argyroxiphium sandwicense.*

Simaruba Or maruba, *Quassia* (=*Simaruba*) *amara,* trop. Am., a commercial timber: source of Jam. b., a bitter tonic.

Simpoh (Burma) *Dillenia obovata,* handsome forest tree, ls. 30 cm long, brown when young, yellow fls., 15 cm diam.

Simsim Sesame, *Sesamum indicum:* **hard —,** *Hyptis spicigera,* trop. Af., cult.

Singhara nut *Trapa bispinosa,* trop. As., aquatic, w. starchy ed. seeds.

Sink gardening Or trough gardening constitutes a form introduced by the Royal Horticultural Society in the 1920s, which soon became very popular. Old fashioned kitchen or scullery sinks or stone pig troughs are used for growing dwarf pls., when a large number can be accommodated. If the trough or sink be raised weeding is easy, while slugs and other pests are seldom troublesome.

Sinningia Low growing hairy herbs, S.Am., cult., usu. called gloxinias.

Siris Or E.I. walnut, *Albizia lebbek,* a good timber: **red —,** *A. toona,* Ind.

Sisal or **sisal hemp** *Agave sisalana,* cult. trop., l. fibre much used for cordage, rope, matting and other purposes.

Sisso *Dalbergia sissoo,* Ind., a commercial wood, hard and durable.

Sisyrinchium Anns. or perens., N. & S.Am., cult. gdn. fls. (satin fl.).

Sitka cypress *Chamaecyparis nootkatensis:* — **spruce,** *Picea sitchensis.*

Six weeks grass *Poa annua,* a troublesome gdn. weed.

Skeleton weed *Lygodesmia juncea,* N.Am.: *Eriogonum* spp., N.Am.

Skeletonizing The skeletonizing of ls. and frs. was a popular hobby in Victorian times. This implies the removal of all soft tissue leaving only the fibrous framework or skeleton. Some ls. become naturally more or less skeletonized when they fall. This may often be seen in the case of beech ls. in a beech wood.

There are various chemical methods of skeletonizing, not always easy to apply. If the ls. are left too long in the skeletonizing solution not only the soft tissue but the fibrous tissue as well becomes attacked. Different kinds of l. require different lengths of time for successful treatment. This has to be ascertained by trial and error. A simple chemical method is as follows. Well matured ls. are placed in boiling water for two minutes then placed in a hot solution of permanganate of potash and kept heated. After an hour or two the cellular tissue may be removed with a soft brush. The skeletons are then bleached.

A method popular in Victorian days, without chemicals, was as follows. A few leaves of cabbage (or other brassicas) are placed in a bowl of rain water. Once decomposition or putrefaction has commenced, due to bacteria or other organisms, the ls. or frs. to be skeletonized are immersed in the bowl. Periodical inspection, after a week or two, will show when the ls. are ready to be removed and the softened tissue removed with a soft brush. Some ls. take only a few weeks, others may take many weeks. The object of the decaying cabbage ls. is to build up colonies of bacteria that will attack the ls. to be skeletonized when they are immersed. As evil smells are likely to accompany the process it is best done in an outhouse or shed. A household bleach may be used to bleach the skeletons afterwards.

Skewerwood *Cornus sanguinea* (= *Swida sanguinea*), Eur., dogwood, long used for skewers by butchers.

Skimmia *S. japonica*, cult. orn. evergreen shrub.

Skin irritants *See* dermatitis.

Skirret *Sium sisarum*, dahlia-like tuberous rs. ed., used less now than formerly, eaten like salsify.

Skull-cap *Scutellaria galericulata* and other spp.: some cult.: *S. coccinea* a trop. bedding pl.

Skunk cabbage *Symplocarpus foetidus*, N.Hemisph.: — **currant**, *Ribes glandulosum:* — **grass**, *Eragrostis megastachya:* — **meadow rue**, *Thalictrum revolutum:* — **weed**, *Croton texensis*.

Sky flower *Duranta repens* (=*D. plumieri*), cult., blue fls.: — **pilot**, *Polemonium viscosum*, N.Am.

Slangkop (S.Af.) *Urginea burkei*, a bulbous pl. w. med. uses.

Sleepy daisy *Xanthisma texanum*, N.Am.

Slipper flower *Calceolaria* spp.: — **orchid**, *Cryptostylis ovata*, Aus.: — **plant**, *Pedilanthus tithymaloides*, C.Am., W.I., a hedge pl.: — **wort**, *Calceolaria* spp.

Slippery elm bark *Ulmus fulva*, N.Am., mucilaginous inner b. used med., demulcent.

Sloe *Prunus spinosa*, Eur. inc. Br., fr. ed. but sour, used for sloe gin: **black** —, *Prunus umbellata*, S. U.S., fr. used for jam and jelly.

Slough grass *Spartina pectinata*, N.Am.: *Beckmannia syzigachne*, N.Am.

Smartweed (Am.) *Polygonum* spp.

Smilax Widespread climbers, rs. of some yield sarsaparilla, used less now than formerly. The name 'smilax' is popularly used for other pls., e.g. *Asparagus* spp., *Medeola asparagoides* a cult. orn. climber, N.Am.

Smithiantha (=*Naegelia*) C.Am. perens. w. showy fls. in terminal leafless panicles, cult.

Smoke tree or **bush** *Cotinus coggygria* (=*Rhus cotinus*), Ind. sumac, cult. orn., striking 'smoke-like' inflor.: *Dalea spinosa*, N.Am.

Smoking mixtures, herbal Herbal smoking mixtures, as substitutes for tobacco, are popular in some quarters. Most of the herbal smoking mixtures used in Br. have a large proportion of coltsfoot leaf (*Tussilago farfara*), doubtless because the hairy undersurface of the l. helps to bind or keep the mixture together. It may also assist it to burn well. Often there is a large proportion of driêd clover fls., which, being brown, assist in giving the mixture a tobacco-like appearance. One recipe for a smoking mixture is as follows – coltsfoot l. 50 %: clover fls. 30 %: lavender fls. 10 %: rose petals 10 %. Various other ingredients or ls. are known to be used in some mixtures. These include – agrimony, balm, bog myrtle, calamint, catmint, centaury, clematis, dead nettle, eyebright, hyssop, mallow, marjoram, meadow sweet, melilot, mullein, pellitory, raspberry, restharrow, rosemary, spikenard, sweet briar, thyme, woodruff, woundwort, yarrow.

Smooth bark apple *Angophora costata*, E.Aus.

Smother weed *Kochia* ($=Bassia$) *hyssopifolia*, E.Eur., W.As.

Snail flower *Arisaema; Phaseolus caracalla*, cult. orn. trop. climber, mauve fls. w. twisted keel: — **orchid**, *Pterostylis pyramidalis*, Aus.: — **seed**, *Cocculus carolinus*.

Snake berry *Actaea rubra*, N.Am.

Snake bush *Hemiandra pungens*, W.Aus.: — **climber**, *Bauhinia anguina*, Ind., orn. climber w. peculiar snake-like stem: — **flower** (S.Af.), *Ornithogalum* spp., *Monsonia speciosa*, *Bulbine asphodeloides* and other spp.: — **gourd**, *Trichosanthes cucumerina* ($=T.$ *anguina*) trop. veg.: — **grass**, *Eragrostis*, N.Am.: —**'s head**, *Fritillaria meleagris* and other spp.; *Hermodactylus tuberosus; Chelone* spp., N.Am.: — **lily** (S.Af.), *Haemanthus natalensis:* — **nut**, *Ophiocaryon paradoxum*, the seed has a peculiar embryo which becomes contorted to resemble a coiled snake: — **orchid**, *Diuris pedunculata*, Aus.: —**'s mouth orchid**, *Pogonia ophioglossoides:* — **plant**, *Dracunculus vulgaris:* — **tongue**, *Ophioglossum:* — **weed**, *Polygonum bistorta; Xanthocephalum:* — **wood**, a name for several woods w. markings suggesting a snake skin, e.g. *Piratinera guianensis* ($=Brosimum aubletii$), trop. Am., a commercial timber; *Colubrina arborescens*, W.I.; *Strychnos colubrina*, Ind.

Snakebite remedies Many plants, all over the world, are used by local inhabitants in treating snakebite but it is doubtful whether they are of much value. Such pls. include spp. of *Actaea*, *Arisaema*, *Aristolochia*, *Bignonia*, *Bragantia*, *Casearia*, *Cimicifuga*, *Cissampelos*, *Eryngium*, *Erythroxylum*, *Foeniculum*, *Gentiana*, *Jatropha*, *Leonotis*, *Liatris*, *Machaerium*, *Polygala*, *Tacca*.

Snakeroot A name used for several pls. in the U.S. **black** —, *Cimicifuga racemosa*, med. uses.; *Zigadenus densus*, *Actaea spicata*, *Sanicula marylandica:* **button** —, *Eryngium yuccifolium*, *Liatris* spp.: **Senega** —, *Polygala senega:* **Texas** or **Virginian** —, *Aristolochia serpentaria:* **white** —, *Eupatorium rugosum*.

Snapdragon *Antirrhinum majus*, a popular gdn. fl., numerous named vars. or cultivars.

Snapweed *Impatiens* spp.

Sneezeweed *Achillea ptarmica*, Eur. inc. Br.: *Helenium tenuifolium*, N.Am.

Sneezewood *Ptaeroxylon obliquum* ($=P. utile$), S.Af., a wood of great durability, resists termites, may cause sneezing when worked.

Snoutbean *Rhynchosia* spp., N.Am.

Snow ball tree *Viburnum opulus* var., cult.: — **berry**, *Symphoricarpos*

rivularis, orn., N.Am. shrub, naturalized in Br.; *Chiococca alba,* W.I.:
— **berry, creeping,** *Gaultheria* spp.; *G. hispidula* (=*Chiogenes serpyllifolia*),
N.Am.: — **bush,** *Ceanothus velutinus,* N.Am.; *Calocephalus brownii,* Aus.;
Breynia disticha forma *nivosa,* W.I., variegated ls., young ones may be quite
white: —**drop,** *Galanthus nivalis,* Eur., wild and cult.: —**drop, Barbados,**
Zephyranthes tubispatha, W.I.: —**drop tree,** *Halesia carolina,* cult. fl. tree;
Chionanthus virginica, cult.: *Diospyros whyteana,* S.Af.: —**flake,** *Leucojum ver-*
num (spring), *L. aestivum* (summer), *L. autumnale* (autumn): —, **glory of the,**
Chionodoxa luciliae, As. Minor: — **grass,** *Poa* spp.; *Festuca eriopoda,* Aus.;
Chionochloa flavescens, N.Z.: —**y mespilus,** *Amelanchier vulgaris,* Eur., cult.:
— **on the mountain,** *Euphorbia marginata,* cult., whitish foliage: — **pear,**
Pyrus ussuriensis: — **in summer,** *Cerastium tomentosum,* Eur.: — **tree,**
Pyrus nivalis.

Snuffbox fern *Thelypteris palustris:* — tree, *Oncoba spinosa,* trop. & S.Af.,
calabash-like fr. used, pulp ed.

Soap plants At least 500 genera of pls., many belonging to the families
Sapindaceae and Caryophyllaceae, are known to contain saponin to some
extent, a substance which has the property of foaming with water. For this
reason some of these pls. have been used as substitutes for soap. Among the
better known soap substitutes are the seeds or frs. of spp. of *Sapindus* (soap
berry trees) such as the Ind. or As. spp. *S. trifoliatus* and *S. mukorossi* and
the W.I. or C.Am. sp. *S. marginatus.* In Kashmir the *Sapindus* soap berry
is said to be preferred to ordinary soap for washing the famous Kashmir
shawls. Pls. used in C. & N.Am. include the bulbous rs. of *Chlorogalum*
pomeridianum, frs. of *Ceanothus* spp. *Shepherdia canadensis* and some spp. of
Yucca or *Agave.*

   In Eur. the common soapworts *Saponaria officinalis* and *S. calabrica* are
used, as are the rs. of *Gypsophila struthium* in various Medit. countries.
Saponaria officinalis is said to be still used for cleaning tapestry to restore
original colours without damaging the fabric. During the Boer War in S.Af.
the Boers are said to have used *Pretrea zanguebarica* (=*Dicerocaryum*) as a
soap substitute. The only one of these soap substitutes to be used commercially
is the b. of the S.Am. soap bark tree *Quillaja saponaria,* Andes, Peru and Chile.
The dried inner b. contains about 9 % saponin which finds use in fire extin-
guishers and in other ways.

Sobralia C. & S.Am. orchids, large showy fls., cult.

Socket wood *Daphnandra micrantha,* Aus. rain forest tree.

Sodom-apple *Solanum aculeatissimum,* N.Am.

Soft-grass, creeping *Holcus mollis,* Eur., N.Am., a common grass, often a
weed, variegated form cult. for orn.

Softwood The term 'softwood' is commonly applied to any wood that is
coniferous, i.e. belongs to the Coniferae, regardless of its actual physical
texture. The principal softwoods in Br. are pine, fir, spruce, larch, cypress,
cedar, sequoia and yew.

Soja bean *See* soy or soya bean.

Sola pith plant *Aeschynomene aspera,* pith used for sun helmets: *A. indica*
may be similarly used.

Solandra Trop. Am. climbing shrubs w. large showy fls., cult.

Solanum Includes several trop. and subtrop. orn. shrubs and climbers and many economic pls., e.g. potato and tomato. *S. capsicastrum* is a favourite orange-red fruited Christmas pot pl. in Br.

Solazzi A trade name (Italian) for liquorice.

Soldier plant *Calliandra*, W.I.: —'s **plume**, *Habenaria psychodes*, N.Am.

Solomon's seal *Polygonatum multiflorum:* **false** — —, *Smilacina racemosa:* **lesser** — —, *P. odoratum:* **two-leaved** — —, *Maianthemum canadense*, N.Am.

Soncoya *Annona purpurea*, C.Am., ed. fr.

Sonerila As. herbs or shrubs, cult., *S. margaritacea*, from Java, is a beautiful pl. w. curiously marked ls. and pink fls.

Sonora gum *Larrea americana*, N.Am., a resinous substance obtained from the twigs, med.

Sophora Several spp. are cult. orn. shrubs or trees in warm cli.: some yield useful timber.

Sophronitis Trop. S.Am. orchids w. showy scarlet or violet fls., cult.

Sorbus Trees or shrubs, N.Hemisph., cult. for orn.: some yield useful grained woods.

Sorghum Great millet, Guinea corn, kaffir corn, dari etc., *Sorghum* spp. bicolor cultivars. One of the World's imp. cereals and of age-old cult. in the Old World, esp. in Africa, Ind. and China and now much grown in the Americas. The relatively small, spherical, white-brown grains are a much used human, animal and poultry food. The sorghum pl., in its numerous vars., is able to withstand drier conditions than maize and so is favoured in the drier parts of the trop. and subtrop. In S.Af. sorghum is used for a breakfast food, or porridge and in the preparation of infant food. It is much used by the Bantu in making their native beer. Sweet-stemmed forms of sorghum may be grown for forage, for making syrup or for chewing. The dried inflor. of broom-corn (*Sorghum dochna* var. *technicum*) are used for brooms and brushes. [Doggett, H. (1970) *Sorghum*, 403 pp.]

Sorrel *Rumex scutatus, R. acetosa.* Ls. may be cooked as spinach or used in salads: much used in France for soup and for flavouring omelettes and sauces. **French** —, *Rumex scutatus:* **ladies** —, *Oxalis* spp.: **maiden** —, *Rumex montanus:* **mountain** —, *Oxyria digyna:* **red** —, *Hibiscus sabdariffa* (roselle): **sheep** —, *Rumex acetosella:* **tree** —, *Oxydendrum arboreum*, E. N.Am., ls. have pleasant acid taste: **violet wood** —, *Oxalis violacea*, N.Am.: **wood** —, *Oxalis acetosella.* Many other spp. of *Oxalis* may be called sorrel.

Sotol *Dasylirion* spp., SW. U.S., Yucca-like desert pls.: polished l. bases sold as curios.

Sauari nut *See* swarri nut.

Souchong Kind of Chinese tea made from coarse l.

Sour cherry *Prunus cerasus, see* cherry: — **grass**, *Panicum conjugatum*, trop. pasture grass: — **greens**, *Rumex venosus*, N.Am.: — **gum**, *Nyssa* spp., N.Am.: — **orange**, *Citrus aurantium:* — **plum**, *Ximenia caffra*, S.Af.; *Owenia acidula*, Aus.: —**sop**, *Annona muricata*, cult. fr.: **mountain** —**sop**, *Annona montana*, N.Am.: — **wood**, *Oxydendrum arboreum.*

Southernwood *Artemisia abrotanum*, S.Eur., cult., fragrant shrub.

Sowa Or lechi-caspi, *Couma macrocarpa*, trop. S.Am., yields a gutta-percha type material used for chewing gum.

Sow-bane *Chenopodium murale* and *C. hybridum:* — **bread,** *Cyclamen hederifolium:* — **thistle,** *Sonchus oleraceus* and other spp.

Soya bean or **soy bean** *Glycine max* (= *G. soja*). An all-imp. pulse or food crop in the Orient, used as human food in innumerable ways. The beans also yield an ed. oil: oil and meal have numerous industrial applications. Cult. in other countries, esp. the U.S.

Spaghetti This well known Italian food product is made from hard or flint wheat.

Spathalia (Cyprus) *Genista sphacelata*, a hedge pl.

Spanglegrass *Uniola* spp., N.Am., useful for pasture.

Spanish arbour vine *Merremia tuberosa*, cult. warm. cli., yellow fls. — **ash,** *Lonchocarpus benthamianus*, W.I.: — **bayonet,** *Yucca aloifolia* and other spp.: — **berries,** *Rhamnus infectorius:* — **bluebell,** *Scilla hispanica:* — **broom,** *Spartium junceum*, fibre from stems once exploited: — **buttons,** *Centaurea nigra:* — **cane,** *Arundo donax:* — **chestnut,** *Castanea sativa:* — **dagger,** *Yucca aloifolia:* — **elm,** *Cordia gerascanthus*, W.I., a furniture wood: — **esparcet,** *Hedysarum coronarium:* — **iris,** *Iris xiphium:* — **larkspur,** *Gilia rubra*, N.Am.: — **lime,** *Melicoccus bijugatus*, ed. fr.: — **liquorice,** *Glycyrrhiza glabra:* — **moss,** *Tillandsia usneoides*, S. U.S., trop. Am., forms long hanging tufts on trees, used for packing or upholstery: — **needle,** *Agave* spp., *Yucca* spp., *Bidens* spp.: — **oak,** *Quercus palustris*, N.Am., *Inga laurina*, W.I.: — **origanum,** *Thymus capitatus:* — **oyster pl.,** *Scolymus hispanicus:* — **plum,** *Spondias* spp., W.I.: — **reed,** *Arundo donax*, used for musical instruments and for screens.

Sparaxis *See* harlequin flower.

Sparkleberry *Vaccinium arboreum*, N.Am.

Sparmannia *S. africana*, S.Af., cult. orn. shrub, large ls. and white fls.

Sparrow grass Corruption of *Asparagus.*

Spathiphyllum Trop. orn. pls., cult. *S. wallisii*, w. a green or white spathe, sold by florists in Br.

Spathoglottis As. and Aus. orchids, cult.

Spatterdock *Nuphar* spp., N.Am.

Spear grass Name for various spinous grasses, esp. *Stipa* and *Poa* spp.: *Heteropogon contortus*, Ind.: *Pentapogon quadrifidus*, Aus.: also used for the spiny umbellifer *Aciphylla squarrosa* in N.Z.: — **mint,** *Mentha spicata*, Eur., oil distilled from ls. used for flavouring, esp. chewing gum: — **wood,** *Acacia homalophylla*, Aus.: —**wort,** *Ranunculus lingua* and other spp.

Speedwell *Veronica* spp. **common** —, *V. officinalis:* **field** —, *V. agrestis:* **germander** —, *V. chamaedrys:* **ivy-leaved** —, *V. hederifolia*, and many more.

Spekboom (S.Af.) *Portulacaria afra*, a valuable fodder or browse pl. w. small succulent ls.

Spelt *Triticum spelta*, a form of wheat.

Spices Most of the spices that are of importance commercially are the product of trop. pls. Examples are pepper, ginger, cloves, nutmeg and cinnamon, all native to the As. trop., while pimento and vanilla originated in the New

World trop. Spices have been used and esteemed from the earliest times, esp. in countries such as Ind., E.I., Medit. countries and Egypt. They attracted the attention of early explorers and traders and much of the wealth of certain cities, such as Venice, was due to their trade in spices. They were brought by the tedious overland route through Arabia and As. Minor, the journey taking many months. With the discovery of the sea route to The East via the Cape, spices became imp. cargo for the E. Indiamen. Fantastic prices were paid for oriental spices during the early days of this trade by sea and attempts were made to establish monopolies in the East, esp. by the Dutch. The urge to obtain trop. spices actually played an imp. part in world history. It stimulated exploration, including the discovery of the Americas.

It is not a simple matter to differentiate between spices, condiments, flavourings and certain culinary herbs. Some of the imp. condiments are from pls. of temp. or subtemp. origin such as mustard, angelica and caraway. Many hundreds of different pls. are known to be used in the flavouring of food in different parts of the world. In the case of curries alone a number of different spices or flavourings may be used, although turmeric is a characteristic and perhaps esst. ingredient. Apart from their uses in domestic cookery spices are of great importance to the food manufacturer, esp. for such commodities as pickles, chutneys and sauces. The world's most imp. spice in terms of quantity used is pepper, followed probably by mustard. The following are among the more imp. spices (*see also* flavourings and separate headings): Angelica, anise, caraway seed, cardamom, Cayenne pepper, celery seed, chillies, cinnamon, cinnamon cassia, cloves, coriander seed, cumin seed, curry leaf, dill, fennel, fenugreek, ginger, mace, mustard seed, nutmeg, paprika, pepper, pimento, saffron, sage, star anise, turmeric, vanilla. [Rosengarten, E. (1969) *The book of spices*, 489 pp.]

Spider flower *Cleome* spp. cult. gdn. pls., the long stamens are suggestive of a spider's legs: S.Af. *Ferraria undulata* (= *Tigridia undulata*): **grey** — —, *Grevillea buxifolia*, Aus.: **red** — —, *G. aculeata*, Aus. shrub w. red fls., cult.: **spiny** — —, *Cleome aculeata*, W.I., a trop. weed, esp. in sugar cane in Queensland: — **lily**, *Hemerocallis* spp.: — **orchid**, *Bartholina pectinata*, Aus.; *Caladenia filamentosa*, Aus.; *Ophrys sphegodes*: **tailed** — **orchid**, *Calandrinia caudata*, Aus.: — **orchis**, *Ophrys sphegodes*: — **plant**, *Chlorophytum comosum*, S.Af., a popular, easily grown house pl.: **—wort**, *Tradescantia virginiana* cult., many vars.: **mountain —wort**, *Lloydia serotina*, Eur. inc. Br., but rare.

Spiderling *Boerhavia caribaea*, trop. and subtrop. Am., often a garden weed.

Spigelia *S. anthelmia*, trop. weed w. explosive fr.

Spignel Or bald-money, *Meum athamanticum*, Eur. inc. Br., aromatic ls.

Spike grass *Uniola* spp., N.Am.; *Distichlis spicata*, N.Am.: — **rush**, *Eleocharis palustris* and other spp.: — **oil**, from spike lavender, *Lavandula latifolia*.

Spikenard *Nardostachys jatamansi*, rhizomes fragrant, yield an oil esteemed as a perfume in the East: **Am., false** or **wild** —, *Smilacina racemosa*, N.Am., med.

Spinach *Spinacia oleracea*, a much grown temp. veg., often canned: many different vars. w. much variation in l. characters. — **beet** or Swiss chard, *Beta vulgaris*, fleshy ls. eaten: **Ceylon** or **Ind.** —, *Basella alba:* **N.Z.** —, *Tetragonia tetragonioides* (= *T. expansa*), much grown, esp. in areas too hot or dry for ordinary spinach. Numerous other pls. are used as spinach, esp.

in trop. and subtrop. Many belong to the following genera – *Amaranthus, Atriplex, Bidens, Boerhavia, Celosia, Corchorus, Cucurbita, Grewia, Ipomoea, Talinum.*

Spindle tree *Euonymus europaeus*, Eur. inc. Br., cult. for delicate fol. and red frs.: wood close-grained and tough, used for spindles when spinning was done at home.

Spinks *Cardamine pratensis*, Eur. inc. Br., cuckoo fl., lady's smock.

Spiraea *Astilbe* spp., As., cult. orn. pls., fls. in dense clusters, white, pink or crimson, many hybrids.

Spiranthes Widespread genus of trop. and temp. orchids, some cult.

Spire lily *Galtonia* spp. S.Af., cult. orn. fl.

Spirit-weed *Lachnanthes tinctoria*, E. N.Am., med.: *Aegiphila elata*, trop. Am., W.I.

Spleenwort *Asplenium* spp. **common maidenhair** —, *A. trichomanes:* **mountain** —, *A. montanum*, N.Am.: **sea** —, *A. marinum*, Eur. inc. Br.

Spogel seed *See* ispaghul and psyllium seed.

Spoon flower *Labisia* spp., Malaysia, petals resemble spoons: — **wood**, *Kalmia latifolia*, N.Am., wood hard and close-grained.

Spotted sun orchid *Thelymitra ixioides*, Aus.

Sprangletop *Scolochloa festucacea*, N.Am., an imp. fodder or hay grass: **green** —, *Leptochloa dubia*, Am., imp. grazing grass.

Spring beauty *Claytonia lanceolata*, N.Am.: — **grass**, *Anthoxanthum odoratum*, contains coumarin giving the smell of new-mown hay: — **savory**, *Satureja acinos:* — **snowflake**, *Leucojum vernum:* — **starflower**, *Ipheion uniflorum.*

Spruce *Picea* spp. Imp. conifers widely distributed in the N.Hemisph., esp. China: much used for timber and grown for orn. The **common Eur.** or **Norway** —, *Picea abies* (=*P. excelsa*) is widely grown in C. & N.Eur.: has been grown in Br. for over 400 years: it is the common Christmas tree: the timber is known as 'whitewood', 'white deal' or 'white fir' and is much used in general joinery and for wood pulp (for paper, rayon etc.): there are many orn. vars. Other spruces yielding commercial timber include – **black** —, *P. mariana*, E. N.Am.: **eastern** or **Can.** —, *P. glauca*, N.Am.: **Himal.** —, *P. smithiana:* **red** —, *P. rubens*, N.Am.: **Rocky Mt.** —, *P. engelmannii*, W. N.Am.: **Sitka** —, *P. sitchensis*, W. N.Am., cult., wood like Eur. spruce: **western white** —, *P. glauca* var. *albertina*, N.Am. There are many other spruces grown for timber or orn. The term 'hemlock spruce' refers to *Tsuga.*

Spruce bark The b. of the common Eur. spruce, *Picea abies* (=*P. excelsa*), has long been used commercially in many parts of Eur. in the tanning of leather, esp. C. & N.Eur. A tanning extract is prepared from the b. The tannin content of spruce b. is extremely variable, being affected by location and age of tree, 10–12 % being the average for most areas.

Spur flower *Plectranthus* spp., S.Af.: — **valerian**, *Centranthus ruber.*

Spurge *Euphorbia* spp. **caper** —, *E. lathyrus*, frs. sometimes pickled as capers, not to be recommended; pl. said to 'drive away' moles from a gdn.: **cypress** —, *E. cyparissias*, a gdn. escape: — **flax**, *Daphne mezereum:* **flowering** —, *Euphorbia corollata:* **ipecacuanha** —, *E. ipecacuanha*, N.Am.: — **laurel**, *Daphne laureola:* — **nettle**, *Cnidoscolus stimulosus*, N.Am.

Spurrey Corn spurrey, *Spergula arvensis*, a common cornfield weed: may be used as a fodder catch-crop, esp. var. *sativa* in Holland.

Squash Ed. gourds of various kinds, summer (non-keeping) and winter (keeping), cultivars derived mainly from *Curcurbita maxima* and *C. moschata*.

Squaw berry *Rhus trilobata*, N.Am.: — **grass**, *Elymus triticoides:* — **huckle-berry**, *Vaccinium stamineum*, *V. caesium:* — **root**, *Cimicifuga racemosa*, *Trillium erectum*, *Conopholis americana*, *Caulophyllum thalictroides*, all med.: — **thorn**, *Lycium:* — **vine**, *Mitchella repens:* — **weed**, *Senecio aureus*.

Squill *Urginea maritima* (= *U. scilla*), Medit., the large bulbs used med. or as a rat poison, there being two forms, the red and the white. The action of squill resembles digitalis but is less potent: it has been used in cough mixtures. **Autumn** —, *Scilla autumnalis:* **blue** —, *Scilla natalensis*, S.Af.: **Lebanon** or **striped** —, *Puschkinia scilloides*, cult. orn. pl.: **spring** —, *Scilla verna*.

Squinancy wort *Asperula cynanchica*, Eur. inc. Br., mainly on chalk.

Squirrel corn *Dicentra canadensis*, N.Am.: — **tail grass**, *Hordeum marinum; H. jubatum; Sitanion hystrix*, N.Am.

Squirting cucumber *Ecballium elaterium*, Medit., a purgative (elaterium) prepared from the fr. which squirts out the seeds when ripe. *See Ecballium* in Willis.

St entries *See under* Saint.

Stachys Genus widespread in N.Hemisph., some cult. in fl. gdn.

Stachytarpheta *S. mutabilis*, S.Am., cult. subshrub; red fls., other spp. are trop. weeds.

Staff tree *Celastrus scandens*, E. U.S., orn. climber.

Stag bush *Viburnum prunifolium*, N.Am.

Stagger bush *Pieris mariana*, N.Am., poisons lambs and calves.

Staghorn bush *Daviesia epiphylla*, W.Aus.: — **fern** or **elk's horn**, *Platy-cerium* spp., handsome trop. ferns of distinctive or unique appearance with forked fertile fronds resembling a stag's horn, cult., popular in hot-houses: — **moss**, *Lycopodium clavatum*. *See Platycerium* and *Lycopodium* in Willis.

Stangeria *S. eriopus*, Natal, S.Af., an interesting monotypic cycad (Hottentot's head), cult.

Stanhopea Trop. Am. orchids w. large showy fls., often scented, cult.

Stapelia Succulents w. usu. brownish-red, evil-smelling fls. (carrion fls.) cult.

Staphylea N. temp. orn. shrubs, cult.

Star anise *Illicium verum*, a small evergreen tree cult., S.China: aromatic star-shaped frs. distilled while green, oil used for flavouring, esp. liqueurs, powdered seeds also used med. or as a spice. — **apple**, *Chrysophyllum cainito*, trop. Am., ed. fr., cut across has a star-like appearance: **monkey** — —, *C. perpulchrum*, trop. Af.: **white** — —, *C. albidum*, trop. Af.: — **of Bethlehem**, *Ornithogalum umbellatum* and other spp.: — **cucumber**, *Sicyos*, Am.: — **flower**, *Lithophragma parviflorum*, N.Am.: — **fr.**, *Damasonium alisma:* — **grass**, *Aletris, Hypoxis, Leptochloa:* — **head**, *Centaurea:* — **hyacinth**, *Scilla amoena:* — **jasmin**, *Jasminum gracillimum:* — **of Jerusalem**, *Trago-pogon pratensis:* — **lotus**, *Nymphaea odorata:* — **of the night**, *Clusia rosea*, N.Am.: — **plum**, *Chrysophyllum cainito*, W.I.: — **thistle**, *Centaurea:* — **tree**, *Astronium*, trop. Am.: —**wort**, *Aster, Stellaria, Callitriche*.

Starch Starch is the most prevalent of the non-nitrogenous food reserves of pls.

Starch grains may occur in any part of the pl. but are often most abundant in seeds and tubers. They vary much in size and shape and may possess characters by which they or the parent pl. may be identified. Some are v. large, like those of the ed. canna, other may be v. small.

Commercially starch is prepared largely from maize, potatoes or cassava, also from wheat, rice, sorghum or other cereals. Protein and oil are by-products of starch manufacture. Starch has many uses, esp. in textile and paper manufacture, also as a thickener in food and confections and for adhesives.

Statice *Limonium* spp., cult. orn. fls. of various colours, often used like everlastings.

Stauropsis Trop. As. orchids w. showy fls., cult.

Stave-wood *Quassia* (= *Simarouba*) *amara*, W.I., source of Jam. b.

Stavesacre *Delphinium staphisagria*, S.Eur., cult., seeds once used as an insecticide or parasiticide.

Steekgrass (S.Af.) *Trachypogon spicatus*, considered for paper.

Steeple bush *Spiraea tomentosa*, N.Am., brown felted shoots, pink fls.

Steer's head *Dicentra uniflora*, N.Am.

Stenomesson Showy trop. Am. bulbous pls., cult.

Stenorrhynchos Trop. Am. orchids, cult.

Stephanotis Twining shrubs, Madag., *S. floribunda*, a favoured hot-house climber w. clusters of fragrant, white fls. (wax fl. or Madag. jasmine).

Sterculia Brown —, *Sterculia rhinopetala*: **yellow** —, *S. oblonga*: both W.Af. commercial timbers. Some spp. of *Sterculia* have large showy fls.

Stevia *S. rebaudiana*, a shrubby pl. of Paraguay, remarkable for the sweetening properties of its ls.

Stick-leaf *Mentzelia oligosperma*, N.Am.: — **seed**, *Lappula* spp., N.Am.: — **tight**, *Bidens* spp., seeds stick to clothing and animals.

Sticky flower *Orphium frutescens*, S.Af.

Stifftia Trop. Am. shrubs w. large orange or yellow fl. heads, cult.

Stigmaphyllon Handsome trop. Am. climbing shrubs. cult., esp. *S. ciliatum*.

Stinging hairs These are present on a number of different pls., esp. in the families *Loasaceae*, *Malpighiaceae* and *Urticaceae*, the common stinging nettle, *Urtica dioica*, being one of the best known. Other spp. w. stinging hairs occur in the following genera – *Blumenbachia*, *Dendrocnide*, *Girardinia*, *Chidoscolus*, *Laportea*, *Mucuna*, *Tragia*, *Urera*, *Urtica*.

Stink bush *Boscia foetida*, S.Af.: — **grass**, *Eragrostis megastachya*, N.Am.: — **weed**, *Pluchea camphorata*, N.Am.

Stinking Benjamin *Trillium erectum*, N.Am., cult. orn. pl., fls. ill-scented: — **bush**, *Cassia occidentalis*, trop. and subtrop.: — **cedar**, *Torreya taxifolia*: — **chamomile**, *Anthemis cotula*: — **elder**, *Sambucus pubens*, N.Am.: — fleabane, *Pluchea foetida*, N.Am.: — **gladwin**, *Iris foetidissima*: — **ground pine**, *Camphorosma monspeliaca*, Medit.: — **miss**, *Gynandropsis gynandra*, trop.: — **passion fruit**, *Passiflora foetida*, a trop. weed: — **weed**, *Cassia occidentalis*, *Pluchea camphorata*: — **Willie**, *Senecio jacobaea*, ragwort: — **yew**, *Torreya*.

Stinkwood Name used for various woods, esp. in S.Af., usu. because they emit a strong odour when worked, e.g. *Celtis africana*, S.Af. **Camdeboo** —, a dull white wood: *Coprosma foetidissima*, N.Z.: *Gustavia augusta*, trop. Am.: *Ocotea bullata*, S.Af.: **black** —, a high quality furniture wood, once used for wagons: **red** —, *Prunus* (=*Pygeum*) *africana*, S.Af., also formerly used for wagons.

Stitchwort *Stellaria* spp.

Stock, garden *Matthiola incana*, S.Eur. popular gdn. pl., many cultivars: **hoary** —, *M. incana*: **night-scented** —, *M. bicornis:* — **rose**, *Sparmannia africana:* **sea** —, *Matthiola sinuata:* **Virginia** —, *Malcolmia maritima*.

Stocking tree *Eucalyptus kondinensis*, W.Aus.

Stokesia *S. laevis* (= *S. cyanea*), N.Am., cult. gdn. pl., fls. white, yellow or blue.

Stone-crop *Sedum anglicum, S. acre* and other spp.: — **drop**, *Penthorum* spp., N.Am.: — **foot**, *Collinsonia canadensis:* — **leek**, *Allium fistulosum:* — **orpine**, *Sedum reflexum:* — **pine**, *Pinus pinea:* — **rush**, *Scleria* spp., N.Am.: — **wort**, *Sedum* spp.

Storax A fragrant balsam or oleo-resin, obtained originally from *Styrax officinale*, a small Medit. tree, and used since ancient times: also obtained from the b. of *Liquidambar orientalis* in As. Minor. **American** — or sweet gum is obtained from *Liquidambar styraciflua*. Storax is an ingredient of compound benzoin tincture and of benzoin inhalant.

Storksbill *Erodium* spp. some spp. cult., orn. fls.

Stove plant One which, in a temp. cli., requires a hot-house with constant high temperature and humidity for successful growth.

Strainer vine (W.I.) Sponge gourd or *Luffa*.

Stramonium Or thorn apple, *Datura stramonium*, a cosmop. weed, also cult., ls. and seeds contain hyoscyamine and hyoscine: the drug stramonium consists of the dried ls.: it is used in the treatment of asthma to relieve bronchial spasm.

Strangle weed (Am.) *Cuscuta* spp., dodder.

Stranvaesia Evergreen Chinese shrubs w. showy fls., cult.

Strapwort *Corrigiola litoralis*, Eur. inc. Br.

Strasburg turpentine Or Alsatian turpentine, *Abies pectinata*, silver fir, collected like Canada balsam, med. in Eur.

Straw The dry stems of various cereals, that of wheat being the most imp. and generally useful: has many agric. uses such as bedding for animals, mulching, insulating (clamps) etc.: used for making paper, straw hats, bottle covers etc.

Strawberry *Fragaria* × *ananassa*. One of the best known and popular temp. frs., numerous cultivars and hybrids, much used for jam and canning. **Barren** —, *Potentilla sterilis*, Eur., *Waldsteinia fragarioides*, N.Am.: — **blite**, *Chenopodium capitatum:* — **bush**, *Euonymus americanus:* — **cactus**, *Echinocereus enneacanthus:* **Chiloe** —, *Fragaria chiloensis*, cult. in Andes, has been used in strawberry breeding: — **guava**, *Psidium cattleianum*, cult.: **Hautbois** —, *Fragaria moschata*, temp. Eur., cult.: **Ind.** —, *Duchesnea indica*, As., cult.: fr. insipid: **Mex.** —, *Echinocactus stramineus:* — **raspberry**, *Rubus illecebrosus*, cult.: — **shrub**, *Calycanthus floridus*, N.Am.: **sow-teat** —, *Fragaria vesca* var. *americana:* — **tomato**, *Physalis* sp.: — **tree**, *Arbutus unedo:* **wild** —, *Fragaria vesca:* **Virginian** —, *Fragaria virginiana*.

Strelitzia Orn. S.Af. pls., cult., esp. *S. reginae*, bird of Paradise fl.

Streptocarpus Or Cape primrose, cult. orn. pls., many hybrids: *S. polyanthus,* often cult.

Streptosolen *S. jamesonii,* orn. climbing shrub from Colombia long grown in hot-houses, has attractive clusters of orange fls.

Stretchberry *Smilax bona-nox,* E. N.Am.

Striga *S. lutea* and other spp., semi-parasites, troublesome when crops such as maize and sorghum are attacked. The word striga also signifies a small, straight, hair-like scale.

String lily *Crinum americanum,* N.Am.

Stringybark (Aus.) Spp. of *Eucalyptus* w. fibrous b., some imp. for timber: **brown** —, *E. capitellata:* **red** —, *E. macrorhyncha:* **white** —, *E. eugenioides:* **white-top** —, *E. gigantea:* **yellow** —, *E. acmenoides, E. muelleriana, E. triantha.*

Striped king of the woods *Anoectochilus regius,* a Sri Lanka orchid, cult.: **— squill,** *Puschkinia scilloides,* Orient., cult. orn. fl.

Strobilanthes Old World trop., cult., some have stigmas sensitive to touch.

Strongman's weed (W.I.) *Petiveria alliacea,* shrub w. strong odour, med.

Strophanthus Shrubs of trop. Af. and As.: seeds of several have been used for arrow poisons in Af., some are med., e.g. *S. kombe, S. gratus* and *S. hispidus* and the source of the drug strophanthine which resembles digitalis in its action. Several spp. have attractive or unusual fls., and are cult., petals may have long strap-like appendages. At one time *Strophanthus* seeds were of interest as a possible source of cortisone.

Strychnine This poisonous alkaloid is extracted from the seeds of *Strychnos nux-vomica* and allied spp. It has med. uses and stimulates all parts of the nervous system but as a therapeutic agent it is no longer used. It has been used for poisoning harmful animals (jackals in S.Af.).

Styptic weed *Cassia occidentalis,* trop.

Succory *Cichorium intybus,* chicory, ed. fol.: **gum** —, *Chondrilla* spp.: **swine** or **lamb's** —, *Arnoseris minima,* Eur. inc. Br.

Succulents The term is applied hort. to fleshy pls. The majority of succulents grown by enthusiasts or collectors are from the Old World (mainly Af.) and are frequently grown with cacti from the New World. Both groups exist in the wild state under similar climatic or environmental conditions, dry or semi-desert conditions for much of the year with great heat and high light intensity. S.Af. is particularly rich in small or low-growing succulents from the Karoo, and other desert or semi-desert regions and provides many of the subjects for the collector and cultivator of succulents. The following are some of the genera from which they were derived but there are of course many more – *Aloe, Anacampseros, Argyroderma, Ceropegia, Cheiridopsis, Conophytum, Cotyledon, Crassula, Dinteranthus, Echeveria, Glottiphyllum, Haworthia, Kalanchoe, Lithops, Pleiospilos, Sedum, Sempervivum, Senecio, Stapelia.*

Sucupira Name for commercial timber derived from spp. of *Bowdichia* and *Diplotropis,* also *Ferreira spectabilis* in Braz.

Sudan grass *Sorghum sudanense,* fodder grass of trop. and subtrop., also a weed: **— teak,** *Cordia myxa.*

Sugar apple Name for custard apple, *Annona* spp.: **wild — —,** *Rollinia* *multiflora*, W.I.: — **bean**, *Phaseolus lunatus:* — **berry**, *Celtis occidentalis* and other spp., N.Am.

Sugar beet *Beta vulgaris* var. Nearly half the world's supply of manufactured sugar is derived from sugar beet, an imp. crop in many temp. countries, inc. Br. It is also grown, often w. irrigation, as a cool season crop in some trop. and subtrop. countries, e.g. Ind. and N.Af. Numerous vars. have been bred to suit different conditions. Sugar beet is an imp. source of industrial alcohol in some countries and the leafy tops and pulp from the factory are a much used stock food.

Sugar bowl *Clematis hirsutissima*, N.Am.: — **bush**, *Protea mellifera* and other spp., S.Af., during early days at the Cape the fl. heads were soaked in water to remove the copious nectar and the solution evaporated to an ed. sugar syrup.

Sugar cane *Saccharum* spp. is responsible for more than half the world's supply of manufactured sugar with molasses an imp. by-product. Formerly vars. of *Saccharum officinarum*, or 'noble' canes, were mainly grown, but these were supplanted by more disease-resistant canes bred partly from wild canes or other spp. of *Saccharum*. Rum is another imp. by-product of the sugar cane industry. Baggasse, or the crushed cane after expression of the juice, is mainly used for fuel in the sugar mills, but may also be used for fibre board or some classes of paper.

Sugar grape *Vitis rupestris*, N.Am.: — **huckleberry**, *Vaccinium vacillans*, N.Am.: — **maple**, *Acer saccharum*, E. N.Am., Can., the juice is tapped from mature trees in the early spring when the sap has commenced to rise. It is evaporated down to yield maple sugar and maple syrup: — **orchid**, *Caladenia saccharata*, Aus.: **palm** —, or syrup, obtained from many palms by tapping, e.g. *Arenga, Borassus, Caryota, Cocos, Elaeis, Jubaea, Nypa, Phoenix:* — **pine**, *Pinus lambertiana*, N.Am.: — **plum**, *Uapaca guineensis*, W.Af., fr. ed.: —**protea**, *Protea mellifera*, S.Af.

Sugi Or Jap. cedar, *Cryptomeria japonica*, widely used timber in Jap. soft and light, tree also cult. for orn.

Sulla *Hedysarum coronarium*, Medit., a fodder crop.

Sulphur flower *Eriogonum flavum*, N.Am.: — **root**, *Peucedanum officinale* and *P. ostruthium*, Eur. inc. Br., yellow sap.

Sultan's flower *Impatiens wallerana*, E.Af., a popular house pl., many cultivars.

Sumac *Rhus coriaria*, Medit. The dried ground ls. are a well known commercial tanning material, Sicily being the main producer. It is favoured in the light leather industry and for sheepskin leathers. Its use has declined with competition from other cheaper tans. Good quality sumac has a tannin content of about 26–27%. **Am.** —; Several spp. of *Rhus* in the eastern U.S. have been used in tanning, esp. *R. copallina* and *R. glabra:* **Cape** —, *Colpoon compressum*, S.Af.: **Chinese** —, *Rhus semialata:* **country** —, *Anogeissus latifolia*, Ind.: **fragrant** —, *Rhus canadensis:* **Ind.** —, *Cotinus coggygria* (= *Rhus cotinus*), also known as Turkish, Venetian, Hungarian or Tyrolean sumac, may be used in local tanning: **lemon** —, *Rhus aromatica*, N.Am.: **poison** —, *R. vernix, R. toxicodendron*, E. U.S.: **staghorn** or **velvet** —, *R. typhina*, N.Am., cult. orn. shrub.

Sumbul Or musk root, *Ferula sumbul*, Turkestan, med.

Summer cypress Or burning bush, *Kochia scoparia*, a gdn. pl., fol. turns red in late summer: — **grape**, *Vitis aestivalis*, N.Am.

Sumpweed *Iva ciliata*, N.Am.

Sun berry *Physalis minima*, cult., fr. used in preserves: — **bonnet**, *Chaptalia tomentosa*, N.Am.: — **bush, purple** (Aus), *Helichrysum purpurascens:* —**dew**, *Drosera* spp., insectivorous pls., often in bogs or wet moors (*see Drosera* in Willis): — **drops**, *Oenothera fruticosa*, N.Am.: —**flower**, *Helianthus* spp., cult. for orn.: **giant** —**flower**, *Helianthus annuus*, an imp. oil seed crop: **Mex.** —**flower**, *Tithonia diversifolia*, trop. weed: **prickly** —**flower**, *Berkheya ilicifolia*, S.Af.: **red** —**flower**, *Tithonia rotundifolia:* — or **sunn hemp**, *Crotalaria juncea*, Ind., fibre and green manure crop: — **orchid**, *Thelymitra* spp., Aus.: — **plant**, *Portulaca grandiflora:* — **ray**, *Enceliopsis argophylla*, N.Am.: — **ray everlasting**, *Helipterus floribundus*, Aus.: — **rose**, *Helianthemum* spp., cult. favoured for dry places.

Sunshine flower Name adopted for *Venidio–Arctotis* hybrids.

Supple-jack *Paullinia* spp., *Serjania* spp., W.I., lianas, stems used for walking sticks: **Aus.** —, *Ventilago viminalis:* **N.Z.** —, *Rhipogonum scandens.*

Surette *Bysonima* spp., W.I. useful timber, b. may be used in tanning.

Sutherlandia *S. frutescens*, S.Af., orn. fl. shrub, cult.

Swallow wort *Chelidonium majus*, blossoms at the time the swallow arrives (in Br.): *Asclepias curassavica*, cult. orn. fl.

Swamp bay *Magnolia glauca*, *M. virginiana*, N.Am.: — **blueberry**, *Vaccinium corymbosum*, N.Am.: — **cabbage**, *Ipomoea reptans*, W.I.; *Spathyema foetida*, N.Am.: — **candles**, *Lysimachia terrestris*, N.Am.: — **cottonwood**, *Populus heterophylla:* — **cypress**, *Taxodium distichum*, N.Am., handsome deciduous conifer, widely cult., protuberances or knees arise from the rs. in damp situations: — **foxtail grass**, *Pennisetum alopecuroideum*, Aus.: — **haw**, *Viburnum prunifolium:* — **heath**, *Sprengelia* spp., Aus.: — **hickory**, *Carya aquatica*, N.Am.: — **laurel**, *Kalmia polifolia*, N.Am.; *Magnolia virginiana*, N.Am.: — **lily**, *Crinum americanum:* — **locust**, *Gleditsia aquatica*, N.Am.: — **mahogany**, *Eucalyptus robusta*, *E. botryoides*, Aus.: — **milkweed**, *Asclepias incarnata*, N.Am.: — **millet**, *Isachne globosa*, Aus.: — **oak**, *Casuarina* spp.: — **potato**, *Sagittaria* spp., N.Am.: — **privet**, *Forestiera acuminata*, N.Am.: — **rose**, *Rosa palustris*, N.Am.

Swan flower *Aristolochia gigas* or other large flowered spp., fl. buds more or less resemble a swan: — **plant**, *Asclepias physocarpa*, cult. orn. shrub, inflated seed pods floating in a bowl resemble young swans: — **river daisy**, *Brachycome iberidifolia*, cult.

Swarri nut Or souari nut, *Caryocar nuciferum*, Guyana, a large thick-shelled ed. nut, also called butternut.

Swaziland grass *Digitaria* sp., a lawn grass.

Swede or **Swedish turnip** *Brassica napo-brassica*, an imp. stock food in temp. countries, also eaten as a veg.

Sweet alyssum *Lobularia* (=*Alyssum*) *maritima:* — **balm**, *Melissa officinalis:* — **bark**, =cascarilla b.: — **basil**, *Ocimum basilicum:* — **bay**, *Laurus nobilis:* — **brier**, *Rosa eglanteria:* — **buckeye**, *Aesculus octandra:* — **bush**, *Clethra alnifolia*, N.Am.: — **chestnut**, *Castanea sativa:* — **cicely**, *Myrrhis odorata:* — **clover**, *Melilotus* spp.: — **corn**, *see* maize: — **cumin**, *Pimpinella anisum:*

— **cup**, *Passiflora maliformis:* — **fennel**, *Foeniculum vulgare:* — **fern**, *Comptonia aspleniifolia*, N.Am., has fern-like ls.: — **flag**, *Acorus calamus*, Eur. inc. Br., scented rhizomes: — **gale**, *Myrica gale*, Br.: — **galingale**, *Cyperus longus:* — **grass**, *Glyceria plicata* and other spp.: — **vernal grass**, *Anthoxanthum odoratum:* — **gum**, *Liquidambar styraciflua*, N.Am.: — **heart**, *Talinum triangulare*, W.I., weed and pot-herb: — **lime**, *Citrus aurantiifolia* var. *limetta:* — **marjoram**, *Origanum majorana:* — **Nancy**, *Achillea ageratum:* — **olive**, *Osmanthus fragrans:* — **pea**, *Lathyrus odoratus:* — **pea bush**, S.Af., *Podalyria calyptrata:* — **pepper bush**, *Clethra alnifolia*, N.Am.: — **potato**, *Ipomoea batatas*, widely cult. trop. and subtrop.: — **rocket**, *Hesperis matronalis*, old favourite gdn. fl.: — **scabious**, *Scabiosa atropurpurea:* — **sop**, *Annona squamosa*, trop.: — **sultan**, *Centaurea imperialis*, fragrant orn. fl.: — **verbena tree**, *Backhousia citriodora*, Aus.: — **William**, *Dianthus barbatus*, popular gdn. fl.

Swertia Large widespread genus, a few spp. of gdn. value, such as *S. perennis*.

Swine's cress *Coronopus didymus*, Eur. inc. Br.: — **succory**, *Arnoseris minima*, Eur. inc. Br.

Swiss chard Or spinach beet, *Beta vulgaris*.

Switch cane *Arundinaria tecta*, N.Am.: — **ivy** *Leucothoe editorum*, N.Am.

Swizzlestick *Rauvolfia vomitoria*, trop. Af.: *Quararibea turbinata*, trop. Am.

Sword bean *Canavalia ensiformis* widely cult. trop. veg.: — **fern**, *Nephrolepis polypodium:* — **grass**, *Scirpus americanus*, N.Am.: — **leaf phlox**, *Phlox buckleyi*, N.Am.: — **plant**, *Echinodorus* spp. aquarium pls.: — **sedge**, *Lepidosperma gladiatum*, used to bind sand in Aus.

Sycamore *Acer pseudoplatanus*, Eur., widely naturalized in Br., will grow in exposed situations, esp. near the sea, where other trees fail: called great maple in Scotland. In Am. the planes (*Platanus*) are called sycamore. The wood of sycamore may be used for interior decoration, furniture or cabinet work and is favoured for dairy and kitchen utensils. 'Fiddle-back sycamore' is the figured wood used for veneers and 'harewood' the wood stained grey The fls. are a good early source of nectar for the hive bee. They furnish a greenish-coloured honey.

Sycomore Or mulberry fig, *Ficus sycomorus*, the sycomore of the Bible, fr. ed., light wood was used for coffins for mummies.

Sydney red gum *Angophora costata*.

Symphyandra *S. asiatica*, Orient, showy fls., cult.

Synedrella *S. nodiflora*, a trop. weed.

Syringa Several different pls. are known by this name. *Philadelphus* spp. mock orange, cult. orn. shrub: *Melia azedarach*, Persian lilac, orn. tree, widely cult. in trop. and subtrop.: *Syringa* spp., common lilac, orn. fl. shrub: **night —**, *Cestrum nocturnum*, W.I., evergreen shrub, fls. delicately perfumed at night: **white —**, *Kirkia acuminata*, S.Af., useful furniture wood: **wild —**, *Burkea africana*, S.Af.: **yellow —**, *Cestrum parqui*, Chile, orn. shrub.

T

Tabasheer or **tabishir** White siliceous concretion sometimes found in the hollow stems of bamboo (*Bambusa arundinacea* and other spp.) in trop. As., has local med. uses.

Tabebuia Trop. Am. trees and shrubs, fls. often large and showy, cult.: some spp. .yield good timber.

Tabernaemontana Trop. trees and shrubs, fls. often fragrant and showy, cult.: some spp. contain rubbery latex.

Tabernanthe *T. iboga*, W.Af., rs. med., contain hallucinogenic alkaloids.

Tacamahac *Populus balsamifera*, balsam poplar, N.Am.

Tacamahaca *Calophyllum inophyllum*, trop., resin once used med.

Tacay *Caryodendron orinocense*, Colombia, seeds ed. after roasting.

Tacca arrowroot *Tacca leontopetaloides* (= *T. pinnatifida*), Fiji.

Taccada pith *Scaevola koenigii*, used by Malays and Thais for making artificial fls. and frs.

Tackstem *Calycoseris wrightii*, SW. U.S., a handsome desert ann.

Tagasate *Cytisus proliferus*, Canary Isles, a fodder tree or shrub.

Tagetes Trop. or subtrop. Am. pls., some cult. as gdn. pls., some are weeds.

Taggar *Cinnamosma fragrans*, Madag. a scented wood exported to the Orient.

Tagua Vegetable ivory, *Phytelephas macrocarpa*, trop. Am.

Tahiti arrowroot *Tacca leontopetaloides*: — **chestnut**, *Inocarpus edulis*, seeds eaten raw or cooked.

Tail flower *Anthurium* spp.

Tainia As. orchids w. large- or medium-sized fls., cult.

Tak-out galls Tamarisk galls, rich in tannin.

Talinum *T. triangulare*, a trop. weed and pot-herb.

Talipot palm *Corypha umbraculifera*, trop. As., a tall palm. w. very large ls. used for thatching, tents and umbrellas, one l. can shelter 7 or 8 people. Buddhist or Sinhalese 'books' made from strips of the ls.

Tallerack *Eucalyptus tetragona*, W.Aus.

Tallicona *Carapa guianensis*, andiroba or crabwood.

Tall-oil By-product of the wood pulp industry used in soap manufacture.

Tallow, vegetable Chinese — **tree**, *Sapium sebiferum*: **Jap.** — **tree**, *Rhus succedanea*: **Mafura** —, *Trichilia emetica*, trop. Af.: **Malabar** —, *Vateria indica*: — **nut**, *Ximenia americana*: — **shrub**, *Myrica cerifera* and other spp.: — **tree**, *Pentadesma butyracea*, W.Af.: — **wood**, *Eucalyptus microcorys*, Aus., a commercial timber.

Tamacoari or **tamaquare** *Caraipa psidiifolia*, Braz., seed oil med. (skin diseases).

Tamarack larch *Larix laricina*, N.Am., a commercial timber.

Tamarilla Tree tomato, *Cyphomandra betacea*.

Tamarind *Tamarindus indica*, valued as a shade and orn. tree in the arid trop.: pulp surrounding the seeds eaten in various ways, has laxative properties.

Tamarinds in syrup are a W.I. delicacy. **Coolie** —, *Averrhoa carambola*, W.I.: **cow** —, *Samanea saman:* **horse** —, *Leucaena glauca:* **native** —, *Diploglottis australis*, Aus.: — **plum**, *Dialium:* **Spanish** —, *Vangueria madagascariensis:* **velvet** —, *Dialium guineense*, W.Af.: **wild** —, *Leucaena glauca.*

Tamarisk *Tamarix* spp. Old World trees and shrubs, often maritime, cult, for orn. and for sand or erosion control. **Common** or **English** —, *T. anglica:* **French** —, *T. gallica:* **manna** —, *T. mannifera*, produces a sweet gummy substance as a result of scale insect attack, probably Biblical manna: still collected by the Bedouins.

Tamarisk galls Used in local leather tanning, esp. in Morocco, derived mainly from *Tamarix aphylla* (= *T. articulata*) and *T. gallica:* used in the production of Morocco leather: tanning content 40–45 %: tanning properties similar to sumac. Other names are 'tak-out' or 'teggaout'.

Tamate (W.I.) Tomato.

Tambootie grass or **tambookie** Any tall coarse grass (S.Af.), esp. *Hyparrhenia aucta* and other spp.; *Andropogon* or *Cymbopogon* spp. used for thatching and soil erosion control: — **tree**, *Spirostachys africanus.*

Tampico fibre Or ixtli, *Agave* spp.

Tan *See* tannin.

Tanekaha *Phyllocladus trichomanoides*, N.Z., good quality wood, b. rich in tannin.

Tangelo Citrus fr., cross between tangerine and grapefruit or pomelo: name derives from *tang*erine and pom*elo.*

Tangerine *Citrus reticulata* (= *C. nobilis*). There are numerous vars. of this citrus fr., popular on account of its fine flavour and being so easy to peel.

Tanghin *Cerbera tanghin*, Madag., frs. poisonous, used as ordeal poison.

Tangier pea *Lathyrus tingitanus*, cult. orn. fl.

Tangle legs *Viburnum alnifolium*, N.Am.

Tangor Citrus fr. which is a hybrid between a *tang*erine and an *or*ange.

Tang-shen *Codonopsis tangshen*, China, a ginseng substitute.

Tanner's dock *See* canaigre.

Tannia or **tania** *Xanthosoma sagittifolium*, a trop. r. crop.

Tannins and tanning materials Veg. tanning materials or pls. rich in tannin have been used by man in many parts of the world from the earliest times for converting the skins of animals into leather or rendering them more suitable, more soft and more pliable, for use as clothing, tents, or other purposes. Tannins are very widely distributed in the veg. kingdom, but it is only those pls. that are rich or relatively rich in tannins that are used in the tanning of leather. Most of the commercially imp. tanning materials are products of warm countries. The use of different veg. tanning materials in commercial tanning has changed with the passage of time and with technological advance, for new chemicals and synthetic materials have made great advances. Furthermore, the use of plastics and composition materials for the soles of footwear has had considerable repercussions in the leather and tanning industries.

There are some pl. families in which tannin occurs very freely, such as the *Leguminosae, Ancardiaceae, Combretaceae, Rhizophoraceae, Myrtaceae* and

Polygonaceae. On the other hand there are families in which tannin is absent or rare, such as the *Gramineae* (grasses), *Cruciferae* and *Papaveraceae*. Tannin may occur in almost any part of the pl., b., stem or trunk, ls., frs., rs. and even hairs.

Bs. are perhaps the most imp. commercial sources of veg. tanning materials. The more notable veg. tanning materials that are or have been used in commercial tanning include the following – *Barks:* black wattle or mimosa (*Acacia*), mangrove, oak, spruce, hemlock, eucalyptus, avaram, babul, birch, willow, pine, larch, alder. *Woods:* quebracho, urunday, chestnut, oak, cutch, wandoo, tizra. *Fruits:* myrobalans, valonea, divi-divi, algarobilla, tara, teri pods, sant pods, pomegranate rinds. *Leaves:* sumac, Am. sumac, gambier, dhawa or country sumac. *Roots:* canaigre, docks, Siberian saxifrage, sea lavender. *Plant-galls:* oak galls, Chinese galls, tamarisk galls, pistacia galls. [Howes, F. N. (1953) *Vegetable tanning materials*, 324 pp.]

Tansy *Tanacetum* (= *Chrysanthemum*) *vulgare*, once a herbal remedy and used for flavouring and garnishing.

Tapa cloth *Broussonetia papyrifera*, Pacific islands, made from b.

Tape grass *Vallisneria spiralis*, N.Am.

Tapioca Food product made from cassava starch, *Manihot esculenta*.

Tar Wood tar, as distinct from ordinary coal tar, is obtained from wood distillation, chiefly Scots pine (*Pinus sylvestris*), an imp. industry in parts of N.Eur ('Stockholm tar'): — **flower**, *Bejaria racemosa*, N.Am. has sticky buds: — **weed**, *Grindelia camporum*, N.Am.; *Amsinckia* spp.; *Madia* spp.

Tara *Caesalpinia spinosa*, Peru, pods used in tanning.

Taraire *Beilschmiedia taraira*, N.Z., a fine timber tree.

Tarata *Pittosporum eugenioides*, N.Z.

Tarbush Or varnish bush, *Flourensia cernua*, SW. U.S.

Tare Or vetch, *Vicia sativa* and other spp., cult. as forage, two main kinds — summer and winter.

Taro *Colocasia esculenta* (=*C. antiquorum*), an imp. trop. r. crop, large starchy tubers: **giant** —, *Alocasia indica*, one of the main food pls. of the Gilbert and Ellice Isles.

Tarragon *Artemisia dracunculus*, E.Eur. Fragrant ls. used for seasoning, esp. in Eur., have a bitter-sweet flavour, used in salads, soups, stews, sauces etc.: tarragon vinegar is favoured for fish sauces.

Tartarian aster *Aster tataricus:* — **honeysuckle**, *Lonicera tatarica:* — **lamb**, *Cibotium barometz*, the rhi. of this fern, covered w. yellow down, was the famous Tartarian lamb of 17th Century travellers.

Tasmanian blue gum *Eucalyptus globulus*, widely cult.: — **cedar**, *Athrotaxis* spp., useful timber: — **daisy bush**, *Olearia stellulata:* — **laurel**, *Anopterus glandulosus*, cult. shrub.

Tassel flower *Emilia flammea* (=*Cacalia coccinea*) trop. Am., scarlet fl. heads, cult.

Tasua *Aphanamixis polystachya; Aglaia cucullata*, Thailand, a commercial timber.

Taupata *Coprosma repens*, N.Z., an evergreen shrub.

Tau-saghyz *Scorzonera tausaghyz*, Russia, considered as a source of rubber during the Second World War.

Tawa *Beilschmiedia tawa*, N.Z., a fine timber tree.

Taya *Xanthosoma peregrinum*, W.I.

Tchirish *Asphodelus ramosus*, powdered rs. contain gum used by book binders in Turkey.

Tea *Camellia sinensis* (= C. *thea*, *Thea sinensis*). Tea has been grown in China and other As. countries from very early times and is now cult. in many other parts of the world. 'Assam' and 'China' teas are the two main groups under cult., the many vars. being termed 'jats'. In the wild state tea is a small tree but under cult. it is kept low by pruning to facilitate harvesting when young shoots w. 3 or 4 young ls. are plucked. Withering, rolling, fermenting and firing are the processes then carried out to produce the familiar black tea that is consumed. Green tea is not fermented. Black tea grades are known as Orange Pekoe, Pekoe, Pekoe Souchong, Souchong and Siftings. Green tea is graded into Young Hyson, Hyson, Siftings and Gunpowder. Brick teas are made for special markets. **Abyssinian** —, *Catha edulis:* **Algerian** —, *Paronychia* spp.: **Arabian** —, *Catha edulis:* **Ayapana** —, *Eupatorium triplinerve*, trop. Am.: **Bahama** —, *Lantana camara:* — **balm**, *Melissa officinalis:* — **berry**, *Gaultheria procumbens:* **Bogota** —, *Symplocos theiformis:* **Braz.** —, *Stachytarpheta dichotoma:* **bush** — (Aus.), *Ocimum sanctum;* (S.Af. or Cape), *Cyclopia* spp.: **chamomile** —, *Matricaria recutita:* **Colombian** —, *Capraria biflora* var. *pilosa:* **Faham** —, *Angraecum fragrans:* **Greek** —, *Salvia triloba:* **Guayusa** —, *Ilex guayusa:* **Matura** —, *Cassia auriculata:* **Mex.** —, *Chenopodium ambrosioides:* **mountain** —, *Gaultheria procumbens:* **New Jersey** —, *Ceanothus americanus:* — **olive**, *Osmanthus fragrans:* **Oswego** —, *Monarda didyma:* **Paraguay** —, *Ilex paraguensis*, *Lycium barbarum:* **Salvador** —, *Gaultheria:* — **seed oil**, *Camellia sasanqua:* — **tree**, *Leptospermum* spp. Aus.: **W.I.** —, *Capraria* spp.: **Yaupon** —, *Ilex vomitoria:* **yerba maté** —, *Ilex paraguensis*.

Teak *Tectona grandis*, Ind., Indomalaysia, cult., imp. commercial timber, strong and durable, much used in shipbuilding, numerous other uses. **Af.** —, *Oldfieldia africana:* **Aus.** —, *Flindersia australis:* **bastard** —, *Butea superba:* **Borneo** —, *Intsia bijuga:* a commercial timber: **Brunei** —, *Dryobalanops* spp., a commercial timber: **grey** —, *Gmelina* sp.: **Malacca** —, *Afzelia palembanica:* **Rhodesian** —, *Baikiaea plurijuga:* **Sudan** —, *Cordia myxa*.

Teasel Or fuller's teasel, *Dipsacus fullonum* subsp. *sativus*, the fruiting heads with their hooked bracts used for raising the nap on cloth, cult.: **wild** —, *D. fullonum* subsp. *fullonum* (*D. sylvestris*), similar but bracts not hooked although spiny.

Tecoma Trop. or subtrop. trees, shrubs or climbers: some trop. Am. spp. yield strong durable timber.

Tecomaria *T. capensis* (= *Tecoma capensis*), S.Af. orn. climber, cult.

Tectaria Old World trop. ferns, cult.

Teff grass *Eragrostis tef*, Ethiopia, seeds used as a cereal in Ethiopia: cult. in other countries, esp. S.Af., as fodder: two forms red- and white-seeded.

Teggaout *See* tamarisk galls.

Telegraph plant *Desmodium gyrans*, the lateral leaflets exhibit a rotary motion during the day if the temperature is suitably high. *Heterotheca subaxillaris*, SW. U.S., has tall, straight stems like telegraph poles.

Tellicherry bark *Holarrhena antidysenterica*, Ind., seeds and b. med.

Telopea Attractive evergreen shrubs, Aus., *T. speciosissima* is the waratah.

Temple tree or **flower** *See* frangipani.

Teosinte *Euchlaena mexicana*, Mex., resembles, and is a close relative of maize, used as a cereal in C.Am. and as a fodder.

Tepa *Laurelia serrata*, Chile, a commercial timber.

Tepary bean *Phaseolus acutifolius* var. *latifolius*, C.Am., very drought-resistant, many vars.

Tephrosia Trop. and subtrop., some are orn. and cult., some grown as cover crops or green manures: *T. vogelii* is a fish poison in trop. Af. and *T. macropoda* in S.Af.

Terblans *Faurea macnaughtonii*, S.Af., prettily marked wood.

Terebinth *Pistacia terebinthus*, Medit., small tree, source of Chian turpentine and tannin galls.

Teree touri (W.I.) *Caesalpinia digyna*, Ind., a good defensive hedge, has thorns.

Teri pods *Caesalpinia digyna*, used for tanning in Ind. and Burma, contains 22–27 % tannin.

Terminalia Many trop. spp. yield durable and attractive woods.

Terra japonica Old name for gambier.

Terrell grass *Elymus virginicus*, N.Am.

Tesota *Olneya tesota*, S.W. U.S., ed. seeds.

Tetracarpidium *See* conophor oil.

Tetranema *T. mexicanum*, Mex. foxglove, cult. orn. fl.

Tetter bush *Lyonia lucida*, N.Am.

Texas blue grass *Poa arachnifera:* — **grass**, *Vaseyochloa multinervosa*. — **millet**, *Panicum texanum*, a fodder grass: — **snakeroot**, *Aristolochia serpentaria*.

Thalictrum Perens., mainly N. temp., some w. attractive fls., cult., some have local med uses.

Thatching, palms used for *Attalea, Brahea, Calamus, Caryota, Cocos, Copernicia, Euterpe, Geonoma, Hyphaene, Livistona, Manicaria, Maximiliana, Nypa, Pandanus* (not a true palm), *Sabal, Socratea, Johannesteijsmannia, Thrinax, Washingtonia* etc.

Theetsee *Melanorrhoea usitata*, Burmese varnish tree.

Thelocactus Small very spiny cacti w. large fls., cult.

Thelymitra Orchids, mainly Aus., cult.

Theobroma oil Or cocoa butter, a fat extracted from crushed cocoa beans, *Theobroma cacao:* used pharmaceutically in making suppositories, pessaries, bougies etc.: cocoa butter melts at about the temperature of the human body: *theobromine*, an alkaloid obtained from cocoa bean husks, used med.

Thibaudia Handsome shrubs from the Andes, scarlet fls., cult.

Thimble berry *Rubus occidentalis*, N.Am. and other spp.: — **weed**, *Anemone virginiana*, N.Am. and other spp.

Thin-tail *Pholiurus incurvus*, SE. Eur., a grass naturalized in N.Am.

Thingadu (Burma) *Parashorea stellata*, a commercial timber.

Thingan (Burma) *Hopea odorata*, an imp. timber tree, also yields resin.

Thinwin (Burma) *Pongamia glabra*, tree of river banks and tidal areas, seed oil used as illuminant.

Thistle Name used for a range of herbaceous pls. usu. with prickly ls. or stems and fls. in heads. Often they are weeds and cosmop. **Blessed** —, *Cnicus benedictus:* **bull** —, *Cirsium vulgare:* **Can.** —, *Cirsium arvense:* **carline** —, *Carlina vulgaris:* **creeping** —, *Cirsium arvense:* **cotton** —, *Onopordon acanthium:* **distaff** —, *Atractylis cancellata:* **globe** —, *Echinops* spp.: **golden** —, *Scolymus hispanicus:* **hedgehog** —, *Echinocactus* spp.: **holy** —, *Silybum marianum:* **marsh** —, *Cirsium palustre:* **melon** —, *Melocactus* spp.: **milk** —, *Silybum marianum:* **musk** —, *Carduus nutans:* **ornamental** —, *Onopordon acanthium:* **plumed** —, *Cirsium* spp.: **red** —, *Emilia coccinea:* **Russian** —, *Salsola kali:* **saffron** —, *Carthamus* spp.: — **sage,** *Salvia carduacea:* **Scottish** —, *Onopordon acanthium, Carduus nutans, Cirsium vulgare:* **sow** —, *Sonchus* spp.: **spear** —, *Cirsium vulgare:* **star** —, *Centaurea calcitrapa:* **swamp** —, *Cirsium muticum:* **weather** —, *Carlina vulgaris:* **welted** —, *Carduus crispus:* **yellow** —, *Argemone mexicana.*

Thitka (Burma) *Pentace burmannica,* tall forest tree w. useful light timber.

Thitmin *Podocarpus neriifolius,* Himal., a much used yellowish timber.

Thorn Thorns or spines are stiff pointed outgrowths which may be of various origins, e.g. epidermal (*Prosopis, Rosa*) branches (*Carissa, Colletia, Crataegus, Gleditsia, Haematoxylon*): fl. stalk (*Mesembryanthemum*): l. (*Azima, Cactaceae, Citrus*): leaflet (*Cycadaceae*): midrib (*Astragalus, Fouquieria*): ovary (*Tetragonia, Tribulus*): petiole (*Astragalus*): r. (*Acanthorhiza, Iriartea*): stipules (*Acacia, Paliurus, Polygala, Zizyphus*). **Thorn** = hawthorn: — **apple,** *Datura stramonium:* **black**—, *Prunus spinosa:* **buffalo** or **Cape** —, *Zizyphus mucronata:* **Christ's** —, *Paliurus spina-christi; Zizyphus spina-christi:* **cockspur** —, *Crataegus crus-galli:* **devil** —, *Tribulus terrestris:* **garland** —, *Paliurus spina-christi:* **goat's** —, *Astragalus tragacantha:* **Jerusalem** —, *Parkinsonia aculeata:* **kaffir** —, *Lycium afrum:* **kangaroo** —, *Acacia* spp.: — **pear,** *Scolopia zeyheri,* S.Af.: **sallow** —, *Hippophae rhamnoides:* **sweet** —, *Acacia karroo:* — **tea,** *Cliffortia ilicifolia:* — **tree,** *Acacia* spp., S.Af.: **Washington** —, *Crataegus cordata.*

Thorow-wax (Am.) *Bupleurum rotundifolium:* — **wort,** *Eupatorium perfoliatum,* med.

Thread-foot *Podostemum ceratophyllum,* N.Am.: — **plant,** *Nemacladus glanduliferus,* SW. U.S. a small, profusely branched, desert pl.

Three-square *Scirpus americanus,* N.Am.

Thrift Or sea pink, *Armeria maritima,* wild (in Br.) and cult.: **Jersey** —, *A. arenaria.*

Throat wort *Trachelium caeruleum, Scrophularia nodosa, Geum virginianum.*

Thrum wort *Damasonium alisma,* rare in Br.

Thuja *See* arbor-vitae.

Thunbergia The genus includes many beautiful pls. often climbers w. large fls., widely cult. in warm cli., e.g. *T. alata,* black-eyed Susan, fls. w. black eyes: *T. fragrans* and *T. grandiflora.*

Thuya Trade name for wood of *Tetraclinis articulata,* N.Af., usu. in the form of burrs (not to be confused with the generic name *Thuja,* formerly *Thuya*).

Thyme, common or **garden** *Thymus vulgare,* is a widely used flavouring herb, its pleasant aromatic flavour blending with most dishes, esp. meats, poultry and fish. Lemon-thyme with its different flavour may be similarly used, also basil-thyme, *Satureja calamintha*, Medit. Many spp. of *Thymus* are cult., esp. in the rock gdn., inc. cultivars of the wild or creeping thyme, *Thymus praecox*.

Thymol Camphor-like substance prepared from thyme or other oils or prepared synthetically. It is a bactericide and has med. uses: may be used in toothpastes, gargles etc.

Tibouchina Handsome trop. Am. shrubs, mainly Braz., cult., esp. *T. semidecandra*, w. large purple fls. or velvety sheen.

Tick bush *Kunzia ambigua*, Aus.: — **clover**, *Desmodium* spp., N.Am.: — **seed**, *Coreopsis* spp., N.Am.: — **sunflower**, *Bidens coronata*, N.Am.: — **trefoil**, *Desmodium* spp., N.Am.: — **weed**, *Verbesina virginica*.

Tidy tips *Layia platyglossa*, W. N.Am., handsome ann., cult.

Tie palm *Cordyline indivisa*, N.Z., cult. fol. pl. w. attractive narrow ls.

Tiger cocoa *Theobroma bicolor:* — **claw**, *Martynia diandra:* — **flower**, *Tigridia pavonia*, C.Am. cult. orn. pl., ed. corm: — **grass**, *Thysanolaena*, trop. As.: —**'s jaws**, *Faucaria tigrina*, S.Af.: — **lily**, *Lilium tigrinum:* — **nut**, *Cyperus esculentus*, the dried tubers are eaten out-of-hand like nuts; used in parrot seed mixtures; a drink is prepared from them in Spain: — **orchid**, *Diuris sulphurea*, Aus.: — **wood**, *Lovoa klaineana*, W.Af.; *Machaerium schomburgkii*.

Tilia *See* lime.

Tillandsia Many spp. in this large trop. Am. genus cult. as orn. pls., handsome fls. and foliage. *See Tillandsia* in Willis.

Timber Physical characteristics are imp. in determining the value of wood or timber and the best uses to which it may be put, esp. its working qualities with different wood-working tools. Other imp. points are strength, nature of grain, resistance to splitting and ability to take nails, durability (outdoors), resistance to abrasion (flooring, counter-tops etc.), resistance to insect attack, ability to take polish, stains, preservatives etc. Most of the timber that enters commerce is from conifers (pine, spruce, larch, etc.) and large quantities are used for paper pulp. The lightest commercial wood is probably balsa and the heaviest lignum vitae, about eighty times as heavy as balsa. The number of timbers used throughout the world runs into many thousands, the genus *Eucalyptus* alone consisting of some 500 spp.

The following are some of the more imp. commercial timbers, the names used being the trade names or recommended trade names. Botanical names may be seen under separate headings.

Afara: Af. blackwood, canarium, celtis, ebony, mahogany, padauk, pencil cedar, satinwood, walnut: afrormosia: afzelia: agba: alan: albizia, W.Af.: alder, common, grey: alerce: alpine fir: alstonia: amabilis fir: Amboyna: Am. ash, beech, cherry, chestnut, hop-hornbeam, mahogany, pitch pine, plane, red gum, red oak, walnut, white oak: Andaman gurjan, marblewood, padauk, pyinma: angelin: antiaris: apitong: apple: ash, Am., Aus., white, Eur., Jap.: aspen, Can., Eur.: Aus. blackwood, cedar, silky oak, white ash: Austrian pine: avodire: ayan.

Balau, red: balsa: balsam fir: banak: basralocus: basswood: bean, black: beech, Am., Eur., Jap., negro-head, silver: berlinia: bicuiba: billian: bintangor: binuang: birch, Eur., paper, western paper, yellow: black bean, cottonwood, poplar, willow: blackbutt: blackwood, Afr., Aus.: bluegum, southern: bombax, W.Af.: bombway, white: box, brush: boxwood, E. London, Eur., Knysna., Maracaibo, San Domingo: Braz. rosewood, tulip wood: brazilwood: Br. Honduras yellowwood: brown sterculia: brush box: bubinga: bunya pine: Burma cedar, gurjun, padauk, tulip wood.

Camphorwood, E.Af.: camwood: Can. aspen, poplar: canarium, Af., Ind., Malayan: Caribbean pitch pine: cativo: cedar, Af. pencil, Am., Burma, C.Am., incense, Port Orford, S.Am., southern white, Virginian pencil, western red, white, yellow: ceiba: celtis, Af.: C.Am. cedar: Ceylon ebony, satinwood: chan: cherry, Am., Eur.: chestnut, Am., sweet: chickrassy: Chile pine: Chilean laurel: chuglam, white: chumprak: coachwood: cocobolo: cocuswood: coigne: common alder: copalwood Rhodesian: Corsican pine: cotton-wood, black, eastern: courbaril: crabwood: cricket bat willow: cypress, southern,

Dahoma: dalli: danta: dark red meranti: dark seraya: dau: degame: dhup, red: dogwood: Douglas fir: Dunkeld larch: Dutch elm.

E.Af. camphorwood, cypress, olive: eastern Can. spruce, cottonwood, hemlock: ebony Af., Ceylon, Macassar: ekki: elm, Dutch, English, Jap., rock, smooth-leaved, white, wych: eng: Engelmann spruce: Erimado: Eur. ash, aspen, beech, birch, boxwood, cherry, hop-hornbeam, horse chestnut, larch, lime, oak, plane, spruce, walnut.

Field maple: Fijian kauri: fir, alpine, amabilis, balsam, Douglas, grand, noble, silver: freijo: fustic: futui: gaboon: gedu nohor: gonçalo alves: grand fir: greenheart: grevillea: grey alder, poplar: guarea: gum, Am. red, saligna, southern blue, spotted: gurjun, Andaman, Burma, Ind.

Hackberry: haldu: hemlock eastern, western: hickory: holly: holm oak: Honduras rosewood: hoop pine: hop-hornbeam, Am., Eur.: hora: hornbeam: horse chestnut, Eur., Jap.: huon pine: hura: indigbo: ilomba: imbuya: incense cedar: Ind. canarium, gurjun, laurel, rosewood, silver-grey wood: Indon. kapin, keruing: iroko: ironbark.

Jacareuba: jack pine: Jap. ash, beech, elm, horse chestnut, larch, lime, maple, oak, walnut: jarrah: jelutong: jequitiba: jongkong: kanluang: kapong: kapur, Indon., Malayan, North Borneo, Sarawak: Karri: katon: katsura: kaunghmu: kauri, E.I., Fijian, N.Z., Queensland: keruing, Indon., Malayan, North Borneo, Sarawak: kiaat: kingwood: Knysna boxwood: kokko: krabak: kumbuk.

Laburnum: lancewood: larch, Dunkeld, Eur., Jap., Siberian, tamarack, western: lauan, red, white: laurel Chilean, Ind.: light red meranti, seraya: lignum vitae: limba: lime, Eur., Jap.: lingue: lodgepole pine: lolagbola: loliondo: louro, red, inamui, preto: lunumidella: Macassar ebony: mafu: magnolia: mahogany, Af., Am.: maidu: makore: Malayan canarium, kapur, keruing: manio: mansonia: maple, field, Jap., Pacific, Queensland, rock, soft: Maracaibo boxwood: marblewood, Andaman: maritime pine: matai: mengkulang: meranti, dark red, light red, white, yellow: merawan: merbau: mersawa: miro: missanda: moabi: mora: mugongo: muhimbi: muhuhu: mukunari: mulga: mussinga: musino: musizi: mvule: myrtle, Tasman.

Nazingu: negro-head beech: New Guinea walnut: N.Z. Kauri, white pine: niangon: noble fir: nongo: North Borneo kapur, keruing: Norway maple: nyatoh: oak, Am. red, white, Eur., holm, Jap., Persian, Tasm., Turkey:

TIMBO

obeche: odoko: odum: ofram: okoume: okwen: olive, E.Af.: olivillo: omu: opepe.

Pacific maple: padauk, Afr., Andaman, Burma: paldao: palosapis: panga-panga: paper birch: Parana pine: partridge wood: pear: pecan: pencil cedar, Af., Virginian: peroba, white, rosa: Persian oak: persimmon: pillarwood: pine, Am. pitch, Austrian, bunya, Can. red, Caribbean pitch, Chile, Corsican, hoop, huon, jack, lodgepole, maritime, N.Z. white, Parana, pitch, ponderosa, radiata, Scots, Siberian yellow, sugar, western white, yellow.

Rosewood, Honduras, Ind.: sal: saligna gum: salmwood: San Domingo boxwood: sandalwood: Santa Maria: sapele: sapodilla: Sarawak kapur, keruing: satiné: satinwood, Af., Ceylon, W.I.: Scots pine: Selangan red, batu: sempilor: sen: sepetir: sequoia: seraya, dark red, light red, white, yellow: Siberian larch, yellow pine: silky oak, Aus.: silver beech, fir: silver grey wood, Ind.: simaruba: sissoo: Sitka spruce: smooth-leaved elm: snake-wood: soft maple: S.Am. cedar: southern blue gum, cypress, white cedar: spotted gum: spruce, eastern Can., Engelmann, Eur., Sitka, western white: sterculia brown, yellow: stringybark, yellow: sucupira: sugar pine: sugi: sweet chestnut: sycamore.

Tallow wood: tamarack larch: Tasm. myrtle: tasua: teak: teak, Rhodesian: tepa: thingadu: thingan: thitka: thuya: totara: tulip wood, Braz., Burma: tupelo: Turkey oak: turpentine: ulmo: utile: verawood: vinhatico: Virginian pencil cedar.

Wallaba: walnut, Af., Am., Eur., Jap., New Guinea, Queensland: wandoo: wawa: wellingtonia: wenge: W.Af. albizzia, bombax: W.I. satinwood: western hemlock, larch, paper birch, red cedar, white pine, white spruce: white ash Aus., bombway, cedar, southern, chuglam, elm, lauan, meranti, oak, Am., peroba, pine N.Z., poplar, seraya, spruce western, willow: white-beam: whitewood, Am.: willow, black, cricket bat, white: wych elm.

Yang: yellow birch, cedar, meranti, pine, seraya, sterculia, stringybark: yellowwood, Honduras: yew: yomkin: yon: zebrano.

Timbo Name for *Lonchocarpus nicou* and other fish poison pls. in S.Am.

Timothy grass *Phleum pratense*, a valuable agric. grass, cult. for pasture or hay in most temp. countries: many vars. or strains: first grown on a field scale in Am. and introduced to England by one Timothy Hansen in 1780, and hence its name.

Tim-tom bush *Chiococca alba*, W.I., snowberry, a scandent shrub.

Tingo fibre *Pouzolzia hypoleuca*, trop. Af., used for fishing nets.

Tinker's weed *Triosteum perfoliatum*, N.Am.

Tinospora Several spp. used med. in trop.

Tirucalli *Euphorbia tirucalli*, trop. & S.Af., cult. as a hedge, latex harmful to the eyes, contains rubber. The presence of this shrub or small tree is often an indication of former native burial grounds.

Tisso flowers Or pulas, *Butea superba* (=*B. frondosa*), Ind., the source of a yellow or orange dye.

Tithonia Or Mex. sunflower, cult.

Titi *Oxydendrum arboreum*, N.Am., sorrel tree, ls. scarlet in autumn; *Cliftonia monophylla*, N.Am., a good honey source: **black —**, *Cyrilla racemiflora*, N.Am.

Titoki *Alectryon excelsum*, N.Z., tough wood used for tool handles.

Tizra *Rhus pentaphylla*, a well known tanning material of Morocco and other parts of N.Af., rs. and heartwood mainly used, employed in the production of Morocco leather.

Toadflax *Linaria* spp. **common** —, *L. vulgaris:* **bastard** —, *Thesium humifusum*, a r. parasite: **blue** —, *Linaria canadensis:* **ivy-leaved** —, *Cymbalaria muralis*, common on walls in Br.: **old field** —, *L. canadensis:* **purple** —, *L. purpurea:* **striped** —, *L. repens.*

Toad lily *Tricyrtis* spp. E.As. cult.: — **plant**, *Stapelia* spp., S.Af.: — **rush**, *Juncus bufonius*: — **shade**, *Trillium sessile*, N.Am.: — **tree**, *Conopharyngia elegans*, fr. skin resembles that of a toad.

Toatoa *Phyllocladus glaucus*, N.Z., a handsome tree.

Tobacco *Nicotiana tabacum*, S.Am., *N. rustica*, C.Am., cult., trop. and subtrop., most of the tobacco smoked throughout the world is derived from *N. tabacum*, that from *N. rustica* is of a coarser, poorer quality but the pl. is hardy, matures quickly and is grown in some countries, esp. Ind. Numerous vars. of *N. tabacum* have evolved or been bred in different tobacco growing regions. The pl. is an ann. and the fl. head is removed to encourage full l. development. The ls. are collected when they commence to change texture or turn yellow and are dried and fermented or 'cured'. In the manufacture of cigarettes and pipe tobacco much blending of different tobaccos takes place. The waste tobacco from tobacco factories (and *N. rustica*) is used in the preparation of nicotine, used as an insecticide. Some other spp. of *Nicotiana* are known to have been smoked by Amerindians. **Aztec** —, *Nicotiana rustica:* **Cavendish** — **plug** or **cake** — made w. the addition of treacle or syrup: **Havana** —, var. *havanensis*, a special large-leaved var.: **honeydew** —, light-coloured cake tobacco: **Hottentot** —, *Tarchonanthus camphoratus*, S.Af.: **Ind.** —, *Lobelia inflata:* **Latakia** —, special flavour due to curing w. smoke from green oak wood: **mountain** —, *Arnica montana:* **negro head** — = Cavendish —: **rock** —, *Primulina tabacum*, China, a tobacco-scented primula-like pl.: **shag** —, a finely cut dark pipe tobacco: **Turkish** —, a specially flavoured cigarette tobacco: **twist** —, prepared like Cavendish w. treacle and popular for chewing: **Virginia** —, var. *virginiana*, the popular cigarette tobacco: **wild tree** —, *Acnistus arborescens*, trop. Am.: **winged** —, *Nicotiana alata.*

Tobira (Jap.) *Pittosporum tobira*, orn. tree, cult., subtrop.

Toddy Or palm wine (fermented palm juice): the juice may be obtained from various palms, e.g. *Arenga pinnata, Beccariophoenix madagascariensis, Borassus, Caryota urens, Cocos nucifera, Elaeis guineensis, Hyphaene, Nypa fruticans, Phoenix* spp., *Raphia* spp. etc.

Toddy palm Name for *Caryota urens*, trop. As., a source of palm wine and palm sugar, also fibre (kitool fibre).

Todea *T. barbara*, N.Z., a cult. fern.

Toetoe *Arundo conspicua*, N.Z.

Tokay Celebrated Hungarian wine, production centred on the town of Tokaj.

Tola wood (W.Af.) *See* agba.

Tollon *Heteromeles arbutifolia*, Calif., branches w. bright red berries, used like holly at Christmas.

Tolu balsam *Myroxylon balsamum* (= *M. toluiferum*), C. & S.Am., the tree is widely distributed but most balsam comes from Colombia and Venezuela.

TOMATILLO

It is used in the perfume industry and med. as a pleasant ingredient in cough mixtures.

Tomatillo *Physalis ixocarpa*, Mex., husk tomato, fr. ed., used in stews.

Tomato *Lycopersicon esculentum*, S.Am., the common tomato is now world-wide in cult. and usage, being used esp. in salads, soups, stews, with meat and as a veg. There are numerous vars., some with yellow frs. It is also extensively canned, as is the juice: the seeds contain an ed. oil. **Bitter —**, *Solanum incanum:* **cherry —**, *Lycopersicon esculentum* var. *cerasiforme:* **children's —**, *Solanum anomalum*, trop. Af., frs. eaten cooked: **tree —**, *Cyphomandra betacea*, S.Am., long cult. in Peru, ed. red egg-shaped frs. used in jams and preserves.

Tonga N.Z. medicine, the main ingredient being *Epipremnum mirabile*.

Tongue grass *Lepidium* spp., N.Am.: **— orchid**, *Cryptostylus* spp.; *Dendrobium cucumerinum*, Aus.: **— violet**, *Schweiggeria fruticosa*.

Tonka bean or **tongue bean** *Dipteryx odorata*, trop. Am., the fragrant kernels, 'cured' in rum, are used in perfumery, esp. for tobacco or snuff.

Tongking cane *Arundinaria amabilis* var. *sativa*, S. China, a strong bamboo w. little taper, used for 'split-cane' fishing rods and many other purposes.

Tooart Or towart, *Eucalyptus gomphocephala*, Aus.

Toolsi (Ind.) *Ocimum sanctum*, cult., held in veneration by the Vishnu Brahmins.

Toon Or red cedar, *Toona ciliata*, Ind., imp. forest tree, cult., timber much used, esp. for tea and cigar boxes.

Toonsi *See* toolsi.

Toothache grass *Ctenium* spp., N.Am.: **— tree**, *Zanthoxylum americanum*, N.Am., pungent b. chewed as a toothache remedy.

Toothbrush tree *Salvadora persica*, trop. Af. and As. (*See* chewstick).

Toothwort *Lathraea squamaria*, *Dentaria* (= *Cardamine*) *bulbifera*, Br., both pls. have tooth-like scales.

Toowoomba canary grass *Phalaris tuberosa*, an imp. fodder grass.

Topi-tamboo (W.I.) Or topinambour, *Calathea allouia*, a gdn. r. crop.

Topiary Or the art of growing and training shrubs or small trees into special shapes, often resembling birds or animals, is very old and goes back many centuries. It probably originated in Italy for it was practised by the Romans, the word topiary being derived from *ars topiaria* (ornamental gardening). It is not known when topiary was first practised in Britain but it reached its zenith in the 16th and 17th Centuries. Topiary was appropriate with the formal type of gardening that was popular for so long. The pls. most commonly used have always been yew (*Taxus baccata*) and box (*Buxus sempervirens*). Reasons for this are their long life and the fact that they are evergreen, have small ls., and stand continuous clipping well. Close attention over many years is necessary to obtain good specimens. [Hadfield, M. (1971) *Topiary or ornamental hedges*, 100 pp.]

Topinambour *See* topi-tamboo.

Toquilla (Ecuador) *Carludovica palmata*, Panama hat pl.

Tora Or Ind. rape, *Brassica napus* var. *dichotoma*, an imp. cold weather oil seed crop.

Torch lily *Kniphofia* spp., S.Af., cult.: **— thistle**, *Cereus peruvianus*, S.Am., cult.: **— wood**, *Amyris balsamifera*, W.I.; *Bikkia mariannensis*, Polynesia, small tree, wood ignites easily; *Balanites maughamii*, S.Af.; *Casearia icosandra*, C.Am.: *Jacquinia arborea*, W.I.

Torenia Cult. orn. pls., trop., esp. *T. fournieri*, blue fls., winged calyx.

Tori *See* tora, Ind. rape.

Tormentilla *Potentilla erecta* (= *P. tomentilla*) common on heaths and acid soils in Br.

Torpedo grass *Panicum repens*, a sugar cane weed in Taiwan.

Torreya *T. nucifera*, Jap., oily seeds ed.

Tortoise bush (S.Af.) *Zygophyllum* spp.

Totara *Podocarpus totara*, N.Z., a well known commercial timber.

Totem pole The totem poles of the N.Am. Amerindians were often made of western red cedar (*Thuja plicata*), easily worked w. good lasting properties.

Tou-fou Or tau-foo, a Chinese food: a bean curd or 'cheese' made from soy bean.

Touch-me-not *Impatiens* spp., balsam, ripe frs. may 'explode' if touched: **common** — — —, *Impatiens noli-tangere*, Br.: **pale** — — —, *I. pallida*, N.Am.: **spotted** — — —, *I. capensis*, many colour vars.: **S.Af.** — — —, *Melianthus major*.

Tough weed *Sida elliotii*, N.Am., used for baskets by Amerindians.

Tournesol Or turn-sole, *Chrozophora tinctoria*, Medit., a dye.

Tous-les-mois *Canna edulis*, W.I., ed. canna, a r. crop.

Tow The short waste fibrous material left after fibre preparation, esp. of flax and hemp.

Towai bark *Weinmannia racemosa*, N.Z., b. rich in tannin.

Towel gourd *Luffa*.

Tower mustard *Arabis glabra*, N.Am.

Trac *Dalbergia cochinchinensis*, Indo-China, timber.

Trachelium *T. caeruleum*, a pot pl. w. large heads of small blue fls.

Trachelospermum Climbing shrubs, As., N.Am., cult.; *T. jasminoides* (Chinese jasmin) an evergreen climber w. white fls.

Tradescantia *T. virginiana*, N.Am. and other spp., cult. (spiderworts).

Tragacanth, gum *Astragalus* spp., several spp. growing wild in Iran, Turkey, Afghanistan and adjacent Russia are the commercial sources of gum tragacanth, the gum being obtained by tapping the stems or rs. The hydrophilic and colloidal properties of the gum are of value in the manufacture of ice-cream, liquors, lotions, sizings and other industrial products. Pharmaceutically it is much used, esp. for suspending resinous tinctures and heavy insoluble, powders, also for medicated toilet creams and jellies. **Af.** —, *Sterculia tragacantha* and other spp.: **Ind.** —, *Sterculia urens*.

Tragasol *Ceratonia siliqua*, carob seed gum, a tragacanth substitute.

Trailing bean *Phaseolus helvolus* (= *Strophostyles umbellata*) and other spp., N.Am., used in fixing sand dunes.

Tramp's trouble *Smilax bona-nox*, N.Am.

Transvaal daisy Or Barberton daisy, *Gerbera jamesonii*, cult., many cultivars, **— gardenia**, *Gardenia spathulifolia* and other spp.

Trap flowers Fls. which entrap insect visitors, e.g. *Aristolochia*, *Arum*, *Asarum*, *Ceropegia*, *Coryanthes*, *Cypripedium*, *Magnolia* etc.

Trapa Old World aquatics, some cult. for ed. frs., e.g. water chestnut *Trapa natans*, Singhara nut, *T. bispinosa*, ling, *T. bicornuta*.

Traveller's joy *Clematis vitalba:* — —, **Aus.**, *C. aristata:* — **palm** or **tree**, *Ravenala madagascariensis*, Madag. cult. orn., palm-like, seed has a blue aril.

Treacle mustard *Erysimum cheiranthoides*, Eur. inc. Br., a weed.

Tread softly *Cnidoscolus stimulosus*, trop. Am.

Treasure flower *Gazania rigens*, S.Af.

Treculia *T. africana*, trop. Af., Af. bread fr. tree, seeds ed.

Tree cotton *Gossypium arboreum:* — **fern**, spp. of *Brainea, Cibotium, Cnemidaria, Culcita, Cyathea, Dicksonia*, etc. have tree-like trunks some v. tall and handsome as in N.Z.: — **fuchsia**, *Fuchsia* spp., N.Z.; *Schotia brachypetala*, S.Af.: — **mallow**, *Lavatera arborea:* — **medick**, *Medicago arborea:* — **nails** or **trenails**, wooden pegs of oak and other woods of various sizes, small used by cobblers, large for railway chairs and in shipbuilding: — **of chastity**, *Vitex agnus-castus:* — **of Heaven**, *Ailanthus altissima*, orn. Chinese tree: — **of life**, *Thuja* spp.: — **of sadness**, *Nyctanthes arbor-tristis:* — **peony**, *Paeonia suffruticosa:* — **peach**, *Trema australis*, Aus.: — **tomato**, *Cyphomandra betacea*, S.Am., widely cult. subtrop., used mainly for preserves.

Trefoil Usu. clover-like pls., often valuable constituents of pasture: **bean** —, *Anagyris foetida*, Medit.: **bird's foot** —, *Lotus corniculatus:* **hop** —, *Trifolium campestre:* **milk** —, *Cytisus:* **moon** —, *Medicago arborea:* **pitch** —, *Psoralea bituminosa:* **prairie** —, *Lotus americanus:* **scented** —, *Melilotus:* **shrubby** —, *Ptelea trifoliata:* **sweet** —, *Trigonella:* **woolly** —, *Trifolium tomentosum:* **yellow** —, *Medicago lupulina.*

Trianthema *T. portulacastrum*, trop. weed.

Trichilia Trop. Af. or Am. trees, some yielding good timber.

Trichocentrum Trop. Am. orchids, mainly dwarf, relatively large bright fls., cult.

Trichocereus Trop. Am. cacti, fls. nocturnal, scented, cult.

Tricholaena *See* Natal grass.

Trichomanes Ferns w. delicate fronds, cult.

Trichopilia Trop. Am orchids, often showy, cult.

Tricyrtis As. perens., large fls., cult. (toad lilies).

Tridax *T. procumbens*, a bad top. weed.

Trigger plants *Stylidium* spp. Aus., some have an irritable gynostemium. It bends to one side: on being touched it springs to the other side. *S. brevicarpum* has an explosive pollination mechanism.

Trigonidium Curious trop. Am. orchids, cult.

Trillium *Trillium* spp., N.Am., E.As., wood lilies, cult.: **large-flowered** —, *T. grandiflorum:* **nodding** —, *T. cernuum:* **painted** —, *T. undulatum:* **purple** —, *T. erectum*, N.Am., ill-scented.

Trimbleberry *Rubus occidentalis*, E. N.Am., ed. fr., cult.

Trincomalee wood *Berrya cordifolia* (=*B. ammonilla*) trop. As., a valuable timber, tough, dark red.

Trinidad pansy *Torenia* sp., W.I., cult.

Tritonia S.Af. bulbous pls. w. handsome fls., cult.

Trollius Or globe flower, *Trollius* spp., cult. orn. fls.

Trompillo *Guarea guidonia*, trop. Aus., a strong and durable wood.

Tropaeolum *See* nasturtium.

Trough garden *See* jardinière, sink gardening.

Trout lily *Erythronium americanum*, N.Am.

Trumpet flower Name used for various pls. w. large tubular or trumpet-shaped fls. **Chinese — —**, *Campsis chinensis* (= *Tecoma grandiflora*): **— climber, creeper** or **vine**, *Campsis radicans*, cult. trop.: **huntsman's —** *Sarracenia flava* and other spp., N.Am. **Nepal — —**, *Beaumontia grandiflora*, cult.: **— reed**, *Arundo:* **— tree**, *Cecropia peltata*, W.I., of rapid growth, hollow stems, used as trumpets by Amerindians, sometimes occupied by fierce ants. *See Cecropia* in Willis.: **— weed**, *Eupatorium fistulosum*, N.Am.: **yellow — —**, *Tecoma* (= *Stenolobium*) *stans*, trop. Am., cult.

Tsubaki oil (Jap.) *Camellia japonica*, oil from seeds used as a hair oil.

Tuba root Derris root.

Tuberose *Polianthes tuberosa*, a gdn. pl.: fl. oil used in perfumery.

Tuckahoe *Peltandra virginica*, N.Am., green arrow arum, cult. orn.

Tucuma palm *Astrocaryum tucuma*, Braz., frs. yield oil resembling coconut oil, l. fibre used for hammocks, bags etc.

Tule Am. name for bulrush.

Tulip *Tulipa* spp. The **garden —** reached Europe from Turkey in the 16th Century and has always been much grown in Holland where many new types or vars. have been bred. **Cape —**, *Homeria* spp., poisonous to livestock: **lady —**, *Tulipa clusiana:* **wild —**, *T. sylvestris*, fls. yellow, naturalized in Br.

Tulip tree *Liriodendron tulipifera*, E. N.Am., cult. orn. tree, large tulip-like fls. which may produce much nectar, a teaspoonful being recorded for one fl.: the wood 'canary white wood' is a commercial timber, light and soft. **Af. — —**, *Spathodea campanulata:* **Chinese — —**, *Liriodendron chinense:* **Ind. — —**, *Thespesia populnea*, cult. trop.

Tulipwood *Physocalymma scaberrimum*, Braz., a handsome reddish wood w. dark zones; *Dalbergia frutescens* and *D. cearensis*, Braz.: **Aus. —**, *Harpullia pendula*, close-grained handsome wood; *Owenia* spp.: **Burmese —**, *Dalbergia oliveri*.

Tulsi (Ind.) *Ocimum basilicum*, basil.

Tumble grass *Schedonnardus paniculatus; Eragrostis spectabilis*, N.Am.: **— mustard**, *Sisymbrium altissimum*, N.Am.: **— weed**, *Amaranthus albus*, N.Am.

Tuna *Opuntia* spp., trop. Am., ed. frs.

Tunbridge filmy fern *Hymenophyllum tunbrigense*, Br.

Tung oil Or Chinese wood oil, *Aleurites* (= *Vernicia*) *fordii*, *A. montana*, China, cult., seeds yield a drying oil w. uses like those of linseed oil. *A. cordata* of S. Jap. yields a similar oil.

Tupelo *Nyssa sylvatica*, E. N.Am., and other spp., a commercial timber, fls. provide bee pasturage: **Chinese —**, *N. sinensis*.

Tupistra Perens., As., bell-shaped fls. cult.

Turf An association in which grasses predominate: applied in horticulture to thin grass sods laid on the surface of prepared ground to form a lawn.

Turkey berry *Solanum* sp., W.I.; *Mitchella repens*, N.Am.: **— beard**, *Xerophyllum asphodeloides*, N.Am.: **— corn**, *Dicentra canadensis* and other spp., N.Am.: — **oak**, *Quercus cerris*, Medit., cult. orn. evergreen tree: — **red**, *Peganum harmala*, Medit., dye from seeds.

Turk's cap *Malvaviscus arboreus*, Mex. shrub, cult., red fls. resemble a fez. — — **cactus**, *Melocactus communis*, trop. Am.: — — **gourd**, *Cucurbita:* — —**lily**, *Lilium superbum:* — —**mallow**, *Malvaviscus arboreus:* — **turban**, *Clerodendrum* spp.: **Turkish galls**, *see* oak galls: **Turkish hazel**, *Corylus colurna*.

Turmeric *Curcuma longa*, trop. As., widely cult, the rhi., dried and ground, is an imp. spice and an esst. curry powder ingredient: also an orange or yellow dye at one time much used for dyeing silk and wool, also carpets: — **root**, *Hydrastis canadensis*.

Turnip *Brassica rapa*, the origin of the turnip is uncertain, having been cult. for centuries: **Ind.** —, *Arisaema* spp., N.Am., corms starchy, fls. believed to be pollinated by snails: **prairie** —, *Psoralea esculenta*, N.Am., r. ed.: **Swede** —, *Brassica napo-brassica*, a much used stock food: — **wood**, *Dysoxylum muelleri*, Aus., red wood used in cabinet making.

Turnsole Or tournesol, *Chrozophora tinctoria*, Medit., a dye pl.

Turpentine *Pinus* spp., a product of the distillation of the oleo-resin of various pines, obtained by tapping, chiefly long leaf pine *P. palustris*, S. U.S. and maritime pine, *P. pinaster*, Eur., esp. France. **Chian** —, *Pistacia terebinthus:* **Jura** —, *Picea excelsa:* — **oil** or **purified** — is used in pharmacy: **Strasbourg** —, *Abies pectinata:* **tree** —, *Pistacia terebinthus*, Medit.; *Bursera* spp., W.I.; *Syncarpia* spp., Aus.; *Pittosporum crassifolium*, N.Z.: **Venice** —, *Larix decidua*, Eur. larch: — **wood**, or 'luster', a trade name for the wood of *Syncarpia* spp., Aus.: **wood** —, obtained by distilling the rs. and stumps of pine trees.

Turpeth root or **turpethum** *Ipomoea turpethum*, trop. As., med., a purgative: **Spanish** — —, *Thapsia garganica*.

Turtle grass *Thalassia testudinum* (Fam. *Hydrocharitaceae*) SE. U.S., submerged marine aquatic: — **head**, *Chelone* spp.

Turwad bark Avaram bark.

Tussock grass *Deschampsia caespitosa* (tufted hair grass), N.Hemisph. inc. Br., a common coarse grass of little use; *Poa* spp., *Chionochloa* spp., N.Z.: *Poa labillardieri* and other spp., Aus. *Poa flabellata*, Falkland Isles: **serrated** — —, *Nassella trichotoma*, Aus.

Tutsan *Hypericum androsaemum*, Eur. inc. Br., cult.

Tutu (N.Z.) *Coriaria arborea* and other spp., poisonous shrub, can give rise to poisonous honey.

Twayblade *Listera ovata*, Br.: **lesser** —, *Listera cordata*, Br., terrestrial orchids; *Liparis* spp., N.Am. *See Listera* in Willis.

Twig rush *Cladium* spp., N.Am.

Twin berry *Mitchella repens*, N.Am., fls. in pairs developing into twin drupes, used for winter decoration: — **flower**, *Bravoa geminiflora*, Mex., cult., fls. in twos; *Dolichos biflorus; Linnaea borealis*, N. Hemisph. inc. Br.: — **leaf**, *Jeffersonia diphylla*, N.Am., ls. deeply divided into two lobes: — **spur**, *Diascia* spp., S.Af.: — **veined wattle**, *Acacia binervata*, Aus.

Twisted stalk (Am.) *Streptopus:* — **leaf garlic**, *Allium obliquum:* **twistwood**, *Viburnum lantana*, Eur.

Twitch Or couch grass, *Agropyron repens,* a bad weed of cult. land, pieces of rhizome quickly develop into new pls.

Two-eyed berry *Mitchella repens,* N.Am.

Tyle berry (W.I.) *Jatropha multifida.*

Tylophora *T. asthmatica,* Ind., ls. a substitute for ipecacuanha.

Typhonodorum *T. lindleyanum,* E.Af., Madag., a giant aroid of fresh water swamps, seeds eaten after boiling.

U

Uba cane Var. of sugar cane once extensively grown, esp. in Natal, S.Af.: no longer cult. for sugar production because of disease susceptibility: a good fodder cane: believed to be of Indian origin.

Ucahuba or **ucuhuba** *Virola surinamensis,* trop. S.Am., seeds yield a solid fat, as do those of *V. sebifera* (virola fat) and *V. venezuelensis* (cuojo).

Udo *Aralia cordata,* Jap. veg., rs. and young shoots eaten.

Uganda grass *Cynodon transvaalensis,* a lawn grass.

Ugli Citrus fr. hybrid: cross between grapefruit and tangerine: exported from W.I., good flavour, like a giant coarse-skinned tangerine.

Uli rubber *Castilla* spp. (=*Castilloa*), C.Am.

Ulla grass (Ind.) *Themeda gigantea,* considered for paper.

Ullucu *Ullucus tuberosus,* a starchy r. crop of the Andes.

Ulmer pipes Tobacco pipes made from the tough heavy wood of the hedge maple (*Acer campestre*).

Ulmo *Eucryphia cordifolia,* Chile, a commercial timber.

Ulva marina Or alva marina, *see* grasswrack (*Zostera marina*).

Umbrella bush *Melaleuca cardiophylla,* W.Aus.: — **fern,** *Sticherus* spp., Aus.: — **fir,** *Sciadopitys verticillata,* Jap.: — **grass,** *Fuirena* spp., N.Am. (sedges): — **handle bamboo,** *Thyrsostachys siamensis,* trop. As.: — **leaf,** *Diphylleia cymosa,* N.Am.: — **pine,** *Sciadopitys verticillata,* Jap.: — **plant,** *Cyperus alternifolius,* E.Af., Madag., a popular house pl.; *Eriogonum* spp., N.Am.: *Peltiphyllum peltatum,* N.Am.: — **sedge,** *Cyperus alternifolius,* naturalized in W.I.: — **tree,** name used for several trees, e.g. *Cussonia* spp., S.Af.; *Magnolia tripetala,* N.Am., has v. large ls.; *Musanga smithii,* trop. Af., *Polyscias murrayi,* Aus.; *Schefflera actinophylla,* Aus.

Umburana Or amburana, *Amburana* (= *Torresia*) *cearensis,* Braz., seeds scented like tonka beans (coumarin), seeds and b. med.: tree yields useful timber.

Umzimbeet *Millettia grandis* (=*M. caffra*), S.Af. ebony-like heartwood, a favourite wood for Zulu knobkerries or clubs.

Unicorn plant *Martynia, Proboscidea* spp., N.Am., horned fr.: — **root,** or colic root, *Aletris farinosa,* N.Am. med.; *Chamaelirium luteum,* N.Am., med,

Uniola *U. virgata,* W.I., used for ropes: *U. paniculata,* N.Am.: used for fixing sand, fl. heads for decoration.

Upas tree *Antiaris toxicaria,* trop. As., latex poisonous, used in arrow poison. At the end of the 18th Century curious or fanciful stories were circulated about the tree and its deadly effect on men and animals, probably due to the tree growing in some volcanic valleys where exudations of carbon dioxide occurred.

×**Urceocharis** Intergeneric hybrids between *Eucharis* and *Urceolaria*, cult.

Urceolina S.Am. bulbous pls. w. long-stalked fls., cult.

Urd (Ind.) *Phaseolus mungo*, black gram, a pulse crop.

Urena fibre Or aramina fibre, *Urena lobata*, Braz., a jute-like fibre used for coffee bags.

Urera Trop. trees and shrubs w. powerful stinging hairs.

Urn flower *Urceolina urceolata*, Peru, cult. orn. fl.: — **plant**, *Aechmea urceolata*, S.Am., cult. orn. pl., a room pl.

Ursinia *U. versicolor*, S.Af. and other spp. orange or yellow daisy-like fls. w. a long season, cult.

Urucury (Braz.) *Maximiliana regia*, *Orbignya martiana*, large palms.

Urunday *Astronium urundeuva* and other spp., heartwood yields a tannin extract similar to quebracho.

Utile Trade name adopted for mahogany-like timber of *Entandrophragma utile*, W.Af., a furniture wood, resembles sapele.

Utricularia *See* bladderwort.

Uva grass (W.I.) Or wild cane, *Gynerium sagittatum*, trop. Am. a tall reed, 3–6 m high, stems used as laths and in mat and basket making, ls. for thatch.

Uva-ursi *Arctostaphylos uva-ursi*, N.Hemisph., bearberry, dried ls. med., also used as tea and in local tanning.

Uvalha (Braz.) *Eugenia uvalha*, small tree, wild and cult., juicy yellow or orange frs. used for drinks.

Uvularia Rhizomatous perens., N.Am., spring fl., cult., esp. *U. perfoliata*.

V

Vacao *Pandanus utilis*, sacks made from ls. in Reunion.

Vaccinium Shrubs or small trees, widely distributed, some cult. for orn. or ed. fr. (cranberry, bilberry, blueberry etc.).

Vahy *Landolphia madagascariensis*, Madag., has been exploited for rubber.

Valerian Common —, *Valeriana officinalis*, Eur. inc. Br., As., cult., rhis. med., has been used in hysteria and other nervous disorders. **Af.** —, *Fedia cornucopiae*, N.Af., a pot-herb: **cat's** —, *V. officinalis*: **edible** —, *V. obovata*, N.Am.: **garden** or **red** —, *Centranthus ruber*, white and pink flowered forms exist: **Greek** —, *Polemonium coeruleum*, N.Hemisph., cult. gdn. pl.: **Ind.** — *Valeriana wallichii*: **Jap.** —, *V. officinalis* var. *latifolia*: **lesser** —, *V. dioica*: **swamp** —, *Valeriana uliginosa*, N.Am.

Vallaris *V. heynei*, Ind., cult. orn. shrub.

Valonea International name for the dried acorn cups or cupules of the valonea oak (*Quercus aegilops* and allied spp.) of E.Medit. and As.Minor, which are used in leather tanning. 'Palamut' is the Turkish name. Valonea extract is prepared in Turkey or Greece and is valued chiefly for the production of high grade, heavy leather.

Vanda Showy orchids, mainly trop. As., cult.

Vangueria Frs. of some trop. Af. spp. ed.

Vanilla *Vanilla fragrans* (= *V. planifolia*), trop. Am., a climbing orchid, cult., esp. in Madag. and Mex., the nearly ripe pods, after curing, constitute the vanilla beans of commerce, a much used spice, esp. for flavouring puddings, confectionery and ice cream. The spice is produced synthetically but the natural product is preferred for many purposes. **Bahia** or **Braz. —**, *Vanilla gardneri:* — **grass**, *Hierochloe odorata*, Euras., has vanilla-like odour: — **lily** or **plant**, *Sowerbaea juncea*, *S. laxiflora*, Aus.

Vanillons Name once applied to vanilla pods from wild or uncult. plants in C.Am.

Varnish tree *See* lacquer tree.

Vegetable gold Or gold thread, *Coptis trifolia*, N.Am., slender yellow rhis. med.: — **hair** or 'horsehair', *Chamaerops humilis*, Medit. ('crin vegetal') used in upholstery; *Carex brizoides*, Eur., used as packing material: — **ivory**, *Phytelephas macrocarpa*, trop. Am., hard kernels used for orn. objects or buttons: — **ivory substitute**, *Hyphaene thebaica*, N.Af.: — **marrow**, *Cucurbita pepo*, var.: — **oyster**, *Tragopogon porrifolius:* — **pear**, *Sechium edule:* — **sheep**, *Raoulia mammillaris* and other spp., N.Z.: — **sponge**, *Luffa aegyptiaca:* — **tallow**, *see* oils: — **wax**, *see* waxes.

Vellozia Lily-like pls. w. woody or fibrous stems, showy fls., African species referable to *Talbotia* and *Xerophyta*, cult., esp. *T. elegans*, S.Af.: stem sections used as scrubbing brushes.

Veltheimia Bulbous pls., S.Af., cult.

Velvet bean *Mucuna* (= *Stizolobium*) *deeringiana*, trop. fodder crop, seeds eaten: — **bur**, *Priva lappulacea*, W.I.: — **bush**, *Lagascea mollis*, W.I.: — **fl.**, *Sparaxis tricolor*, S.Af.; *Amaranthus caudatus:* — **grass**, *Holcus lanatus*, Eurasia, pasture, cult.: — **leaf**, *Cissampelos pareira*, trop., med.; *Lavatera arborea:* — **seed**, *Guettarda*, W.I.: — **tamarind**, *Dialium guineense*, W.Af.: — **tree**, *Gynura aurantiaca*, W.I.

Veneers Thin layers of fancy or ornamental woods superimposed on less valuable woods.

Venetian whisk *See* broom corn.

Venezuelan rose *Brownea grandiceps*, cult. orn. trop. tree w. balls of rhododendron-like red fls.

Venice turpentine *Larix decidua*, an oleo-resin obtained by boring holes in larch trees: used in veterinary medicine.

Venidio–Arctotis A name adopted for handsome gdn. pls. developed by crossing *Venidium fastuosum* with certain spp. of *Arctotis* (S.Af.): not winter hardy in Br. but easily grown from cuttings.

Venidium *V. fastuosum*, hort., S.Af., large sunflower-like fls., cult.

Ventilago *V. madraspatana*, Ind., woody vine, rs. yield a red dye.

Venus' comb Or shepherd's needle, *Scandix pecten-veneris*, Eur. inc. Br., fr. w. long beak: — **fly trap**, *Dionaea muscipula*, SE. U.S. (*See* Dionaea in Willis): — **hair fern**, *Adiantum capillus-veneris:* — **looking glass**, *Legousia hybrida* (= *Specularia speculum*), Eur. inc. Br., cult. gdn. pl., several vars.: — **navel-wort**, *Omphalodes linifolia*, SW.Eur.

Veratrin *Schoenocaulon officinale*, C.Am., alkaloid from seeds.

Verawood *Bulnesia arborea*, Venezuela, a commercial timber, resembles lignum-vitae and has similar uses, also called Maracaibo lignum vitae.

VERBASCUM

Verbascum Or mullein, *Verbascum thapsus* or other spp., med., ls. used in asthma cigarettes: some spp. cult. gdn. pls., favoured for cottage and wild gdns.

Verbena *Verbena* spp., cult. orn. fls., mainly of Am. origin, popular for edging, bedding and window-boxes. **Lemon-scented** —, *Lippia citriodora*, Chile, cult. for fragrant ls., distilled for verbena oil, used in perfumery.

Vermicelli This typically Italian food product is made from hard or flint wheat.

Vermilion wood (Am.) Name for padouk.

Vermouth This white wine of France and Italy is flavoured with absinthe, gentian or other herbs.

Vernal grass, sweet *Anthoxanthum odoratum*.

Veronica Large variable genus, many spp. cult. for orn.

Verticordia Aus. ericoid shrubs of various fl. colours except blue: many W.Aus. spp. are considered very promising hort. for other countries with suitable cli.

Vervain *Verbena officinalis*, almost cosmop. herb w. various med. uses. **Am. blue** —, *V. hastata:* **bastard** —, *Stachytarpheta jamaicensis*, W.I., a common weed: **hoary** —, *Verbena stricta:* **rose** —, *V. canadensis:* — **sage**, *Salvia verbenaca:* **white** —, *V. urticifolia*, N.Am.

Vetch *Vicia sativa* and other spp.; Eur. inc. Br., wild and cult. for fodder, 'winter' and 'summer' vetch. **Am.** —, *V. americana:* **bitter** —, *V. ervilea*, Medit., cult. for forage and seeds: **bush** —, *V. sepium:* **chickling** —, *Lathyrus sativus:* **French** —, *V. narbonnensis:* **hairy** —, *V. villosa:* **horse-shoe** —, *Hippocrepis comosa:* **Hungarian** —, *Vicia pannonica:* **kidney** —, *Anthyllis vulneraria:* **milk** —, *Astragalus glycyphyllos:* **Russian** —, *Vicia villosa:* **scarlet** —, *V. fulgens:* **Siberian** —, *V. villosa:* **tufted** —, *V. cracca:* **wood** —, *V. sylvatica*.

Vetchling Common or **meadow** —, *Lathyrus pratensis*, Eur. inc. Br.: **hairy** —, *L. hirsutus:* **tuberous** —, *L. tuberosus:* **yellow** —, *L. aphaca*.

Vetiver Or khus-khus grass, *Vetiveria zizanioides*, Ind., cult. for its fragrant rs. yielding vetivert oil, used in perfumery. Fans and screens made of the rs. in Ind. produce fragrance when wetted.

Viburnum Shrubs, mainly N. temp., cult., some w. fragrant fol. or ed. fr.: **maple-leaved** —, *V. acerifolium*, N.Am.: **sweet** —, *V. lentago*, N.Am.

Victoria *V. amazonica* (= *V. regia*), trop. Am., giant water lily, cult. *See* Willis.

Vijgie (S.Af.) *Lampranthus* spp. ed. fig-like frs.

Vine Any pl. w. a climbing or running stem: word often used for the grape. **Balloon** —, *Cardiospermum halicacabum*, trop., med. in Ind.: **pipe** —, *Aristolochia* spp.: **poison** —, *Rhus radicans, R. toxicodendron:* **potato** —, *Ipomoea pandurata:* **silk** —, *Periploca graeca*.

Vinegar Usually made on a commercial scale from malt and in wine making countries from poor or sour wine. For culinary purposes it is often flavoured with other substances such as tarragon.

Vinhatico *Plathymenia reticulata*, Braz., a commercial timber.

Viola Cult. orn. fls.; this large genus (over 500 spp.) includes the violas, pansies and violets of the fl. gdn. *See Viola* in Willis.

Violet Name has been applied to many pls. besides the true violets, *Viola* spp., esp. in olden times. **Af.** —, *Saintpaulia ionantha:* **Aus.** —, *Hybanthus:*

WAHLENBERGIA

bog —, *Viola palustris, Pinguicula vulgaris:* **Calathian** —, *Gentiana pneumonanthe:* **Can.** —, *Viola canadensis:* **Cape** —, *Ionidium capense:* **common** — (Br.), *Viola riviniana:* **confederate** —, *V. papilionacea:* **dame's** —, *Hesperis matronalis:* **dog** —, *Viola canina, V. riviniana:* **dog's tooth** —, *Erythronium dens-canis:* **essence of** —, *Iris florentina,* oil distilled from violet-scented rhi. (orris root): **fen** —, *Viola stagnina:* **green** —, *Hybanthus* spp.: **S.Af.** — **trees** *Securidaca longipedunculata,* violet-scented fls.: **sweet** —, *Viola odorata:* **Usambara** —, *Saintpaulia ionantha:* **water** —, *Hottonia palustris:* **yellow** —, *V. pubescens.*

Violin wood Name used for spruce wood: other woods used for violins are beech, poplar, cedar and birch. Stradivarius violins are said to have been made of alder.

Viper's bugloss *Echium vulgare,* Eur., med.: **— gourd,** *Trichosanthes cucumerina,* trop., snake-like fr.: **— grass,** *Scorzonera hispanica.*

Virgilia *V. capensis,* S.Af., orn. fl. tree or shrub, cult.

Virgin oil Trade name used for the best grade of olive oil.

Virginia(n) bluebell *Mertensia virginica:* **— cowslip,** *M. virginica:* **— creeper,** *Parthenocissus tricuspidata* (= *Ampelopsis veitchii*), As.: *P. quinquefolia,* both widely cult.: **— date plum,** *Diospyros virginiana:* **— mallow,** *Sida hermaphrodita:* **— mountain mint,** *Pycnanthemum virginianum:* **— rose,** *Rosa virginiana:* **— silk,** *Periploca graeca:* **— stock,** *Malcolmia maritima:* **— snakeroot,** *Aristolochia serpentaria:* **— strawberry,** *Fragaria virginiana:* **— tuckahoe,** *Peltandra virginica:* **— willow,** *Itea* spp.

Virgins' bower *Clematis cirrhosa,* Medit., cult.: *C. virginiana,* N.Am.

Virola *V. surinamensis* and allied spp., trop. Am., yield commercial timber, also called 'dalli': seeds of some trop. Am. spp. are sources of ed. fat.

Viscaria Cult. orn. fls. allied to *Lychnis,* some w. sticky stems.

Visnaga *Ammi visnaga,* Medit., med., has been used for angina and asthma. Fr. stalks used as tooth picks.

Vitex A large genus (250 spp.) of shrubs and trees, some yield useful timber.

Vitis Mainly climbers: many cult. for autumn colouring and covering large pergolas: many spp. yield ed. grape-like frs.

Vittaria Trop. ferns w. hard grass-like fronds, cult.

Volatile oils *See* oils.

Vomit nut *Jatropha curcas,* trop.

Vriesea Trop. Am. bromeliads w. showy fls., cult., esp. *V. splendens* a popular room or house pl.

W

Waddy wood *Acacia peuce,* Aus., a comparatively rare wattle.

Wafer ash *Ptelea trifoliata,* N.Am., orn. shrub or tree, cult.

Wagon tree or **waboom** *Protea* spp., used in wagon making by early settlers in S.Af.

Wahlenbergia Some spp. w. orn. fls. cult., allied to *Campanula.*

Wahoo *Euonymus atropurpurea*, N.Am.: *Ulmus alata*, N.Am., winged elm.

Waitzia Aus. anns. w. 'everlasting' fl. heads, cult.

Waiuatua *Rhabdothamnus solandri*, N.Z., fl. shrub, cult., fls. orange, striped purple.

Wake robin *Trillium grandiflorum*, N.Am., and other spp., cult. orn. pl.: *Arum maculatum*.

Walking fern or **leaf** *Camptosorus rhizophyllus*, so-called because the l. tips form new pls. where they touch the soil, causing the pl. to spread rapidly.

Walking sticks The popularity of walking sticks varies with fashion. Almost any wood may be used. Among those often favoured are the following – ash, a strong tough stick, used for 'alpenstocks': bamboo, various kinds used, esp. Jap. 'wangee' and 'whampoa' (*Phyllostachys*): cherry: chestnut (coppice shoots) called 'Congo sticks' in the trade: ebony: elm: knobwood (*Zanthoxylum*): lancewood: Malacca cane (*Calamus*): palmyra palm: partridge wood: Penang lawyer (*Licuala*), palms w. small stems: rajah cane (*Eugeissona*): rattan cane: reed (*Arundo*): Tongking cane (*Arundinaria amabilis*).

Wall bedstraw *Galium parisiense:* — **cress**, *Arabis* spp.: — **fern**, *Polypodium vulgare:* — **germander**, *Teucrium chamaedrys:* — **lettuce**, *Mycelis* (= *Lactuca muralis:* — **pennywort**, *Umbilicus rupestris:* — **pepper**, *Sedum acre:* — **rocket**, *Diplotaxis muralis:* — **rue**, *Asplenium ruta-muraria:* — **speedwell**, *Veronica arvensis:* — **spleenwort**, *Asplenium ruta-muraria:* — **whitlow grass**, *Draba muralis*.

Wall plants Term used for two classes of pls., (1) those planted against a wall for protection against winter cold or wind or to benefit from the extra heat from the sun in the growing season due to the presence of the wall, and (2) those planted for orn. in a dry wall, i.e. one built using soil not mortar or cement. For the latter tufted or mat-forming pls. are favoured, e.g. spp. of *Arabis, Dianthus, Aubrieta, Sedum, Sempervivum*, ferns etc.

Wallaba (Guyana) *Eperua falcata, E. grandiflora*, trop. S.Am., a commercial timber.

Wallaby grass (Aus). *Danthonia* spp.: **swamp** — —, *Amphibromus neesii*.

Wallapatta (Sri Lanka) *Gyrinops walla*, small tree, fibrous inner b. used for ropes.

Wallflower *Cheiranthus cheiri*, Eur., naturalized in Br., popular spring fl. gdn. pl., many vars.: **alpine** or **fairy** —, *Erysimum* spp.: **Siberian** —, *Cheiranthus* × *allionii; Erysimum perofskianum:* **western** —, *E. asperum*, N.Am.

Walnut The **Eur.** —, *Juglans regia*, also called 'English' or Persian walnut, has been used in Eur. for food from the earliest times. The quantities of nuts produced or entering commerce are exceeded only by the almond and Brazil nut. France and the S. U.S. are imp. producers.

Walnut wood is a valuable hardwood, much used for high class furniture and interior decoration. At the beginning of the 18th Century ('the walnut age') it was much in favour but gave way to mahogany, less vulnerable to woodworm. The wood has long been favoured for gun or rifle stocks as it does not warp and stands rough usage. The main commercial timbers termed 'walnut' are – **Af.** —, *Lovoa klaineana:* **Am.** or **black** —, *Juglans nigra:* **Eur.** or **Persian** —, *Juglans regia:* **Jap.** —, *J. sieboldiana:* **New Guinea** —,

Dracontomelum spp. (also called Pacific or Papuan —): **Queensland** or **Aus.** —, *Endiandra palmerstonii.*

A number of other trees w. woods or frs. resembling walnut, which are referred to as 'walnut' include – **Braz.** —, *Phoebe porosa:* **country** or **Otaheite** —, *Aleurites triloba:* **E.I.** —, *Albizia lebbeck:* **satin** —, *Liquidambar styraciflua* (heartwood).

The name Carpathian walnut has been adopted for those vars. or cultivars of the Eur. walnut, *Juglans regia,* that grow successfully as nut producers in the colder climates, as in N.Eur.

Wampee *Pontederia cordata,* N.Am., pickerel weed, bright blue fls., cult.

Wampi *Clausena lansium* (= *C. wampi*), trop. As., cult., ed fr., resembles a small lime.

Wand flower *Dierama pendulum,* S.Af.: *Galax aphylla,* N.Am., orn. ls., cult.

Wandering Jew Name applied to various pls. w. prostrate rooting stems, often grown as house pls., e.g. *Tradescantia fluminensis, T. albiflora, Zebrina pendula, Saxifraga stolonifera:* — **sailor,** *Cymbalaria muralis,* Eur., inc. Br., many vars., cult., used for walls and hanging baskets.

Wandoo *Eucalyptus redunca* var. *elata* (= *E. wandoo*), W.Aus., a hard, heavy commercial timber, rich in tannin which was at one time extracted, like quebracho, and used in tanning leather.

Wangee cane Or whangi, *Phyllostachys nigra,* Orient, a bamboo, stems black when mature.

Wapato *Sagittaria cuneata, S. latifolia,* N.Am., rs. eaten cooked.

Wara *Calotropis gigantea,* trop. As., yields a strong b. fibre; seed floss used for stuffing pillows etc.

Waratah *Telopea speciosissima,* Aus. evergreen shrub w. handsome coral-red fls.: the floral emblem of New South Wales: **Gippsland** —, *T. oreades,* **Tasmanian** —, *T. truncata:* **tree** —, *Oreocallis wickhamii.*

Wardian case The Wardian case, invented by Dr Nathaniel Ward (1791–1868), is in effect a small portable greenhouse. It consists essentially of a stout wooden box with handles at either end and a sloping glass roof to admit light to the potplants it contains, which are tightly packed in peat or coconut fibre and made secure with laths. It is used for sending plants long distances by sea, being stored in a light place on the ship and watered from time to time if need be, often by arrangement with the shipping authorities.

Another type of Wardian case, basically similar but much smaller and orn. has been used in living rooms for ferns, esp. filmy ferns, and other pls., in order to maintain a moist atmosphere.

Warrea Trop. Am. orchids, large showy fls., cult., esp. *W. tricolor.*

Wart cress *Coronopus squamatus* and other spp.: —**ed gourd,** *Cucurbita pepo* var. *verrucosa:* — **weed,** *Euphorbia helioscopia.*

Washiba wood (Guyana) *Tabebuia serratifolia,* a commercial timber.

Washington thorn *Crataegus phaenopyrum* (= *C. cordata*) handsome frs. and autumn colouring, cult.

Wasp flowers Some fls. are visited by wasps and they no doubt assist in pollination, e.g. spp. of *Cotoneaster, Hedera, Scrophularia* and *Symphoricarpus:* — **orchid** (Br.), *Ophrys apifera* var. *trollii.*

Water arum *Calla palustris,* N.Hemisph., naturalized in Br., a pl. of pond

margins: — **ash**, *Fraxinus carolinianus*, N.Am.: — **avens**, *Geum rivale:* — **beech**, *Carpinus caroliniana:* — **berry**, *Syzygium* spp. trop. and S.Af., ed. fr.: — **blinks**, *Montia fontana*, *M. rivularis:* — **caltrop**, *Trapa natans.*
Water chestnut *Trapa* spp., cult. for ed. starchy seeds, *T. bicornis* and *T. bispinosa* in the Orient and trop. As., and *T. natans* in Eur.: **Chinese** — —, *Eleocharis tuberosa*, the starchy tubers are eaten and used in Chinese dishes, canned in China: — **chinquepin** (Am.), *Nelumbo pentapetala* (= *N. luteum*): — **clover**, *Marsilea* spp.
Water cress *Nasturtium officinale*, Eur., an aquatic and much used salad pl. w, a pleasant, slightly pungent flavour, cult., usu. in special water cress beds. **mountain** — —, *Cardamine rotundifolia*, N.Am.: — **crowfoot**, *Ranunculus aquatilis:* — **daffodil**, *Sternbergia* spp.: — **daisy**, *Brachycome cardiocarpa*, Aus.: — **dock**, *Rumex hydrolapathum:* — **dropwort**, *Oenanthe crocata* and other spp.: — **elm**, *Planera aquatica*, N.Am.: — **feather**, *Myriophyllum brasiliense*, much used in aquaria: — **fern**, *Salvinia natans*, *Azolla* spp. (*see Salviniaceae* in Willis): — **figwort**, *Scrophularia auriculata:* — **forget-me-not**, *Myosotis scorpioides:* — **fringe**, *Nymphoides peltata*, an aquarium pl.: — **germander**, *Teucrium scordium:* — **grass**, *Panicum molle*, *Cyperus* spp. bad weeds: — **gum**, *Nyssa biflora*, N.Am.: — **hair**, *Catabrosa aquatica:* — **hawthorn**, *Aponogeton distachyus:* — **hemlock**, *Oenanthe crocata:* — **hyacinth**, *Eichhornia crassipes*, trop. Am., now a troublesome aquatic weed in many parts of the world, blocking water courses and canals or impeding boats and shipping (*see Eichhornia* in Willis): — **hyssop** (W.I.), *Bacopa* spp.: — **lettuce**, *Pistia stratiotes* (*see Pistia* in Willis): — **lily**, *Nymphaea* spp., *Nuphar* spp., numerous spp. in both genera cult. for orn.: seeds and rs. of some are eaten (*see* Willis): — **lily, giant**, *Victoria amazonica*, trop. S.Am., ls. 2 m across, cult. (*see* Willis): **Aus. giant** — —, *Nymphaea gigantea:* — **lobelia**, *Lobelia dortmanna:* — **meal**, *Wolffia* spp.
Water melon *Citrullus lanatus* (= *C. vulgaris*), widely cult. in warm countries: the juicy flesh varies in colour from greenish white to dark red: it is sometimes preserved or used in jam: when young it may be boiled and used like veg. narrow: in some parts of the world, esp. China, the oily seed kernels are eaten after parching or used in soups or stews. — **milfoil**, *Myriophyllum* spp.: — **mint**, *Mentha aquatica:* — **parsnip**, *Sium latifolium:* — **pennywort**, *Hydrocotyle* spp.: — **pepper**, *Polygonum hydropiper:* — **pimpernel** (Am.), *Samolus:* — **plantain**, *Alisma* spp.
Water Rice See wild rice, *Zizania aquatica*, N.Am.: — **sedge**, *Carex aquatilis:* — **shield**, *Cabomba* spp., used as aquarium pls.: — **target**, *Brasenia schreberi:* — **soldier**, *Stratiotes aloides*, Eur.; *Pistia stratiotes*, trop. and subtrop. (*see* Willis): — **speedwell**, *Veronica catenata:* — **spinach**, *Ipomoea aquatica*, SE.As., cult: — **star wort**, *Callitriche stagnalis:* — **thyme**, *Elodea*, *Anacharis:* — **tree**, *Tetracera potatoria*, trop. Af., sap used as a beverage: — **tupelo**, *Nyssa aquatica*, N.Am.: — **vine**, *Doliocarpus*, W.I.: — **violet**, *Hottonia palustris:* — **weed, Am.**, *Elodea canadensis:* — **weed, Can.**, *Elodea canadensis:* — **weed, Chilean**, *Elodea ernstiae* aquarium pls.: — **wisteria**, *Synnema triflorum*, a popular aquarium pl.: — **wort**, *Elatine* spp.: — **yam**, *Dioscorea alata.*
Watsonia S.Af. cormous pls. resembling *Gladiolus*, several spp. and hybrids cult.

Wattle The word signifies a flexible rod or hurdle but has come to be used for Aus. spp. of *Acacia* of which there are over 600. Many are cult. for timber, shelter, wind breaks, hedges, sand binding, for orn. and for wattle b. for tanning. The latter, b. or extract, also called 'mimosa' in the trade, is one of the world's most extensively used tanning materials, derived from the black wattle, *Acacia mearnsii* (=*A. decurrens* var. *mollis*) grown in plantations. Tannin content of b. (air dried) is 35–39% and of the extract 60–65%. **Black** —, *Acacia mearnsii:* **blue** —, *A. dealbata:* **blue-leaved** —, *A. cyanophylla:* **Cootaminda** —, *A. baileyana:* **golden** —, *A. pycnantha:* **golden rain** —, *A. prominens:* **green** —, *A. decurrens:* **hairy** —, *A. pubescens*, cult. v. orn.: **Mudgee** —, *A. spectabilis:* **silver** —, *A. dealbata:* **sunshine** —, *A. botrycephala:* **swamp** —, *A. elongata:* **Sydney golden** —, *A. longifolia.*

Wawa (Ghana) *See* obeche.

Wax flower *Hoya carnosa*, Queensland, pinkish white wax-like fls., a popular house pl.: — **flower, Aus.**, *Chamaelaucium* spp., W.Aus.; *Crowea saligna; Eriostemon lanceolatus, E. verrucosus:* — **flower, Am.**, *Chimaphila* spp.: — **gourd**, *Benincasa hispida* (= *B. cerifera*), trop, ed. wax-coated gourd: **Jap.** —, *Rhus succedanea:* — **lip orchid**, *Glossodia major*, Aus.: — **myrtle**, *Myrtus* spp., the wax coating on the frs. has been exploited, mainly for candles, in several countries, e.g. U.S., *M. cerifera;* Mex., *M. mexicana;* Colombia, *M. arguta;* S.Af., *M. cordifolia:* **palm** —, *Ceroxylon andicola*, Peru, waxy substance on trunk: — **plant**, *Hoya carnosa:* **Sumatra** —, *Ficus variegata*, wax used in batik work: — **tree, Jap.**, *Rhus succedanea:* **Urucury** —, *Syagrus coronata:* — **weed, blue**, *Cuphea petiolata*, N.Am.

Wax, vegetable Wax occurs widely among pls., usu. as a thin layer or coating on the l., stem, fr. or other organ. Its function may be to assist in preventing water loss. This waxy coating may be seen in many desert pls. and in a number of everyday pls. Examples are the ls. of the carnation, cabbage and certain conifers. Wax is also conspicuous on the frs. of many vars. of grape and plum as they ripen. Only a few veg. waxes are exploited or used commercially, such as carnauba and uricury wax from the young ls. of Braz. palms, esparto or fibre wax obtained from the ls. of esparto grass while they are being processed for paper making. Wax occurs on floral organs in some instances and internally in *Balanophora* (in the ·stem parenchyma). In *Ficus* and *Brosimum* wax is present in the milky sap or latex. In cotton (*Gossypium*) wax is present on the lint or seed fibres, in minute quantities. Nevertheless it is imp. in certain manufacturing processes with cotton and has a direct bearing on the 'feel' of cotton as estimated by the grader. Some so-called veg. waxes, such as Jap. wax (*Rhus*) and myrtle or berry wax (*Myrica*), although wax-like in appearance, are not true waxes but are really fats (glycerides).

The main use of veg. waxes is in polish manufacture, particularly polishes for footwear. They are frequently mixed with other waxes such as beeswax and mineral or synthetic waxes. A value characteristic of veg. waxes, particularly of carnauba (*Copernicia cerifera*), which is the most extensively used vegetable wax, is their relatively high melting point. Even a small quantity mixed with softer waxes (beeswax or mineral waxes) may cause a marked rise in the melting point of the resulting mixture. A well known use for veg. waxes is in the manufacture of 'carbon' or copying papers for typewriters. Only small quantities of veg. wax are now used for candles, these being for special purposes, e.g. for church use, on account of the agreeable odour they give out when burning.

The following are among the better known or interesting veg. waxes. **Balanophora wax**, dried rhis. used for illumination in Indon.: **banana —**, from wild banana ls. in Malaysia: **candelilla —**, *Euphorbia cerifera*, a stem wax, Mex.: **carnauba —**, *Copernicia cerifera*, from young ls., Braz.: **esparto** or **'fibre' —**, from esparto grass: **flax —**, a by-product from flax preparation: **palm —**, *Ceroxylon andicola:* **raffia —**, from leaves of raffia palm: **sugar cane —**, a by-product from sugar cane mills.

Wayfaring tree *Viburnum lantana*, Eur. inc. Br., orn. shrub, cult.

Weasel's snout *Misopates* (=*Antirrhinum*) *orontium* Br., pink fls.

Weather plant *Abrus precatorius:* — **thistle**, *Carlina acaulis*.

Weaver's broom *Spartium junceum*, Eur., naturalized in Br.

Wedding flower Or bridal wreath, *Francoa sonchifolia*, Chile, pinkish fls. in loose clusters, cult., fls. last well: — **bush**, *Ricinocarpos pinifolius*, *R. bowmanii*, Aus.

Wedge pea *Gompholobium* spp.

Weeds A brief and apt description of a weed is 'a pl. out of place'. Weeds have no doubt been a problem to man from the time when he gave up the nomadic life for a settled existence and began to grow his own food pls. They still constitute a major problem in all parts of the world and large sums of money are spent on their control in mechanical and hand weeding and by means of herbicides. Weeds may consist of local spp. and spp. accidentally introduced from other lands. The early trade routes were responsible for the introduction of weeds from one country to another. With the discovery of the Americas a mutual exchange of weed spp. between the Old and the New World began and has more or less continued ever since. Many of the troublesome cornfield weeds of Eur. soon established themselves in N.Am. and plagued the early settlers there. Am. spp. that have become troublesome in many parts of the Old World included such pls. as prickly pear (*Opuntia*), the water hyacinth (*Eichhornia crassipes*) and many more. Modern intercontinental air travel, ever wider in its ramifications, may well be a likely cause for further accidental weed introduction. Mud on passengers' shoes is a likely place for weed seeds, as would be mud on the landing wheels of aircraft picked up from runways.

Some weeds are notoriously difficult to subdue when growing among crops or in the fl. or veg. gdn., good examples being gout weed, couch grass and bindweed or convolvulus. Some tuberous rooted weeds such as spp. of *Oxalis* and *Cyperus* (notably nut grass) are particularly difficult to control, even with modern herbicides. Nut grass (*Cyperus rotundus*) is a bad weed pest in practically all sugar-cane-growing countries. The so-called 'hormone weedkillers' or auxin-type growth regulators, such as MCPA and 2,4-D revolutionized weed control by their ability to kill many ann. and peren. weeds without harming cereals and other graminaceous crops or the grasses of lawns. Low cost and ease of application were in their favour. These herbicides are absorbed by both rs. and shoots (*see* herbicides). There are several good examples of successful biological control of troublesome weeds by means of insects. In introducing insects for this purpose great care has to be exercised to make sure that the insect will not also attack any economic or cult. pl. One of the best known examples of successful control of a weed by an introduced insect is the control of the prickly pear (*Opuntia*) in Aus. Other examples are the control of the trop. weed *Lantana camara* in Hawaii and elsewhere and of St John's wort (*Hypericum perforatum*) in the western U.S. [King, L. J. (1966) *Weeds of the World*, 526 pp.; Salisbury, E. (1964) *Weeds and aliens*, 384 pp.]

Weed-wind Old name for bindweed, *Convolvulus arvensis*.

Weeping grass (Aus.) *Microlaena stipoides:* — **love grass**, *Eragrostis curvula*, S.Af., name used in Am., a pasture grass, succeeds in poor sandy soil, used in erosion control: — **willow**, *Salix alba* cv. Tristis; (Bible) *Populus euphratica*.

Weigela Deciduous shrubs, E.As., cult., attractive fls. change colour after fertilization.

Weights, seeds used as Crab's eyes, *Abrus precatorius*, Ind.: carob, *Ceratonia siliqua*, Medit.

Weinmannia B. of several spp. used in local tanning.

Weld *Reseda luteola*, Eur., Medit., As., an imp. source of yellow dye in olden times: the Romans used it esp. for wedding garments: also called dyer's rocket or dyer's weed.

Wellingtonia Name sometimes used for the Calif. bigwood or mammoth tree, *Sequoiadendron giganteum* (= *Wellingtonia* or *Sequoia gigantea*). See *Sequoiadendron* in Willis.

Welsh onion Name used for two different kinds of onion, *Allium fistulosum* and *A. cepa* var. *perutile:* — **poppy**, *Meconopsis cambrica*.

Welwitschia *Welwitschia mirabilis* (= *W. bainesii*), SW.Af., a remarkable desert pl. that lives for at least a century (*see* Willis).

Wendock *Brasenia schreberi*, trop.

Wenge *Millettia laurentii*, Congo, a commercial timber.

West Indian arrowroot *Maranta arundinacea:* — — **birch**, *Bursera gummifera:* — — **boxwood**, *Tabebuia pentaphylla:* — — **cedar**, *Cedrela odorata:* — — **cherry**, *Malpighia punicifolia:* — — **ebony**, *Brya ebenus:* — — **elemi**, *Bursera gummifera:* — — **gherkin**, *Cucumis anguria:* — — **gooseberry**, *Pereskia aculeata:* — — **locust-tree**, *Hymenaea courbaril:* — — **mahogany**, *Swietenia mahagoni:* — — **redwood**, *Guarea guidonia:* — — **sandalwood**, *Amyris balsamifera*.

Western rose (Aus.) *Diplolaena dampieri*.

Westland pine (N.Z.) *Dacrydium colensoi*, mottled wood esteemed for furniture.

Weymouth pine *Pinus strobus*, N.Am., eastern white pine.

Whale bone tree *Streblus brunonianus*, Aus. rain forest tree.

Whangee cane *Phyllostachys nigra* and other spp., a Chinese bamboo.

Wharariki (N.Z.) *Phormium colensoi*.

Wheat Common or **bread** —, *Triticum aestivum* (= *T. vulgare*, *T. sativum*): origin uncertain, cult. from ancient times, used for breads, bakery goods, breakfast foods etc. of all kinds. Wheat flour or starch has various other uses as has wheat straw. The numerous vars. of wheat may be divided into (1) spring or winter wheat, (2) hard or soft wheat, (3) bearded or non-bearded, (4) red or white wheat. [Petersen, R. F. (1965) *Wheat*, 422 pp.]. **Club** —,

Triticum compactum: **cine** —, *T. turgidum,* cult. mainly in Medit.: **Dinkel** — =spelt: **durum** —, *T. durum:* **einkorn** —, *T. monococcum:* **emmer** —, *T. dicoccum:* **flint** —, *T. durum:* **hard** —, *T. durum:* **macaroni** —, *T. durum,* used for macaroni and similar foods: **miracle** —, *T. turgidum* var.: **mummy** —, *T. turgidum:* **Polish** or **Astracan** —, *T. polonicum:* **Poulard** —, *T. turgidum:* **rivet** —, *T. turgidum:* **spelt** —, *T. spelta,* cult. in Eur., esp. Germany: **wild** —, *Elymus triticoides:* — **grass** (Am.), *Agropyron* spp., some are useful pasture or fodder grasses.

Wheel-of-fire tree *Stenocarpus sinuatus,* Aus. rain forest tree.

Whimberry Or cowberry, *Vaccinium vitis-idaea,* Br., common on moors.

Whin Or gorse, *Ulex europaeus,* Eur. inc. Br.

Whisk French —, *Chrysopogon gryllus,* Medit., used for brushes: **Mex.** —, *Epicampes macroura* or broom root: **Venetian** —, *Sorghum dochna* var. *technicum* or brown corn.

Whistlewood *Acer pennsylvanicum,* N.Am.: **whistling acacia,** *Acacia drepanolobium,* E.Af., small holes in the swollen spine bases, caused by ants, cause a whistling noise in the wind: **whistling pine,** *Casuarina equisetifolia.*

White alder *Platylophus trifoliatus,* S.Af.: — **ash,** *Fraxinus americanus:* — **asparagus,** *Asparagus albus,* N.Af.: — **bark,** *Endiandra compressa,* Aus.: — **beam,** *Sorbus aria* and other spp.: — **beard** (Aus.), *Leucopogon lanceolatus:* — **box,** *Eucalyptus hemiphloia, Tristania conferta:* — **bryony,** *Bryonia dioica:* — **cedar,** *Thuja occidentalis, Calocedrus decurrens, Chamaecyparis thyoides:* — **chandon wood,** *Dalbergia hupeana,* China: — **chuglan wood,** *Terminalia bialata,* Andamans, good furniture wood: — **cinnamon**=canella bark: — **clover,** *Trifolium repens:* — **cypress,** *Taxodium distichum:* — **dammar,** *Vateria indica:* — **dead nettle,** *Lamium album:* — **elm,** *Ulmus americana:* — **fairy orchid,** *Caladenia paniculata:* — **fir,** *Abies amabilis* and other spp.: — **gourd,** *Benincasa hispida:* — **grass** (Am.), *Leersia* spp.: — **gum,** *Eucalyptus viminalis* and other spp.: — **haiari,** *Lonchocarpus densiflorus,* S.Am.: — **hellebore,** *Veratrum album:* — **horehound,** *Marrubium vulgare:* — **horse,** *Portlandia,* W.I.: — **ironbark,** *Eucalyptus leucoxylon,* Aus.: — **lime,** *Tilia tomentosa:* — **lupin,** *Lupinus albus:* — **mahogany,** *Eucalyptus robusta:* — **mangrove,** *Avicennia, Laguncularia:* — **maple,** *Acer saccharinum:* — **melilot,** *Melilotus alba:* — **mignonette orchid,** *Microtis alba,* Aus.: — **mulberry,** *Morus alba:* — **mullein,** *Verbascum lychnitis:* — **mustard,** *Sinapis alba:* — **oak,** *Quercus alba:* — **pear,** *Apodytes dimidiata,* S.Af., useful timber: — **pepper,** *Piper nigrum:* — **pine,** *Pinus strobus* and other spp.; *Podocarpus* spp.: — **poplar,** *Populus alba:* — **Sally,** *Eucalyptus pauciflora,* Aus.: — **spruce,** *Picea glauca:* — **stem tree,** *Boscia albitrunca:* — **sweet clover,** *Melilotus alba:* — **thorn**=hawthorn; *Acacia albida,* S.Af.: — **thunbergia,** *Thunbergia erecta,* W.Af., cult.: — **willow,** *Salix alba:* — **wood,** name used for several timbers, e.g. *Abies alba,* Eur.; *Atalaya hemiglauca,* Aus.; *Liriodendron tulipifera,* Am. (canary white wood); *Maerua triphylla,* S.Af.; *Triplochiton scleroxylon,* trop. Af.; *Tilia americana,* N.Am.: — **yam,** *Dioscorea alata:* — **zapote,** *Casimiroa edulis.*

Whitlow grass *Draba* spp.: **wall** — —, *D. muralis,* common on old walls.

Whortleberry Bilberry or blueberry, *Vaccinium myrtillus,* Eur. inc. Br., on heaths, ed. blue frs. used for jam or stewed: **bog** —, *V. uliginosum:* **Caucasian** —, *V. arctostaphylos:* **Madeira** —, *V. padifolium.*

Wickerwork *See* osier.

Wig tree Or smoke tree, *Cotinus coggygria* (= *Rhus cotinus*), Medit., cult. orn. shrub.

Wild allspice *Lindera benzoin:* — **almond**, *Brebejum stellatifolium*, S.Af.: — **angelica**, *Angelica sylvestris:* — **asparagus**, *Asparagus officinalis:* — **barley**, *Hordeum murinum:* — **bean**, *Apios tuberosa*, N.Am.: — **bergamot**, *Monarda fistulosa:* — **buckwheat**, *Eriogonum tomentosum:* — **chamomile** (Am.), *Matricaria* spp.: — **cherry**, *Prunus avium:* — **cherry bark**, *Prunus serotina*, N.Am., med.: — **comfrey**, *Cynoglossum virginianum:* — **date**, *Phoenix sylvestris:* — **ginger**, *Asarum canadense:* — **hyacinth**, *Camassia* spp., N.Am.: — **Irishman**, *Discaria toumatou*, N.Z., spiny shrub or small tree: — **lettuce**, *Lactuca* spp.: — **liquorice**, *Astragalus glycyphyllos:* — **mignonette**, *Reseda lutea:* — **oat**, *Avena fatua:* — **olive**, *Elaeagnus* spp.: — **parsnip**, *Pastinaca sativa:* — **pineapple**, *Tillandsia fasciculata*, N.Am.

Wild rice Or Canada rice, *Zizania aquatica*, grain resembles ordinary rice, but is longer, an imp. food of Amerindians at one time, collected from canoes; still gathered in Manitoba and Minnesota as a commercial enterprise for modern culinary use in Can. and the U.S.; a good wild fowl food: — —, **W.Af.**, *Oryza barthii:* — **rye**, *Elymus* spp., N.Am.: — **sage**, *Salvia horminoides:* — **senna**, *Cassia marylandica:* — **service tree**, *Sorbus torminalis:* — **snowball**, *Ceanothus americanus:* — **Spaniard** or spear grass (N.Z.), *Aciphylla* spp.: — **tamarind**, *Lysiloma latisiliqua*, W.I.: — **thyme**, *Thymus praecox:* — **wheat**, *Elymus triticoides:* — **yam**, *Dioscorea spinosa* and other spp.

Wilga (Aus.) *Geijera parviflora* and other spp., a fodder tree: — **boom** (S.Af.), willow tree.

Willow *Salix* spp., widely distributed, esp. in N.Hemisph.: many cult. for orn. such as weeping willow, *S. alba* cv. Tristis, or for their wood such as the cricket bat willow, *S. alba* cv. caerulea (= *S. caerulea*), from which the best bats are made: inferior bats are made from the white willow, *S. alba*, or crack willow, *S. fragilis*. Willow wood is light and is used for artificial limbs and brake blocks as it does not catch fire with friction. **Almond** —, *S. triandra:* **balsam** —, *S. pyrifolia:* **basket** —, *S. viminalis:* **black** —, *S. nigra*, imp. for timber: **brittle** —, *S. fragilis:* **Cape** —, *S. capensis:* **creeping** —, *S. repens:* **desert** or **flowing** —, *Chilopsis linearis*, N.Am.: **eared** —, *Salix aurita:* **goat** —, *S. caprea:* — **grass**, *Polygonum amphibium:* — **herb**, *Epilobium* spp.: **Ind.** —, *Salix tetrasperma:* — **myrtle**, *Agonis flexuosa*, Aus., a much used street tree in Aus.: **native** —, *Pittosporum phylliraeoides*, *Oxylobium*, Aus.: — **oak**, *Quercus phellos:* **prairie** —, *Salix humilis:* **river** —, *S. capensis*, S.Af.: **sand dune** —, *S. syrticola*, N.Am.: **water** — (Am.), *Dianthera americana:* **weeping** —, *S. alba* cv. Tristis: **weeping** —, **Kilmarnock**, *S. caprea* var. *pendula.*

Willow bark Long used in N.Eur. in local leather tanning, esp. *Salix viminalis* and *S. caprea*, the b. being a by-product of some other enterprise: yields a light-coloured, yellowish-brown leather that is soft and flexible, favoured for gloves at one time. Extracts formerly the basis of aspirin.

Wimmera rye (Aus.) *Lolium rigidum*, an imp. pasture grass in Aus.

Wind flower *Anemone* spp.; *Anemonella thalictroides*, E. N.Am.; *Dierama* spp., S. & E.Af.

Windmill grass *Chloris truncata* and other spp.: **windmills**, *Allionia incarnata*, SW. U.S.

Window bearing orchids *Cryptophoranthus* spp., trop. Am., sepals w. slits, windows, through which insects pass: — **plants**, succulent pls. buried in the soil or sand except for the top of the l. which is transparent and admits light: occur in the Karoo, S.Af.

Wine berry Chilean —, *Aristotelia maqui:* **Jap.** —, *Rubus phoenicolasius*, cult.: **N.Z.** —, *Aristotelia fruticosa* and *A. serrata:* — **cups**, *Geissorhiza rochensis*, S.Af.: — **palms**, *Jubaea chilensis*, Chile, and spp. of *Borassus*, *Caryota*, *Elaeis*, *Mauritia*, *Phoenix*, *Raphia* etc.

Winter aconite *Eranthis hyemalis* cult., yellow fls.: —**'s bark**, *Drimys winteri*, S.Am., b. med., stomachic: — **berry**, *Ilex verticillata*, N.Am., and other spp.: — **bloom**, *Hamamelis virginiana*, N.Am.: — **buds**, shortened and crowded vegetative shoots found in *Potamogeton* and some other genera: — **cherry**, *Physalis alkekengi*, *Solanum capsicastrum*, orn. red berries, a room pl.: — **cress**, *Barbarea* spp.: — **crookneck**, *Cucurbita moschata*, fr. ed.: — **grape**, *Vitis vulpina*, N.Am.: — **green**, *Gaultheria procumbens*, E. N.Am., low shrub, ls. distilled for med. oil; *Pyrola* spp.: — **heliotrope**, *Petasites fragrans:* — **sweet**, *Chimonanthus praecox* (= *C. fragrans*), China, cult. shrub, fragrant fls.; *Acokanthera spectabilis*, cult. orn. shrub: — **vetch**, *Vicia villosa*.

Wire grass Various tough grasses have acquired the name, spp. of *Andropogon*, *Aristida*, *Cynodon*, *Eleusine*, *Poa*, *Sporobolus* etc.

Wissadula Mainly trop. Am., some spp. yield b. fibre, used locally.

Wisteria Climbing shrubs, cult., esp. *W. sinensis*, w. showy scented fls. **Am.** —, *W. frutescens:* **Aus.** or **native** —, *Hardenbergia comptoniana*, *Millettia megasperma:* **Jap.** —, *Wisteria floribunda:* **Rhodesian** or **S.Af.** —, *Bolusanthus speciosus:* **water** —, *Synnema triflorum*, a popular aquarium pl.

Witch alder *Fothergilla gardeni*, N.Am.: —**'s broom** or **knot**, the abnormal dense growth of shoots seen on some trees, e.g. birch, due usu. to fungi, sometimes to mites, as in willows, or bacteria: — **elm** = wych elm: — **grass**, *Agropyron repens*, couch or twitch; *Leptoloma cognatum*, N.Am.: — **hazel**, *Hamamelis virginiana*, N.Am., med. oil distilled from b., much used in ointments and rubifacients; b. of *Betula lutea* var. *macrolepis* used similarly; **Af.** — **hazel**, *Trichocladus ellipticus*, orn. fl. shrub: —**'s hobble**, *Viburnum alnifolium*, N.Am.: —**'s thimble**, *Silene maritima:* — **weed**, *Striga lutea* and other spp. semi-parasites troublesome with some crops, e.g. maize and sorghum.

Withy, withe or **wyth** Flexible stick or rod, commonly used for willow, *Salix fragilis*, *S. viminalis:* **black** —, *Trichostigma fruticosum*, W.I., used for baskets: **cat's claw** —, *Macfadyena uncata*, W.I.: **poison** or **scratch** —, *Cissus sicyoides*, W.I.

Witloof Well known variety of chicory.

Woad *Isatis tinctoria*, for centuries the imp. dye of Eur., displaced by natural indigo (1631) and then the synthetic product (1890). The ls. were crushed, made into balls and dried; before use, they were mixed with water and fermented: the last small English factory ceased to operate in the late 1930s. Woad was the blue body paint of the ancient Britons: mixed w. other dyes in dyeing cloth it gave different colours, e.g. green with weld. [Hurry, J. B. (1930) *The woad plant and its dye*, 328 pp.].

Wolffia *W. arrhiza*, an aquatic; claimed to be the world's smallest fl. pl. only 1 mm long, ovoid: in still water in S. England.

Wolf berry *Symphoricarpus occidentalis*, W. N.Am.: **Arabian —**, *Lycium arabicum:* **Chinese —**, *Lycium chinense*.

Wolf's bane *Aconitum napellus*, aconite or monk's hood, *A. lycoctonum*, cult.: **trailing — —**, *A. reclinatum*, N.Am.: **— milk**, *Euphorbia* spp.

Woman's tongue *Albizia lebbek*, trop., cult. tree, ripe pods may rattle incessantly in the wind.

Wombat berry (Aus.) *Eustrephus latifolius*.

Wonder-berry *Solanum nigrum*, a large-fruited cultivar, used in preserves: **— boom** (S.Af.), *Ficus pretoriae*, a famous tree (or group of trees originating from one) near Pretoria: **— tree** (N.Z.), *Idesia polycarpa*, Jap., China, in N.Z. and elsewhere has proved a valuable street tree, bears masses of crimson berries untouched by birds.

Wonga-wonga vine (Aus.) *Pandorea pandorana*, cult. orn.

Wood *See* timber.

Wood apple *Feronia limonia* (= *F. elephantum*), trop. As., small tree, cult., aromatic fr. pulp ed. and used in jellies and sherbets, tree yields a gum. **— avens**, *Geum urbanum:* **— barley**, *Hordelymus europaeus:* **—bine**, *Lonicera periclymenum*, honeysuckle: **— cow wheat**, *Melampyrum sylvaticum:* **— flour**, prepared by grinding various woods and used in plastic compounds: **— flowers** or 'flores de palo', *Loranthus* spp.; *Phoradendron* spp., curious fl.-like structures, the remains of dead woody growth of the parasites, sometimes sold as curiosities: **— garlic**, *Allium ursinum:* **— groundsel**, *Senecio sylvaticus:* **— grass**, *Sorghastrum nutans*, N.Am.: **— lily**, *Lilium philadelphicum:* **— melick**, *Melica uniflora:* **— millet**, *Milium effusum:* **— mint**, *Blephilia hirsuta*, N.Am.: **— nettle**, *Laportea canadensis*, N.Am.: **— oil**, *Aleurites* (= *Vernicia*) *fordii*, tung oil, China: **— pear**, *Xylomelum* spp., Aus., woody pear-like frs.: **— poppy**, *Stylophorum diphyllum*, N.Am.: **— rose**, *Ipomoea tuberosa:* **— rush**, *Luzula* spp.: **— sage**, *Teucrium scorodonia:* **— shamrock**, *Oxalis montana*, N.Am.: **— sorrel**, *Oxalis acetosella:* **— vetch**, *Vicia sylvatica*, *V. caroliniana*.

Woodfordia *W. floribunda*, trop. As. orn. shrub, scarlet fls., cult.

Woodruff *Galium odoratum*, Eur., used for flavouring.

Woodsia Tufted ferns of cold cli., cult.

Woodwardia N. temp. ferns (chain ferns), cult.

Woody nightshade *Solanum dulcamara*, Eur. inc. Br., scarlet frs. poisonous.

Wool flower (S.Af.) *Lanaria plumosa*, white woolly fls.: **— spider** (S.Af.), *Harpagophytum procumbens* (*see* grapple plant).

Woolly beard (S.Af.) *Protea* spp.: **— buckthorn**, *Bumelia lanuginosa*, N.Am., ed. fr., b. used as chewing gum: **— butt** (Aus.), *Eucalyptus miniata* and other spp.: **— pod milkweed**, *Asclepias eriocarpa*, SW. U.S., yields fibre: **— tree**, *Ochroma pyramidalis* (= *O. lagopus*), yields balsa wood.

Worm grass *Spigelia marylandica* or Ind. pink, orn. gdn. pl.: **— seed**,

Chenopodium ambrosioides, cult., oil a well known vermifuge: **Levant —
seed**, *Artemisia cina:* **—wood**, *Artemisia absinthium*, cult., oil med. once
used in liqueurs: **Roman —wood**, *Artemisia pontica:* **sea —wood**, *A.
maritima.*
Wound wort *Stachys arvensis* and other spp.; *Anthyllis vulneraria.*
Wrightia Some trop. As. spp. yield useful light-coloured timber.
Wulfenia Rock gdn. pls. w. blue or purple fls.
Wych elm *Ulmus glabra*, so named because the wood was used for making
chests or 'wyches' of old writers.

X

Xantheranthemum *X. igneum*, Peru, cult. orn. fl.
Xanthium Includes several cosmop. weeds, burweeds, some v. troublesome,
esp. *X. spinosum*, frs. become embedded in the wool of sheep.
Xanthoceras *X. sorbifolium*, China, orn. deciduous tree, cult.
Xanthophyllum *X. lanceolatum*, Indon. seeds yield ed. oil.
Xanthorrhoea *Xanthorrhoea* spp., Aus., known as 'grass trees', 'black boys'
(stems blackened by fires) or 'yaccas' in Aus.: a characteristic feature of the
veg. in some areas: a resin may be collected from the l. bases known as
'acaroid resin' or 'gum acaroides': it may be obtained from the following
spp. – *X. hastilis*, *X. arborea*, *X. tateana*, *X. preissii*, and is classified as
'red' or 'yellow'. Its main use has been in varnishes for metals. It has been
exploited mainly in S.Aus., esp. in Kangaroo Is.
Xanthosma Trop. Am. aroids, orn. ls., cult.: corms of some spp. ed.
Xeranthemum 'Everlasting' fls., SW.Medit., cult. and prized for winter
decoration.
Xerophyllum N.Am. liliaceous pls., orn., cult., ls. used for baskets by
Amerindians.
Xerospermum Trop. As. trees, some yield useful hard and durable wood.
Ximenia Trop. or subtrop. trees or shrubs, some yield ed. frs. or ed. kernel
oil, e.g. *X. americana.*
Xylia Trees of Old World trop., w. hard heavy wood, esp. *X. dolabriformis*,
Ind., 'pyinkado'.
Xylobium Trop. Am. orchids, cult.
Xylocarpus Trees of Old World coastal trop., mangroves, b. used for tanning
or dyeing.
Xylomelum Aus. trees or shrubs w. woody frs. resembling pears, called
wooden pears, cult.
Xylopia Trees or shrubs of Old World trop., some yield useful timber:
X. aethiopica, W.Af., has spicy frs. used as a condiment, 'negro pepper'.
Xyris Rush-like pls., trop. and subtrop., some w. local med. uses.
Xysmalobium *X. heudelotianum*, trop. Af., ed. roots.

Y

Yacca *Xanthorrhoea* spp., Aus.; *Podocarpus urbanii* (Jam.).

Yachan (Argen.) *Chorisia insignis*, down or floss from the frs. resembles and is used like kapok: was used by the Matico Indians for padding garments to make them arrow-proof.

Yacon (Peru) *Polymnia edulis*, cult. in the Andes for ed. tubers.

Yagé (Colombia) *Banisteriopsis caapi*, ls. and twigs are the source of a drug causing hallucinations, cult.

Yam The true yams are derived from spp. of *Dioscorea* and are imp. r. crops in many parts of the trop. and subtrop., esp. W.Af., the Caribbean and south Pacific regions. *D. composita* (Mexico) is an important source of diosgenin (q.v.). The most imp. and widely grown sp. is *Dioscorea alata*, of which there are numerous vars. or cultivars. Two other widely grown spp., also with numerous vars., are *D. rotundata* and *D. cayenensis*. Wild yams are commonly eaten as emergency or famine foods, usually after treatment to remove harmful substances. Under cult. yams are grown as anns., and, being climbers, need some form of support. Tubers are usually 1–5 kg in weight but with special cult. or treatment may attain enormous dimensions and weight, as in New Guinea, where yams are specially grown for ceremonial or religious purposes and when 20 kg is common and some even reach 60 kg. In yam competitions in the W.I. similar weights may be obtained.

These very large yams are laborious to harvest. The crop, as usually grown, is unsuitable for mechanical harvesting (with present machinery) and is mainly a peasant cultivar's crop. The following are some of the many common names (English) of yams – **Acom** —, *Dioscorea bulbifera*: **aerial** —, *D. bulbifera*: **Asiatic** —, *D. esculenta*: **bitter** —, *D. dumetorum*: **Carib.** —, *D. cordata*: **Chinese** —, *D. opposita*: **cush-cush** —, *D. trifida*: **cut-and-come-again** —, *D. cayenensis*: **greater** —, *D. alata*: **Ind.** —, *D. trifida*: **negro** —, *D. rotundata*: **Otaheite** —, *D. bulbifera*: **potato** —, *D. bulbifera*: **ten months** —, *D. alata*: **twelve months** —, *D. cayenensis*: **water** —, *D. alata*: **white** —, *D. alata, D. rotundata*: **white Guinea** —, *D. rotundata, D. alata*, **white-winged** —, *D. alata*: **wing-stalked** —, *D. alata*: **yampi** —, *D. trifida*: **yellow** —, *D. cayenensis*.

The term yam is also used, erroneously perhaps, for certain other r. crops. In parts of the southern U.S. it is used for the sweet potato (*Ipomoea batatas*) and in W.Af., as cocoyam, for the dasheen (*Colocasia*). [Coursey, D. G. (1967) *Yams*, 230 pp.].

Yam bean *Pachyrhizus tuberosus*, trop. Am., cult., young pods ed. also tubers: *Sphenostylis stenocarpa*, W.Af., seeds and tubers eaten.

Yampa *Perideridia gairdneri*, N.Am.

Yang *Dipterocarpus* spp., mainly *D. alatus* and *D. turbinatus*, Burma, a commercial timber.

Yaragua grass *Andropogon rufus*, trop. Am., a trop. fodder grass.

Yard grass *Eleusine indica*, a common trop. grass, used for fodder, sometimes a weed: — **long bean**, *Vigna sesquipedalis*, v. long pods eaten as a veg., trop.

Yareta (Chile) *Azorella yareta* and other spp., cushion-like pls., an imp. source of fuel in the Andes.

Yarroconalli (Arawak) *Tephrosia toxicaria*, S.Am., cult. by Amerindians as a fish poison pl.: used as a green manure or cover crop.

Yarrow *Achillea millefolium*, used in herbal remedies, other spp. cult. orn. pls.

Yate (Aus.) *Eucalyptus cornuta*, a v. tough wood.

Yaupon or **yapon** *Ilex vomitoria*, N.Am., ls. commonly used as tea.

Yautia *Xanthosoma* spp. trop. Am., some, e.g. *X. sagittifolium* and *X. jacquinii*, cult. for starchy ed. corms or rhi.

Yaw root (Am.) *Stillingia sylvatica*: — **weed**, *Morinda umbellata*, As.

Yawa *Vigna sinensis* var. *textilis*, W.Af., a var. of cow-pea cult. for fibre, used for ropes, fishing nets etc.

Yeenga (Aus.) *Geodorum pictum*, this orchid is an article of diet with the aboriginals.

Yegoma oil (Jap.) *Perilla ocymoides*, seed oil used in waterproofing paper umbrellas.

Yeheb nut (Somalia) *Cordeauxia edulis*, NE.Af., nut kernels of this small tree much eaten by Somalis, taste not unlike chestnuts.

Yellow ash *Cladrastis lutea*, N.Am.: — **bark**, *Cinchona calisaya*: — **bark oak**, *Quercus velutina*, N.Am.: —**berries**, *Rhamnus infectoria*, S.Eur., frs. yielded green and yellow dyes: — **birch**, *Betula lutea*: — **bird's nest**, *Monotropa* spp.: — **box**, *Eucalyptus hemiphloia*: — **bugle**, *Ajuga*; — **cedar**, *Chamaecyparis* spp.: — **cress**, *Barbarea praecox*: — **cypress**, *Chamaecyparis* spp.: — **deal**, *Pinus sylvestris*: —**eyed grass**, *Xyris* spp.: — **flag**, *Iris pseudacorus*: — **gentian**, *Gentiana lutea*: — **Hercules** (W.I.), *Zanthoxylum* spp.: — **horned poppy**, *Glaucium flavum*: — **loosestrife**, *Lysimachia vulgaris*: — **lupin**, *Lupinus luteus*: — **mombin**, *Spondias lutea*: — **monkey fl.**, *Mimulus luteus*: — **morning glory**, *Ipomoea tuberosa*: — **oleander**, *Thevetia peruviana* (= *T. neriifolia*), trop.: — **pimpernel**, *Lysimachia nemorum*: — **pine**, *Pinus echinata*: — **poplar**, *Liriodendron tulipifera*: — **poui** (W.I.), *Tecoma serratifolia*: — **puccoon root**, *Hydrastis canadensis*: — **rocket**, *Barbarea praecox*: — **root**, *Hydrastis canadensis*: — **sapote**, *Lucuma salicifolia*: — **shak-shak**, *Crotalaria retusa*: — **star flower**, *Sternbergia lutea*: — **sultan**, *Centaurea moschata*: — **sweet clover**, *Melilotus officinalis*: — **toadflax**, *Linaria vulgaris*: — **water lily**, *Nuphar luteum*: —**weed, dyer's**, *Reseda lutea*: — **wood**, many woods inc. spp. of *Podocarpus*, *Chlorophora*, *Cladrastis*, *Flindersia*, *Zanthoxylum*, *Ochrosia* etc.: —**wort**, *Blackstonia perfoliata* (=*Chlora perfoliata*), Br.: — **yam**, *Dioscorea cayenensis*.

Yerba buena *Micromeria chamissonis* (and other spp.) W. N.Am., fls. purple, ls. fragrant, cult., a rock gdn. pl.: — **de la feche**, *Sebastiania bilocularis*, SW. U.S. an arrow and fish poison: — **maté**, *Ilex paraguensis*, see maté: — **reuma**, *Frankenia grandifolia*, shrub, N.Mex., ls. med. (rheumatism): — **de la sangre**, *Cordia globosa*, trop. Am., shrub, ls. med.: — **santa**, *Eriodictyon californicum*, SW. U.S. and N.Mex. ls. med. and used as tea.

Yercum fibre Or madar, *Calotropis gigantea*, Ind., a strong b. fibre.

Yew Common or **Eur.**, *Taxus baccata*, cult. orn. tree, many vars., much used for hedges and in topiary: long-lived and a symbol of eternity, much planted near churches and in graveyards: formerly the traditional wood for archers' bows: sapwood narrow and white, heartwood dark, once used for furniture, now mainly for small fancy articles. **Am.** —, *T. canadensis:* **Calif.** —, *T. brevifolia:* **Can.** —, *T. canadensis:* **Chinese** —, *T. celebica:* **English** —, *T. baccata:* **golden** —, *T. baccata* var. *aurea:* **Irish** —, *T. baccata* var. *fastigiata:* **Jap.** —, *T. cuspidata; Cephalotaxus:* **jointed** —, *Athrotaxis:* **Lord Harrington's** —, *Cephalotaxus:* **Pacific** —, *T. brevifolia:* **Prince Albert's** —, *Saxegothaea:* **stinking** —, *Torreya.*

Yiel-yiel (Aus.) *Grevillea hilliana.*

Ylang-ylang Or cananga oil, *Cananga odorata*, trop. As., the large greenish fls. of this quick-growing tree are the source of the perfume.

Yoco (Colombia) *Paullinia yoco*, source of a beverage containing caffein.

Yohimbe bark *Pausinystalia johimba*, W. trop. Af., the source of the alkaloid yohimbine, a local anaesthetic.

Yom hin *Chukrasia velutina*, Thailand, a commercial timber.

Yon *Anogeissus acuminatus*, Thailand, Burma, Ind., a commercial timber.

Yorkshire fog Also called soft grass or velvet grass (Am.), *Holcus lanatus*, a widespread grass in pasture, often a weed.

Yorrell *Eucalyptus griffithii*, W.Aus. orn. cult.

Yoruba indigo *Lonchocarpus cyanescens*, W.Af., used in dyeing native cloth.

Youngberry Hybrid berry fr. of Am. origin believed to be a cross between a loganberry (*Rubus loganobaccus*) and dewberry (*Rubus flagellaris*).

Young fustic Name for a yellow dye obtained from Turkish or Venetian sumac, *Cotinus coggygria* (= *Rhus cotinus*).

Yuca (S.Am.) = cassava, *Manihot esculenta.*

Yucca *Yucca* spp., C.Am., many spp. cult. for orn., inc. several hybrids and cultivars: some yield a strong l. fibre (e.g. *Y. aloifolia*, *Y. elata*, *Y. filamentosa*, *Y. recurvifolia*), used for ropes by early settlers: the fibre has never become of commercial importance probably because of its short staple: frs. of some spp. eaten raw or cooked (*Y. baccata*, *Y. macrocarpa*). For details of interesting pollination *see* Willis.

Yulan Or lily tree, *Magnolia conspicua*, China, large white fls., a beautiful fl. tree, cult.

Z

Zacate *Nolina longifolia*, Mex., tough ls. used for brooms, baskets, thatch etc.

Zahlbrucknera *Z. paradoxa*, Eur. Alps., related to saxifrage: of interest in that petals of the small greenish fls. are smaller and narrower than the sepals.

Zakaton Or Mexican whisk, *Epicampes macroura*, cleaned rs. of this grass used for brooms, brushes etc., exported.

Zalacca *Z. edulis*, Malaysia, a more or less stemless palm w. large reddish brown fls., pleasantly sweet when ripe, young frs. pickled.

Zalil *Delphinium sulphureum* (= *D. zalil*), W.As., fls. yield a yellow dye, used esp. for silk.

ZALUZIANSKYA

Zaluzianskya Orn. S.Af. pls., cult.

Zambac Arabic name for *Jasminum sambac*, Arabian jasmine, fls. used for scenting tea, cult.

Zambesi redwood *Baikiaea plurijuga*, 'Rhodesian teak'.

Zamia Trop. Am. cycads, cult. for orn., stems a local source of starchy food, seeds also ed.

Zamioculcas *Z. zamiifolia*, an E.Af. aroid w. large ls., cult. for orn.

Zantedeschia *See* arum lily.

Zanthorhiza *Z. apiifolia*, N.Am., rs. yield a yellow dye.

Zanthoxylum Trees or shrubs, trop. and subtrop., often aromatic or spiny or w. spine-tipped knobs on the trunk, hence the name 'knobthorn': many have local med. uses or spicy seeds and are used as condiments (e.g. *Z. bungei*, *Z. piperitum*, Chinese pepper): some spp. yield useful timber (yellowwood).

Zanzibar redheads Trade name for cloves from Zanzibar.

Zapetero Or W.I. boxwood, *Gossypiospermum praecox* (=*Casearia praecox*), the principal boxwood of commerce.

Zapote *See* sapodilla.

Zauschneria Dwarf perens., W. N.Am., scarlet fuchsia-like fls., cult.

Zawa *Lophira alata*, W.Af., kernel oil used for cooking and other purposes: hard, heavy wood difficult to work (red ironwood).

Zazil Persian name for *Delphinium*.

Zebra orchid (Aus.) *Caladenia cairnsiana:* — **plant**, *Calathea zebrina*, Braz., orn. foliage; *Aphelandra squarrosa*, Braz., orn. fol., popular room pl.: — **wood**, *Astronium fraxinifolium*, Braz.; *Centrolobium robustum*, trop. Am.; *Diospyros kurzii*, Ind.; *D. marmorata*, Andamans; *Microberlinia brazzavillensis*, W.Af.

Zebrano *Microberlinia brazzavillensis* and *M. bisulcata*, W.Af., commercial timbers.

Zebrina *Zebrina* spp., C.Am., orn. fol. pls. cult., some are popular house or room pls., e.g. *Z. pendula* ls. striped w. silver and green, purple below, var. *quadricolor* has ls. w. green, red and cream markings on upper surface.

Zedoary *Curcuma zedoaria*, SE.As., allied to turmeric, rhizomes used as a condiment or tonic, also in cosmetics.

Zelkova Trees and shrubs, E.Eur. and As., cult., esp. *Z. carpinifolia* Caucasus and *Z. serrata*, Orient, a much used timber in Japan.

Zenobia *Z. speciosa*, SE. N.Am., cult. orn. shrub.

Zephyr lily *Zephyranthes atamasco*, N.Am. cult.: bulbs eaten by Amerindians.

Zeuxine Trop. and subtrop. Old World orchids: *Z. strateumatica* is an As. terrestrial orchid now naturalized in Florida.

Zeyheria *Z. tuberculosa*, S.Am., seeds yield oil.

Zezegany (W.I.) =sesame, *Sesamum indicum*.

Zigadenus Bulbous orn. pls., N.Am., As. cult.

Zimbabwe creeper *Podranea brycei* (= *Tecoma brycei*), trop. and S.Af. a strong grower w. large scented fls.

Zingana *Microberlinia brazzavillensis*, W.Af., a commercial timber.

Zingiber *Zingiber* spp. trop. As., several spp. cult. for orn. may have aromatic rhis. used as condiments (see ginger).

Zinnia *Z. elegans*, Mex. is the parent of the many gdn. forms cult. w. fls. scarlet, crimson, rose, buff, white or striped: **desert** —, *Z. pumila*, SW. U.S.

Ziricote *Cordia dodecandra*, C.Am., a handsome dark wood: rough ls. used as sandpaper: fr. ed.: cult.

Zizania Aquatic grasses. *Z. latifolia*, ls. used for mats in Jap.: *Z. aquatica* is Can. wild rice (*see* wild rice).

Zizyphus Trees or shrubs, frs. of some spp. ed. *See* jujube.

Zomicarpa Braz. aroids w. tuberous rhis., cult.

Zostera *See* eel grass.

Zoysia (=*Zoisia*). Old World maritime grasses, some used for lawns, e.g. *Z. japonica* (Jap. lawn grass), *Z. matrella* (Manila grass), *Z. tenuifolia*, a lawn grass in the U.S.

Zygopetalum Trop. Am. orchids, cult.

Zygophyllum Old World shrubs, often prostrate, ls. often fleshy, fls. white or red: *Z. fabago* is the Syrian bean caper.

SOME USEFUL REFERENCE WORKS

Bailey, L. H. (1939) *The standard cyclopedia of horticulture*, 3629 pp.

Bailey, L. H. (1949) *Manual of cultivated plants*, 1116 pp.

Bean, W. J. (1970) *Trees and shrubs hardy in the British Isles*, 4 vols., eighth edition.

Blomberg, A. M. (1967) *A guide to Australian native plants*, 481 pp.

British Standards Institution (1955) *Nomenclature of commercial timbers including sources of supply*, 144 pp.

Burkill, I. H. (1935) *A dictionary of the economic products of the Malay Peninsula*, 2402 pp.

Cassidy, F. G. and Le Page, R. B. (1967) *Dictionary of Jamaican English*, 489 pp.

Dallimore, W. and Jackson, A. B. (1966) *A handbook of Coniferae*, fourth edition, 729 pp.

Dalziel, J. M. (1955) *The useful plants of West Tropical Africa*, 612 pp.

Eliovson, S. (1960) *South African flowers for the garden*, third edition, 306 pp.

Emboden, W. (1972) *Narcotic plants*, 168 pp.

Everett, T. H. (1971) *Living trees of the world*, 315 pp.

Fairall, A. P. (1970) *West Australian native plants in cultivation*, 253 pp.

Graf, A. B. (1963) *Exotica: pictorial cyclopedia of exotic plants*, 1826 pp.

Gray, A. (1950) *Gray's Manual of Botany*, eighth edition, 1632 pp.

Greenway, P. J. (1940) *A Swahili-botanical-English dictionary of plant names*, 208 pp.

Harrison, S. G. and Masefield, G. B. (1969) *The Oxford book of food plants*, 206 pp.

Hay, R. and Synge, P. M. (1969) *The dictionary of garden plants in colour*, 373 pp.

Hill, A. F. (1952) *Economic botany, a textbook of useful plants and plant products*, 592 pp.

Hosie, R. C. (1969) *Native trees of Canada*, seventh edition, 380 pp.

Howard, A. (1934) *A manual of the timbers of the world, their characteristics and uses*, 672 pp.

Hubbard, C. E. (1954) *Grasses*, 428 pp.

Indian Government (1948) *The wealth of India, a dictionary of Indian raw materials and industrial products*, several vols.

REFERENCES

Jardin, C. (F.A.O.) (1967) *List of foods used in Africa*, 320 pp.

Jeffrey, C. (1968) *An introduction to plant taxonomy*, 128 pp.

Jepson, W. L. (1923) *A manual of the flowering plants of California*, 1238 pp.

Jex Blake, A. J. (1957) *Gardening in East Africa*, fourth edition, 414 pp.

Keble Martin, W. (1965) *The concise British flora in colour*, 254 pp.

Meikle, R. D. (1963) *Garden flowers*, 479 pp.

Meyer, H. (1933) *Buch der Holznamen*, 564 pp.

Morley B. D. and Everard, B. (1970) *Flowers of the world*, 432 pp.

Mors, W. B. and Rizzini, C. T. (1966) *Useful plants of Brazil*, 166 pp.

Palmer, E. and Pitman, N. (1961) *Trees of South Africa*, 352 pp.

Parodi, L. R. (1964) *Enciclopedia Argentina de Agricultura y Jardineria*, 1408 pp.

Pharmaceutical Society (1963) *British Pharmaceutical Codex*, 1433 pp.

Polunin, O. and Huxley, A. (1965) *Flowers of the Mediterranean*, 257 pp.

Popenoe, W. (1938) *Manual of tropical and subtropical fruits*, 474 pp.

Purseglove, J. W. (1968–72) *Tropical crops*, four volumes.

Record, S. J. and Mell, C. D. (1924) *Timbers of tropical America*, 610 pp.

Rehder, A. (1956) *Manual of cultivated trees and shrubs hardy in North America*, second edition, 996 pp.

Robinson, W. (1956) *The English flower garden*, sixteenth edition.

Royal Horticultural Society (1956) *Dictionary of gardening*, 2316 pp.; supplement, second edition (1969), 555 pp.

Russell Smith, J. (1935) *Tree crops, a permanent agriculture*, second edition, 333 pp.

Salmon, J. T. (1963) *New Zealand flowers and plants*, 204 pp.

Smith, C. A. (1966) *Common names of South African plants*, 642 pp.

Taylor, P. (1960) *British ferns and mosses*, 231 pp.

Thomas, A. (1966) *Gardening in hot climates*, 207 pp.

Uphof, J. C. Th. (1968) *Dictionary of economic plants*, second edition, 591 pp.

van Wijk, G. H. L. (1911) *A dictionary of plant names*, 1696 pp.

Williams, R. O. (1949) *The useful and ornamental plants of Zanzibar and Pemba*, 497 pp.

Winton, A. L. and Winton, K. B. (1932) *The structure and composition of foods*, 905 pp.